UNIVERSITY OF STRATHCLYDE
30125 00469966 5

D1614637

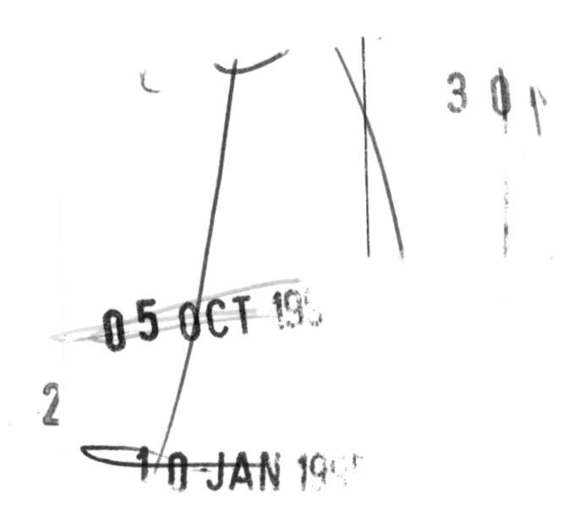
3 0
05 OCT 19
2
10 JAN 19

Electronics Manufacturing Processes

PRENTICE HALL INTERNATIONAL SERIES
IN INDUSTRIAL AND SYSTEMS ENGINEERING

W. J. Fabrycky and J. H. Mize, Editors

ALEXANDER *The Practice and Management of Industrial Ergonomics*
AMOS AND SARCHET *Management for Engineers*
AMRINE, RITCHEY, MOODIE, AND KMEC *Manufacturing Organization and Management,* 6/E
ASFAHL *Industrial Safety and Health Management,* 2/E
BABCOCK *Managing Engineering and Technology*
BADIRU *Expert Systems Applications in Engineering and Manufacturing*
BANKS AND CARSON *Discrete-Event System Simulation*
BLANCHARD *Logistics Engineering and Management,* 4/E
BLANCHARD AND FABRYCKY *Systems Engineering and Analysis,* 2/E
BUSSEY AND ESCHENBACH *The Economic Analysis of Industrial Projects,* 2/E
BUZACOTT AND SHANTHIKUMAR *Stochastic Models of Manufacturing Systems*
CANADA AND SULLIVAN *Economic and Multi-Attribute Evaluation of Advanced Manufacturing Systems*
CHANG AND WYSK *An Introduction to Automated Process Planning Systems*
CHANG, WYSK, AND WANG *Computer Aided Manufacturing*
CLYMER *Systems Analysis Using Simulation and Markov Models*
EBERTS *User Interface Design*
ELSAYED AND BOUCHER *Analysis and Control of Production Systems,* 2/E
FABRYCKY AND BLANCHARD *Life-Cycle Cost and Economic Analysis*
FABRYCKY AND THUESEN *Economic Decision Analysis,* 2/E
FRANCIS, MCGINNIS, AND WHITE *Facility Layout and Location: An Analytical Approach,* 2/E
GIBSON *Modern Management of the High-Technology Enterprise*
HALL *Queuing Methods: For Services and Manufacturing*
HAMMER *Occupational Safety Management and Engineering,* 4/E
HUTCHINSON *An Integrated Approach to Logistics Management*
IGNIZIO *Linear Programming in Single- and Multiple-Objective Systems*
IGNIZIO AND CAVALIER *Linear Programming*
KROEMER, KROEMER, AND KROEMER-ELBERT *Ergonomics: How To Design for Ease and Efficiency*
KUSIAK *Intelligent Manufacturing Systems*
LANDERS, BROWN, FANT, MALSTROM, AND SCHMITT *Electronics Manufacturing Processes*
MUNDEL AND DANNER *Motion and Time Study: Improving Productivity,* 7/E
OSTWALD *Engineering Cost Estimating,* 3/E
PULAT *Fundamentals of Industrial Ergonomics*
SHTUB, BARD, GLOBERSON *Project Management: Engineering Technology and Implementation*
TAHA *Simulation Modeling and SIMNET*
THUESEN AND FABRYCKY *Engineering Economy,* 8/E
TURNER, MIZE, CASE, AND NAZEMETZ *Introduction to Industrial and Systems Engineering,* 3/E
WOLFF *Stochastic Modeling and the Theory of Queues*

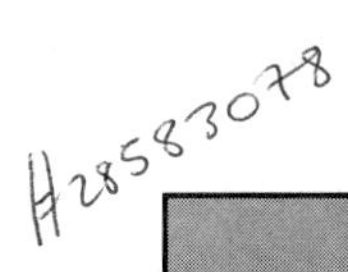

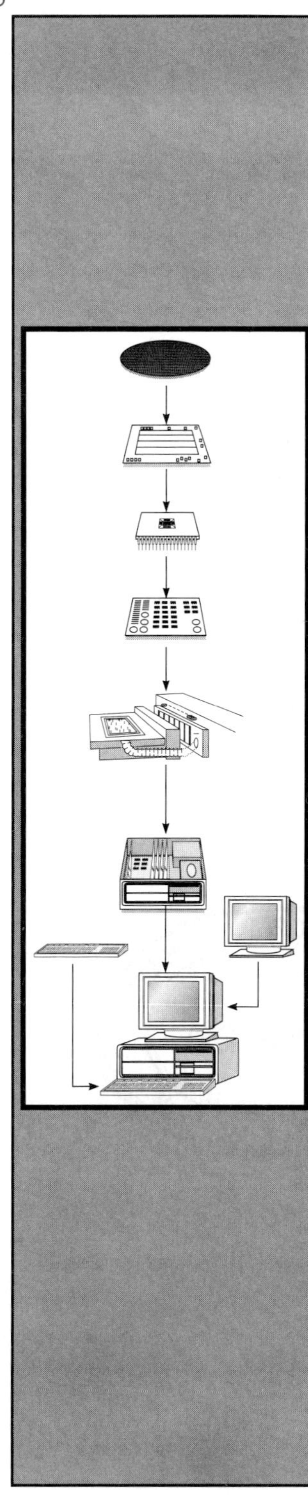

ELECTRONICS MANUFACTURING PROCESSES

Thomas L. Landers
William D. Brown
Earnest W. Fant
Eric M. Malstrom
Neil M. Schmitt

University of Arkansas

PRENTICE HALL
Englewood Cliffs, New Jersey 07632

Library of Congress Cataloging-in-Publication Data

Electronics manufacturing processes/by Thomas L. Landers . . . [et al.].
p. cm.—(Prentice Hall International series in industrial and systems engineering)
Includes bibliographical references and index.
ISBN 0-13-176470-5
1. Electronic industries—Management. 2. Production engineering. 3. Electronic industries—Automation. I. Landers, Thomas L. II. Series.
TK7836.E467 1994
621.381—dc20 93-5980
CIP

Acquisitions editor: Marcia Horton
Editorial/production supervision: Merrill Peterson
Interior design: Joan Stone
Copy editor: Nikki Herbst
Cover designer: Wanda Lubelska Design
Production coordinator: Dave Dickey
Editorial assistant: Dolores Mars

Credits appear on p. 556, which constitutes a continuation of the copyright page.

Printed in the United States of America

10 9 8 7 6 5 4 3 2 1

PRENTICE-HALL INTERNATIONAL (UK) LIMITED, *London*
PRENTICE-HALL OF AUSTRALIA PTY. LIMITED, *Sydney*
PRENTICE-HALL CANADA INC., *Toronto*
PRENTICE-HALL HISPANOAMERICANA, S.A., *Mexico*
PRENTICE-HALL OF INDIA PRIVATE LIMITED, *New Delhi*
PRENTICE-HALL OF JAPAN, INC., *Tokyo*
SIMON & SCHUSTER ASIA PTE. LTD., *Singapore*
EDITORA PRENTICE-HALL DO BRASIL, LTDA., *Rio de Janeiro*

Contents

Preface

Electronics technology is pervasive in modern society, and the global electronics industry is vital to our economic progress. Scarce resources, fierce global competition, and rapid technological advancement in the electronics industry require continuous improvement in quality and productivity. Product life cycles are becoming shorter, and there is increasing pressure to concurrently design products and processes. The concurrent engineering approach promotes rapid development of high-quality, low-cost products through close cooperation among engineering disciplines. The goal of this book is to promote concurrent engineering in the electronics industry by orienting engineers of all disciplines to electronic technology and manufacturing processes.

For many years there have been books available that surveyed the products and processes of the metal-working industries. These books have oriented manufacturing engineers to the processes of fabricating and assembling mechanical products. These books have also provided the design engineer a broad perspective of the industry and the capabilities and limitations of available production processes. This book seeks to meet the same objectives for the electronics industry. Engineers with interests in manufacturing will learn about the products and processes associated with electronics. Conversely, engineers with backgrounds or interests in the design of electronics will gain a better understanding of the processes by which electronics products are made.

This book is intended for a wide range of users in all engineering disciplines:

- Engineering students (upper-level undergraduate and graduate students)
- Engineering technology students
- Practicing engineers and engineering technologists in the electronics industry who

seek a broader perspective of their industry and how their job functions interrelate with others in the concurrent engineering environment

In addition to teaching and research in the industrial and electrical engineering disciplines, the authors have several years of experience in the electronics industry. We have worked in design, development, manufacturing, and test functions at Texas Instruments, Rockwell Collins Radio, Sandia Laboratories, the Naval Avionics Center (Naval Air Warfare Center), and the Air Force Aeronautical Systems Division. Because of this experience, we have sought to provide a practical approach to the topic of electronics manufacturing processes.

There are five parts of the book. In Part One (Perspectives) we discuss the global competitive environment, trends in the electronics industry, and fundamental principles of production. Part Two (Electronics Fundamentals) describes electronic hardware, including passive and active devices, printed wiring boards, through-hole and surface-mount technologies, and methods of interconnection. We discuss both the devices and the fabrication processes by which they are made. Printed wiring board solderability and cleaning processes are emphasized. Part Three (Automatic Assembly) introduces principles of process automation, including robotics, machine vision, sensors, and advanced inspection technologies (x-ray and laser). We describe insertion assembly processes for through-hole technology and placement assembly processes for surface-mount technology.

Part Four (Life-Cycle Engineering) orients engineers to concurrent engineering approaches which optimize the total life-cycle cost of a product: quality assurance, reliability, testability, and environmental stress screening and testing.

Part Five (Manufacturing Systems) provides an overview of principles for efficient manufacturing, including design of production facilities, workplace layout and material handling, systems for planning and control of production and inventory, and methods for economic evaluation of production alternatives. Chapter 14 on facilities and material handling includes discussions of three topics of keen current interest in the electronics industry: electrostatic discharge control, clean-room environmental control, and worker ergonomics, including cumulative-trauma disorders. Chapter 15 includes the topics of material requirements planning and just-in-time production. Chapter 16 introduces the theory of product-improvement curves, which several electronics manufacturers have used to achieve competitive advantages through cost reduction and strategic pricing.

This book is the culmination of a program funded in large part by the AT&T Foundation, through the Manufacturing Technology Grants Program. We gratefully acknowledge the support that the AT&T Foundation provides to the general infrastructure of engineering and particularly to revitalization of the manufacturing sector. Ann Alexander, Maureen D'Elia, Guido Schlesinger and W.D. Seelig have provided leadership for their project through the AT&T Foundation. Several current and former employees of AT&T have given their time to provide information and critical reviews of the draft manuscripts for this book. The support and leadership provided by Clell Callaway, Wyck Seelig, Joseph Eisenhauer, George Foo, Alan Marney, and Nick Parsons of the AT&T Little Rock Operations Center have made this book possible. We also appreciate the valuable technical comments of Patsy Ball, Jerry Barnes, Gregg Bess, Roger Boyer, John Burlie,

Kirbit Cole, Tom DeGreve, Bill Godfrey, Dick Klein, Ann Jarrett, Brian Martin, Terry Medal, Perry Mitchell, Paul Parker, and Cathy Webb.

Several employees of Texas Instruments have also made valuable contributions to this work by providing case study and technical information and by reviewing the draft manuscript. Louis Boudreaux and Robert Hawiszczak have provided encouragement and have committed company resources to the project. Thanks to Jeff Brueggeman, David Stark, Pat Thomas, and Scott Webb for their contributions and helpful comments. The following TI employees provided design guidelines and case study information for Chapter 10: Vicki Allen, Ann Asberry, Chris Beczak, Carol Cantrell-Hodges, and Laura Turner.

Kay Fowler has done an excellent job of typing the manuscript. Her task has not been easy, working with five coauthors, but she has always come through! Bonnie Swayze and Genevie Payne have also assisted in typing from time to time. Several graduate students have assisted in the project through research, documentation of references, and preparation of the manuscript and illustrations. Special appreciation is due Gary Lee, Malik Sadiq, and Kevin Martin, for major assistance in research and manuscript preparation for Parts Two, Three, and Four, respectively. John Morelock has done an outstanding job of preparing illustrations and securing permissions for use of material from other sources. Kay Beavers, John Mann, and Eric Webb have provided valuable assistance in preparation of exercises, discussion questions, and illustrations.

Two colleagues at the University of Arkansas assisted by reviewing and commenting on chapters. Dr. John English (Industrial Engineering) and Dr. Rick Couvillion (Mechanical Engineering) provided valuable reviews of Chapter 11 (Quality and Reliability) and Chapter 13 (Environmental Stress Screening), respectively.

We thank Professor Masood Baig and his students at the University of Massachusetts, Lowell, for testing the manuscript in the classroom. Their comments were especially helpful, since they are practicing engineers involved in both industry and academia. Our colleague L. Ken Keys at Louisiana State University also provided very useful suggestions. We also appreciate comments from Muhammad A. Khaliq, Mankato State University, and Nick Parsons, AT&T.

Finally, we express our appreciation to our editor, Marcia Horton, to our production coordinator, Merrill Peterson, and to professors Joe Mize and Wolter Fabrycky for their valuable suggestions and interest in this book.

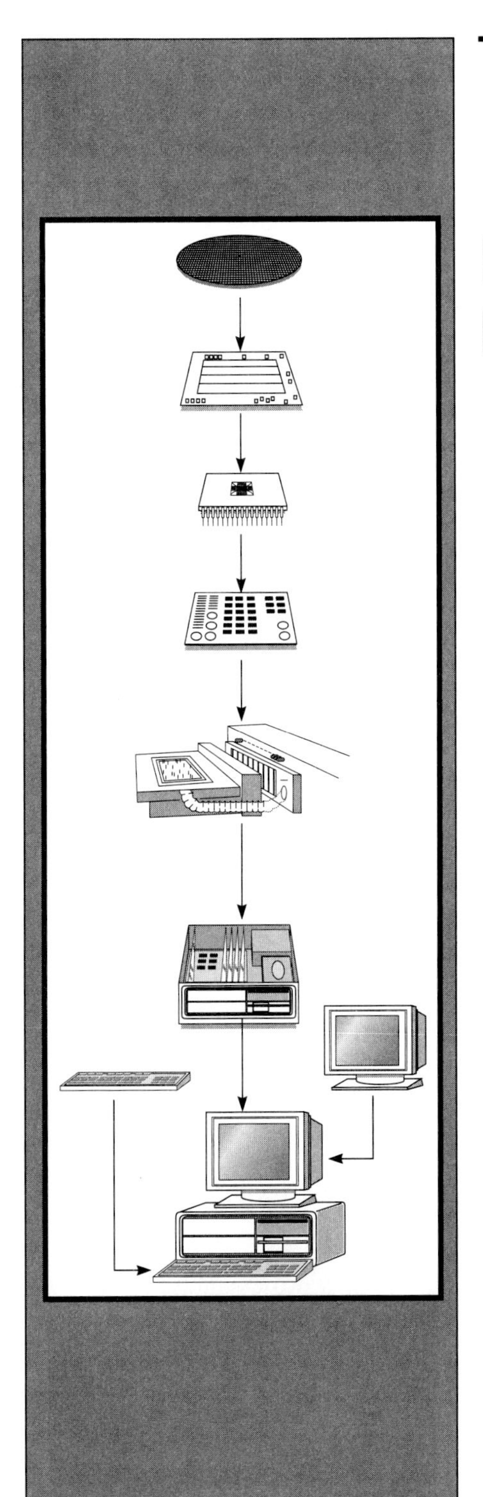

PART ONE
PERSPECTIVES

1 Introduction to the Electronics Industry

1.1 INTRODUCTION AND DEFINITIONS

Manufacturing has accounted for at least 30 percent of the United States (U.S.) Gross National Product (GNP) since World War II. It is important to note that while 55 percent of the United States work force was in manufacturing in 1945, today less than 20 percent of the work force is involved in manufacturing. This trend reflects substantial improvements in productivity, and the electronics manufacturing industry has both contributed to and benefited from the increased efficiencies. However, global competition continues to escalate, particularly in the electronics industry.

1.1.1 Definitions

We begin our study of electronics manufacturing processes by defining terms, as follows:

- *Electronics:* Electronics is the branch of engineering that pertains to the control of electrons for useful purposes. Electron flow can occur in a vacuum, gas, or liquid, as well as in solid materials that allow either restricted (semiconductor), nearly unrestricted (conductor), or totally unrestricted (superconductor) flow.
- *Manufacturing:* Electronics manufacturing may be defined as the processes of design, development, fabrication, assembly, and testing of electronic parts, tools, technologies, components, and systems.

1.1.2 Building a Successful Product

Building a successful product requires an understanding of the marketplace and customer requirements of the product, including cost and quality. In electronics and most other markets, the competition is now global. In the domestic U.S. market, American manufacturers must meet the high standard being set by global competitors. Conversely, when selling products in other countries, U.S. companies must understand and deal with diverse cultures, customs, and ethical standards. The administrative procedures involved in international trade tend to make global manufacturing and marketing very challenging tasks.

With the increasing competition for higher quality and performance at lower cost, customer expectations continue to rise. Manufacturers are finding it necessary to focus on customer needs and carefully define the requirements of customer satisfaction very early in the design process. For example, if the design engineer creates a product that is perceived by the customer as too complex, then the customer may be afraid to buy it or may be disappointed in its use. On the other hand, if the product is perceived as too simple, then competitive products may attract the customer. There are at least nine major considerations in building a competitive product, as follows:

- *Component Selection:* The design engineer must consider component price, availability, and reliability. Size, weight, power consumption, and environmental restrictions may apply. Performance is key, but the design engineer should select the minimum components that will do the job. For example, the designer should not specify a 16-bit microprocessor if a 4-bit microprocessor will accomplish the objective.
- *Fabrication and Assembly Techniques:* The methods used to manufacture the product have a major impact on quality and price competitiveness. The engineer must determine the optimum approach.
- *Materials Management:* Materials management involves purchasing, inventory control, storage, movement, packaging, and shipping. Many of the recent advances in decreasing costs have been in the area of material handling and inventory control.
- *Standards:* There are normally minimum standards a product must meet (for example, safety and environmental impact standards) in order to be allowed on the market. These standards will vary widely among countries. The product may be required to have interface compatibility with other products in accordance with industry standards.
- *Manufacturability:* Manufacturability is a term used to describe a product designed to be producible at high quality and low cost and includes the principles of design for assembly. Manufacturability requires design engineers to understand the capabilities and limitations of the production process and to work closely with manufacturing personnel.
- *Facility Functionality:* In addition to the product itself, the production facility in which it is manufactured must be designed properly. There will be an optimum size and layout for the facility that minimizes manufacturing time while maximizing quality. The facility should provide both productivity and employee satisfaction.

The design should also allow for a free flow of materials throughout the manufacturing process.

- *Management Team-Approach:* While the production process might be highly automated, ultimate control is always in human hands. Management must select programs and procedures that enhance productivity and foster a cooperative rather than confrontational attitude among all people in the company. U.S. companies are now employing quality-improvement techniques long used in Japan. Ironically, these techniques were initially developed in the United States by visionaries such as Deming and Juran but were neglected by the leaders of U.S. companies. The objectives and tools of total quality management were enthusiastically adopted by international competitors.
- *Product Life-Cycle Management:* Generally the electronic product life cycle (the time a product will be useful) is decreasing, while the design time required to bring a product to market is increasing. Figure 1-1 illustrates what is happening in the industry. The use of computer-aided design (CAD) has reduced product design time in many cases, thereby allowing the rapid recovery of the product development costs. The efficient company will hold design time to a fraction of the product life cycle and will design multiple products simultaneously. Business success requires skilled management of all phases in the product life cycle, including all steps from product design to receipt of raw materials to shipping of the completed product. Increasingly, a major competitive goal is to reduce the time from product conception to customer delivery. This goal necessitates the use of the computer in all phases of the project. CAD systems now exist for virtually every type of electronics manufacturing. Computer-aided manufacturing (CAM) automates and monitors each stage of the production process, thereby improving delivery performance to customers.
- *Cost Control:* The competitive manufacturer must understand production economics and make sound decisions regarding both investment in plant and equipment and management of operating expenses. Japanese industry has demonstrated that long-term competitiveness requires dedication to continuous improvement.

1.2 HISTORY OF ELECTRONICS MANUFACTURING

The history of electronics manufacturing can be divided into three eras, as follows:

Vacuum Tube Era	1920–1950
Transistor Era	1950–Mid 1960s
Integrated Circuit Era	Mid 1960s–Present

The microprocessor is a major technological advancement in the era of integrated circuits. Microprocessors, which are programmable integrated circuits, appeared on the market early in the 1970s and continue to increase in importance. Personal computers

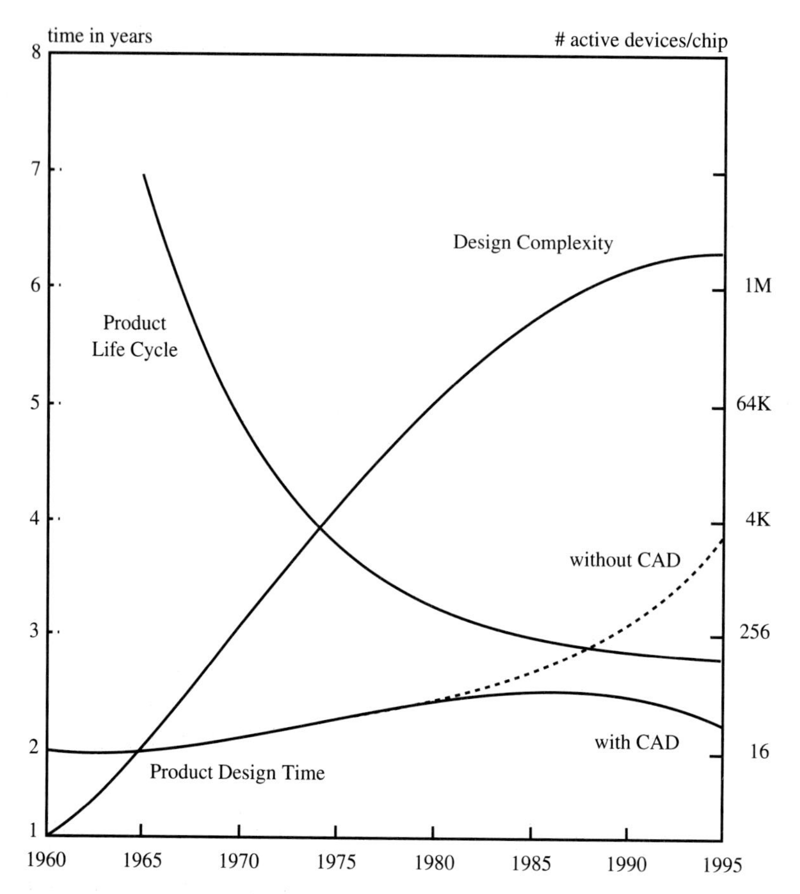

Figure 1-1 Product life cycle vs. design time.

(*Source:* Reprinted from *Industrial Engineering,* February 1986, p. 36. Copyright 1986, Institute of Industrial Engineers, 25 Technology Park/Atlanta, Norcross, Georgia 30092.)

were an immediate outgrowth of the microprocessor. They consist of various integrated circuits, including the microprocessor, random-access memory, programmable memory, and special-purpose devices such as math coprocessors and clock circuits.

1.2.1 Industry Growth

During the time frame from 1920 to the present, the global electronics market has experienced substantial growth. In 1927, the industrywide sales volume was approximately $200 million. By 1980 the industry had grown to over $200 billion, and by 1990 sales had exceeded $650 billion. The projection for industry sales in the year 2000 is over $1.4 trillion.[1] Even accounting for inflation, the real growth of the electronics industry has been impressive.

[1]Frances Stewart, "A Look at the Ups and Downs," *Circuits Assembly,* 3, 9 (1992), pp 24–26.

1.2.2 Market Share by Product Type

Figure 1-2 shows the breakdown of market share for several product groupings in 1980 and 1990 and projections for 2000. Business, retail, and computers form the largest product group (40 percent market share by 2000), with computers being the major product.

It is perhaps surprising that the consumer electronics market share is relatively small (3 percent of the U.S. electronics industry). Three products account for nearly 70 percent of the U.S. consumer electronics market: color televisions (30 percent), video cassette recorders (25 percent), and car audio systems (14 percent).[2]

All markets are observed to be expanding but maintaining relative market share, except for the government and military. With the apparent ending of the arms race between the United States and the former Soviet Union, this market segment is expected to drop by about 50 percent between 1990 and 2000.[3] U.S. companies that depended on military contracts are aggressively merging and seeking other market segments in which to compete.

1.3 EMERGENCE OF THE GLOBAL MARKET

The electronics market has truly become global in nature. Technology, including communications and transportation, has contributed to creation of a world market with enormous opportunity but dramatically increasing competition. Consumer demands for higher-quality and lower-cost products are creating opportunities for world-class manufacturers to gain market share. New global markets are also emerging as consumers in developing countries begin to demand products available from other countries. The recent opening of the eastern European nations to outside interests is creating new markets, and highlights the need for engineers to think globally.

1.3.1 The History

From the end of World War II (1945) to the mid-seventies, the United States dominated the global arena and dictated economic events for most markets, including electronics. In the decade between 1975 and 1985, the United States was challenged by the Pacific Rim countries. As a result of losses in this competition, since 1985 the United States has become a follower rather than a leader in the world market.

Between 1985 and 1990 the U.S. share of the global electronics market decreased by 21 percent, and its worldwide production share decreased by 33 percent. In a 30-month period from September, 1989, to April, 1992, the industry lost approximately 250,000 jobs. These statistics were used by Mr. R. J. Iverson, president and CEO of the American Electronics Association, to conclude, "the U.S. electronics industry at mid-1992 is considerably less healthy than it was a decade or even five years ago."[4]

[2]Johnson A. Edosomwan and Arvid Ballakur, *Productivity and Quality Improvement in Electronics Assembly* (New York: McGraw-Hill Book Company, 1989), pp. 4, 6.

[3]Stewart, "A Look at the Ups and Downs," p. 25.

[4]Stewart, "A Look at the Ups and Downs," p. 24.

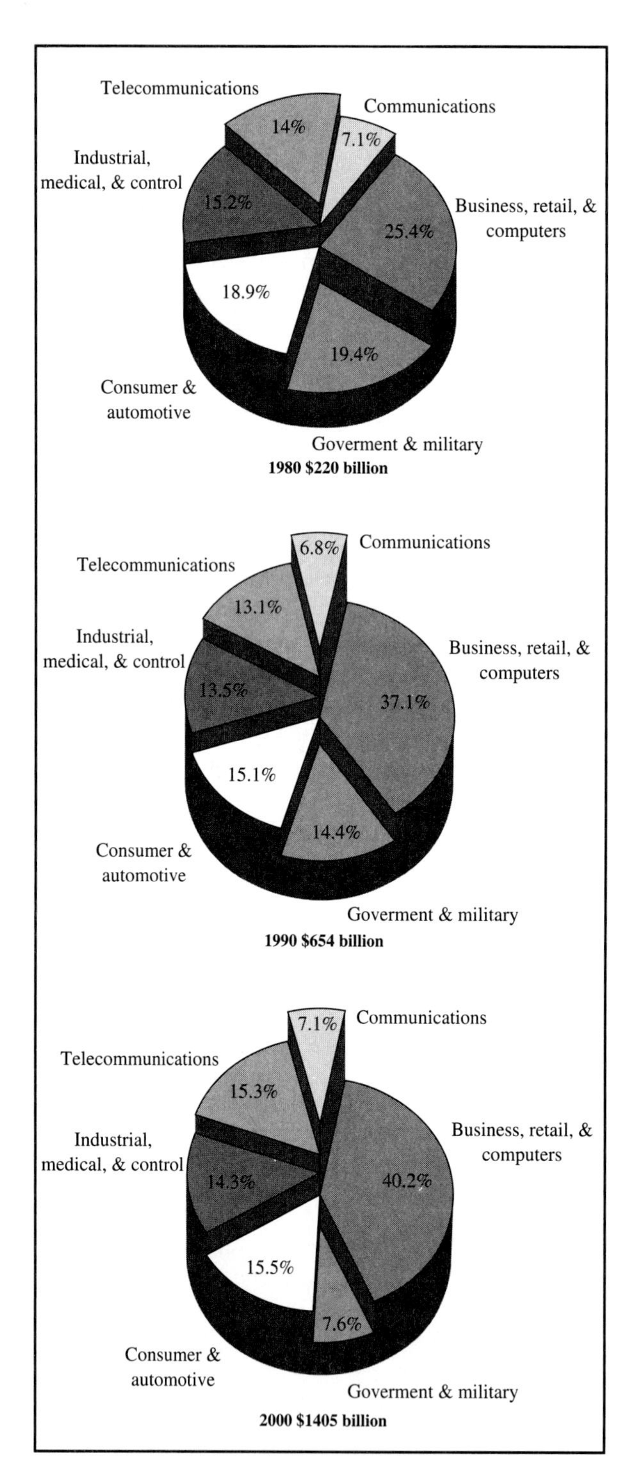

Figure 1-2 Global electronics markets.

(*Source:* Frances Stewart, "A Look at the Ups and Downs," *Circuits Assembly,* September 1992, p. 25 (Courtesy of BPA Technology Management, Inc.).)

1.3.2 The Measure

In 1980, the United States was the world's largest creditor nation ($106 billion surplus). However, by 1988, the United States had become the world's biggest debtor nation ($368 billion deficit), and continues to be the biggest debtor nation today.

1.3.3 The Reasons

There are three primary reasons why the United States has declined as an industrial leader:

- There have been inequities in trade policies creating what is sometimes called the unlevel playing field.
- U.S. management has paid little attention to global business issues or opportunities.
- U.S. interests have arrogantly ignored other cultures, languages, and customs.

1.3.4 The Potential

In 1990, the United States had a worldwide trade deficit of about $80 billion. Between 1989 and 1990, the worldwide consumption of electronic equipment rose by over $40 billion to over $650 billion. By capturing one-fourth of the new electronics equipment market, the United States could have reduced the deficit by 12.5 percent. In 1990, the United States retained 36 percent of the global semiconductor market of $48 billion.[5] By increasing semiconductor market share 20 percent from 36 percent to 56 percent, the United States could have gained another $10 billion in exports, for a total deficit reduction of 25 percent from the electronics industry alone.

1.3.5 The Major Markets and Competitors

In the foreseeable future, there will be three major geographic groupings in the world electronics market. The North American region, including Canada, Mexico, and the United States, will undergo significant trade adjustments. In the 1990s, a three-nation free-trade zone is emerging. In the short term, these three countries should all mutually benefit from more open markets, although there may be some displacements in the labor forces. In the long term, the free-trade zone may well improve the competitiveness of North America in the world economy.

Asian countries, including Japan, Taiwan, Korea, Singapore, Malaysia, and Hong Kong, will continue to assert leadership in world-class manufacturing. Increasing standard of living, higher costs, and an aging work force are making Japan increasingly vulnerable to other competitors in the Pacific Rim. Hong Kong is scheduled to be annexed by the People's Republic of China in 1997, and as a result its role in the future world economy is uncertain.

The European economic community (EC) will function much like the United States,

[5]Samual Weber, "Holding Steady as the '90s Begin," *Electronics,* G3, 1 (1990), pp. 53–55.

with one currency, free trade, unified standards, a common approach, and no transportation hindrances. West Germany has for three decades been a leading exporter. With the reunification of Germany, limited resources may be strained in the short term by the development of eastern Germany. However, in the foreseeable future, Germany could displace Japan as the most competitive nation in electronics and other industries. Germany enjoys a clear strategic advantage in development of the rapidly expanding markets of eastern Europe.

1.3.6 The Developing Markets

The breakup of the Soviet Union into separate republics, and the resulting increasing freedom and openness, will create some new markets. However, a rapidly deteriorating economy in most of the republics and a fragmented political structure will prevent any of the republics other than Russia and the Ukraine from being competitive economic challenges in the world arena for many years.

China is an important, low-cost supplier, but theirs is not yet a substantial market. However, with over one billion people, China is a huge but unpredictable potential market. In the foreseeable future, China will likely be a major competitor in quality high-technology products.

South America has inexpensive, though largely unskilled, labor; and some countries such as Brazil are aggressive in economic development. The large population base creates a potential market. Political instability in the region will likely continue to hinder major improvements. Similarly, the African continent is a possible long-term market because of population. Political instability and absence of a manufacturing base will prevent Africa from making a major competitive challenge well into the future.

1.3.7 The Pattern

Industry observers often cite Japan as a role model for manufacturing excellence. Japan overtook the United States in less than a decade, but the Japanese had planned and laid the groundwork for their success beginning shortly after World War II. When Dr. W. Edwards Deming found little interest for his concepts in U.S. industry, he joined the postwar rebuilding effort in Japan and found an enthusiastic audience. He and others educated Japanese managers and engineers in techniques developed in the Unites States. There were at least eight strategies involved in Japan's successful effort:

- Use of low-cost labor to gain a foothold in targeted markets
- Heavy investment in new processes and technologies
- Aggressive pricing strategies that promote early sales and facilitate economies of scale
- Focus on elimination of waste, through quality and productivity
- Strong marketing and distribution initiatives
- Constant development of new and related products

- Government assistance via MITI (Ministry of International Trade and Industry)
- Willingness to invest in the short term to gain long-term return, at the expense of short-term profits

1.3.8 The Steps to Improved Competitiveness

In order to regain global competitiveness, the United States must accomplish the following objectives:

- Convert to the metric system
- Recognize the role of bartering
- Implement strategic thinking and planning
- Produce world-class products and services
- Erase arrogance and become more tolerant of other cultures and customs
- Form multinational strategic alliances

The metric system of measurement is the standard throughout the world, with the exception of a few countries including the United States and Nigeria. During the 1970s, the United States attempted unsuccessfully to convert to the metric system; and since that time the United States has continued to lose ground in international competition. Recently, there has been renewed interest in the metric system, since its adoption would make sales and customer support around the world much easier.

Barter is a mode of trade that has existed since the earliest human history, and it is still common practice in many cultures. U.S. manufacturers and the U.S. government are finding that barter arrangements can open new markets for products, where monetary exchanges and/or credit arrangements are impractical. For example, Boeing Commercial Airplane Company recently traded aircraft to Poland in exchange for Polish ham.

U.S. manufacturers need management with a long-range perspective. There is a need for more technology forecasting to anticipate changes and emerging trends. There must also be a willingness to invest in the short term, at the possible expense of immediate profits, in order to gain market share and profitability in the long term.

U.S. products are now judged around the world to have a relatively low quality, much like Japanese products were perceived after World War II. The United States is capable of making high-quality products at low cost, but all participants in the company, including management, labor, and suppliers, must be dedicated to continuous improvement as the Japanese manufacturers have been since the early 1950s.

Americans have long tended to hold an arrogant attitude that the United States is always the best, always right, and always the idol of others. When American business people travel overseas and complain openly about accommodations and other conditions, their hosts become offended and less cooperative. American business people need much more education about world geography, politics, and culture. Through better understanding, they will become more sensitive to and tolerant of cultural differences.

1.4 SUMMARY

We have reviewed the history, status, and foreseeable trends in the global electronics industry. Continued increasing pressure to deliver low-cost, high-quality products, in ever-compressing product life cycles, is anticipated. This text is devoted to providing the engineer an orientation to the technologies and management approaches to enhance competitiveness as the electronics industry advances.

EXERCISES

1. Define electronics and electronics manufacturing.
2. Identify and discuss the elements involved in building successful electronics products.
3. Sketch and discuss the life cycle/design curve.
4. List the three major eras in electronics manufacturing.
5. Why has the United States had difficulty competing in a global market?
6. List the strategies Japan used to become a major player in the global market.
7. What can the United States do to become more competitive in the global market?
8. Discuss the potential of the global market in the electronics equipment industry and why it is important to the United States.
9. What geographic regions and major countries are the major producers of electronic products? What regions and countries represent important future markets?

2 Principles of Production

2.1 INTRODUCTION

In Chapter 1, we described the history, status, and trends in the electronics industry. We emphasized the importance of an approach integrating marketing, design, manufacturing, and support functions to compete successfully in the world electronics market. The purpose of this chapter is to introduce you to the manufacturing function.

In this chapter, we set forth definitions for some important terminology (Sec. 2.1). A life-cycle concept for products is introduced in Sec. 2.2; then in Sec. 2.3, we discuss production processes and the relationships between product and process design. We show that there is a variety of methods used in industry for organizing production processes, and that the appropriate process arrangement is dependent upon the product mix and product sales volumes. Finally, Sec. 2.4 discusses productivity, that is, the concept of engineering efficiency in the context of production processes.

2.1.1 Manufacturing Processes

If we define **production** as the transformation of resource inputs to create useful outputs, then production results in products, services, or a combination of the two. Manufacturing processes are those production processes that convert raw materials and purchased components into products (goods). This text is devoted to electronics manufacturing processes. Figure 2-1 summarizes the major stages in the manufacturing of an electronic product (a desk-top personal computer). Component fabrication results in a packaged part which performs an electrical or electronic function. These parts are then assembled onto a **printed wiring board** (PWB), and PWBs are, in turn, assembled into a system. Part Two of

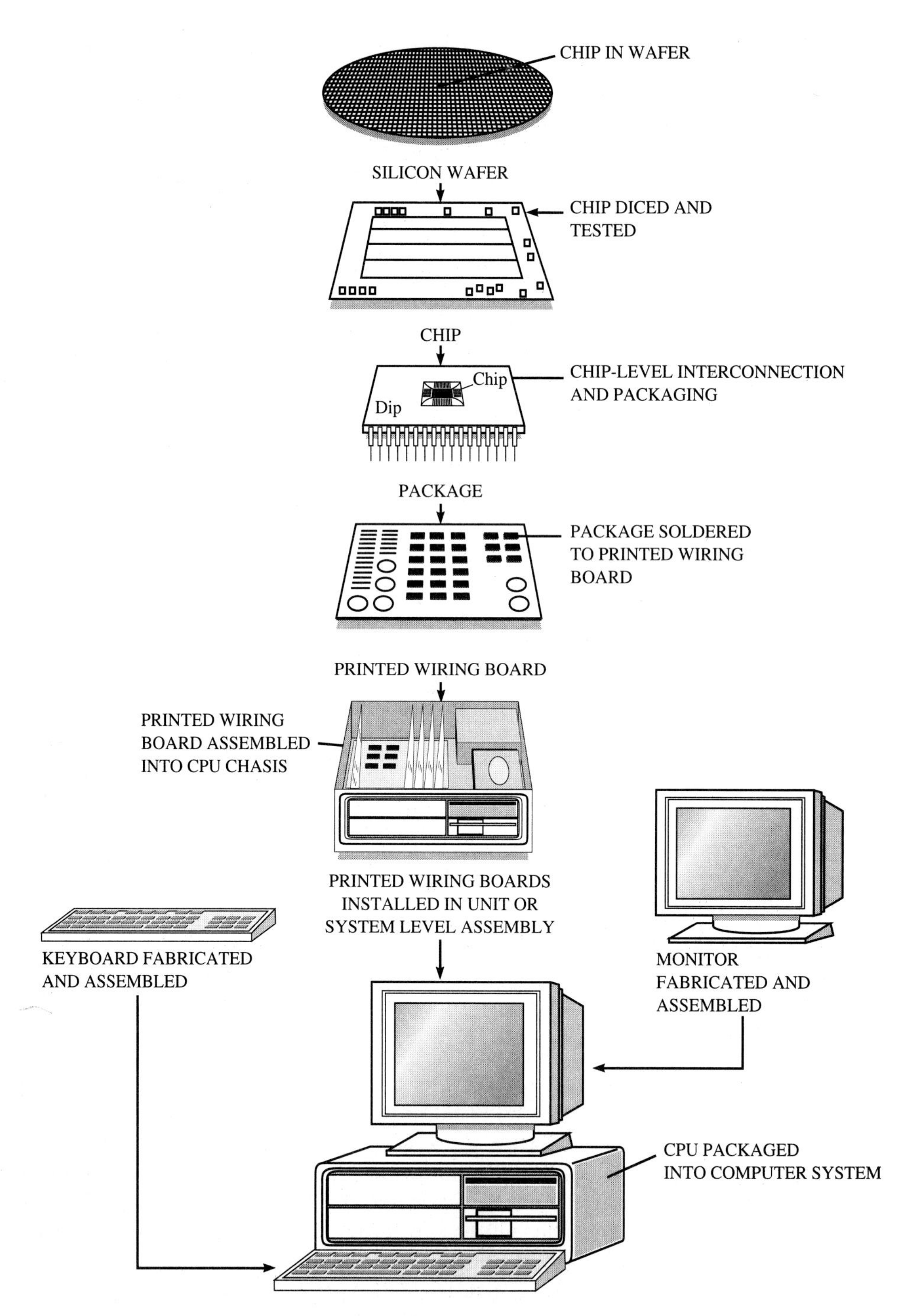

Figure 2-1 Electronic products.

this text introduces electrical/electronic components and the processes of component fabrication. Part Three introduces the principles of automation and the processes of assembling components onto PWBs.

The typical manufacturer performs some, but not all, stages of the production process shown in Fig. 2-1. For example, a manufacturer of populated boards might purchase bare PWBs and component parts from outside suppliers. A manufacturer of personal computers might then purchase the populated boards from that supplier and insert the boards into a chassis assembly, resulting in a modular central processing unit (CPU). Finally, a distributor might purchase the CPU, monitor, keyboard, and other peripherals from different manufacturers and package the system for shipment to retail stores or for direct mailing to individual customers.

2.1.2 Productivity

Productivity is a measure of the efficiency by which a production process transforms resource inputs into outputs. In Sec. 2.4, we discuss the concept of productivity and some tools and methods for productivity improvement. Additionally, Part Four (Life-Cycle Engineering) and Part Five (Manufacturing Systems) discuss key principles and processes for achieving manufacturing productivity.

2.1.3 Concurrent Engineering

Historically, technical specialists have tended to independently design the products and the manufacturing processes without adequate cooperation. Progressive companies in the electronics industry have broken down the organizational walls between design (electrical, mechanical, and computer systems) engineering and manufacturing (industrial) engineering through a concept called **concurrent engineering,** defined as follows:

> Concurrent engineering is a systematic approach to the integrated, concurrent design of products and the related processes, including manufacturing and support. This approach is intended to cause the developers, from the outset, to consider all elements of the product life cycle from conception through disposal, including cost, schedule and user requirements.[1]

To understand the importance of concurrent engineering, we must realize that the design of a product profoundly affects the productivity of the production process. A study by General Electric concluded that 75–90 percent of manufacturing costs are determined by decisions made in the product design process.[2] In the past, most of those decisions were made by the design engineers with minimal input from production or maintenance people.

Design engineers often assume that labor is the primary component of product cost, and that labor productivity is a problem to be dealt with by industrial engineers and

[1]R. I. Winner and others, *The Role of Concurrent Engineering in Weapons System Acquisition* (Alexandria, Va.: Institute for Defense Analysis, December 1988), p. 2.

[2]J. T. Vesey, "Meet the New Competitors: They Think in Terms of Speed to Market," *Industrial Engineering,* 22, 12 (1990), pp. 20–26.

management. However, material cost is becoming increasingly important. The General Electric study also noted that for typical electrical and electronic products, material cost constitutes 65–80 percent of total product costs. A plant producing personal computers reported that material was 80–90 percent of product cost. Design engineers select the materials and components, therefore determining the materials cost for the product. Design engineers also make many decisions affecting the labor costs of product fabrication and assembly.

A similar case exists for the problem of product quality. Design engineers have tended to treat product quality as the responsibility of production workers, quality inspectors, and manufacturing engineers. When a machine operator sets the machine improperly, as few as one defective part out of thousands may result. But when the design engineer makes an error, a defect rate as high as 100 percent may result!

Clearly, engineers of all disciplines must work more closely to ensure a competitive product. Consumer demands for increasing functionality and miniaturization necessitate greater product design cooperation among mechanical, electrical, and software engineers. **Mechatronics** is the union of mechanical, electronic, and computer systems engineering to design and manufacture compact and sophisticated products such as the electronic 35mm camera,[3] the palm-sized video camcorder, and the new generation of sensing and control devices in automotive applications, such as fuel-injection systems.[4]

Concurrent engineering means that in the electronics industry, electrical, mechanical, and software design engineers will increasingly work on product development teams where industrial engineers and other manufacturing engineers are integral participants, to ensure that the product requirements and the process capabilities are compatible. The progression toward concurrent engineering and the development of CAD tools will also result in design engineers performing functions, such as producibility analysis or process planning, which have traditionally been performed by industrial or manufacturing engineers.

The traditional organizational structure, shown in Fig. 2-2, arranges human resources functionally along departmental lines. This arrangement promotes cooperation within a function. For example, within an engineering department junior engineers can efficiently learn from more senior engineers. However, the functional organization inhibits cross-functional integration. With the trend toward shorter product lives and concurrent engineering, new organizational structures are needed.

The organizational trend in the electronics industry has been toward the formation of interdisciplinary product teams. These teams are formed on a temporary basis and are freed from many of the normal bureaucratic procedures. Such teams have demonstrated extraordinary innovation and success in compressing the product development cycle. This virtual organization is distinct from the project organization, where a company is permanently organized by projects (such as in the construction industry). The electronics industry is too dynamic and product life cycles too short for the project teams to become

[3]L. Ken Keys and Charles M. Parks, "Mechatronics, Systems, Elements and Technology: A Perspective," *IEEE Transactions on Components, Hybrids, and Manufacturing Technology,* 14, 3 (1991), pp. 457–61.

[4]L. A. Berardinis, "Mechatronics: A New Design Strategy," *Machine Design,* April 26, 1990, pp. 50–58.

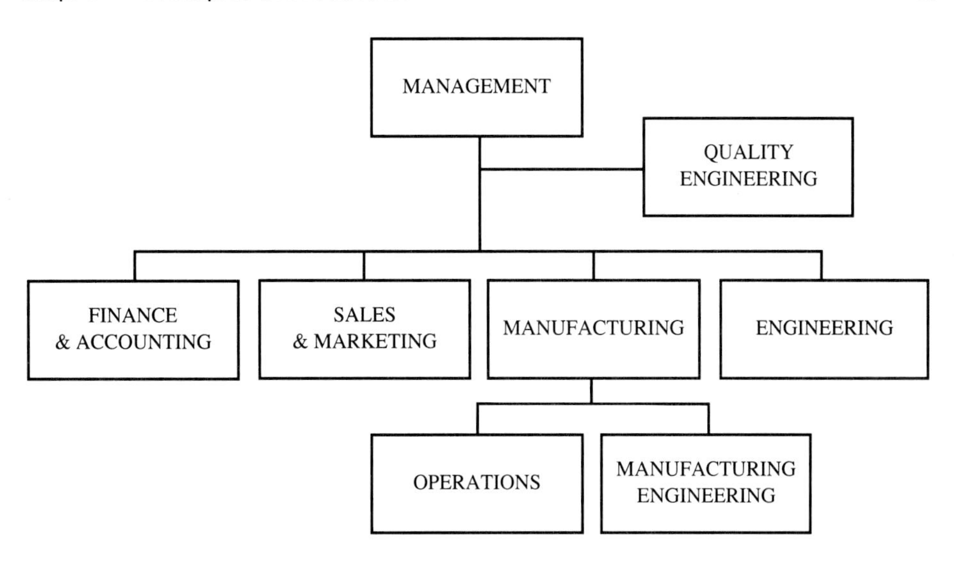

Figure 2-2 Traditional organizations.

permanent. The virtual organization is also different from the matrix management organization, where a person simultaneously reports to a project manager and a home-office functional manager over a prolonged period of time. Matrix management is flawed in that an engineer has multiple supervisors and may receive conflicting direction.

2.2 PRODUCT LIFE CYCLE

Suppose you buy a personal computer (PC) such as the one illustrated in Fig. 2-1. If we were to trace the history of your PC, we would find that it passes through a **product life cycle,** consisting of the stages shown in Fig. 2-3(a). First, engineers design and develop the PC model. Then the PC is produced in a manufacturing process. After an interval of time in a distribution and retail sales network, you purchase the PC. The PC is used, and when failures occur maintenance is performed. Finally, either because of technological obsolescence or excessive frequency of repair, the PC reaches the end of its useful life.

In Sec. 2.1.3, we stressed that designers must consider the requirements to be met by the product during the life-cycle phases following design and development. The product should be designed so that it is easy to manufacture in the plant and easy to repair in the field. With environmental protection becoming increasingly important, design engineers are also beginning to recognize that a well-designed product is made of materials that do not harm the environment upon disposal at the end of the product's life. Engineers must also be aware of the environmental impacts of scrap and waste by-products of the production process such as packaging material and processing chemicals.

From the viewpoint of the manufacturer, a product model or product family progresses through the life cycle shown in Fig. 2-3(b). The manufacturer introduces a product model. As the model increases in popularity, sales and production volumes increase during

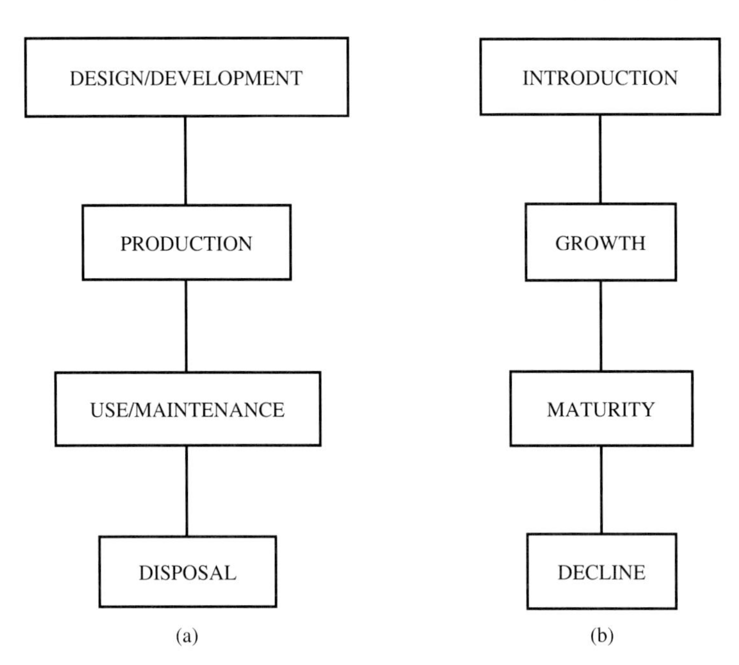

Figure 2-3 Product life cycle from the (a) customer's viewpoint, and (b) producer's viewpoint.

a period of growth. Eventually, the market becomes saturated and sales volume reaches a peak. Finally, the product begins to decline in popularity, perhaps because of technological obsolescence or competitive pressure.

Figure 2-4 depicts the product life cycle in terms of popularity, where sales rate is plotted versus time. Early in the product life cycle, only a few units are produced for development testing and for limited market tests. If the development and marketing tests are successful, then the manufacturer begins to produce and distribute the product in larger quantities, such as for an immediate geographic market or for established customers. If the product continues to increase in popularity, a growth period is experienced with sales rate following an S-curve. Early in the growth phase, sales grow at an increasing rate. Sales continue to grow, but at a decreasing rate, through the maturity phase and up to the saturation point. Eventually, the product declines to obsolescence.

When designing products and processes, we must recognize the dynamic nature of the product life cycle. Different processes are required in the different stages of the product life cycle. For example, early in the introduction and growth phases, volume is small and we are uncertain about the future potential. For these reasons, we tend to use low-cost and general-purpose equipment which reduces the financial risks and maximizes the operational flexibility. As the sales volume increases, it may become necessary to invest in mechanization and automation to increase productivity, thereby reducing sales price and further stimulating sales while deterring competition. However, the economic studies to justify the investment in mechanization and automation must consider the finite nature

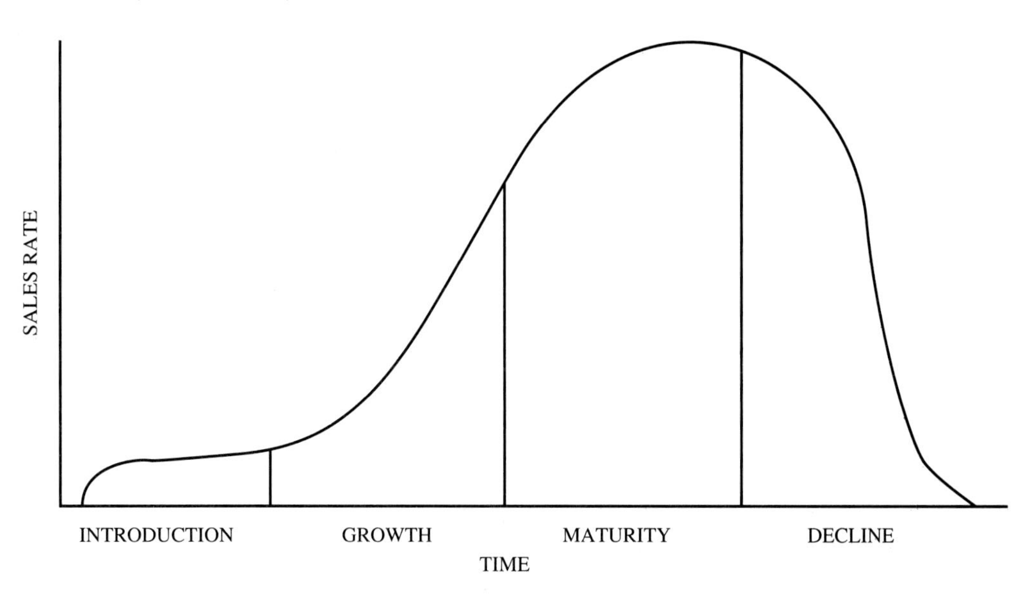

Figure 2-4 Product life-cycle curve.

of the product life cycle. If the product enters the declining phase before the investment is recovered, then the company will sustain heavy financial losses.

The product life cycles in the electronics industry are becoming shorter. In some products such as personal computers, the life cycle can be as short as 9 to 18 months. Compression of product life cycles requires that the manufacturer be very efficient in product introduction. The objective is to accelerate product and process development in order to produce quality products in high volume, thereby capturing market share and gaining the financial benefits of the maturity phase. If a company gets its product to market a few months late, it misses out on a major fraction of the potential profit to be gained over the product life cycle.

Electronics manufacturers that can compress the time for getting a product to market gain a strong competitive advantage. Since the early 1980s, Japanese companies have focused their attention on increasing the productivity of the engineering function. Their traditional strengths in teamwork and group problem-solving have promoted concurrent engineering of the product and the process.

2.3 PRODUCT AND PROCESS RELATIONSHIPS

How should we organize production for a given product or group of products? Every product has unique features requiring special processes. Typically, though, many of a company's products have a great deal in common, and we can usually determine the right general form of process by studying the products, product sales volumes, and product mix. Product sales volumes can range from a quantity of one (one-of-a-kind products, such as custom-built test equipment) to quantities of thousands per day (such as microcircuits). Typical volumes for PWBs are in the range of fewer than 100 per day to a few hundred per day. **Product mix** refers to the diversity of products produced by a manufacturer. Some manufacturers produce only one or a few products. An example is the integrated circuit fabrication plant making a particular scale of memory chips. Other manufacturers (such as those making custom industrial electronic products) manufacture hundreds of different products. The typical assembler of populated PWBs produces a mix of boards ranging from a few dozen to a few hundred, and a day's production might include fewer than 20 different board types.

Successful experience in industry has shown that one of several general types of processes is well suited to a given product mix. For example, if a plant produces a high mix (wide variety) of products, most of them in low volume, then a batch process is effective. In a batch process, we take advantage of the volume in common parts across the product mix to produce in economical lot sizes. We do a setup operation (that is, we prepare the machines in the workplace to produce a particular product); then we produce a batch of the product. For example, a manufacturer of microcomputers produces several different models of computers. However, many of the models use the same memory board. So a PWB assembly machine is set up to produce that board. This means that the components (memory microcircuits) are loaded onto the insertion machine and the machine is programmed to efficiently insert the memory chips into an array of sockets on the board.

2.3.1 Product-Process Matrix

Figure 2-5 is the product-process matrix, which summarizes the typical processes, according to the product mix and volume relationships. Note that the batch process is often used in board shops and in industrial electronics, where electronic products are made for other manufacturers. Subcontractors that make avionics equipment for aerospace contractors or controller units for manufacturers of machine tools are examples of companies in industrial electronics.

The four types of processes are as follows:

- *Project:* A project shop produces one-of-a-kind products. An example is the laboratory, or model shop, where engineering prototypes are built.
- *Batch:* A batch shop makes products in lots (batches) of quantity greater than one. Production is intermittent and very complex. The process is difficult to control and is characterized by high levels of work-in-process inventory.
- *Flow:* A flow shop takes advantage of economies of scale in high volume and low mix. Because of the high volume, it is feasible to invest in engineering time and specialized equipment to make the process very efficient. The classic example of a flow shop is the assembly line, where products flow continuously through the process. The high volumes in consumer electronics can justify the investment in an assembly line.
- *Process:* The true process shop produces one or a few products continuously, in very high volume. The chemical plant, where a raw material is continuously trans-

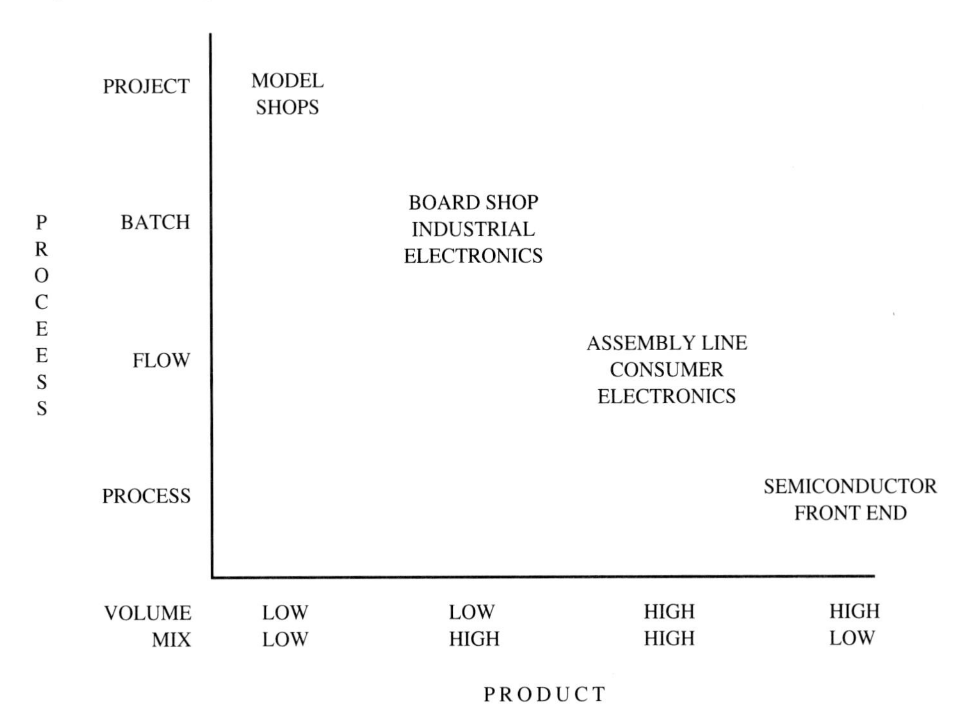

Figure 2-5 Product-process matrix.

formed into a product or family of products and by-products, is an example of a process shop. It is difficult to cite examples of process shops in the electronics industry. The semiconductor foundry and front end approaches the process shop, in that a commodity (one product) is manufactured in a large quantity, almost continuously.

Table 2-1 summarizes some other features of the four typical processes.

2.3.2 Layout

Layout refers to the spatial arrangement of physical resources such as equipment. This topic is introduced here and discussed in greater detail in Chapter 14 (Facilities and Materials Handling). Figure 2-6 illustrates the following four main types of layouts:

- *Fixed:* The product is built in one place, such as on a workbench in the laboratory. All materials, tools, and skills are brought to the location. The fixed layout is typical in the construction industry.
- *Functional:* In the functional layout, equipment is arranged by process or function performed. This layout is very flexible and therefore consistent with the needs of a batch shop where many different products are made in low volumes.

TABLE 2-1 PROCESS CHARACTERISTICS

Process	Layout	Equipment	Control
Project	Fixed	Specialized	Project Scheduling
Batch	Functional	Conventional, General-Purpose Equipment; Flexible Automation	Material Requirements Planning (MRP)
Fow	Group	Conventional General-Purpose Equipment; Flexible Automation	Just-In-Time (JIT)
Process	Product	Hard Automation	Capacity Management

- *Product:* The product layout arranges equipment in an efficient straight-line flow optimal for a high-volume commodity product.
- *Group:* The group layout blends the flexibility of a functional layout with the efficiency of a product layout. The group layout is made possible by identifying a product family (group) with common processing requirements. The grouping creates the greater volume needed to justify dedicating a work center to a specific process. This analytical technique of defining product families is called **group technology.** The U-shaped arrangement allows for some flexibility in the sequence in which operations are performed for different products in the family.

Figure 2-7 summarizes the general types of layouts suited to product-process combinations. A specific application might vary somewhat from the layout recommendations in Fig. 2-7. Also, a specific plant may include a mixture of layout designs for a variety of product-process combinations. Fixed and functional layouts tend to be process-focused, since they are not optimized for the product. Conversely, the group and product layouts tend to be product-focused, since they tend to be specifically designed for a product or family of products. Process-focused layouts are generally more flexible, while product-focused layouts are more efficient.

2.3.3 Equipment

We discuss electronics manufacturing equipment in Parts Two and Three of this text. Table 2-1 identifies the types of equipment used in the different processes. In the electronics industry, the project-type process and fixed layout are used for design and development of engineering prototypes for emerging products. This work is done in a model shop or laboratory, where small-scale equipment, special tools, and precision instruments are used. As a product becomes more defined, the prototypes may be built in the actual production environment, rather than the laboratory, to identify potential manufacturing problems.

In both the batch (functional layout) and flow (group layout) processes, manufacturers tend to use conventional, general-purpose equipment that is economical and flexible.

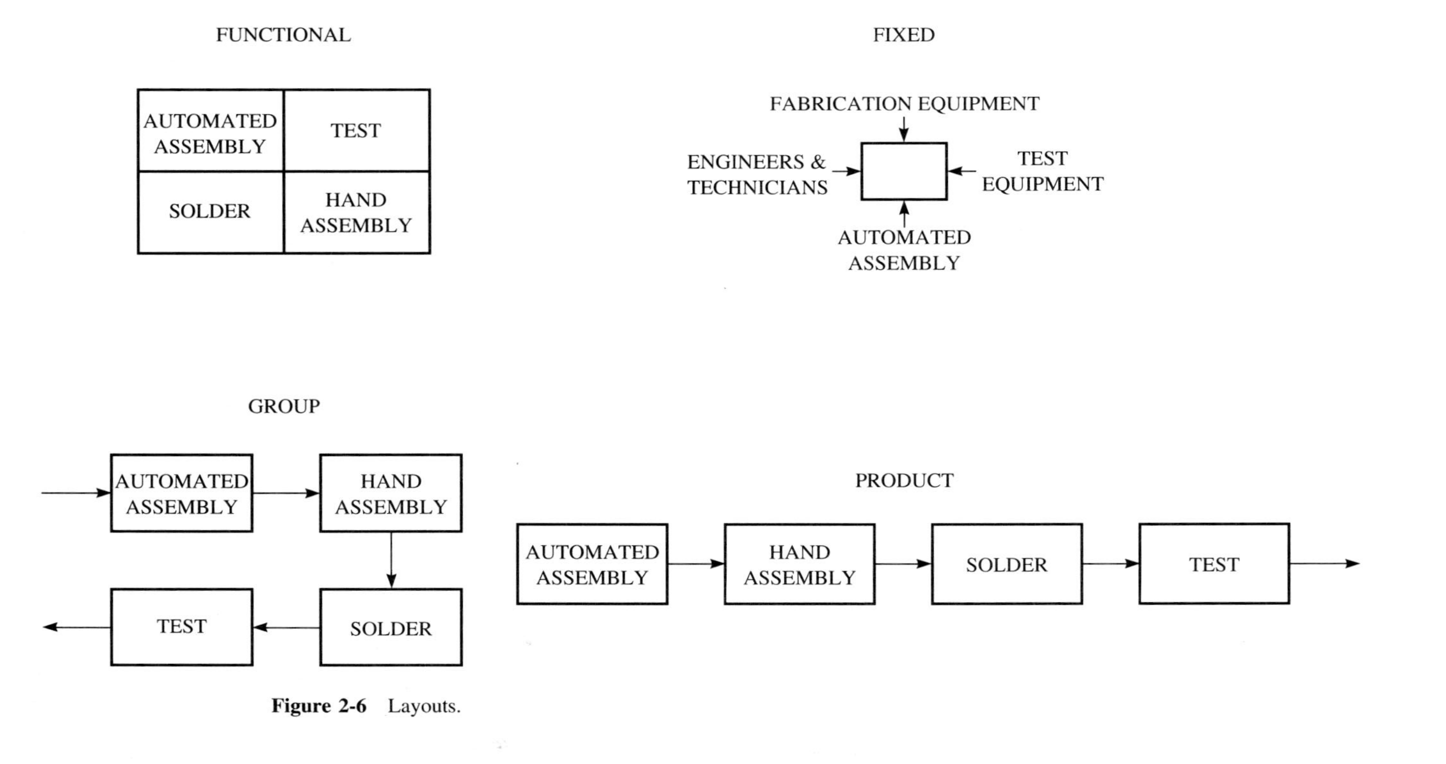

Figure 2-6 Layouts.

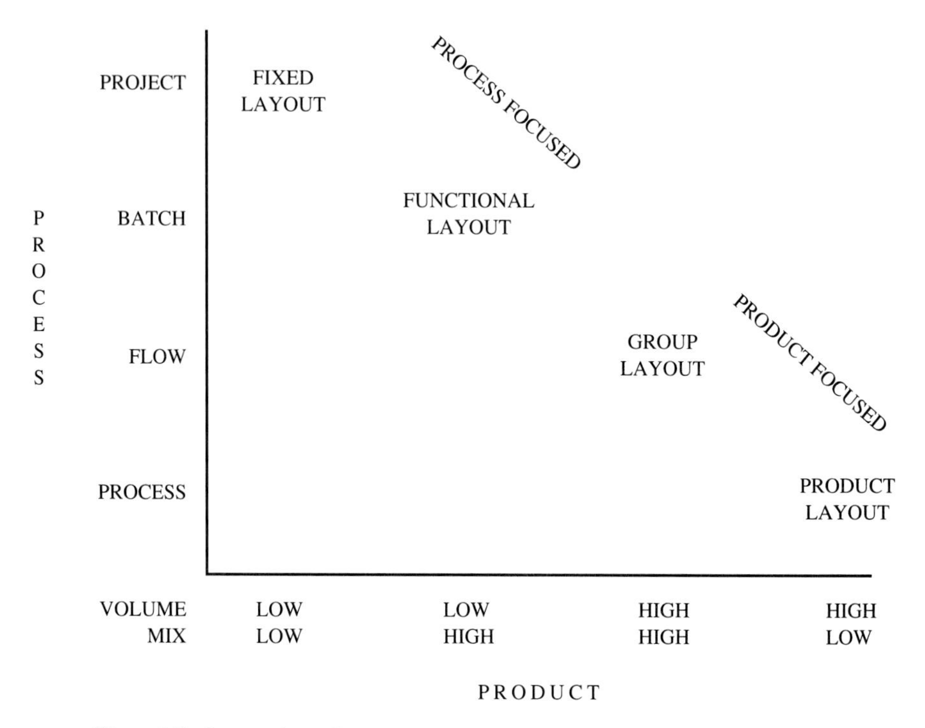

Figure 2-7 Layout alternatives.

There is a tendency to use flexible automation as production volume increases. The only difference in the two processes is in the way the equipment is spatially arranged. In the functional layout, machines (or processes) are clustered based on function; while in the group layout, machines (or processes) are clustered based on the requirements of the product or product family.

In batch and flow shops, manufacturers use flexible automation when the production quantities or other considerations (such as stringent quality requirements or safety hazards) justify the investment. Flexible automation is so named because it is programmable, and changes among different products are accomplished by loading different programs and changing-out tooling. Part Three of this text discusses flexible automation of insertion and placement equipment.

As production quantities increase, or if the products are high-value with large profit margins, then the **flexible manufacturing system** (FMS) may be justified. The FMS automates the storage, handling, and production control in a cellular group layout consisting of flexible, automated machines.

In the process shop, high-volume, low-mix production justifies heavy investment in mechanization and fixed (hard) automation. Hard automation is optimized to a specific product, and processing another product may be difficult, if not impossible. Hard automation is justified based on production quantity and competitive requirements to produce at very low cost. Although highly efficient, hard automation is inflexible and should only be used with caution in the electronics industry, where product lives are short and flexibility is increasingly important.

As a product progresses through its life cycle, the process requirements change. During product development, prototyping is organized on a project basis. Early in the growth phase, volume is low and batch processing in a functional layout, along with other different products, is appropriate. As the product increases in popularity, the flow process is preferred. If the product matures as a high-volume commodity, then the investment in mechanization and hard automation in a continuous process may be justified. Finally, as the product declines, production may once again be in batches.

2.3.4 Production and Inventory Control

Every shop requires a **production and inventory control** (P&IC) system to ensure timely delivery of products to customers, efficient operations, and minimum inventory levels. The P&IC system schedules shop activities and orders of materials and components from suppliers. Industry experience has shown that certain P&IC systems are well suited to the typical processes, as shown in Table 2-1. These systems are defined as follows and further discussed in Chapter 15:

Project Scheduling. Projects consist of activities which must be performed according to certain precedence relationships and completed by specific milestone dates. Project scheduling systems make use of network diagrams to plan and control the activities. The standard systems are Critical Path Method (CPM), Project Evaluation Review Technique (PERT), and Gantt-chart (bar-chart) schedules.

Material Requirements Planning (MRP). MRP is a computer-based system for scheduling production and outside purchases. It focuses on the product structure as defined in the bill of materials (BOM) and combines the material requirements for a given component, considering its usage in all end products. MRP is oriented toward batch production and is common in low-volume, high-mix applications such as batch shops.

Just-In-Time (JIT). JIT is a management philosophy emphasizing minimization of waste. A major part of JIT is a P&IC system which schedules stable, repetitive production for an extended time interval, such as a month. A high-volume, low-mix product line tends to make this approach more feasible. JIT uses very simple procedures for signaling the need to produce components and subassemblies, in response to near-term customer demand. Rather than using scheduling to push production of all items (as in MRP), JIT pulls components and subassemblies through the stages of production and into final assembly and is therefore called a pull system. JIT seeks to minimize inventory levels, and it requires strict disciplines, such as preventive maintenance on equipment, low defect rates, and very efficient setup operations.

Capacity Management. Capacity management is a set of policies, procedures, and systems to determine requirements for production capacity (labor, equipment, and facilities) and control daily utilization of existing capacity. It is an essential function in any manufacturing operation, but it is the most important element of P&IC in the process industries. The product-focused operation tends to involve heavy capital investment in plant and equipment (including inflexible hard automation). Because of the high risks involved, the capacity plan must be accurate. Also, once the capacity is in place, efficient utilization of the capacity determines the efficiency of the operation.

2.4 PRODUCTIVITY

Engineers are familiar with the concept of efficiency, as illustrated in Fig. 2-8. In a production process efficiency is called **productivity** and is calculated as follows:

$$Productivity = Efficiency = \frac{Output}{Input} \tag{2-1}$$

2.4.1 Productivity Measurement

Much of engineering analysis and design seeks to optimize efficiency within the constraints of economics and technology. In the case of an amplifier circuit, the amplification process converts an input signal to an output signal. To measure amplifier efficiency, we calculate the following ratio:

$$Gain = \frac{Output\ Amplitude}{Input\ Amplitude} \tag{2-2}$$

In a manufacturing system, it is difficult to measure productivity, because there are multiple input resources and multiple outputs (Fig. 2-9). We have noted that product cost is heavily dependent upon material costs. The other input factors in Fig. 2-9 affect the

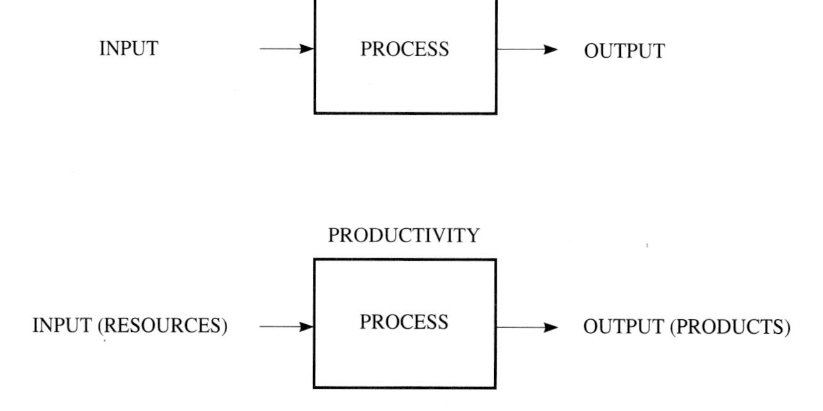

Figure 2-8 Efficiency and productivity.

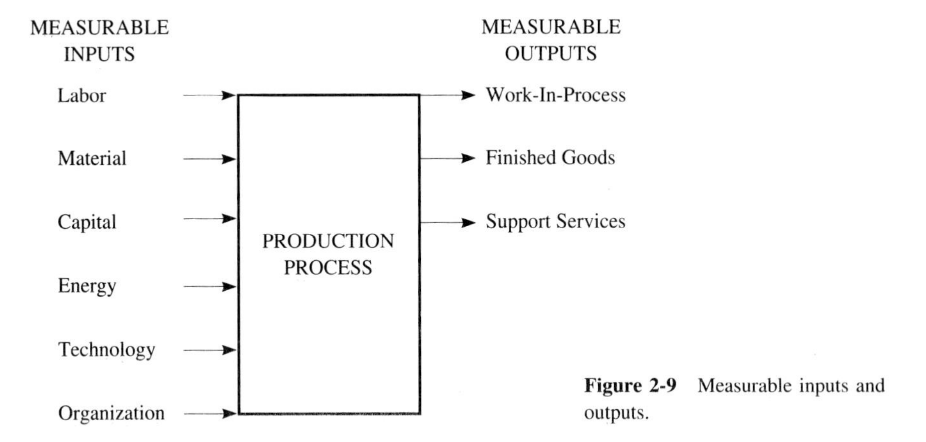

Figure 2-9 Measurable inputs and outputs.

product cost indirectly through the processing costs incurred in fabrication, assembly, and testing. Material selection also affects processing costs, since some materials are more difficult to process than are others.

There are two basic ways to measure productivity, as follows:

- *Partial Productivity:* The ratio of total output to one class of input
- *Total Productivity:* The ratio of total output to all input factors

Partial productivity is easier to measure, and historically the most common measure has been labor productivity (measured in output, such as boards, per unit of labor input, defined as direct labor hours). However, labor productivity is an inadequate and misleading measure. As automation becomes more common and material costs increase as a percentage of total costs, output per direct labor hour is neither a representative nor a consistent basis for comparing among products or process alternatives. Chapter 16 provides guidance for determining manufacturing costs and methods of evaluating economic alternatives.

2.4.2 Productivity Improvement

Productivity measurement helps us to better understand a process and to monitor progress or compare alternatives. However, in a competitive business environment our primary objective is to continually improve productivity. In this section, we will consider some methods for productivity improvement. Part Four of this text is devoted to examining methods of productivity improvement in detail.

First, the engineer should define the process. The process flowchart is a systematic tool for documenting the steps (operations) in a production process. The method uses five symbols, as shown in Fig. 2-10. Tools such as the form shown in Fig. 2-11 aid in the development of flow process charts.

Figure 2-12 illustrates application of the flow process symbols to chart the process for manufacturing of a PWB. The chart shows the operations and includes estimates of the elapsed time for each operation and the distances traveled in a proposed layout.

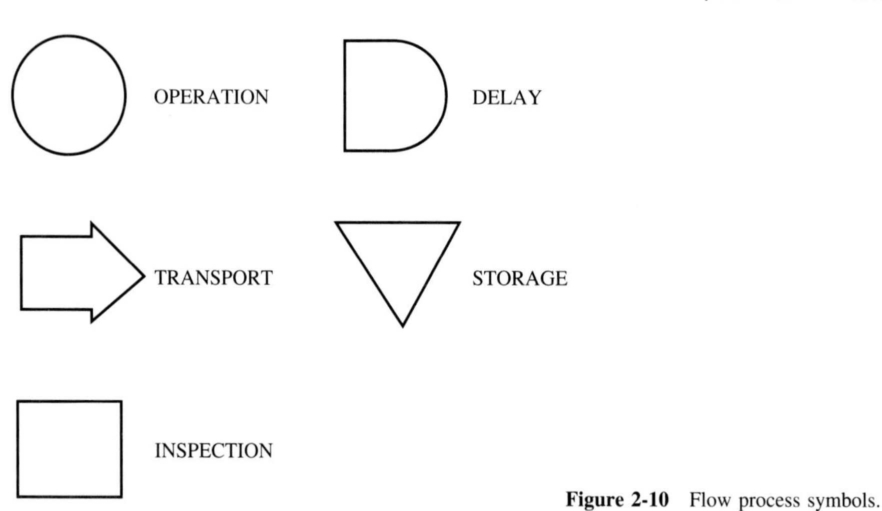

Figure 2-10 Flow process symbols.

A major objective in productivity improvement is to streamline the process as much as possible. In Fig. 2-12, we see that productivity can be improved in several ways, including eliminating an operation, reducing the time required for an operation, and minimizing movement, storage, and delays. World-class manufacturers have demonstrated that the most effective strategy for productivity improvement is to utilize the potential of human imagination. When management encourages participation of all employees in group problem-solving, dramatic improvements are possible. While industrial engineering principles have been neglected in the United States, Japanese companies have made productivity improvement a national crusade. Barnes, in his classic text of work measurement and methods improvement, states the following principles of work simplification:[5]

- Eliminate all unnecessary work
- Combine operations
- Change the sequence of operations
- Simplify the necessary operations

For the example in Fig. 2-12, we see several improvement opportunities. The **design-for-assembly** (DFA) process (Chapter 10) eliminates unnecessary operations and leads to the most efficient sequence of essential operations. The principles of total quality control (Chapter 11) can be used to eliminate the operations, storage, and delays associated with inspection. The principles of layout and material handling (Chapter 14) lead to reductions in the material movement distances and enhanced efficiency of the required handling. The principles of production and inventory control (Chapter 15) can be used to minimize other storage and delays, such as the in-queue times, when the product waits to be processed on a machine. Note that the total throughput time for the product is 8 to 9

[5]Ralph M. Barnes, *Motion and Time Study: Design and Measurement of Work* (New York: John Wiley & Sons, 1984), p. 50.

FLOW PROCESS CHART

NO. ______
PAGE ___ OF ___

ANALYSIS				
WHY?				
WHAT?	WHERE?	WHEN?	WHO?	HOW?

QUESTION EACH DETAIL

SUMMARY

	PRESENT		PROPOSED		DIFFER'CE	
	NO.	TIME	NO.	TIME	NO.	TIME
○ OPERATIONS						
⇨ TRANSPORTATIONS						
□ INSPECTIONS						
D DELAYS						
▽ STORAGES						
DISTANCE TRAVELED		FT.		FT.		FT.

JOB ______

□ PERSON or □ MATERIAL ______

CHART BIGINS ______
CHART ENDS ______
CHARTED BY ______ DATE ______

	DETAILS OF (PRESENT) METHOD	TIME minutes	Operation	Transport	Inspection	Delay	Storage	Distance	Quantity	Time	ACTION: Eliminate	Combine	Segue	Change: Place	Person	Improve	NOTES
1			○	⇨	□	D	▽										
2			○	⇨	□	D	▽										
3			○	⇨	□	D	▽										
4			○	⇨	□	D	▽										
5			○	⇨	□	D	▽										
6			○	⇨	□	D	▽										
7			○	⇨	□	D	▽										
8			○	⇨	□	D	▽										
9			○	⇨	□	D	▽										
10			○	⇨	□	D	▽										
11			○	⇨	□	D	▽										
12			○	⇨	□	D	▽										
13			○	⇨	□	D	▽										
14			○	⇨	□	D	▽										
15			○	⇨	□	D	▽										
16			○	⇨	□	D	▽										
17			○	⇨	□	D	▽										
18			○	⇨	□	D	▽										
19			○	⇨	□	D	▽										
20			○	⇨	□	D	▽										
21			○	⇨	□	D	▽										
22			○	⇨	□	D	▽										
23			○	⇨	□	D	▽										
24			○	⇨	□	D	▽										

Figure 2-11 Flow process chart form.

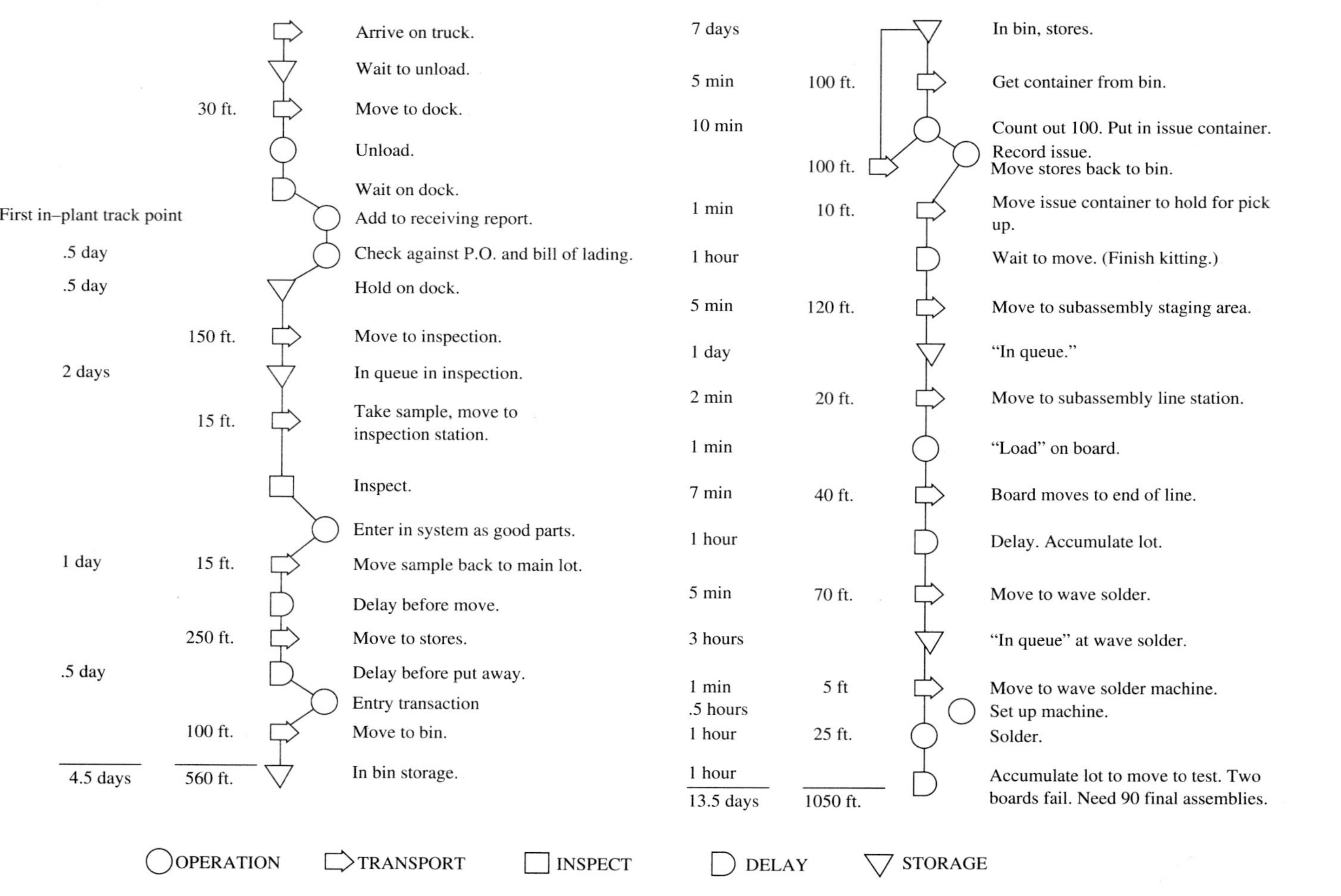

Figure 2-12 Flow process chart for a PWB.

(*Source:* Robert W. Hall, "The Implementation of Zero Inventory/Just In Time," in *The APICS Library of American Production* (Falls Church, Va.: American Production and Inventory Control Society, 1986).)

days. In general, the longer the throughput time, the higher the work-in-process (WIP) inventory level (incomplete units) and the lower the productivity of the material input resource.

Once we have determined the minimum essential operations, we can then consider automation to further improve productivity. However, it is often the case that once we determine the optimum manual process, efficiency is improved to a point that automation is marginally cost-effective. There are still many situations in which automation is essential. We will see one such example in the case of surface-mount technology (SMT) in Chapter 9. SMT involves very small components with high lead-densities requiring handling care and placement accuracy not achievable by manual assembly.

2.5 CASE STUDY

A case study in productivity improvement has been documented for the assembly operations on the Hewlett-Packard (HP) HP9000 computer line.[6] The objective was to implement some principles of JIT productivity improvement, in a low-volume, high-mix environment. Figure 2-13 shows the layout in the early stages of the project. We see that a functional layout was used, but that there was a typical process flow for most of the product (indicated with arrows).

Note the ESS oven in the lower portion of the layout. ESS refers to environmental stress screening, a quality-control screening approach that uses severe environmental stress such as temperature cycling and vibration to reveal defects in materials and workmanship. ESS has been used for several years in the defense industry and is becoming very important in commercial electronics (Chap. 13). In the initial stage of the program, ESS was not fully integrated into the process flow. Several inspection activities were used instead.

PWBs were assembled along a straight-line conveyor called a transporter. Transporters permit assembly to be operator-paced; but as a consequence, a temporary buffer storage carousel is required for the WIP inventory that accumulates due to imbalances in the processing rates. Transporters and carousels are discussed in Chapter 14. These handling and storage devices improve the productivity of labor by reducing manual handling, but they require substantial capital investment and can lead to excessive WIP inventory.

A material requirement planning system was used for production and inventory control (discussed in Chap. 15). MRP is based on forecasts, and if sales demand falls short of forecast, MRP can lead to the scheduling of excessive production and resulting surplus inventories.

In the next phase of the improvement program, HP sought to change the operation into a flow process, by using additional JIT principles. Because the product mix was characterized by many products in small quantities, a continuous-flow process was not feasible. However, Hewlett-Packard focused on reducing quality problems and increasing responsiveness to market demands.

The new layout was still functional, except that U-shaped group layouts were

[6] William Sandras, "Low Volume/High Mix JIT," *Twenty-Ninth Annual International Conference Proceedings,* American Production and Inventory Control Society, October 1986, pp. 295–99.

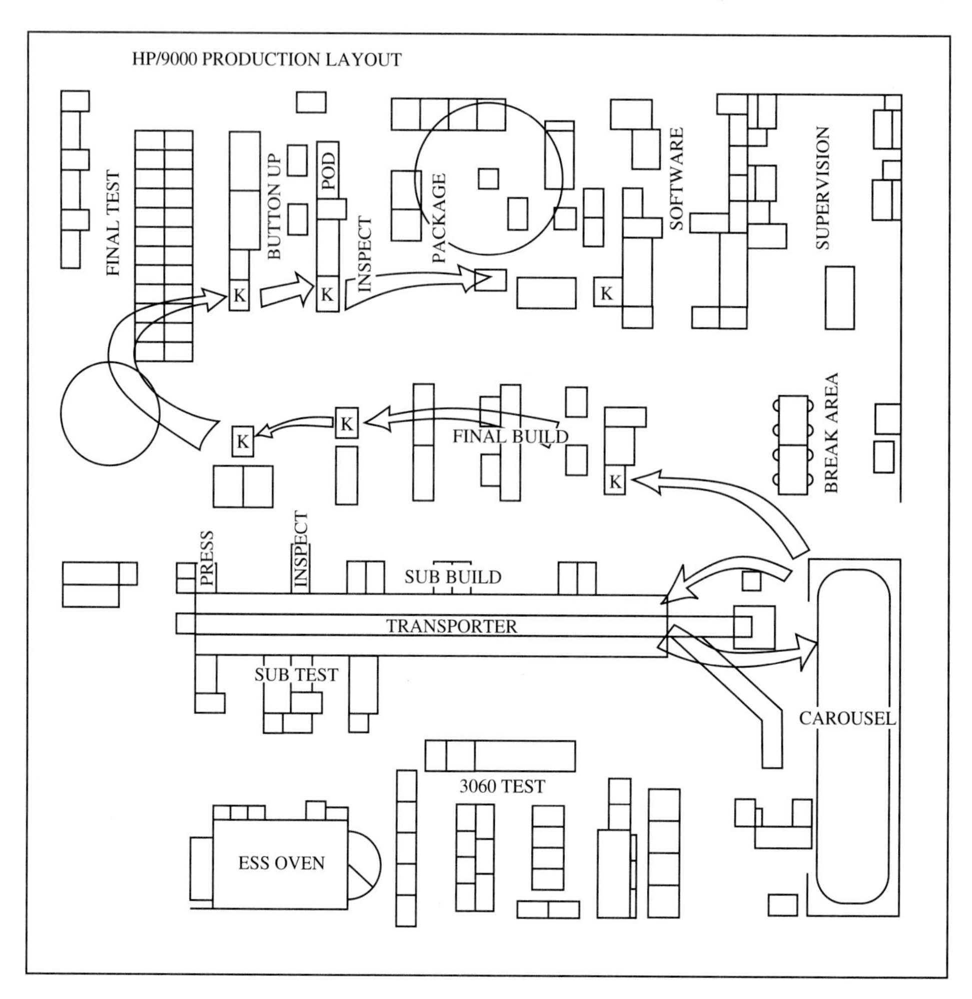

Figure 2-13 HP-9000 production layout.

(*Source:* William Sandras, "Low Volume/High Mix JIT," pp. 295–99.)

used within functional areas to enhance material flow, flexibility, communications, and cooperation. ESS was integrated into the product flow, and as a result of the screen effectiveness, process problems were identified and defects reduced to the point that the inspection steps were eliminated. HP modified the MRP system to focus primarily on management of purchased components. Daily shop-floor control was changed to a JIT signaling system whereby boards were pulled through production only when customer demand created the need for final assembly. This pull system eliminated the need for the transporter and the carousel buffer storage. These changes resulted in the following improvements:

- Inventory reduced more than 70 percent
- Throughput time reduced from 16 days to 2 days
- Scrap and rework reduced 60 percent

Case studies such as this for the HP9000 computer shop illustrate the potential for improving the competitiveness of an electronic manufacturer, through application of the principles discussed in this text.

2.6 SUMMARY

This chapter has introduced you to basic principles of production. In the electronics industry the accelerating pace of product developments, compressing product life cycles, and increasing customer demands for high-performance, high-quality, low-cost products require the concurrent engineering approach. Increasingly, engineers are working in multidisciplinary teams, and engineers who have historically focused only on product design are, with CAD tools, performing many of the functions of the traditional manufacturing engineer. Mechatronics is an approach to product and process design which reflects the increasing interdependency of engineering disciplines.

The production-process design must be compatible with the product mix, and we noted that the production quantities and the diversity of the product mix influence the organization of production and the choice of: (1) layout, (2) equipment and degree of automation, and (3) systems for control of production and inventory.

Production efficiency is defined as productivity. Traditionally, productivity has been measured in terms of output per unit of direct-labor input. With the trend toward automation of many production processes and the increasing share of indirect labor (such as design and manufacturing engineering), direct labor productivity is no longer a suitable measure. The trend is toward total productivity measures, and vigorous efforts toward continuous improvement of productivity in order to remain competitive.

KEY TERMS

Concurrent engineering
Design for assembly (DFA)
Flexible manufacturing system (FMS)
Group technology
Just in time (JIT)
Material requirement planning (MRP)
Mechatronics
Printed wiring board (PWB)
Product life cycle
Product mix
Production
Production and inventory control (P&IC)
Productivity

EXERCISES

1. Discuss the importance of concurrent engineering in electronic manufacturing.
2. Relate the product life cycle to market share and financial benefits.

3. Identify consumer electronic products that are in each of the product life-cycle phases.
4. Discuss how product mix and volumes influence the selection of a facility layout. Address all four types of layouts.
5. Discuss equipment in the context of different processes.
6. Discuss the JIT approach to production and inventory control, and identify the situations in which JIT best applies. Explain how JIT is more than a P&IC system.
7. Define productivity and two ways of measuring it.
8. Why is labor productivity becoming an inadequate and misleading measure of efficiency in electronic manufacturing?
9. What is the most effective strategy for productivity improvement?
10. What issues must a designer consider for a product over its life cycle?
11. Explain why the manufacturer with a high-mix, low-volume product line might chose to produce in batches.
12. Discuss how a group layout promotes flexibility.
13. Some processes, such as assembly lines, are machine-paced, while other processes, such as functional layouts, tend to be operator-paced. Read about transporters and carousels in Chapter 14 and discuss how a transporter and a carousel, such as in the HP9000 case study, permit operator-paced assembly.
14. Make a photocopy of Fig. 2-11 and prepare a flow process chart of a food tray in a dormitory cafeteria, from the time it is dropped off dirty from one meal, until it is dropped off dirty at the next meal.
15. Tour a manufacturing plant, preferably in the electronics industry, and chart the flow process.

BIBLIOGRAPHY

1. ADAM, EVERETT E., AND RONALD J. EBERT, *Production and Operations Management Concepts, Models and Behavior.* Englewood Cliffs, N.J.: Prentice Hall, Inc., 1992.
2. AFT, LAWRENCE S., *Productivity Measurement and Improvement.* Englewood Cliffs, N.J.: Prentice Hall, Inc., 1992.
3. BARNS, RALPH M., *Motion and Time Study: Design and Measurement of Work.* New York: John Wiley & Sons, 1980.
4. EDOSOMWAN, JOHNSON A., AND ARVIND BALLAKUR, *Productivity and Quality Improvement in Electronics Assembly.* New York: McGraw-Hill, Inc., 1989.
5. HUNT, V. D., *Mechatronics.* New York: Chapman and Hall, 1988.
6. SCHONBERGER, RICHARD J., *World Class Manufacturing.* New York: The Free Press, 1986.

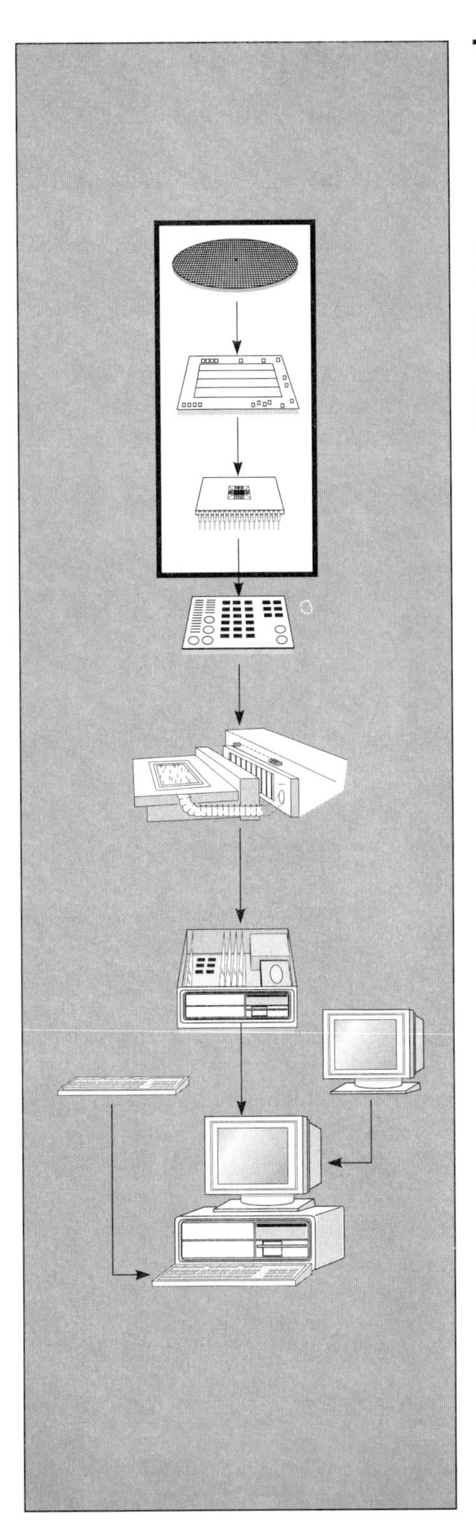

PART TWO
ELECTRONICS FUNDAMENTALS

3 Introduction to Electronic Components

3.1 INTRODUCTION

In electronics manufacturing, the majority of components fall into two main categories: through-hole (TH) components and surface-mount (SM) components. In this chapter, we discuss device types (Sec. 3.2–3.3) and their fabrication and packaging methods (Sec. 3.4–3.9)

3.1.1 Through-Hole Components

Through-hole (TH) components are components that have wire leads which must be inserted through predrilled holes on a PWB. The wire leads are then clinched, trimmed, and soldered. The leads of these components serve the dual purpose of providing circuit connectivity, by being soldered to the circuit paths, and acting as a secure mounting structure to hold the component in place. This has been the dominant method of attaching components to PWBs since the invention of PWB technology, by Dr. Paul Eisner, in 1936.

3.1.2 Surface-Mount Components

Surface-mount components (SMC) or surface-mount devices (SMD) are components that are mounted directly to the surface of the PWB, so that it is not necessary to have holes drilled through the substrate to mount the component. Since the leads do not have to pass through the PWB, the leads can be much smaller than their TH counterparts. Some SMCs are considered leadless, having only a metallized termination on the body of the component for mechanical and electrical connection to the board. The components are initially attached to the board with an adhesive material that contains very small beads of solder. The PWB will later pass through another machine that will melt the solder, thus providing electrical continuity.

Surface-mount technology (SMT) has been used in the electronics industry since the mid-1960s, but only since the early 1980s has it become economically competitive with TH technology. The reasons for this are related to a demand for reduced package sizes, a need for PWB circuit traces (paths) which are both more narrow and closer together, and the evolution of component packages that require much higher numbers of connection leads.

The SMT answers to these demands will become apparent later when component packages are discussed. To ensure a common foundation of knowledge, attention must first be focused on the passive component types, such as resistors and capacitors.

3.2 PASSIVE DISCRETE DEVICES

The first two circuit components to be discussed are passive devices known as resistors and capacitors.

3.2.1 Resistors

Resistors are used in a circuit to control the flow of current and, in the process, produce dissipation of power. The amount of power dissipated is given by the following formula:

$$\text{Power dissipated} = P = V^2/R = I^2R \tag{3-1}$$

where P = power in watts (or milliwatts)
V = voltage in volts
I = current in amps (or milliamps)
R = resistance in ohms

Resistors come in three basic types: carbon composition, wirewound, and film. For carbon composition resistors, shown in Fig. 3-1, the resistive element is a molded plug of ground carbon or graphite combined with a nonconducting filler that is stuck together with an inert binder. The material is molded under high pressure, then cured by baking. The resistor is then treated with a sealant to prevent moisture penetration. Important characteristics of carbon composition resistors include that they are:

- Inexpensive
- Of low inductance and capacitance
- Extremely durable and able to tolerate rough handling
- Not a good choice for low-noise applications, as they are susceptible to thermal and current noise

A wirewound resistor consists of a single layer of high-resistance wire wound around a nonconducting, usually ceramic or fiberglass, core as shown in Fig. 3-2. The winding

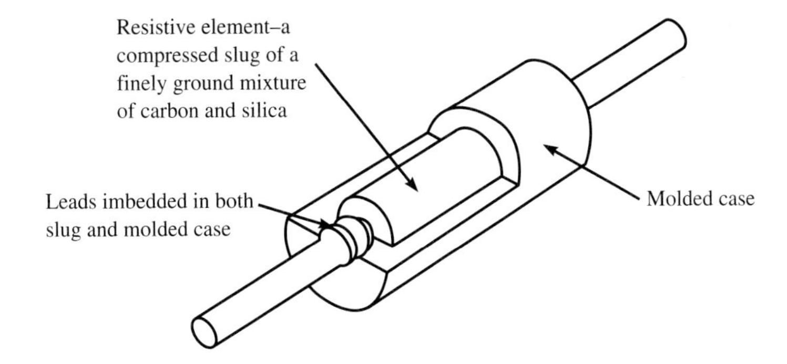

Figure 3-1 Carbon composition resistor construction.

(*Source:* T. H. Jones, *Electronic Components Handbook* (Reston, Va.: Reston Publishing Company, 1978) p. 117).

is then protected by a silicone or epoxy coating or encased in a ceramic or metal tube. Important characteristics of wirewound resistors include that they are:

- Good for high-frequency applications
- Replacements for carbon composition resistors in low-noise and low-resistance applications
- Essentially the only choice for power ratings of 2 watts or more
- Available in a wide variety of types for specific applications

The resistive element in film resistors is a thin layer of material that is applied to the surface of a ceramic or glass rod or cylinder, as can be seen in Fig. 3-3. The material may be tin oxide, carbon, or any of several metals such as nickel-chromium. Film resistors exhibit highly accurate values of resistance ($\pm 0.01\%$), due to spiral laser incising technique (that is, laser trimming). Spiral incising degenerates usefulness above 1MHz due to capacitive reactance.

Resistors are also classified according to resistance tolerance (that is, according to

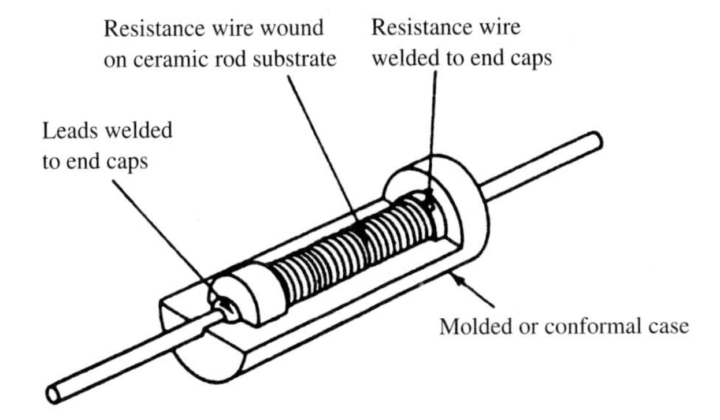

Figure 3-2 Wirewound resistor construction.

(*Source:* Jones, *Electronic Components Handbook,* p. 121.)

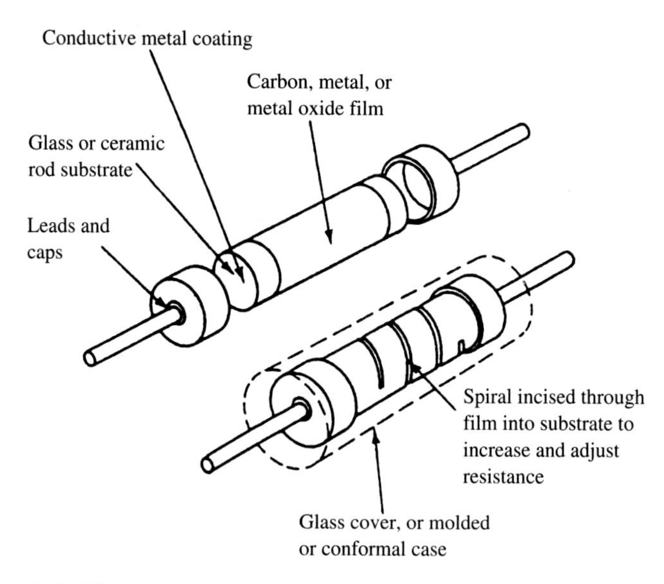

Figure 3-3 Film resistor construction.

(*Source:* Jones, *Electronic Components Handbook,* p. 129.)

how close an average resistor should be expected to come to its rated value). The divisions are as follows:

- General purpose (±20% to ±5%)
- Semiprecision (±5% to ±1%)
- Precision (±1% to ±0.1%)
- Ultraprecision (±0.1% to ±0.01%)

Resistors are available in standard values, as can be seen in Tables 3-1 and 3-2. Fixed resistors are marked for identification.

One method of marking is with the use of color-code bands. Resistors may be marked with three or four bands. The coding scheme is shown in Fig. 3-4. With the band nearest the end held to the left, the first two bands give the first two numbers of resistance value. The next band indicates the number of following zeros, and the fourth band, if present, indicates the tolerance value. If there is no fourth band, the tolerance value is ±20 percent. This coding scheme is the Electronic Industries Association/Military code, commonly known as EIA/MIL.

Additional identifying methods may be used whereby a fifth blue band would indicate a wirewound flameproof resistor, a wide first band indicates the resistor is wirewound rather than carbon composition, or a fifth wide band offset to the right indicates a tolerance of 1 percent (see Fig. 3-5). A more complex set of EIA/MIL coding schemes, which involves the use of alphanumeric codes, is shown in Figs. 3-6 and 3-7.

TABLE 3-1 STANDARD RESISTANCE VALUES FOR GENERAL-PURPOSE RESISTORS

A				±20% and ±10% Values			
1.0†	10	100	1000	10 k	100 k	1.0 M	10 M
1.2†	12	120	1200	12 k	120 k	1.2 M	12 M
1.5†	15	150	1500	15 k	150 k	1.5 M	15 M
1.8†	18	180	1800	18 k	180 k	1.8 M	18 M
2.2†	22	220	2200	22 k	220 k	2.2 M	22 M
2.7	27	270	2700	27 k	270 k	2.7 M	
3.3	33	330	3300	33 k	330 k	3.3 M	
3.9	39	390	3900	39 k	390 k	3.9 M	
4.7	47	470	4700	47 k	470 k	4.7 M	
5.6	56	560	5600	56 k	560 k	5.6 M	
6.8	68	680	6800	68 k	680 k	6.8 M	
8.2	82	820	8200	82 k	820 k	8.2 M	

B				±5% Values			
1.0†	10	100	1000	10 k	100 k	1.0 M	10 M
1.1†	11	110	1100	11 k	110 k	1.1 M	11 M
1.2†	12	120	1200	12 k	120 k	1.2 M	12 M
1.3†	13	130	1300	13 k	130 k	1.3 M	13 M
1.5†	15	150	1500	15 k	150 k	1.5 M	15 M
1.6†	16	160	1600	16 k	160 k	1.6 M	16 M
1.8†	18	180	1800	18 k	180 k	1.8 M	18 M
2.0†	20	200	2000	20 k	200 k	2.0 M	20 M
2.2†	22	220	2200	22 k	220 k	2.2 M	22 M
2.4†	24	240	2400	24 k	240 k	2.4 M	
2.7	27	270	2700	27 k	270 k	2.7 M	
3.0	30	300	3000	30 k	300 k	3.0 M	
3.3	33	330	3300	33 k	330 k	3.3 M	
3.6	36	360	3600	36 k	360 k	3.6 M	
3.9	39	390	3900	39 k	390 k	3.9 M	
4.3	43	430	4300	43 k	430 k	4.3 M	
4.7	47	470	4700	47 k	470 k	4.7 M	
5.1	51	510	5100	51 k	510 k	5.1 M	
5.6	56	560	5600	56 k	560 k	5.6 M	
6.2	62	620	6200	62 k	620 k	6.2 M	
6.8	68	680	6800	68 k	680 k	6.8 M	
7.5	75	750	7500	75 k	750 k	7.5 M	
8.2	82	820	8200	82 k	820 k	8.2 M	
9.1	91	910	9100	91 k	910 k	9.1 M	

† Industrial grade only.

Source: T. H. Jones, *Electronic Component Handbook,* p. 114.

TABLE 3-2 STANDARD EIA/MIL RESISTANCE VALUES FOR ULTRAPRECISION, PRECISION, AND SEMIPRECISION RESISTORS†

*	±1%	±2%	*	±1%	±2%	*	±1%	±2%
1.00	1.00	1.0	1.58	1.58		2.49	2.49	
1.01			1.60			2.52		
1.02	1.02		1.62	1.62	1.6	2.55	2.55	
1.04			1.64			2.58		
1.05	1.05		1.65	1.65		2.61	2.61	
1.06			1.67			2.64		
1.07	1.07		1.69	1.69		2.67	2.67	
1.09			1.72			2.71		2.7
1.10	1.10	1.1	1.74	1.74		2.74	2.74	
1.11			1.76			2.77		
1.13	1.13		1.78	1.78	1.8	2.80	2.80	
1.14			1.80			2.84		
1.15	1.15		1.82	1.82		2.87	2.87	
1.17			1.84			2.91		
1.18	1.18		1.87			2.94	2.94	
1.20			1.89			2.98		
1.21	1.21	1.2	1.91	1.91		3.01	3.01	3.0
1.23			1.93			3.05		
1.24	1.24		1.96	1.96		3.09	3.09	
1.26			1.98			3.12		
1.27	1.27		2.00	2.00	2.0	3.16	3.16	
1.29			2.03			3.20		
1.30	1.30	1.3	2.05	2.05		3.24	3.24	
1.32			2.08			3.28		
1.33	1.33		2.10	2.10		3.32	3.32	3.3
1.35			2.13			3.36		
1.37	1.37		2.15	2.15		3.40	3.40	
1.38			2.18			3.44		
1.40	1.40		2.21	2.21	2.2	3.48	3.48	
1.42			2.23			3.52		
1.43	1.43		2.26	2.26		3.57	3.57	
1.45			2.29			3.61		
1.47	1.47		2.32	2.32		3.65	3.65	3.6
1.49			2.34			3.70		
1.50	1.50	1.5	2.37	2.37	2.4	3.74	3.74	
1.52			2.40			3.79		
1.54	1.54		2.43	2.43		3.83	3.83	
1.56			2.46			3.88		

* ±0.1%, ±0.25%, and ±0.5%.

† Values are given for one decade. All other decades are the same with appropriate multiplier. For each type of resistor, values are subject to minimum and maximum limits.

Source: T. H. Jones, *Electronic Components Handbook,* p. 115.

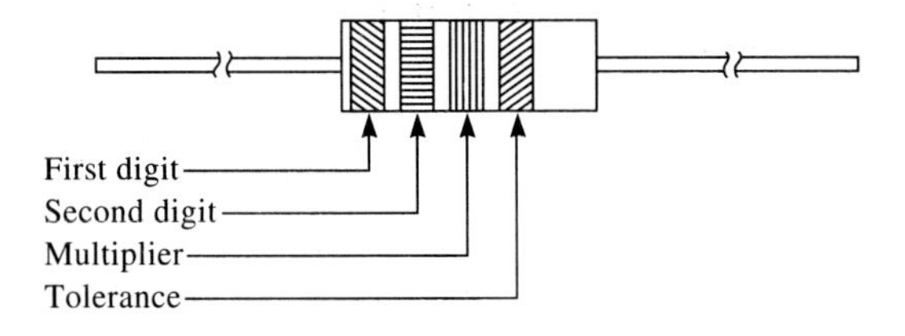

(a) Color code for general-purpose industrial resistors

(b) Color code for general-purpose wirewound resistors

Color	Significant digit	Multiplier	Tolerance
Black	0	1	–
Brown	1	10	–
Red	2	100	–
Orange	3	1000	–
Yellow	4	10,000	–
Green	5	100,000	–
Blue	6	1,000,000	–
Violet	7	10,000,000	–
Gray	8	–	–
White	9	–	–
Gold	–	–	± 5%
Silver	–	–	±10%
No color	–	–	±20%

Example:

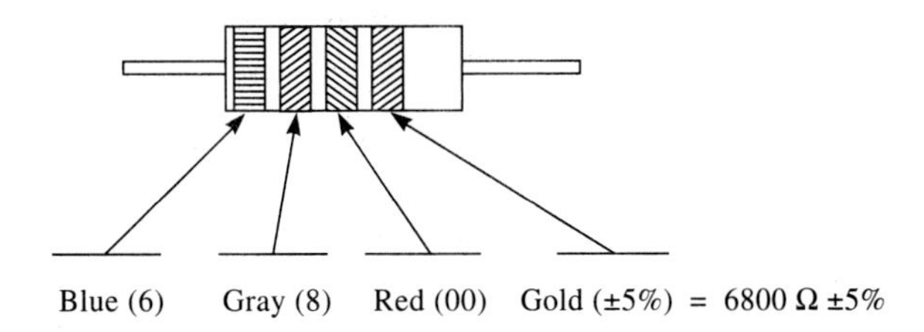

Blue (6) Gray (8) Red (00) Gold (±5%) = 6800 Ω ±5%

Figure 3-4 EIA/MIL color codes for ±5%, ±10%, and ±20% resistors.

(*Source:* Jones, *Electronic Components Handbook,* p. 109.)

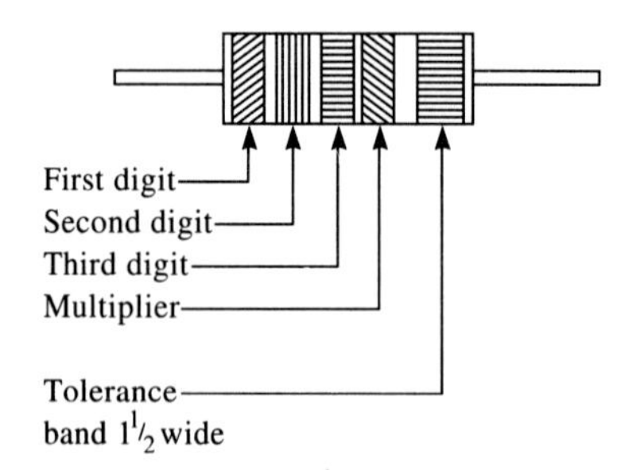

Color code for industrial 1% and better resistors

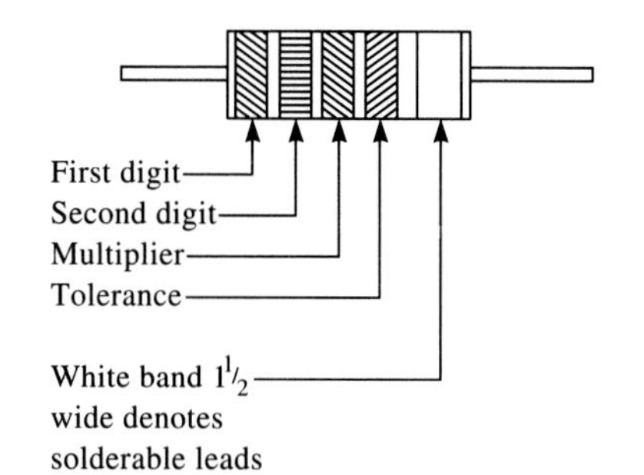

Color code for military resistors

Tolerance code

Black	–
Brown	± 1%
Red	± 2%
Orange	–
Yellow	–
Green	±0.5%
Blue	±0.25%
Violet	±0.1%
Gray	±0.05%
White	–
Gold	± 5%
Silver	±10%

Figure 3-5 Color codes for industrial ±1% and better MIL-SPEC resistors.

(*Source:* Jones, *Electronic Components Handbook,* p. 110.)

3.2.2 Capacitors

A **capacitor** is a circuit component that has the ability to store electrical energy. Capacitors are made up of two parallel-plate conductors separated by a vacuum or an insulating material called a **dielectric.** One of the plates is charged with electrons, and the insulating dielectric prevents the electrons from traveling to the other plate. The electrons on the first plate repel electrons on the second plate, thus causing the second plate to become positively charged. The opposite charges on the two plates attract one another and form a stable system. If the dielectric insulator were not there, the two plates would neutralize each other. The simplest capacitor is made up of two equal-sized plates separated by an air gap.

The measure of the capacitor's ability to store electrical energy, in the form of electrons, is called its capacitance. The units of capacitance are farads in honor of the electrical pioneer, Michael Faraday. A farad is defined as one coulomb of electricity applied at a potential of one volt, where one coulomb is 6.25×10^{18} electrons.

In most applications, a farad is more capacitance than is ever needed, so capacitors

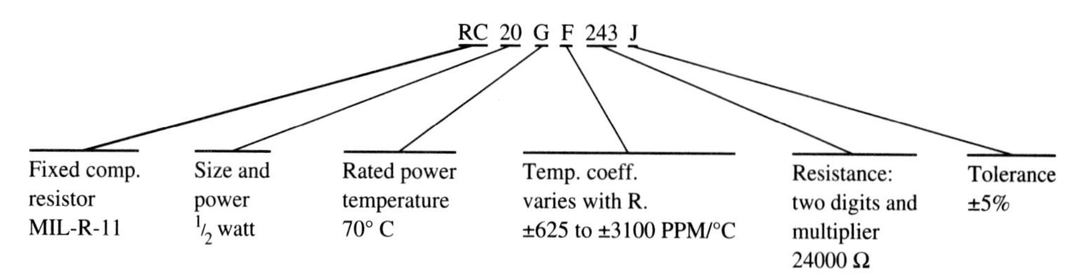

MIL-R-11 Resistor part numbers

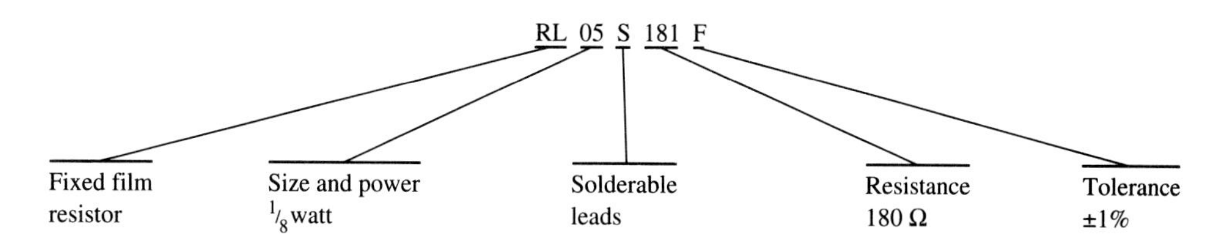

MIL-R-22684 Resistor part numbers (all have ± 200/°C temp. coeff.)

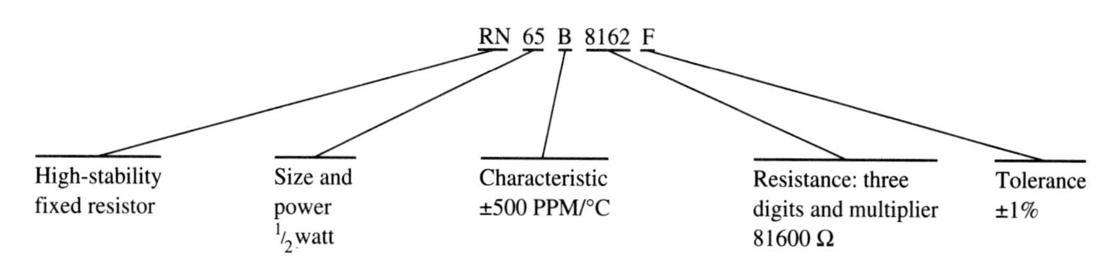

MIL-R-10509 Resistor part numbers

Size and power

05 - 1/8 watt
07 - 1/4 watt
20 - 1/2 watt
32 - 1 watt
42 - 2 watt
} 70° C

50 - 1/10 watt
55 - 1/8 watt
60 - 1/4 watt
65 - 1/2 watt
70 - 3/4 watt
75 - 1 watt
80 - 2 watt
} 70° C

Tolerance

K = ±10%
J = ± 5%
G = ± 2%
F = ± 1%
D = ±0.5%
C = ±0.25%
B = ±0.1%

Note: The MIL-R-26 (fixed wirewound resistors, power type) numbering system is complex and incomprehensible without a copy of the MIL spec.

Figure 3-6 MIL-SPEC part numbering for carbon composition and film resistors.

(*Source:* Jones, *Electronic Components Handbook,* p. 112.)

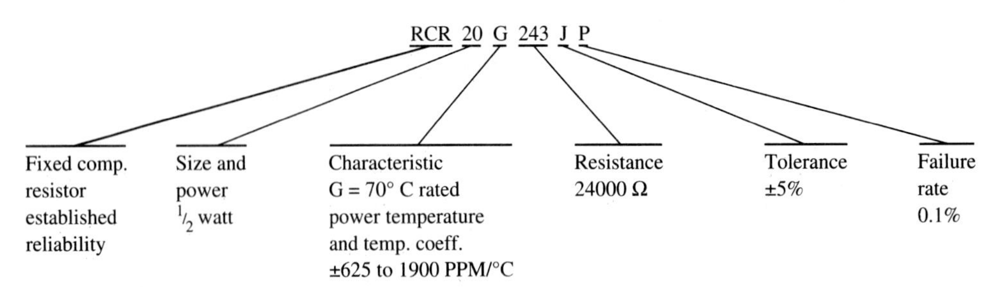

MIL-R-39008 Resistor part numbers

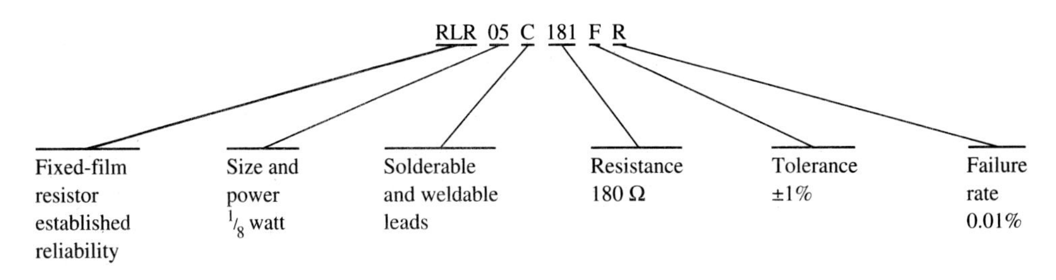

MIL-R-39017 Resistor part numbers

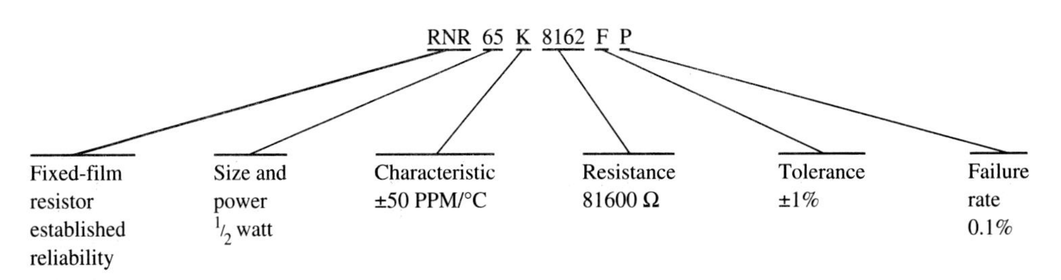

MIL-R-55182 Resistor part numbers

Characteristic (RN)
B = ±500 PPM/°C
C = ± 50 PPM/°C
D = ± 200 or ±500 PPM/°C
E = ± 25 PPM/°C
F = ± 50 PPM/°C

Characteristic (RNR)
H = ± 50 PPM/°C
J = ± 25 PPM/°C
K = ± 100 PPM/°C

Failure rate
per 1000 hours
(60% confidence)
M = 1.0%
P = 0.1%
R = 0.01%
S = 0.001%

Figure 3-7 MIL-SPEC part numbering for established-reliability carbon composition, film, and wirewound resistors.

(*Source:* Jones, *Electronic Components Handbook,* p. 113.)

are usually rated in the microfarad or picofarad range. Factors that affect the value of capacitance include the area of the plates, the thickness of the dielectric (that is, the separation of the plates), and the material from which the dielectric is made. Capacitance is directly proportional to the area of the plates and dielectric constant, but inversely proportional to the distance between the plates.

3.2.3 Dielectric Materials

Dielectric materials are insulating materials that have a dielectric constant associated with them. This constant provides an indicator of the material's ability to improve the energy-storing capability of the capacitor. Capacitance is improved when a material of higher dielectric constant is used.

Some examples of dielectric materials include air, glass, mica, ceramics, plastics, paper, oil, oxides of some metals, and the nonmaterial vacuum. Table 3-3 shows a list of dielectric materials and their dielectric constants. The dielectric constant of a vacuum is arbitrarily considered to be one, and other materials are rated relative to a vacuum.

Capacitance can also be improved by increasing the size of the plates. This method is less useful than improving the dielectric material. Size constraints prevail in all microminiature circuits and generally prohibit increasing capacitance by increasing the size of the plates.

The third way that capacitance can be improved is by decreasing the distance between the plates by making the dielectric material very thin (refer to Fig. 3-8). The problem with this approach is that the mutually attractive forces on the two plates increase as the inverse square of the distance separating them. That is, as the distance between them is halved, the attractive force is quadrupled. A critical distance is soon reached where the plates are so close together that the electrons jump the gap, destroying the dielectric material.

Standard fixed capacitors that are used in the manufacturing of electronic equipment are grouped according to the dielectric material used. A table of capacitance values and available dielectric materials is shown in Fig. 3-9. The construction of various capacitors is shown in Figs. 3-10, 3-11, and 3-12.

TABLE 3-3 DIELECTRIC CONSTANTS

Dielectric Material	Dielectric Constant	Dielectric Material	Dielectric Constant
Air	1.0001	Mica	4.5 to 7.5
Vacuum	1.0000	Glass	6.7
Paper (Impregnated)	4.6	Green Glass	8.3
Polyester	3.0	Quartz	4.2
Polystyrene	2.5 to 2.7	Steatite	5.5 to 6.5
Polycarbonate	3	Titanium Dioxide	80 to 120
Polypropylene	2	Barium Titanate	200 to 16,000
Polysulfone	3	Aluminum Oxide	10
Teflon	2	Tantalum Oxide	11
Polyimide	3.5		

Source: T. H. Jones, *Electronic Components Handbook,* p. 3.

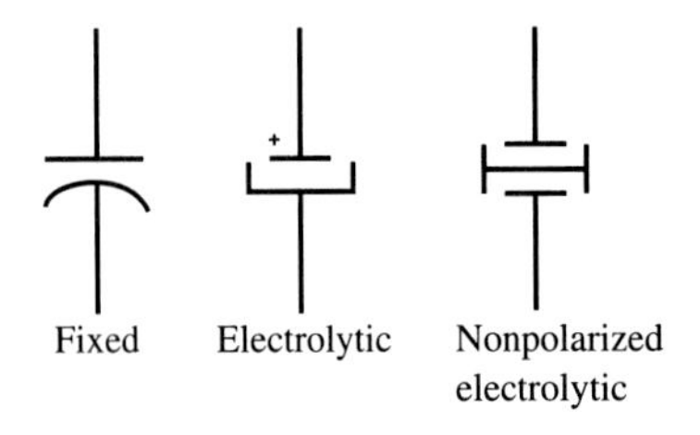

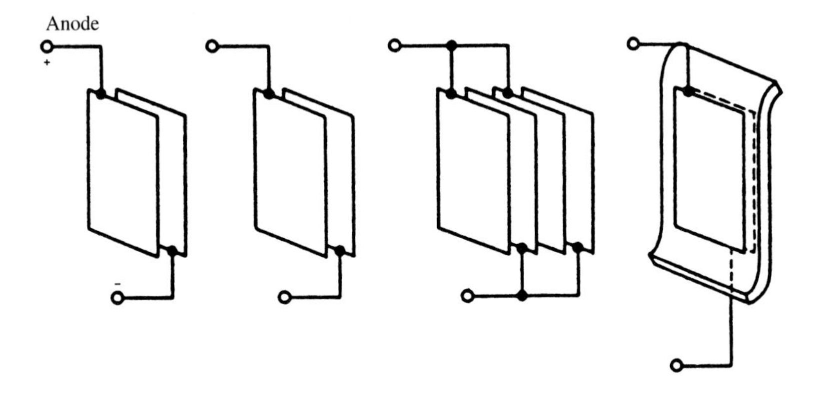

Figure 3-8 Capacitors. (a) Symbols for capacitors. (b) The capacitance of a simple air-dielectric capacitor can be increased by positioning the plates closer together, increasing the number of plates in parallel, or by substituting a material with a higher dielectric constant such as one of the plastic-film dielectrics.

(*Source:* Jones, *Electronic Components Handbook,* p. 3.)

Most capacitors are marked with their capacitance, working voltage, tolerance, and polarity, as can be seen in Fig. 3-13. Another method for marking capacitors is with the use of color-code marking. Figs. 3-14 and 3-15 show how ceramic capacitors and molded mica capacitors are marked with color codes.

3.3 SOLID-STATE DEVICES

Semiconductor devices are electronic components whose main functioning parts are made of a **semiconductor material** such as silicon or germanium. Because they have no moving parts, as do switches, relays, and the like, they are also known as **solid-state devices.**

The main semiconductor devices, upon which most microminiature electronic circuits are based, are diodes and transistors. Since these devices play such a large part in the creation of very complex components, the principle of how they work will be presented here, without delving too deeply into solid-state physics. Plenty of resource material is available if, after this introduction, a deeper understanding of semiconductor electronics is desired.

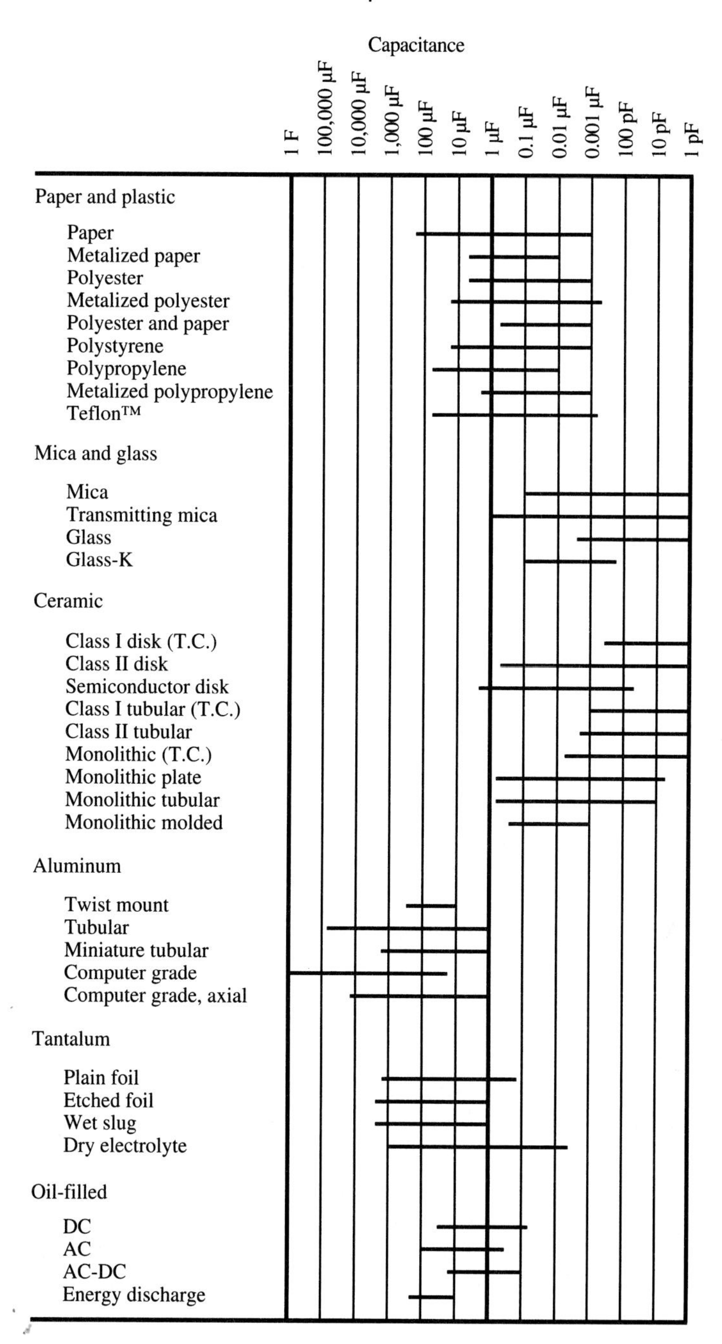

Figure 3-9 Range of capacitances generally available in capacitors on the basis of dielectric material used.

(*Source:* Jones, *Electronic Components Handbook,* p. 7.)

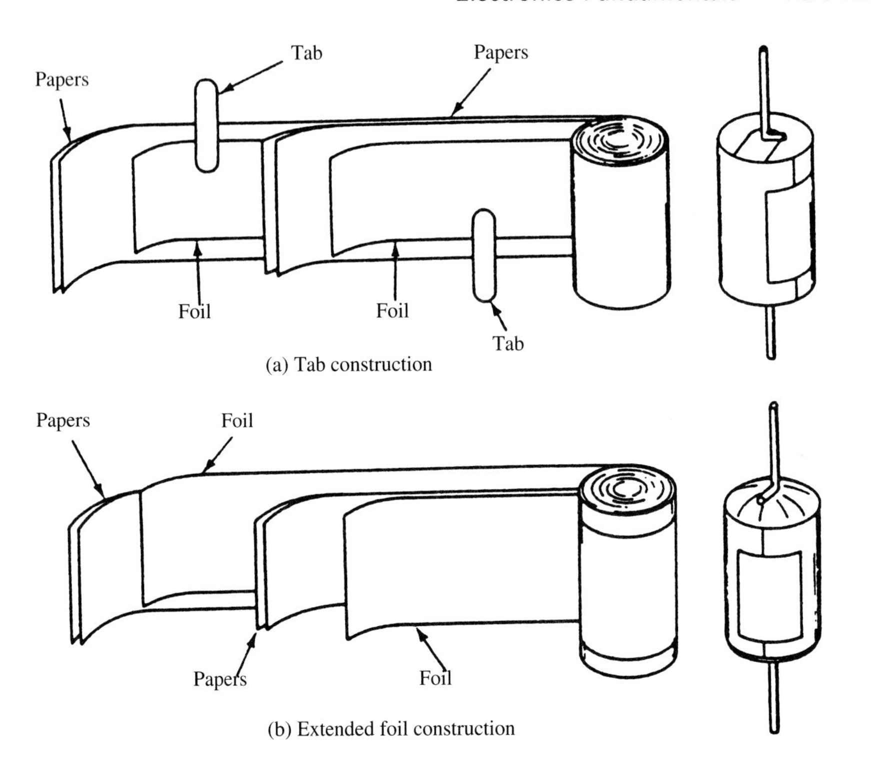

Figure 3-10 Tubular paper construction; film capacitors are constructed in essentially the same manner. (a) Tab construction. (b) Extended-foil construction.

(*Source:* Jones, *Electronic Components Handbook,* p. 12.)

3.3.1 Semiconductor Materials

The two most common materials used in the production of semiconductor devices are crystalline germanium and silicon, although silicon dominates the market. These two materials are neither good conductors nor good insulators. Hence the name, semiconductor. Both these materials possess qualities that offer advantages for specific applications. However, for simplicity, we will focus discussion on silicon. The processes that are performed on germanium are basically the same as those performed on silicon.

The making of a silicon chip begins with the growing of a single-crystal ingot of silicon. A seed crystal of silicon is placed on the end of a rod and dipped into a vat of molten silicon. The rod is slowly withdrawn from the vat, and during this withdrawal process, the silicon that is in contact with the seed crystal slowly cools and the crystal of silicon grows. The end result is an ingot of silicon that is usually not less than 50 cm long and 5.0 cm in diameter. This ingot is then sawed into slices called wafers (see Fig. 3-16(a)), and both surfaces of the wafers are ground and polished to form a smooth finish. The last step of wafer preparation is a chemical etching of one side to remove any final irregularities that are left from the polishing process. The next step is to add the desired

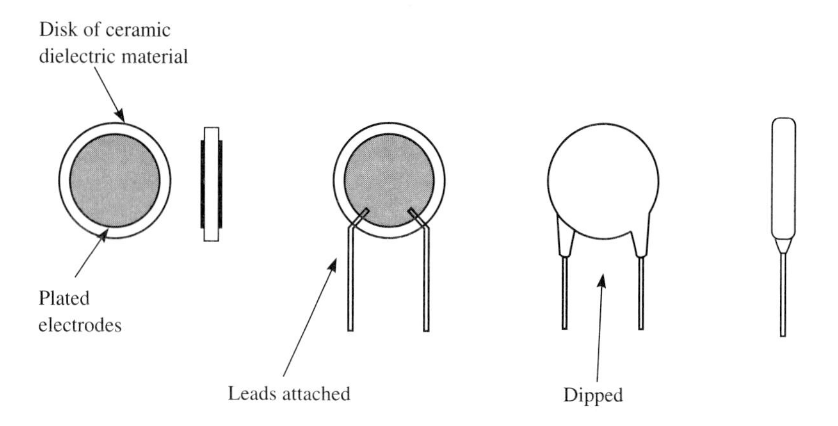

Disk ceramic capacitor construction.

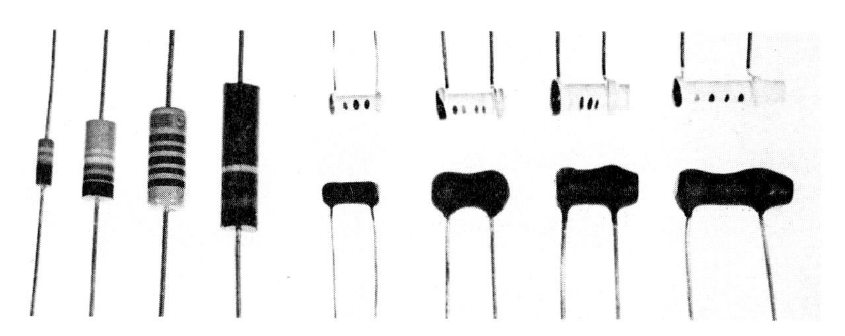

Tubular ceramic capacitors. **Left:** Molded. **Top:** Enamel coated. **Bottom:** Dipped phenolic. Values range from 0.42 pF to about 1000 pF. Capacitors are shown full size. *(Courtesy of Erie Technological Products)*

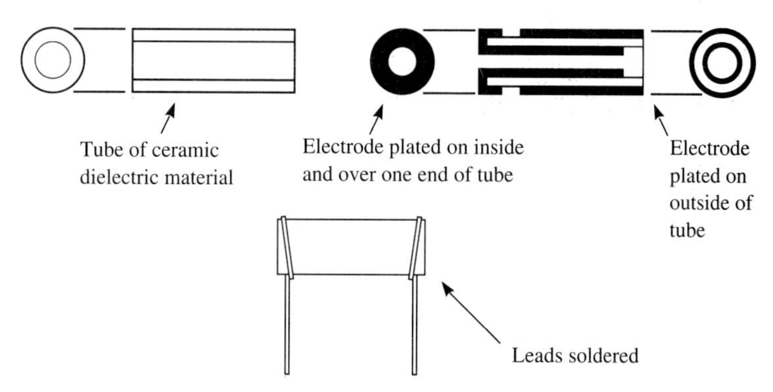

Tubular ceramic capacitor construction.

Figure 3-11 Construction details of disk and tubular ceramic capacitors.

(*Source:* Jones, *Electronic Components Handbook,* p. 35.)

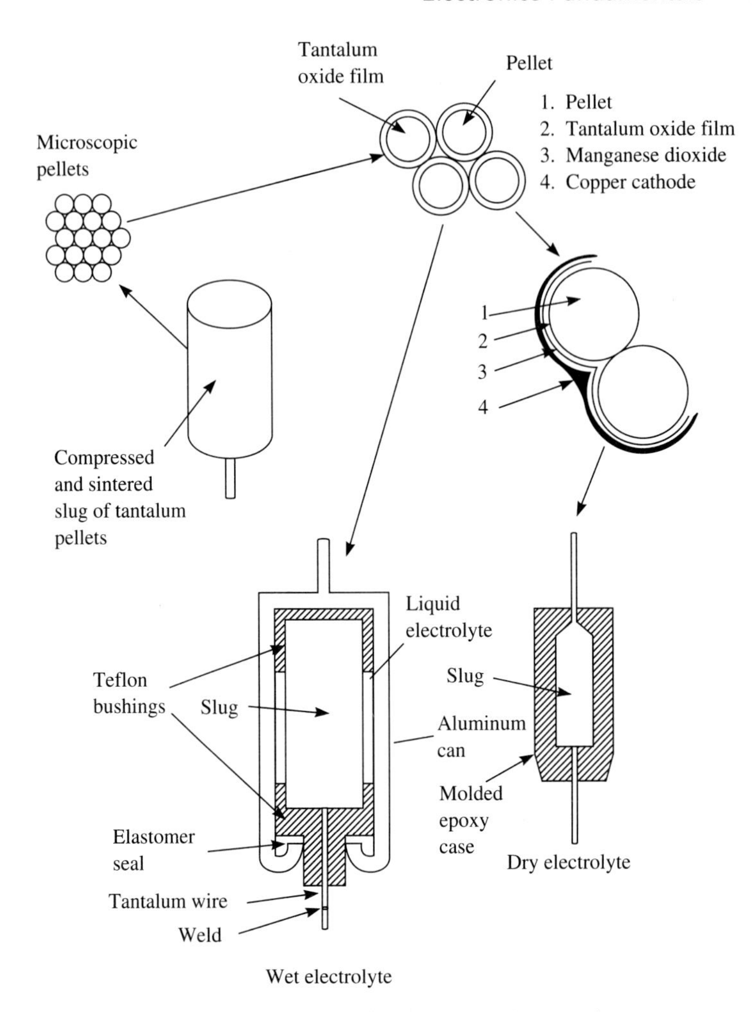

Figure 3-12 Construction details of sintered-anode tantalum capacitors.

(*Source:* Jones, *Electronic Components Handbook,* p. 51.)

electrical characteristics to this semiconductor material, then cut the wafer into many small pieces called dice or chips, and, finally, assemble the chips into a protective package. Figure 3-16(b) illustrates the result of these processes.

3.3.2 Semiconductor Doping

Pure crystals of silicon cannot be used as electronic circuit components such as transistors, because there are no free electrons to move in the crystal. Silicon atoms each have four electrons in their outer shell. When in crystal form, each atom shares these four valence electrons with four neighboring atoms so that the atoms of the crystals are in equilibrium (see Fig. 3-17). In this state, silicon is a poor conductor and not a very good insulator. However, the region just below the surface of a silicon wafer can be selectively altered to give it the attributes of a conductor. This process of altering the silicon is called **doping.**

Doping is the process of taking materials that differ from silicon in the number of electrons that reside in their outer shell, and inserting them into the crystalline lattice of silicon. This results in a crystal lattice that has an abundance or shortage of electrons (see Figs. 3-18 and 3-19).

Figures 3-17, 3-18, and 3-19 compare the conditions that exist in a pure silicon crystal with those that exist after doping, where a donor or acceptor has been added to the lattice. Materials such as phosphorus, antimony, and arsenic are called donors and yield an abundance of electrons, which are relatively free to travel throughout the lattice. Materials such as indium, boron, aluminum, gallium, and thallium are called acceptors and yield a shortage of electrons.

Figure 3-20 shows the nucleus and outer shells of the semiconductors (germanium and silicon), the donor atoms (antimony, phosphorus, and so on), and the acceptor atoms (boron, gallium, and so on), in isolation.

Silicon that has been doped with a donor has an abundance of electrons and is called n-type silicon because of its extra carriers of negative charge. If a battery were placed across the crystal, the extra electrons would migrate toward the positive terminal and enter the battery. Simultaneously, at the negative terminal, electrons would leave the battery and enter the crystal to take the place of those electrons that had left.

Silicon that has been doped with an acceptor has an abundance of **holes** where the acceptor atom lacks a fourth valence electron to pair-bond with its neighboring silicon

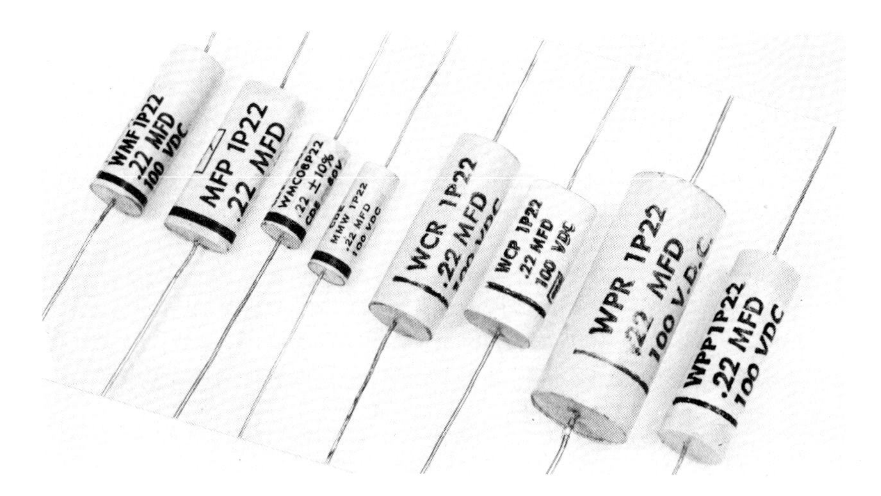

Figure 3-13 Capacitors with specific markings to indicate capacitance, tolerance, and working voltage.

(*Source:* Jones, *Electronic Components Handbook,* p. 17.)

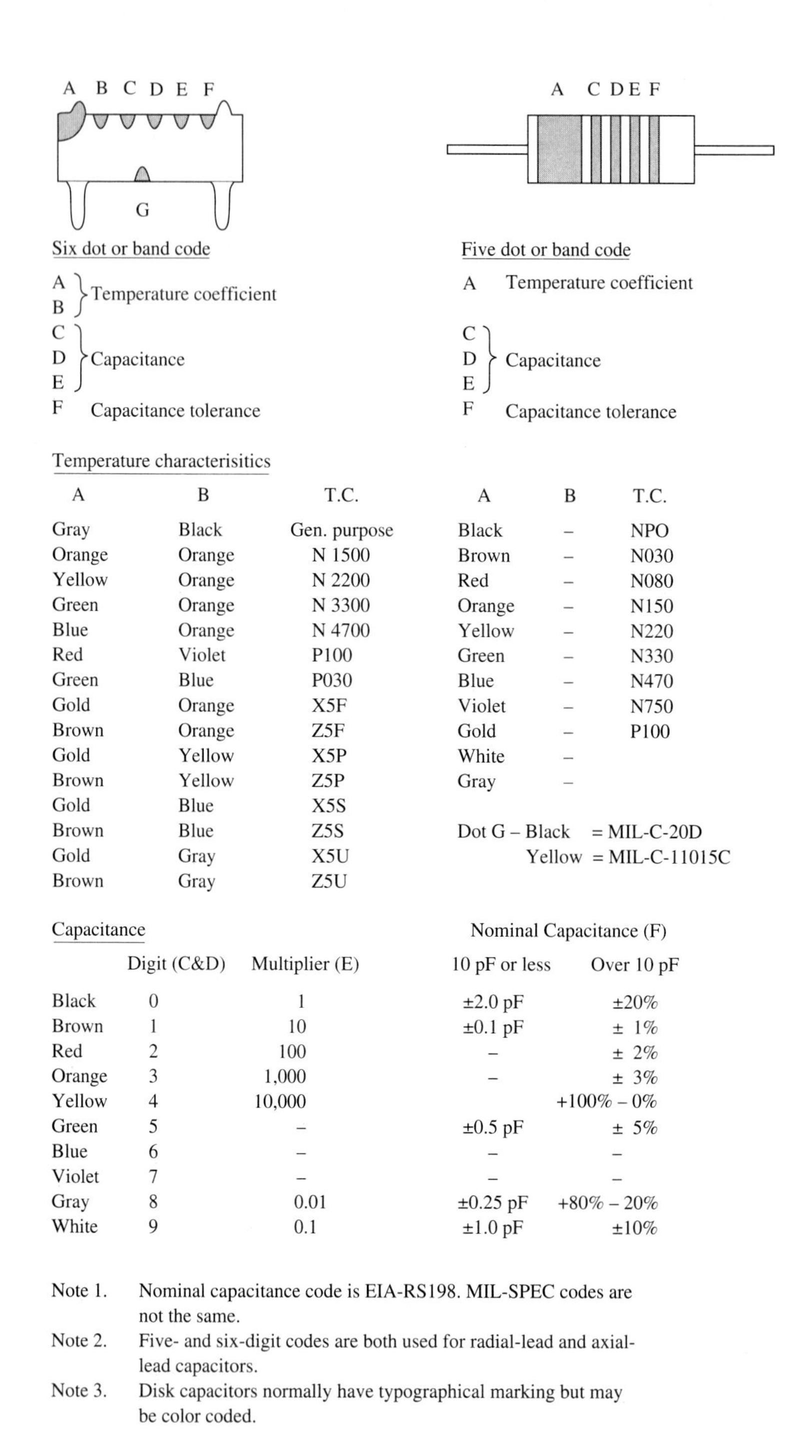

Six dot or band code

A, B	Temperature coefficient
C, D, E	Capacitance
F	Capacitance tolerance

Five dot or band code

A	Temperature coefficient
C, D, E	Capacitance
F	Capacitance tolerance

Temperature characterisitics

A	B	T.C.
Gray	Black	Gen. purpose
Orange	Orange	N 1500
Yellow	Orange	N 2200
Green	Orange	N 3300
Blue	Orange	N 4700
Red	Violet	P100
Green	Blue	P030
Gold	Orange	X5F
Brown	Orange	Z5F
Gold	Yellow	X5P
Brown	Yellow	Z5P
Gold	Blue	X5S
Brown	Blue	Z5S
Gold	Gray	X5U
Brown	Gray	Z5U

A	B	T.C.
Black	–	NPO
Brown	–	N030
Red	–	N080
Orange	–	N150
Yellow	–	N220
Green	–	N330
Blue	–	N470
Violet	–	N750
Gold	–	P100
White	–	
Gray	–	

Dot G – Black = MIL-C-20D
Yellow = MIL-C-11015C

Capacitance

	Digit (C&D)	Multiplier (E)	Nominal Capacitance (F) 10 pF or less	Over 10 pF
Black	0	1	±2.0 pF	±20%
Brown	1	10	±0.1 pF	± 1%
Red	2	100	–	± 2%
Orange	3	1,000	–	± 3%
Yellow	4	10,000		+100% – 0%
Green	5	–	±0.5 pF	± 5%
Blue	6	–	–	–
Violet	7	–	–	–
Gray	8	0.01	±0.25 pF	+80% – 20%
White	9	0.1	±1.0 pF	±10%

Note 1. Nominal capacitance code is EIA-RS198. MIL-SPEC codes are not the same.
Note 2. Five- and six-digit codes are both used for radial-lead and axial-lead capacitors.
Note 3. Disk capacitors normally have typographical marking but may be color coded.

Figure 3-14 Tubular ceramic capacitor color-code markings.

(*Source:* Jones, *Electronic Components Handbook,* p. 10.)

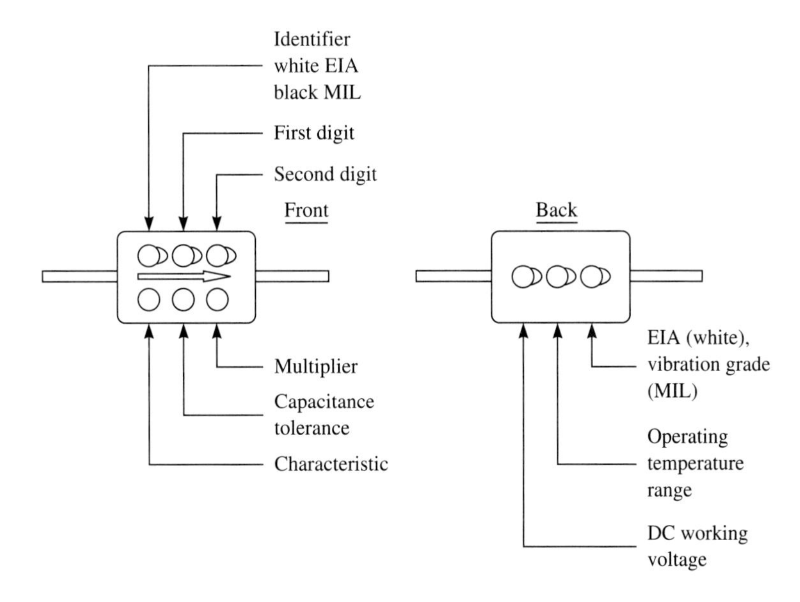

Color	Significant digits	Multiplier	Capacitance tolerance*	Characteristic	DC working voltage	Operating temperature	EIA/ Vibration
Black	0	1	±20%	–	–	–55° to +70° C	10-55 Hz
Brown	1	10	± 1%	B	100	–	–
Red	2	100	± 2%	C	–	–55° to +85° C	–
Orange	3	1,000	–	D	300	–	–
Yellow	4	10,000	–	E	–	–55° to +125° C	10-2000 Hz
Green	5	–	± 5%	F	500	–	–
Blue	6	–	–	–	–	–55° to +150° C	–
Violet	7	–	–	–	–	–	–
Gray	8	–	–	–	–	–	–
White	9	–	–	–	–	–	EIA
Gold	–	–	±0.5%**	–	1000	–	–
Silver	–	–	± 10%	–	–	–	–

* ±1.0 pF
** Or ±0.5 pF, whichever is greater.

Figure 3-15 Color-code markings for molded mica capacitors.

(*Source:* Jones, *Electronic Components Handbook,* p. 11.)

atoms. These holes are considered to be carriers of positive charge, and silicon with an abundance of holes is called p-type silicon because of these positive charge carriers. It has been shown experimentally that these holes are free to move through the crystal as though they were actual particles. In this case, if a battery were placed across the crystal, the holes would migrate toward the negative terminal of the battery, and an electron would enter the crystal and fill (or eliminate) the hole. At the same time, near the positive

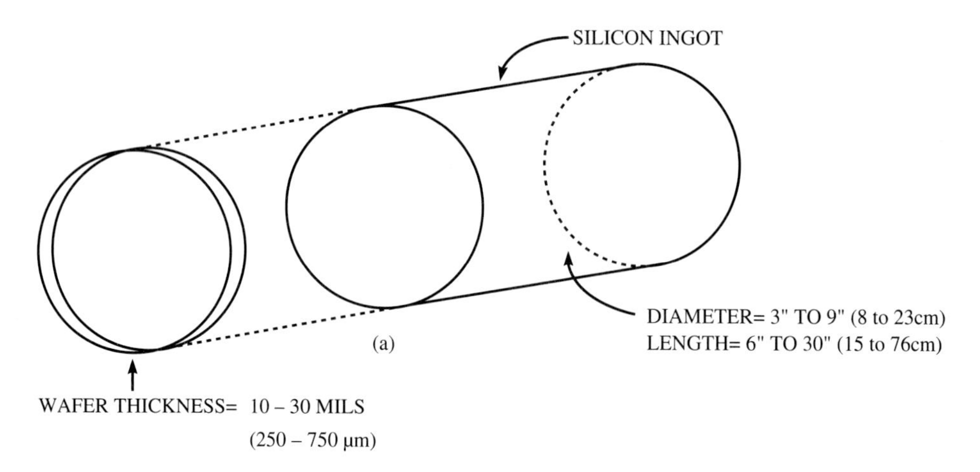

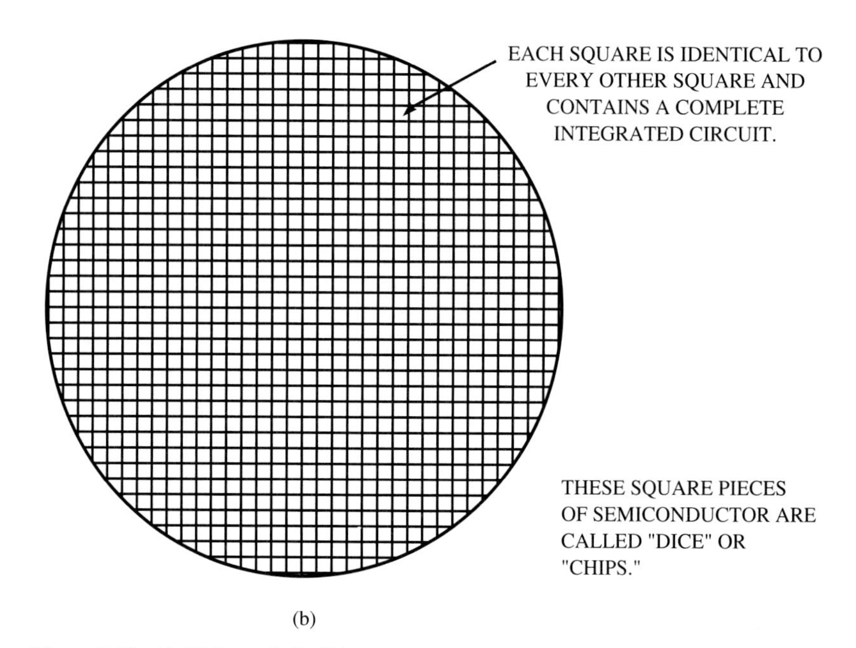

Figure 3-16 (a) Slicing and (b) dicing process.

terminal of the battery, an electron would leave one of the pair-bonds in the crystal, enter the battery, and leave another hole in the crystal.

3.3.3 P-N Junctions

The main building block of semiconductor devices is the **p-n junction.** This is created when p-type silicon is joined with n-type silicon, bringing negative and positive charge carriers into close proximity.

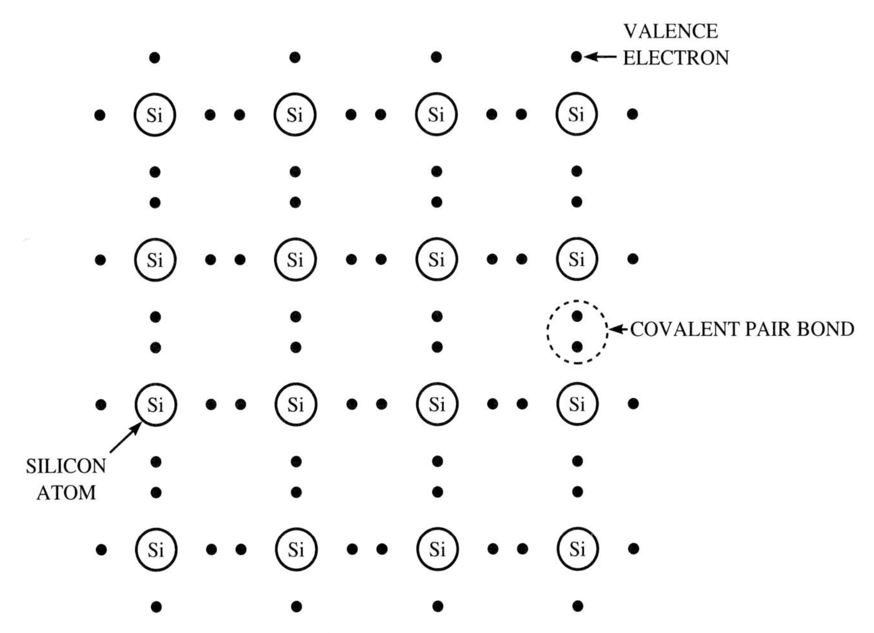

Figure 3-17 Crystal structure of silicon illustrated symbolically in two dimensions. The atoms of the crystal are in equilibrium.

When a junction is formed between p-type and n-type silicon, the resulting electrical component is called a junction **diode.** A diode possesses the qualities of a rectifier, in that it is very easy for electrical charge to flow through the diode in one direction, and very difficult for charge to flow in the opposite direction. This unidirectional flow is the result of potential energy barriers that are created when the p-type and n-type surfaces

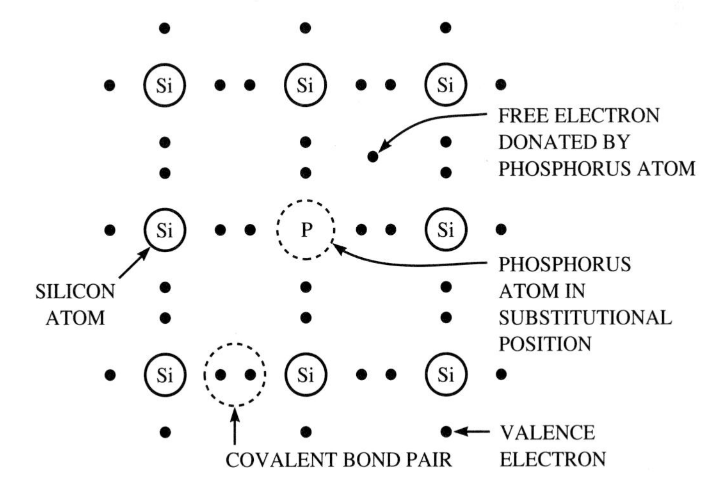

Figure 3-18 Crystal lattice with a silicon atom displaced by a pentavalent (phosphorus) impurity atom. The pentavalent impurities are electron donors.

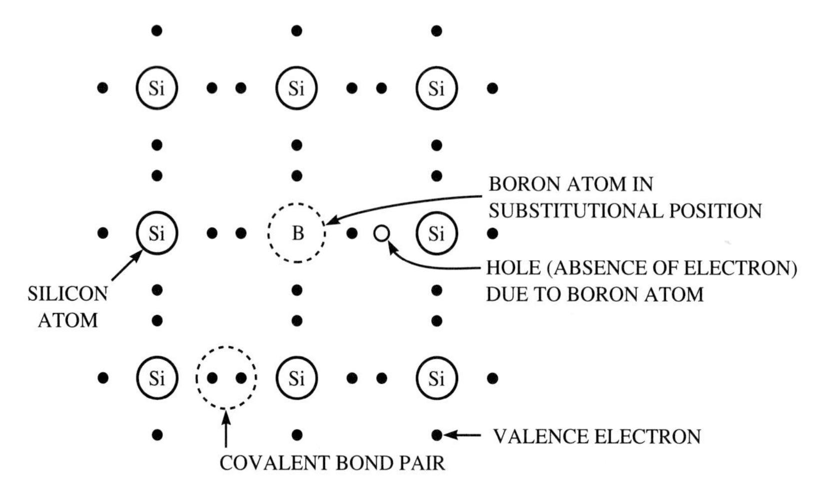

Figure 3-19 Crystal lattice with a silicon atom displaced by a trivalent atom (boron) impurity. The trivalent impurities accept lattice electrons, thereby contributing holes to the lattice.

are brought together.

Figure 3-21 shows a p-n junction in equilibrium. In this picture, the acceptor atoms are indicated by a "−" because they accepted an electron upon entering the silicon lattice, thus becoming negative ions. The donor atoms are indicated by a "+" because these atoms gave up an electron upon entering the crystal lattice, thus becoming positive ions. When the p-type and n-type silicon are brought together, there is an initial exchange of charge. Some holes migrate from the p-type silicon, cross the junction, and combine with electrons. At the same time, some electrons cross the junction from the n-type silicon and combine with holes in the p-type crystal. This mutual migration, known as diffusion current, results in a depletion of charge carriers in the area very near the junction, giving it the names **space-charge region,** or **depletion region.** In the depletion region, the acceptor ions are no longer neutralized by holes, nor are the donor ions neutralized by electrons. These unneutralized ions that reside in the depletion region are referred to as uncovered (or uncompensated) charges.

As charges become uncovered near the junction, an electric field (E) is created whose lines of flux point from right to left. This tends to counteract the flow of diffusion current. Equilibrium is reached when this field gains enough strength to prevent further migration of charge carriers across the junction.

Figures 3-22(a) and (b) also show a p-n junction in equilibrium. Part (c) of the

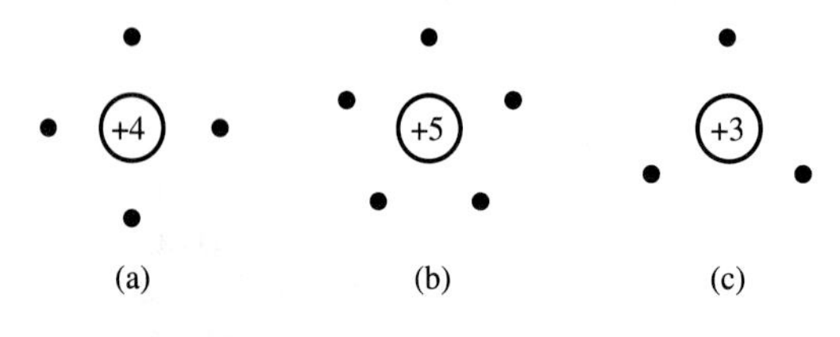

Figure 3-20 Nucleus and outer shells of (a) semiconductors such as silicon and germanium, (b) donor atoms such as atoms such as boron and aluminum.

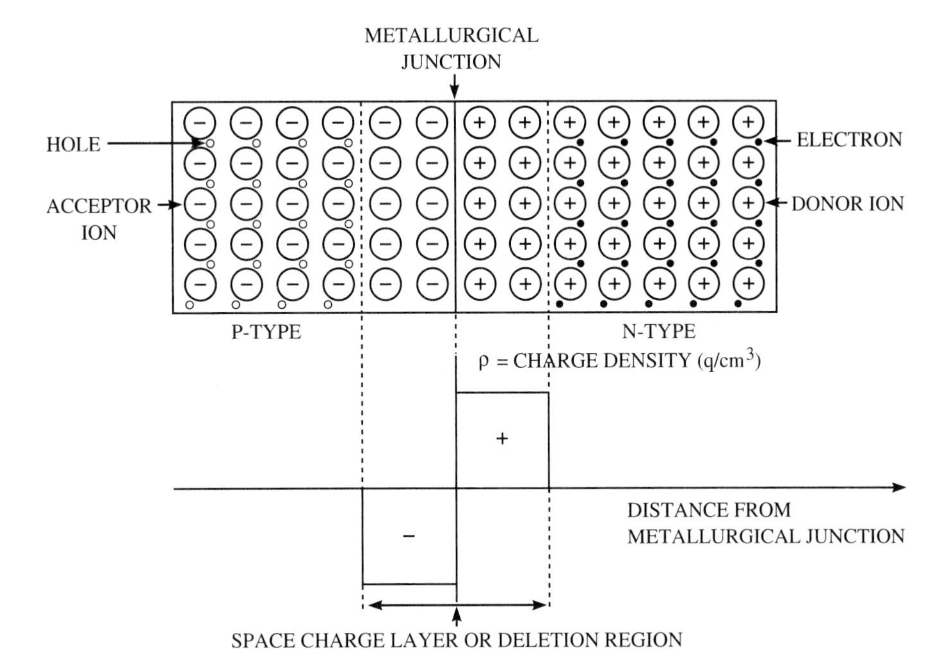

Figure 3-21 A p-n junction and depletion region.

figure plots the electric field intensity as a function of distance from the metallurgical junction. The electrostatic potential that is established due to the creation of the E-field across the junction is shown in Fig. 3-22(d). It is the existence of this electrostatic potential hill that prevents further migration of holes from the p-type silicon to the n-type silicon. Likewise, there is a potential energy hill on the left side of the junction that prevents electrons from diffusing across the junction to the p-type silicon (see Fig. 3-22(e)). These potential energy barriers are what give the p-n junction its rectifier characteristics.

If a battery is placed across a p-n junction, as in Fig. 3-23, so that the negative terminal is connected to the p-type silicon, holes will try to migrate from the p-side to the negative terminal. At the same time, electrons will try to migrate from the n-side to the positive terminal. Migration can only happen for a short time, because in order to continue, holes and electrons would eventually have to migrate across the junction. The battery is connected in such a way as to widen the depletion region and increase the height of the potential energy barriers; the easy flow of charge is restrained from flowing from the n-side to the p-side of the junction. Thus, the p-n junction in Fig. 3-23 is a rectifier, and voltage applied in this direction is known as reverse-bias voltage.

If the battery is connected the opposite way, as shown in Fig. 3-24, so that the positive terminal is connected to the p-side and the negative terminal to the n-side, it creates the opposite effect. Once enough voltage has been applied to overcome the strength of E, free migration of holes and electrons across the junction can occur. Electrons can migrate from the n side, across the junction, and into the positive terminal of the battery. Holes can migrate from the p side, across the junction, and combine with electrons moving

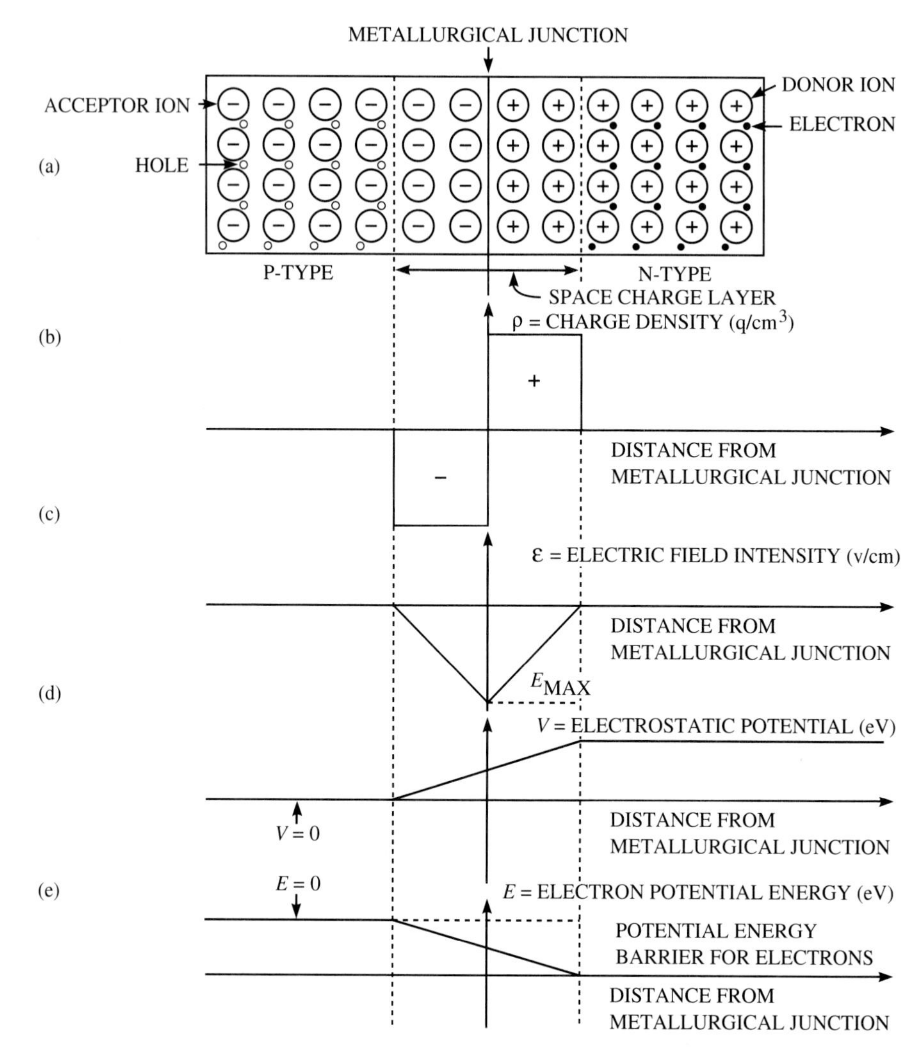

Figure 3-22 A p-n junction. (a) Physical representation. (b) The charge density. (c) Electric field intensity. (d,e) Potential-energy barriers at the junction. Since potential energy = potential + charge, the curve in (d) is proportional to the potential energy for a hole (a positive charge), and the curve in (e) is proportional to the negative of that in (d) (an electron is a negative charge). (Assumption: the diode dimensions are large compared with the space charge region.) Not drawn to scale.

out of the negative terminal of the battery. Connected in this way, the battery has the effect of lowering the potential energy barriers on each side of the junction. Voltage applied as shown in Fig. 3-24 is known as forward-bias voltage.

The amount of voltage required to overcome the strength of E is referred to as the cut-in voltage of the diode. For a silicon diode, the cut-in voltage is about 0.6 volt.

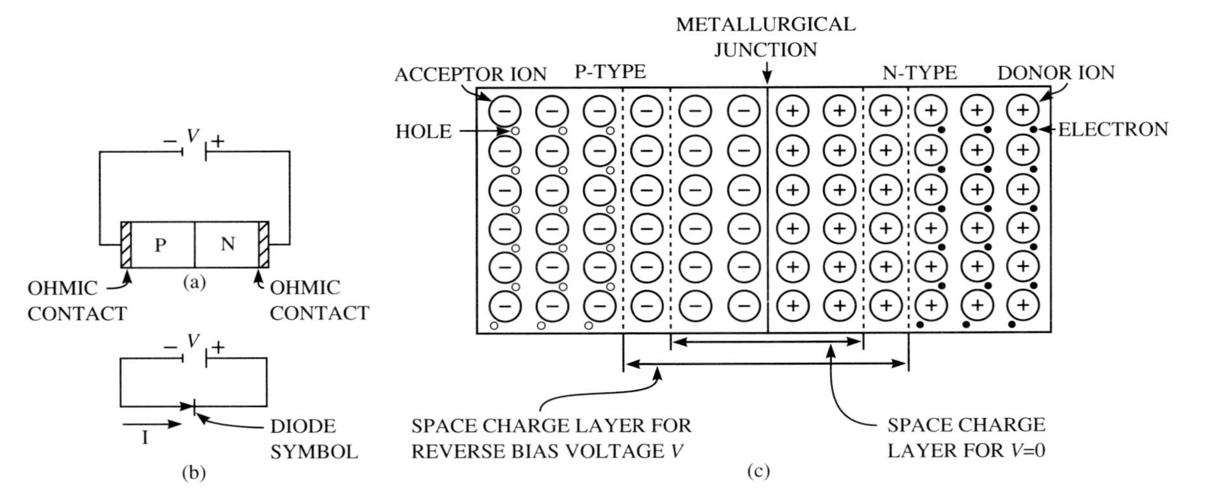

Figure 3-23 (a) A p-n junction biased in the reverse direction. (b) The rectifier (diode) symbol is used for the p-n junction. (c) Physical details of the p-n junction.

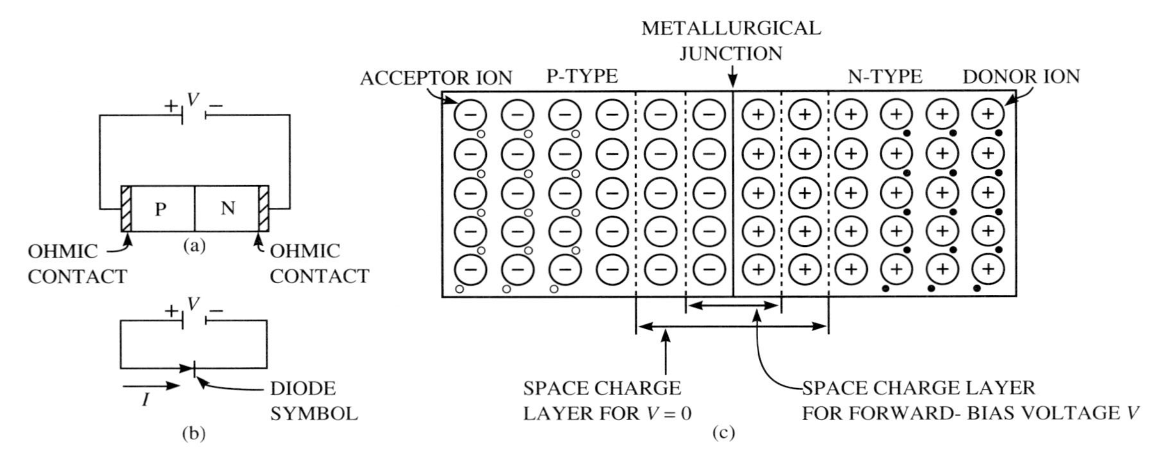

Figure 3-24 (a) A p-n junction biased in the forward direction. (b) The rectifier (diode) symbol is used for the p-n junction. (c) Physical details of the p-n junction.

3.3.4 Transistors

A **triode** is formed when p-type silicon is sandwiched between two pieces of n-type silicon, or when n-type silicon is sandwiched between two pieces of p-type silicon. These electrical components are known as NPN and PNP **transistors,** respectively.

The three sections of a transistor are known as base, emitter, and collector. Figure 3-25 shows the distribution of donors, acceptors, holes, and electrons in an NPN junction transistor under equilibrium conditions, with no external voltages applied. Also shown are circuit symbols for both NPN and PNP transistors. Although not shown in the figure, space charge layers exist at both metallurgical junctions.

The potential energy of electrons, under these conditions, is shown in Fig. 3-26. The electrons are in their position of lowest potential energy, so no current flows. In normal bias conditions, the emitter-base junction is forward-biased and the collector-base junction is reverse-biased, as shown in Fig. 3-27.

The potential energy for electrons then changes to that shown in Fig. 3-28, when no AC signal is applied to the base. Note that the barrier between the emitter and base regions is reduced substantially compared to that in Fig. 3-26.

Forward bias of the emitter-base junction has reduced the potential energy hill at the left p-n junction, causing some electrons to be injected into the base region. Since the base region is relatively thin, most of the electrons pass through the base and go down

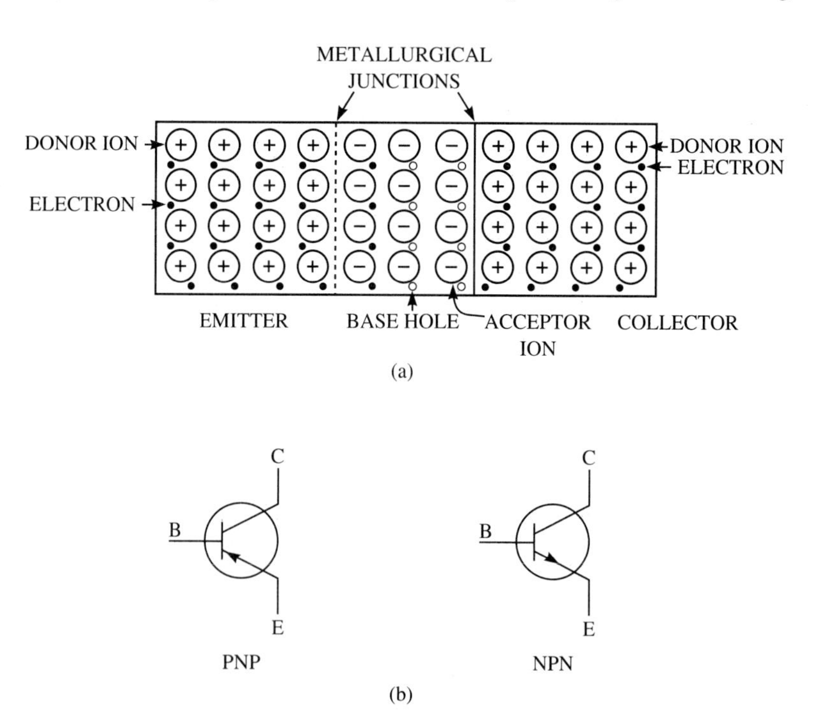

Figure 3-25 (a) Physical details of an NPN junction transistor in a state of equilibrium. (b) Circuit symbols used for PNP and NPN transistors.

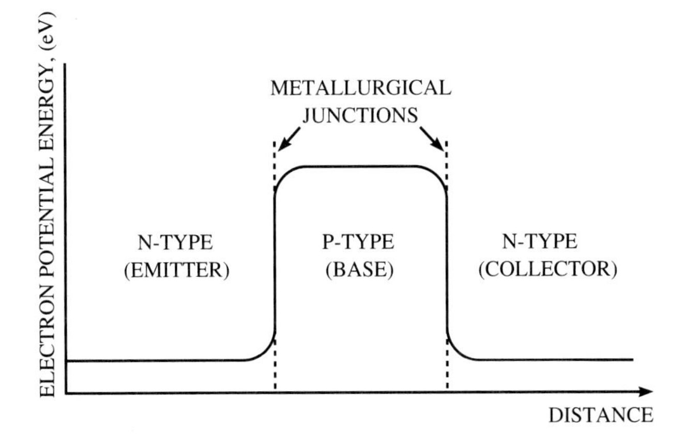

Figure 3-26 Potential energy of electrons in an NPN junction transistor, in a state of equilibrium.

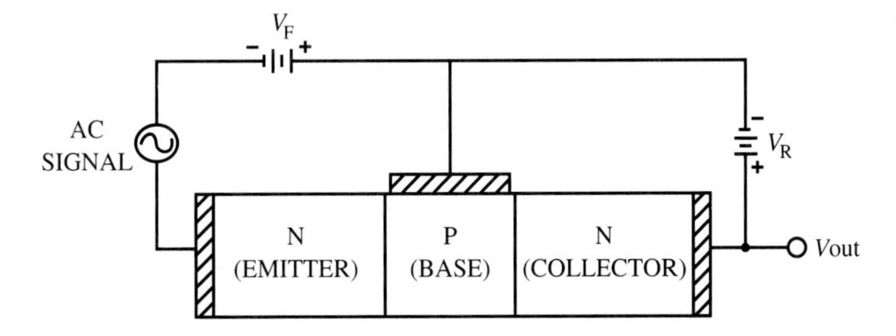

Figure 3-27 An NPN junction transistor under normal operating conditions (biases).

the potential energy slope at the right p-n junction into the collector region, as shown in Fig. 3-28. Electrons that do not enter the collector region recombine with holes in the base region, and both are eliminated.

When the applied signal is aiding the increase of forward bias of the emitter-base junction (see Fig. 3-27), the potential energy hill, between the base and emitter, is reduced further (see Fig. 3-29(a)), causing more electrons to enter the p-type region and pass on through to the collector region. Since the emitter is doped much more heavily than the

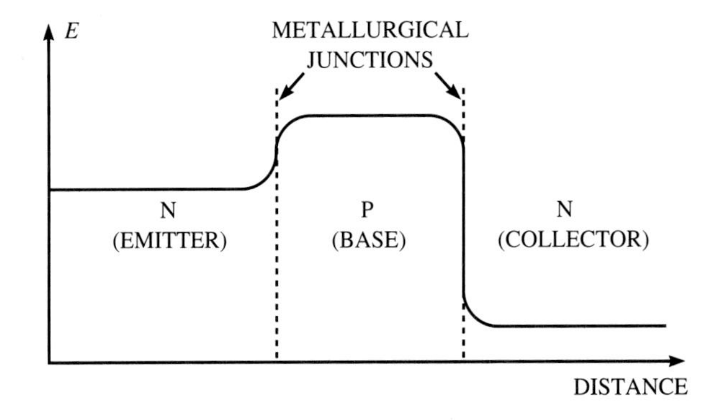

Figure 3-28 Potential energy of an electron with no signal applied to the base.

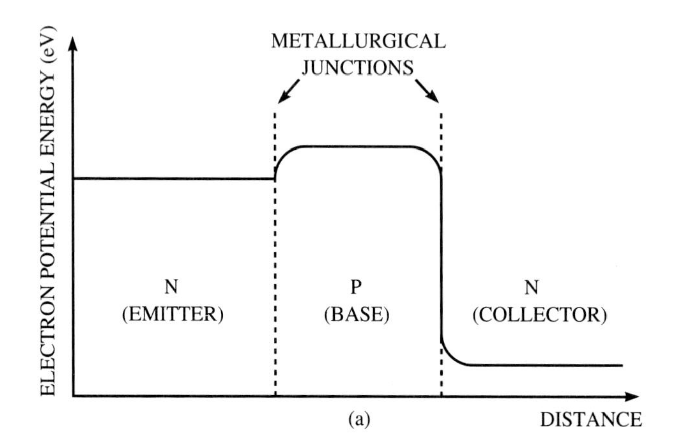

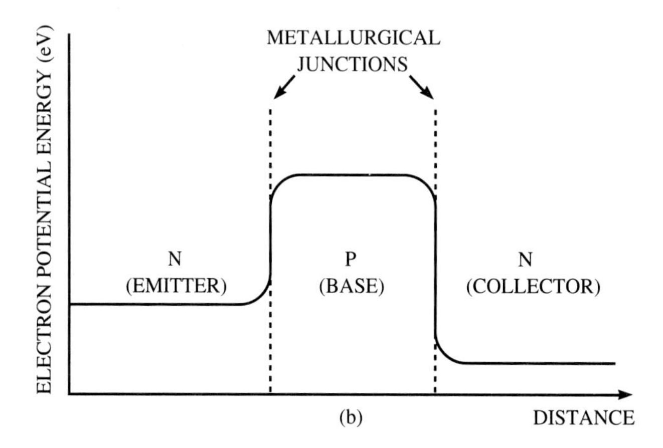

Figure 3-29 (a) An NPN transistor with an AC signal applied so that it increases the forward-bias condition at the emitter(E)-base(B) junction. (b) Signal swings back so that it opposes the forward-bias condition at the emitter-base junction.

base, the injection of holes into the emitter from the base is negligible, and a small base current signal yields an avalanche of electrons. The collector current is then almost equal to the emitter current, because there is very little recombination in the base region.

As the applied signal swings back to oppose forward bias of the emitter-base junction, the potential energy hill takes on the profile shown in Fig. 3-29(b). Few electrons will then climb the hill to enter the p-type region, but of those that do, most will fall into the collector region of low potential energy. To see the effect of the AC signal, the reader should compare Figs. 3-29(a) and 3-29(b) with Fig. 3-28.

3.4 COMPONENT MANUFACTURING AND PACKAGING

The remaining sections of this chapter present and discuss methods by which components such as resistors, capacitors, diodes, and transistors are made, where they are used, and how they are packaged.

The two dominant methods of component manufacturing that are discussed are film technology and planar-diffusion technology.

3.5 FILM TECHNOLOGY

Film technology is the construction of electronic circuits and devices through layering of films on a substrate by vacuum deposition and etching, screening, and a combination of these methods. In general, only passive devices such as resistors, capacitors, and inductors are created by film technology, although thin film transistors exist and are used sparingly. It should be noted that, even though the discussion is of individual components, these components are not connected by leads or other separate methods. The interconnections are an integral part of the process itself, as can be seen in Fig. 3-30.

There are two divisions of film technology: thick film and thin film. The terms **thick** and **thin** are relative but are commonly understood to mean printed-and-fired and vacuum-deposited, respectively. Thin film offers the highest component precision and reliability, but thick films involve lower capital investment because the design and production costs are much lower than for thin film technology.

3.5.1 Thick-Film Technology

Thick-film technology is the science of forming electronic elements or networks by printing liquid or paste through a screen or mask onto a supporting substrate and then firing (heating) it. The primary method used is screen printing, much like the method of silk screening used on T-shirts and sweatshirts (see Fig. 3-31). Film thicknesses in the thick-film method are usually 10 μm or greater. Circuit components that are created with this technique include capacitors, resistors, and conductors.

3.5.2 Thin-Film Technology

Thin-film technology is the science of forming electronic elements or networks by vacuum evaporation, sputtering, or anodization on a supporting substrate. Film thicknesses in the thin-film method are less than 5 μm, and usually on the order of 0.03 to 0.01 μm. Generally, only passive devices are created with this technique, including capacitors and resistors (see Fig. 3-32).

3.5.3 Resistive Elements

Resistor films are described in terms of sheet resistance in units of ohms per square. This is the resistance of a square of the film material, and it varies inversely with the film thickness. The value of resistance for a given film is determined by its length-to-width ratio, as shown in Fig. 3-33, and is given by

$$R = \rho_B L/tW \tag{3-2}$$

where R equals resistance in ohms, ρ_B is bulk resistivity in ohm-cm, L is resistor length

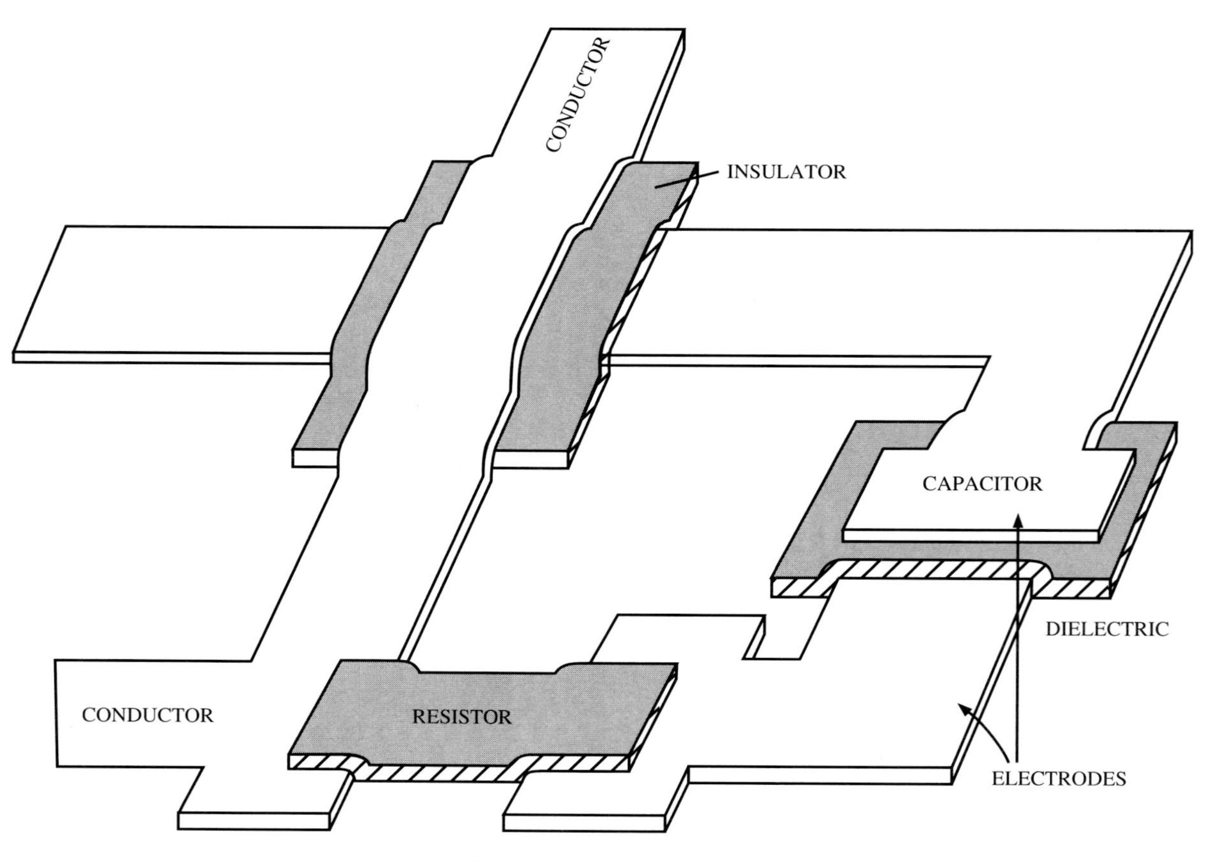

Figure 3-30 Hybrid microcircuit film components.

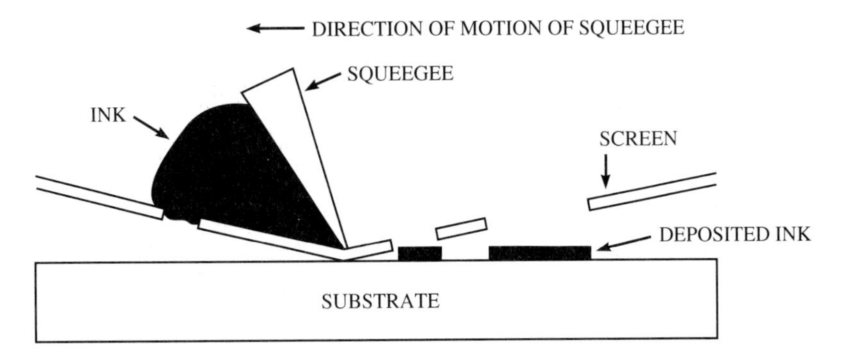

Figure 3-31 Basic screen printing operation for deposition of thick films.

in cm, t is film thickness in cm, and W is resistor width in cm. Let $L/W = n$ (number of squares, dimensionless) so that

$$R = \rho_B n/t \tag{3-3}$$

and sheet resistance can be defined as

$$\rho_s = \rho_B/t = R/n \text{ (ohms/square)} \tag{3-4}$$

When a resistor is formed using film technology, the closest tolerances that are achievable are in the neighborhood of ±15 percent of the desired value. This is not an acceptable tolerance. Therefore, trimming methods have been developed that allow the adjustment of the value of the resistor.

There are three primary methods of resistor trimming currently in use: computer-controlled laser, abrasive trimming, and anodization. The laser may either be a YAG (yttrium-aluminum-garnet) or CO_2 (carbon dioxide) laser. The concentrated energy of the laser is focused onto the substrate film, and the film is vaporized at the point of contact, as shown in Fig. 3-34. Test probes determine when the desired resistance is achieved, and trimming is halted.

The advantages of using lasers are that very fast trimming rates are available (several inches per second) and the laser beam can be concentrated to a very small diameter (on the order of 2.54 mm to 25.4 mm). Laser trimming makes very close resistive tolerances achievable and makes it possible to work within very high-density circuits.

In the abrasive trimming method, very fine alumina particles are directed by a stream of air, through a nozzle, against the substrate film, as shown in Fig. 3-35. Abrasion cuts away the resistive film, allowing adjustment to the required value.

Abrasive trimming is a much older method than is laser trimming, but it is still very popular because of reliability, relatively low cost, and adequate tolerances (on the order of ±0.5 percent). Also, abrasive trimming avoids the formation of microcracks in the substrate, which may occur with laser trimming due to heating of the substrate by the laser.

Anodization is an electrochemical process which is used to grow an oxide in certain thin-film metals. When a tantalum resistor is anodized, parts of the metal are oxidized, causing the metal film to become thinner and thus increasing the resistance. Very accurate control of the process is possible, thus allowing standard tolerances of ±1 percent to be

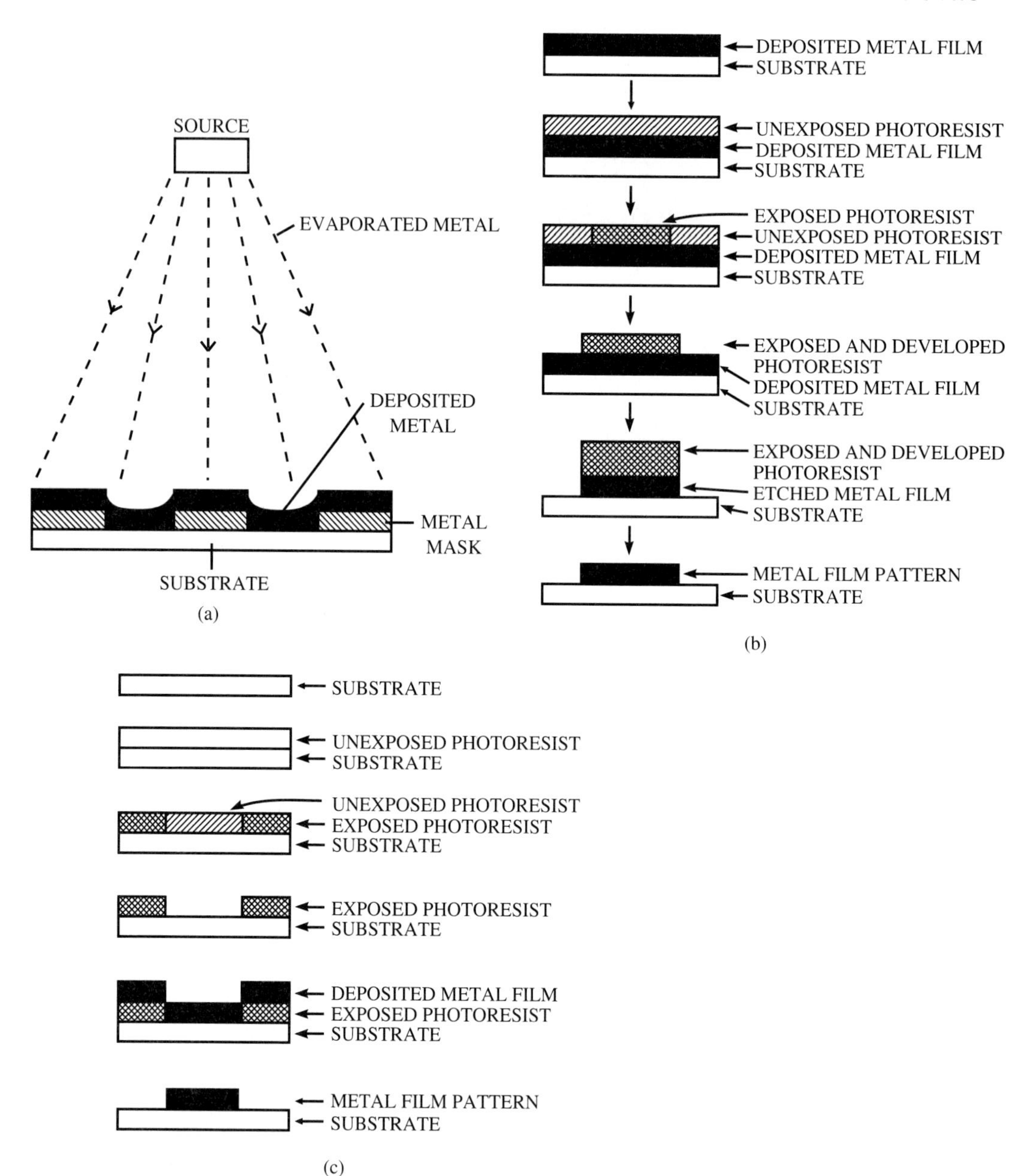

Figure 3-32 Methods of producing thin-film patterns on a substrate: (a) mechanical vacuum-deposition mask, photo-resist, (b) etched-subtractive, (c) vacuum-additive.

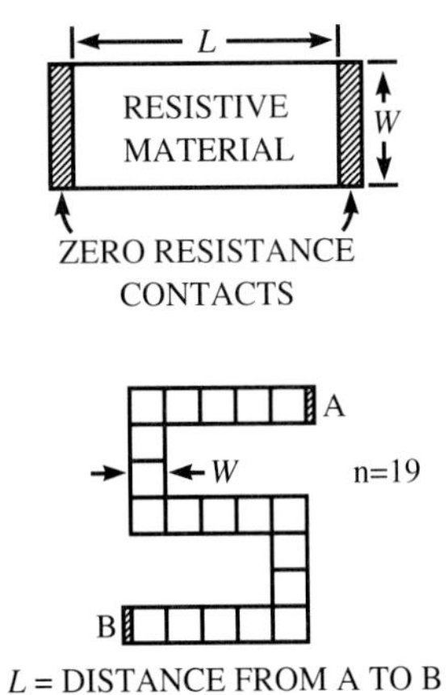

Figure 3-33 Designs for resistive microcircuit elements.

achieved. High-precision adjustments, in the neighborhood of ±0.02 percent to 0.001 percent, are possible. Unique to this method is the advantage that trimming can be carried out regardless of the physical geometry of the resistor.

3.5.4 Capacitive Elements

Deposited-film capacitors are composed of two layers of conductive material separated by a dielectric layer. The use of thick-film capacitors is relatively limited, compared to

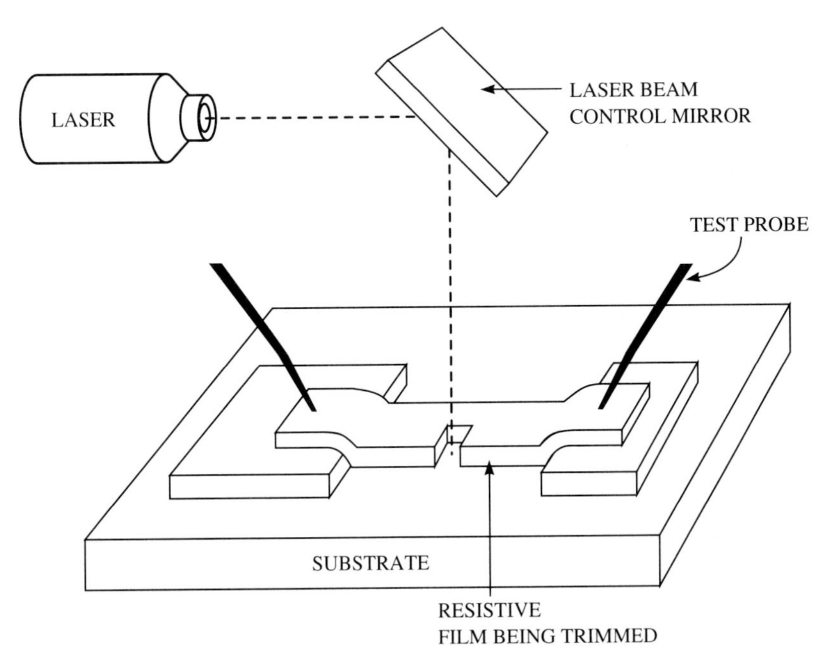

Figure 3-34 Laser method of adjusting the value of thin-film components by reducing the width of the resistive material.

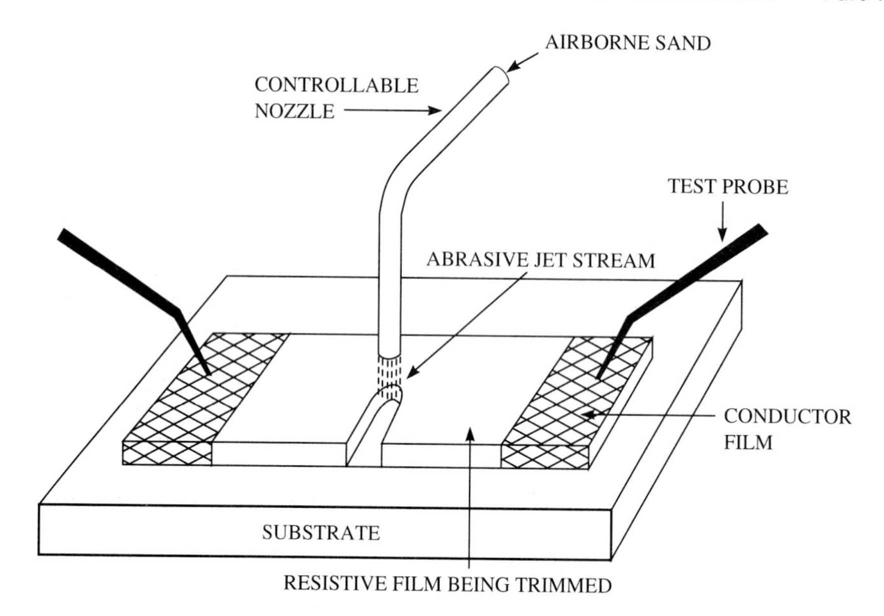

Figure 3-35 Abrasive method of adjusting the value of thin-film resistive components by reducing the width of the resistive material.

the use of thin-film capacitors. This is due to problems with achieving acceptable tolerances, and the fact that thick-film capacitors occupy large substrate areas when high capacitance values are needed.

3.6 PLANAR-DIFFUSION TECHNOLOGY

Planar simply implies that the circuit components are created beneath the surface of a silicon wafer, rather than layered on top of the surface. **Planar-diffusion technology** is a method used to make active components such as diodes and transistors, as well as passive devices such as resistors and capacitors. Several methods are presently used to create these components, but we will only discuss the epitaxial and diffused junction methods.

3.6.1 Epitaxial Method

Epitaxial means to arrange upon or to deposit. In the **epitaxial method,** a monocrystalline wafer is placed into a chamber and heated. A silicon-containing gas, carrying a dopant (impurity), is then admitted into the chamber. The doped silicon gas forms a layer of doped silicon on the surface of the wafer. The newly formed layer may be either n-type or p-type silicon. If a transistor is to be formed, the wafer must next go through a diffusion process.

3.6.2 Diffusion Method

In the **diffusion** process, a doped silicon wafer is thermally oxidized, resulting in a layer of silicon dioxide on the surface, as shown in Fig. 3-36. Next, a large number of windows are opened in the oxide layer using a photolithographic process. Dopants are then introduced into the silicon wafer through these windows.

The windows in the oxide layer are created by a photoetching process that is much

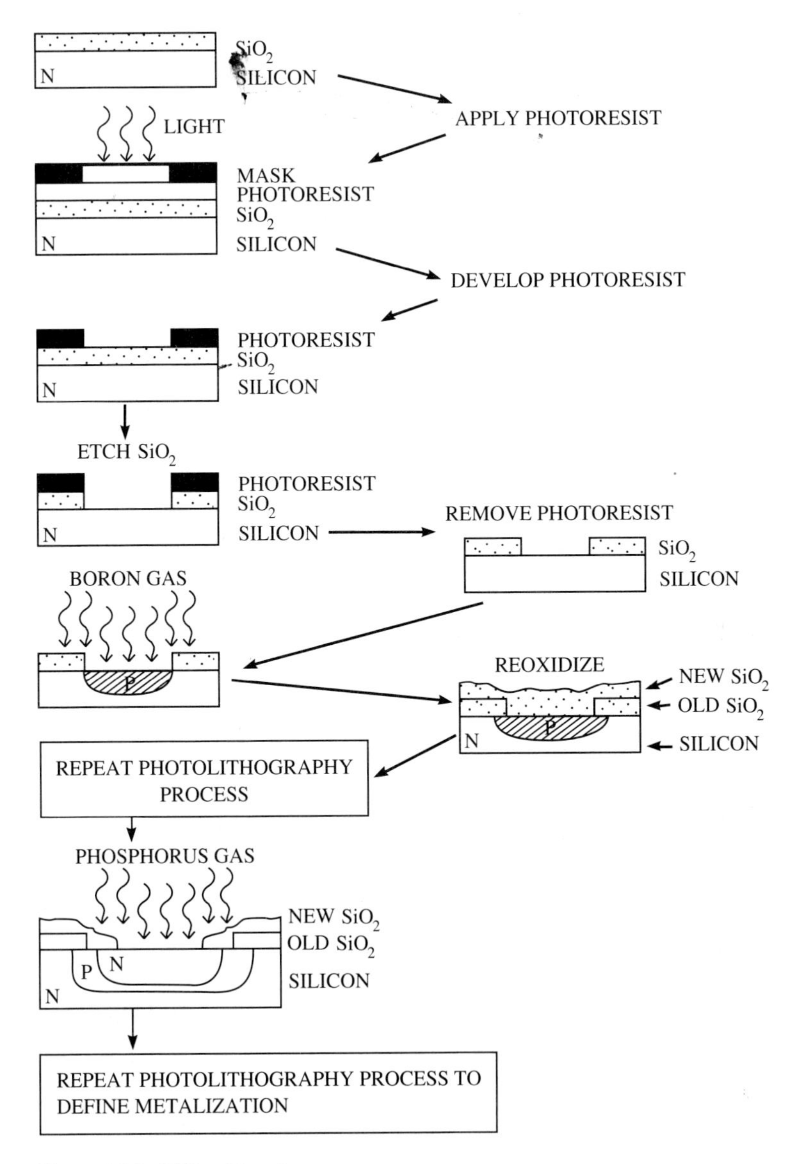

Figure 3-36 Diffused-junction process.

the same process as is used for making circuit traces on PWBs. A photoresist layer is masked off, exposed, developed, and washed away. These processes produce windows in the photoresist mask, allowing the oxide to be etched away in specific places. The wafer is then washed in an etchant, to cut through the oxide.

The windowed silicon wafer is then placed into a furnace, where it is heated to near melting temperature, and a dopant gas is admitted into the furnace. The gas is not able to penetrate the layer of oxide, but where windows have been etched out, the gas penetrates the silicon. If, for example, boron gas is being admitted, a p-type layer will be formed in the wafer.

The wafer may then go through another oxide/mask/etching process so that phosphorus gas can be diffused into the wafer to form the last n-type layer, resulting in the three regions needed for an NPN transistor (refer to Fig. 3-36). The mask contains thousands, or perhaps millions, of windows to create many transistors simultaneously, making the process cost-effective.

3.7 INTEGRATED CIRCUITS

Both planar and film techniques may be used to make either individual circuit components or entire **integrated circuits** (ICs). ICs are simply electronic circuits composed of two or more components, whether they are capacitors, resistors, transistors, or any combination of these or other circuit components.

In microelectronics, ICs usually come in the form of monolithic ICs or hybrid ICs. **Monolithic** (single stone) simply means that the entire circuit is built upon one chip. The chips are normally ten to twenty mils (250 to 500 μm) thick, and they can be up to 0.25 in^2 (1.5 cm^2) in area. The creation of these circuits may be accomplished using only planar techniques, or the fabrication process may incorporate film technology as well.

3.7.1 Monolithic ICs

As electronic products became more complex, integrated circuits came into being, not only to reduce the size of circuits, but more important, to reduce the number of interconnections that are required in a circuit. A primary cause of circuit failures is defective soldered connections, so a reduced number of interconnections dramatically increases circuit reliability.

A good example of improved reliability can be seen in the single-chip circuit of a J-K flip-flop (Fig. 3-37). This single flip-flop, if assembled with discrete circuit components, would require 40 separate discrete circuit elements, 200 connections, 40 hermetic seals, and 300 separate processing operations. Each component or connection is a possible source of failure. When integrated onto one chip, the device requires approximately fourteen connections, and all interconnections are created by a method known as vapor metallization. Instead of 40 hermetic seals there is only one, and the 300 processing steps are reduced to about 30. As can be seen from the dimensions in Fig. 3-38, the IC package takes about 1/1000 the area that is required for the discrete circuit.

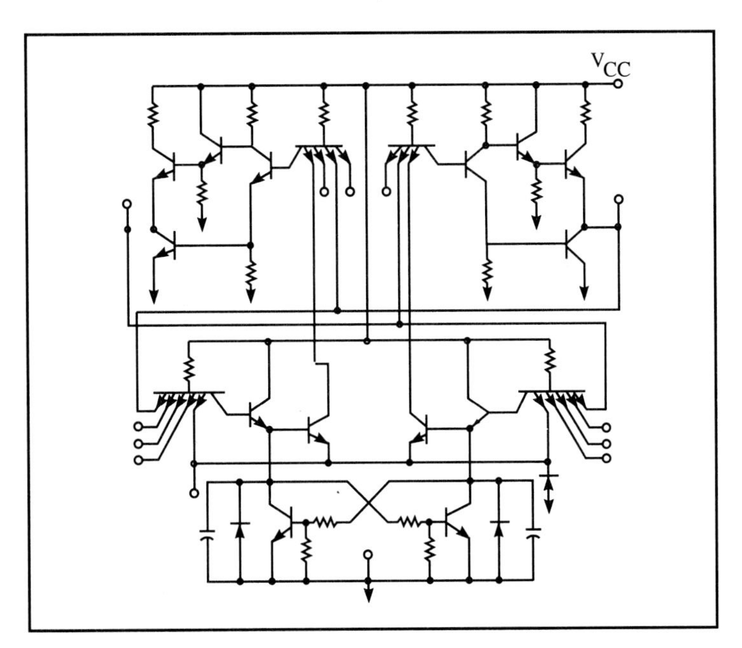

Figure 3-37 Schematic of a J-K flip-flop.

(*Source:* Reprinted with permission, from *Microelectronics* by Clayton L. Hallmark. Copyright 1976 by TAB Books, a Division of McGraw-Hill Book Company, Blue Ridge Summit, Pa., 17294-0850, 800-233-1128.)

3.7.2 Hybrid ICs

Hybrid ICs make use of all the techniques discussed so far. A **hybrid circuit** is normally composed of one or more monolithic ICs and includes many components using thick-film and thin-film techniques. All components and chips are assembled into a relatively small enclosure, or package.

The package contains an insulating substrate, with deposited networks, usually conductors and resistors, to which semiconductor devices, ICs, and passive elements are attached in chip form. The most popular substrate is alumina, because it can easily withstand the high firing temperature required for thick-film processing.

We present a more thorough discussion of hybrid ICs in Chapter 4, where tape automated bonding (TAB) and chip on board (COB) technologies are discussed.

3.8 THROUGH-HOLE COMPONENTS

There are three basic profiles of through-hole (TH) components that are discussed in the following sections: axial-lead, radial-lead, and multiple-lead.

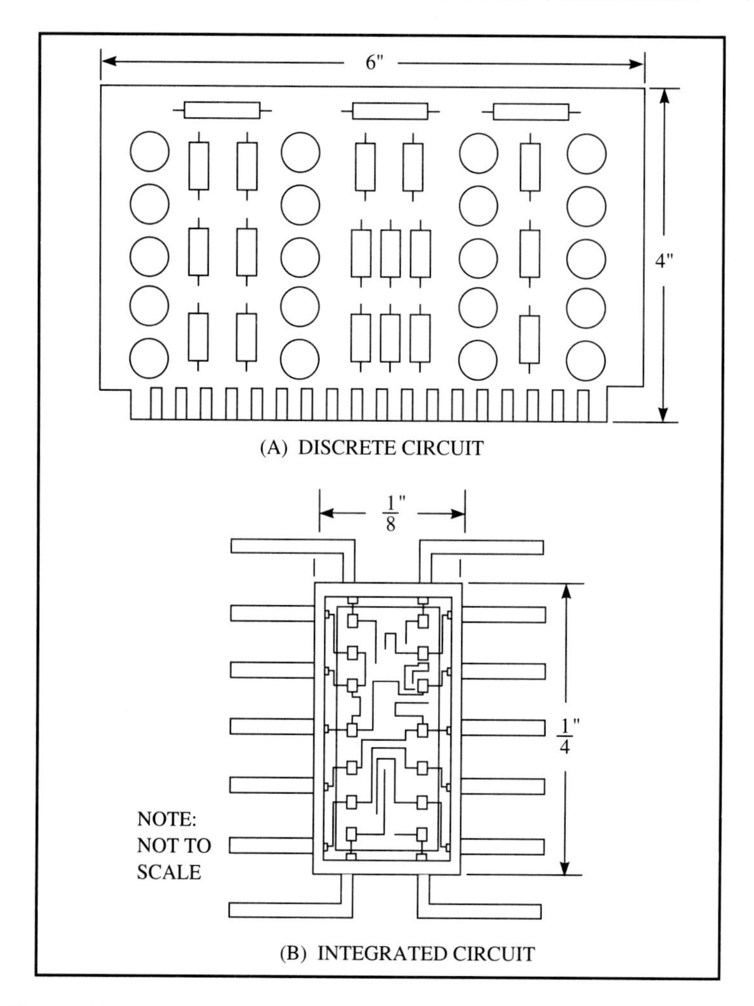

Figure 3-38 Size comparison between discrete and integrated circuit versions of a J-K flip-flop. The discrete circuit is approximately 1,000 times larger.

(*Source:* Hallmark, *Microelectronics.*)

3.8.1 Axial-Lead Components

Axial-lead components are usually cylindrically shaped. The leads run through the central axis of the component and exit from each end of the cylinder (see Fig. 3-39). Examples of axial-lead components include resistors, capacitors, and diodes.

3.8.2 Radial-Lead Components

Radial-lead components come in various shapes, but all have leads exiting from only one end of the component package. Examples of radial-lead components include transistor outline (TO) packages (see Fig. 3-40) and dipped capacitors.

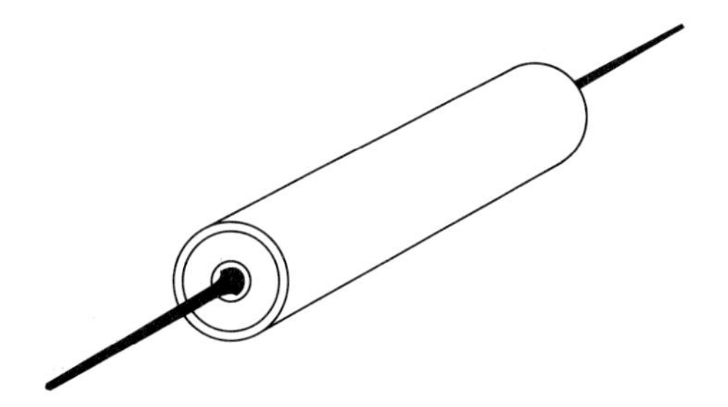

Figure 3-39 Axial-lead component.

Transistor outline packages are hat-shaped, metal cans (TO cans) with leads coming off one end, as shown in Fig. 3-41. TO cans were first popularized as reliable transistor packages, but they are now also used as chip carriers and hybrid circuit packages. Their main advantages are extreme ruggedness and low cost; the disadvantage of TO cans is that the small size restricts the area of the substrate, severely limiting the potential number of terminal leadouts.

The three main TO packages are the small (TO-5), medium (TO-8), and large (TO-3). The number of leads in the TO-5 style package will vary from as few as 4 to as many as 12. The large TO-3 package (see Fig. 3-42) is used for two-pin power transistors and multipin hybrid circuits, with a lead count of 3 to 12 leads. The medium-sized TO-8 can is also used for multipin hybrid circuits, and may have 12 or 16 pins located on 75 or 100 mil (1.91 or 2.54 mm) grid lines, as shown.

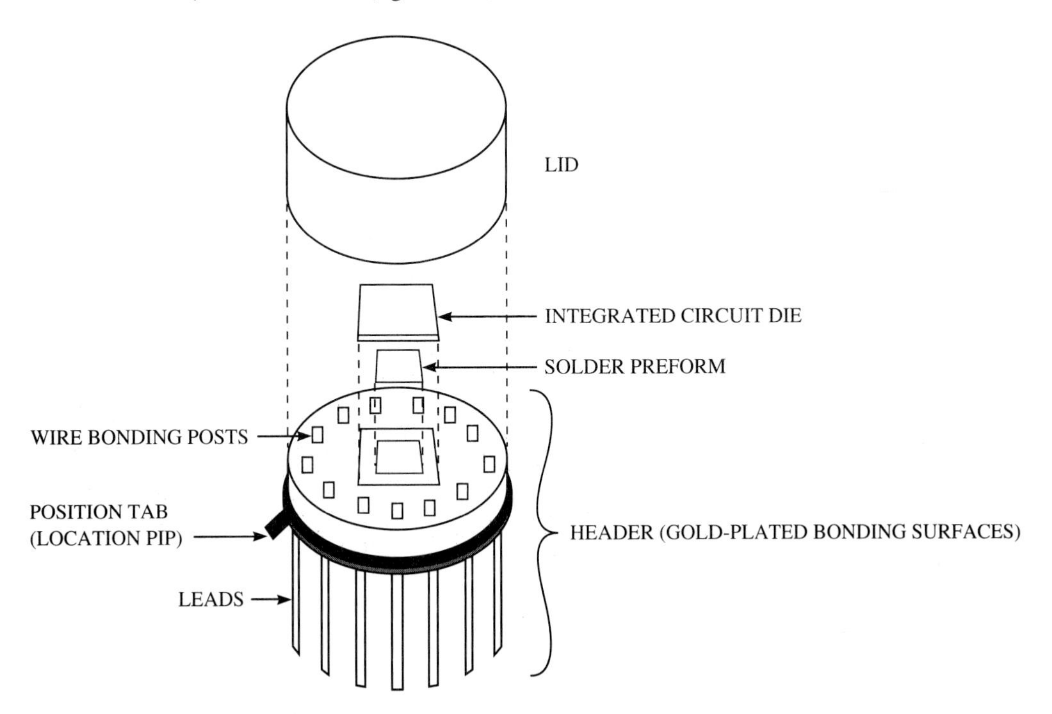

Figure 3-40 Exploded view of modified TO package.

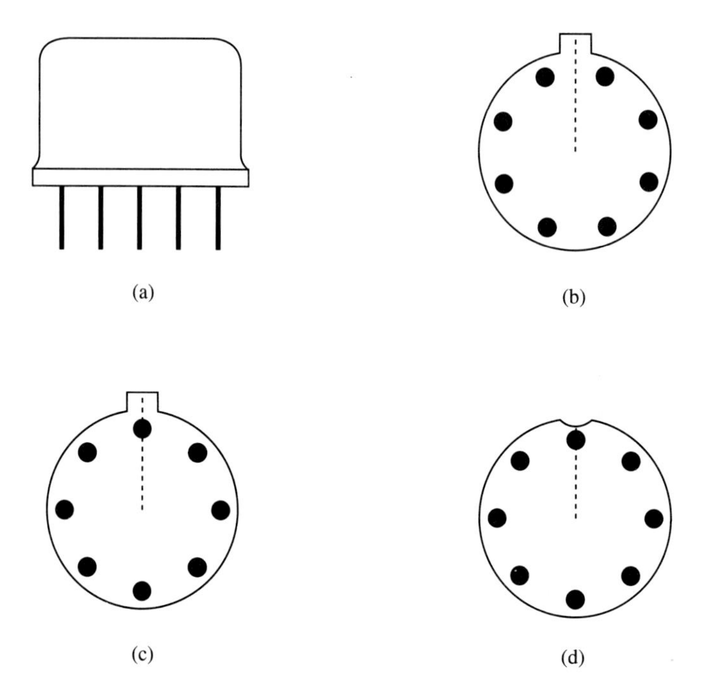

Figure 3-41 Multilead TO-5 packages for hybrid microcircuits: (a) TO-5 outline basic features with (b) lead array symmetrically offset from location pip as in EIA-JEDEC MO-002 standard, (c) leads symmetrical on pip as in EIA-JEDEC MO-006 standard, and (d) plastic pipless MO-009.

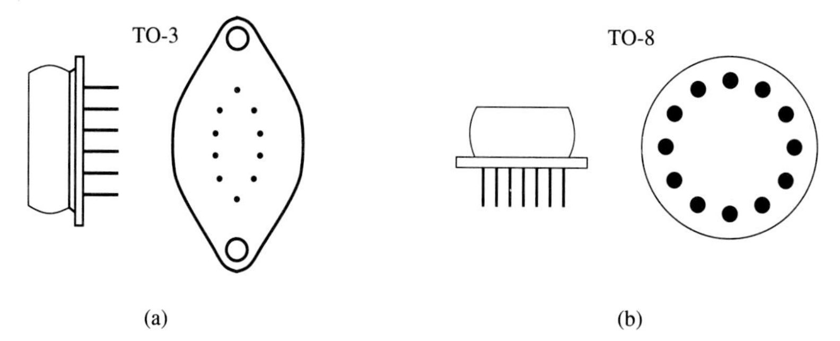

Figure 3-42 Typical hybrid conduction-cooled packages designed for attachment of heatsink: (a) multilead TO-3; (b) multilead TO-8.

3.8.3 Multiple-Lead Components

Multiple-lead components are actually collections of components that are housed in one package and inserted into a wiring assembly as one unit. Examples of multiple-lead components include dual-in-line packages, single-in-line packages, and pin grid arrays (or grid arrays).

Single-in-line (SIL) packages (Fig. 3-43) may have round leads (diameter = 15–

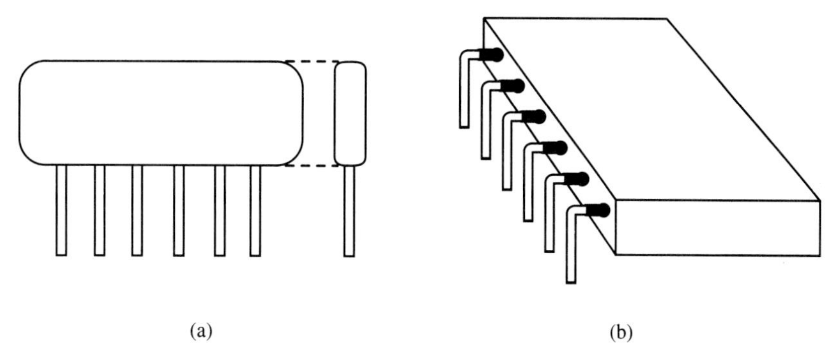

Figure 3-43 The SIL (single-in-line) hybrid package, also called SIP (single-in-line package): (a) stand-off version; (b) flush version.

25 mils or 381–635 μm) or flat ribbon leads of about the same thickness. Leads are normally on a standard center spacing of 0.075, 0.100, 0.125, or 0.150 inch (1.9, 2.5, 3.2, or 3.8 mm). SILs may be mounted in an upright position, or if height is of concern, there is a flush-mounting package that allows the SIL to lie flush to the PWB.

The dual-in-line package (DIP or DIL) is probably the most common package for multiple-lead TH components (see Fig. 3-44). There is essentially one standard DIP lead spacing: 0.100 inch (2.5 mm) centers. Spacing between the rows is most commonly 0.300 inch (0.76 cm) centers, but there are also double- and treble-width packages that have row-widths of 0.600 inch (1.5 cm) and 0.900 inch (2.3 cm), respectively.

The most common DIPs will have 8, 14, 16, 24, or 40 leads, but there are some special dual-in-line packages that may have as few as 4, or as many as 80 leads.

3.9 SURFACE-MOUNT COMPONENTS

3.9.1 Discrete Chip Components

Discrete chip components are essentially miniature axial-leaded components without the leads. They are available both in cylindrical and flat rectangular shapes (see Figs. 3-45 and 3-46).

Almost any type of TH component will have a counterpart **surface-mount component** (SMC), but most choices will be in power ranges of less than one or two watts. SMCs are impractical above this range because of the large physical sizes that are required and because of heat dissipation problems that occur for higher power ranges.

3.9.2 Resistors

Rectangular chip resistors are made by applying thick-film resistor paste to a ceramic substrate and firing it in an oven, much like the glazing process used in ceramics. The chip will then have metallized terminations on the end of the body for electrical contact to the PWB (see Fig. 3-46).

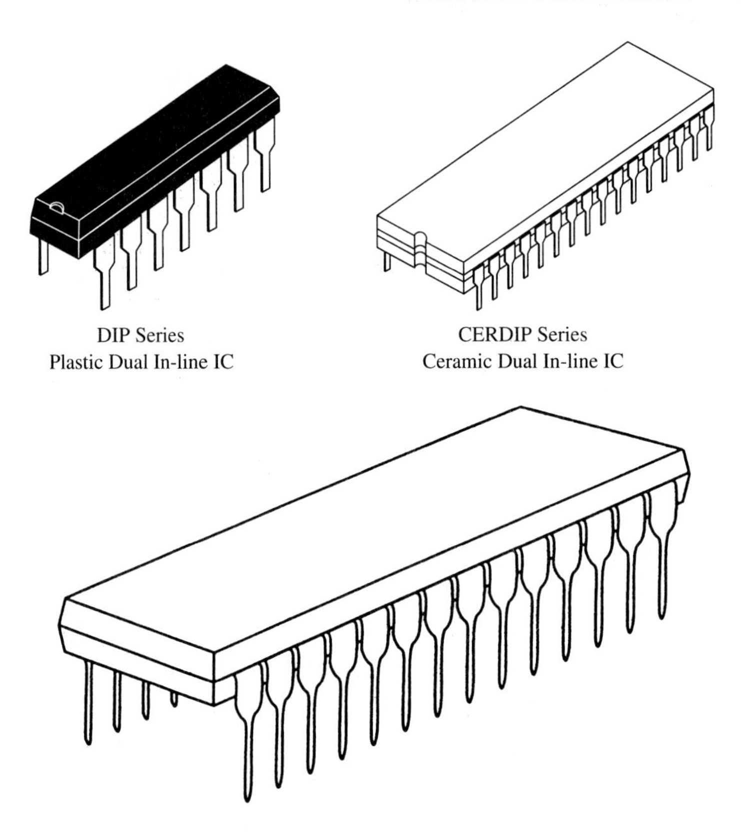

Figure 3-44 Dual-in-line (DIP) packages (courtesy of TopLine Components).

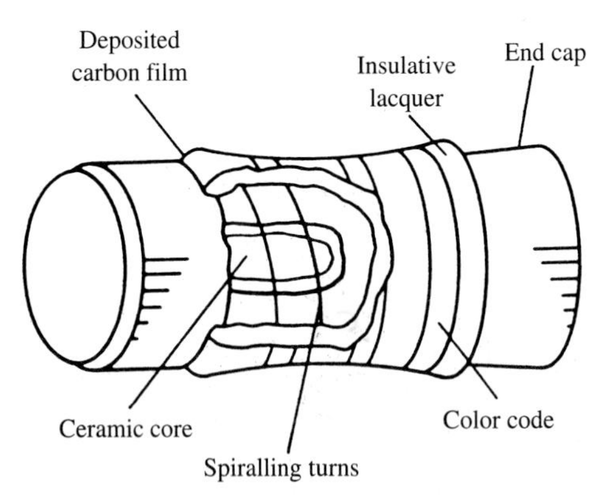

Figure 3-45 Cylindrical chip resistor.

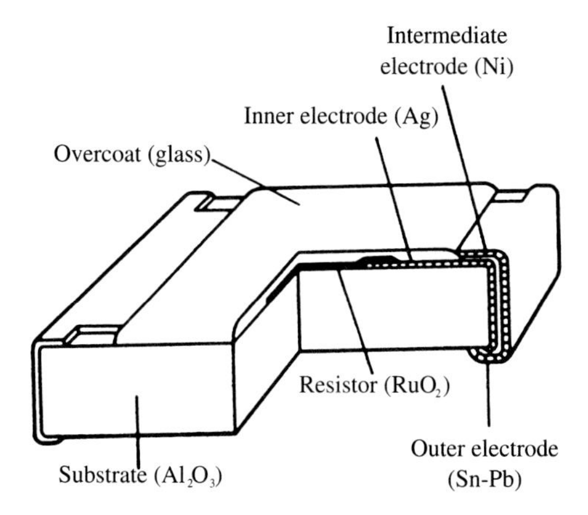

Figure 3-46 Rectangular chip resistors.

The most common body style is the 1206, which refers to its 0.125 inch (3.2 mm) length and 0.063 inch (1.60 mm) width. This style is also known as the 3216, due to its dimensions in millimeters. This thick-film variety is typically in the ¼ watt range and is economically manufactured with tolerances of ±1 percent to ±2 percent. Tighter tolerances are more expensive and are usually attained by the thin-film processes.

The metal electrode leadless face (MELF) resistor is just a conventional resistor without the leads. MELFs are cylindrical with metallized ends for surface mounting (see Fig. 3-47). MELFs take maximum advantage of existing manufacturing technology and are less expensive than rectangular chips.

Until MELFs are soldered in place, they are able to roll out of alignment on the PWB, unless attached with an adhesive. Wave soldering equipment usually has inclined inlet and outlet conveyors, and for this reason, rectangular ceramic chip resistors have been popular in the United States. In Asia, where reflow soldering is popular, MELFs and rectangular chips are both used extensively.

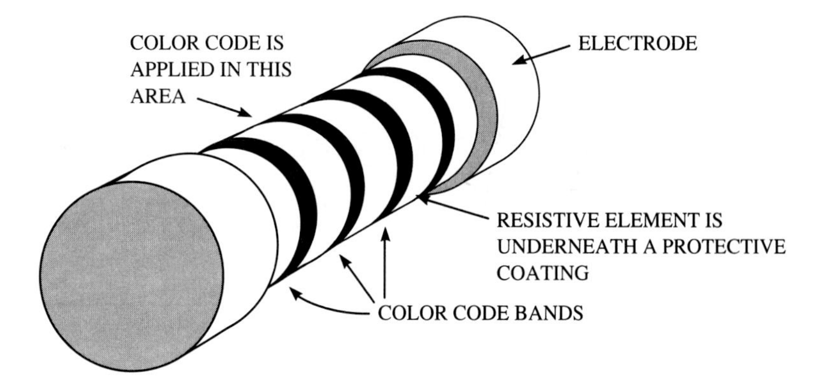

Figure 3-47 Metal electrode face resistor (MELF).

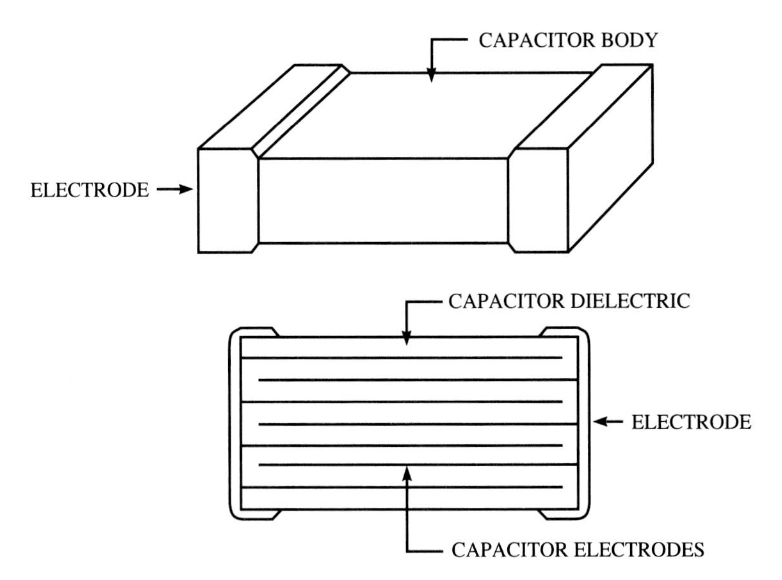

Figure 3-48 Multilayer ceramic chip capacitor: (a) physical appearance; (b) cross-section.

3.9.3 Capacitors

The multilayer ceramic capacitor is the most commonly used surface-mount capacitor. It is made up of a series of parallel precious metal electrodes, separated by layers of ceramic dielectric (see Fig. 3-48). Five standard package sizes have been adopted by the EIA (see Table 3-4), covering the range of capacitance from 10 pF to 2 μF.

Tantalum capacitors (Fig. 3-49) are used when higher capacitance values are needed. Capacitance values up to 30 μF, at an operating voltage of 10 V, are available in this type capacitor. Tantalum capacitors come in a flat-molded body style that makes them especially suited for automatic assembly processes. Tantalum capacitors require a minimum 50 percent voltage derating, or added series resistance, to provide protection from current spiking during turn-on. Otherwise, a steady failure rate of 1,000–2,000 ppm must be tolerated.

TABLE 3-4 EIA STANDARD CERAMIC CAPACITOR SIZES OF SURFACE-MOUNT COMPONENTS

Code designator	Dimensions in mm		
	Length	Width	Height
0805	2.0	1.2	1.2
1206	3.2	1.6	1.5
1210	3.2	2.5	1.7
1812	4.6	3.2	1.7
1825	4.6	6.4	1.7

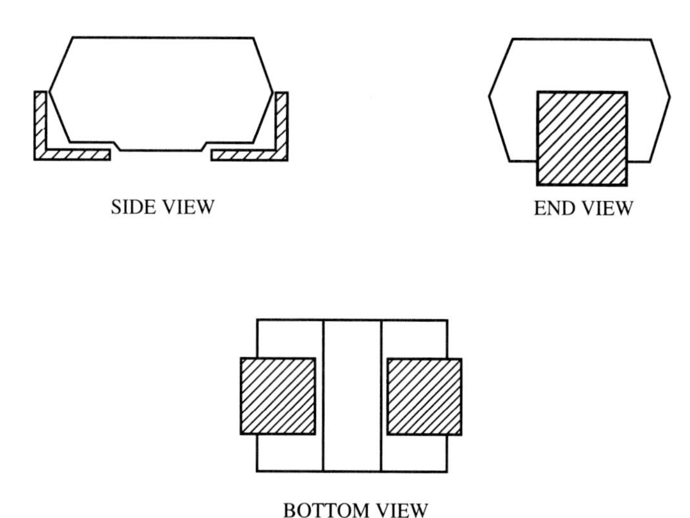

Figure 3-49 Tantalum chip capacitors.

3.9.4 Discrete Semiconductors

Discrete semiconductors are the same devices as are used in TH mounting, except for the package. In TH mounting, the leads offer the primary path for heat removal from the chip. In surface mounting, the leads are smaller than those used in TH mounting, so the thermal resistance of the package must be higher.

3.9.5 Small-Outline Packages

Small-outline packages, as shown in Fig. 3-50, resemble miniature versions of DIPs and may have three or more leads. Their primary advantage is their small size, which makes them the best choice where space and weight are at a premium. Small-outline packages typically occupy 30–50 percent less area and have 70 percent less thickness than an equivalent DIP. The two types of small-outline packages are small-outline transistors (SOTs) and small-outline integrated circuits (SOICs). Both of these types come in a variety of choices. The SOICs will be discussed later when components that come in these packages are covered.

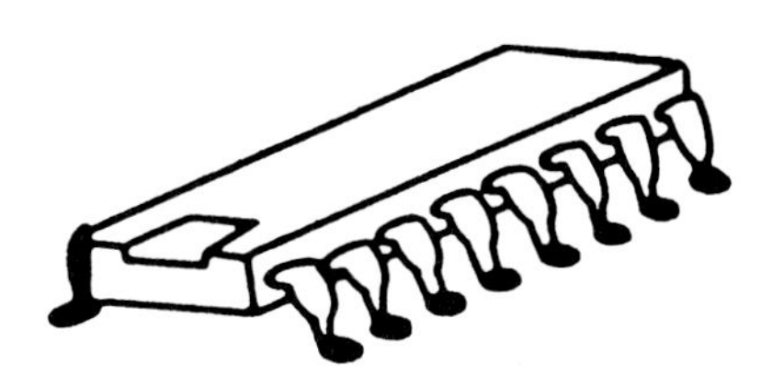

Figure 3-50 Small-outline integrated circuit (SOIC) packages (courtesy of TopLine Components).

3.9.6 Transistors

SOTs are the most popular package for surface-mount transistors (see Fig. 3-51). SOT packages are classified according to the power ranges for which they are designed, and the main choices are as follows:

- SOT-23 (power dissipation ≤ 200 mW)
- SOT-143 (power dissipation ≤ 200 mW)
- SOT-89 (power dissipation ≤ 500 mW)
- DPAK (power dissipation ≤ 1.3 W).

The SOT-89 has a heat-sink tab that solders directly to the PWB to improve thermal conductivity. The SOT-143 is basically a SOT-23 but with four leads instead of three.

For applications above 500 mW, the choices are fairly limited. Figure 3-52 depicts the DPAK. The DPAK is available for power dissipations of up to about 1.3 W, but for applications above 1.3 W, a standard TO-220 package can be used if its leads have been preformed for surface mount.

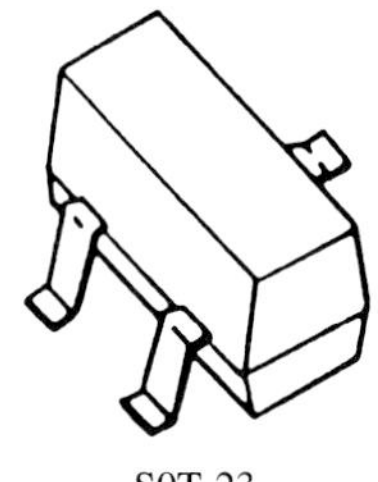

SOT-23

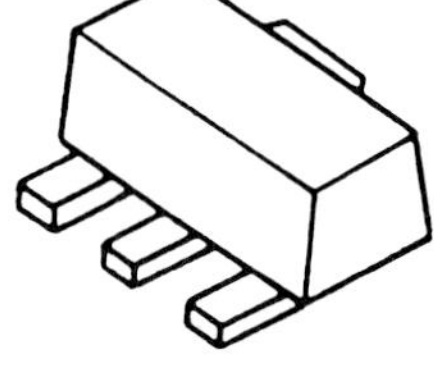

SOT-89

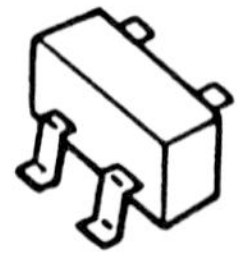

SOT-143

Figure 3-51 Small-outline-transistor (SOT) packages.

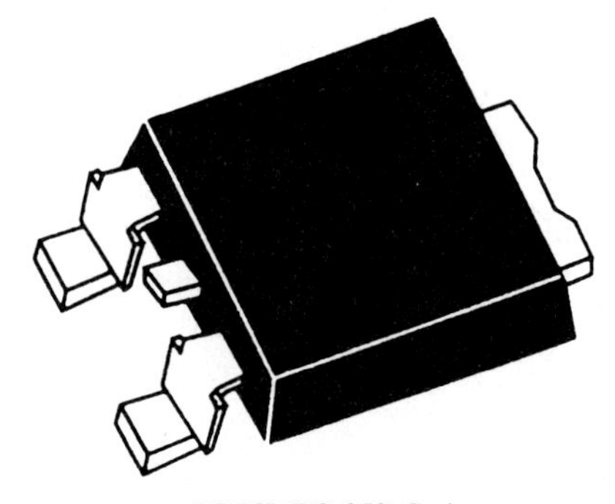

DPAK (TO-252) Series Transistor

Figure 3-52 DPAK (courtesy of TopLine Components).

3.9.7 Diodes

Diodes use much the same packages as are used for transistors. The SOT-23 and SOT-143 are used for low-power single and dual diodes, respectively. A two-terminal version SOT-89 has been designed for medium-power diodes. MELF packages are also sometimes employed for diodes.

3.9.8 Integrated Circuits

As with discrete semiconductors, surface-mount integrated circuits only differ from their TH counterparts by their packaging. Packages tend to vary from inexpensive molded plastic for consumer products to hermetically sealed ceramic packages for military and high-reliability use.

Integrated circuit packages include:

- Small-outline integrated circuits (SOICs)
- Plastic leaded chip carriers (PLCCs)
- Ceramic leaded chip carriers (CLCCs)
- Leadless ceramic chip carriers (LCCCs)
- Quadpacks
- Flat packs

3.9.9 Lead Configurations

Each of the leaded-type packages may be offered in a variety of **lead configurations,** which can be seen in Fig. 3-53. The lead types include gull-wing leads (part (a) of the figure), J-form leads (part (b)), and I-leads (part (c)).

The gull-wing configuration offers the advantage of easy visual inspection of the solder joint. Its disadvantages are that it takes up a lot of room on a PWB and its leads are more susceptible to damage during handling. If the leads are bent out of their seating plan by as little as 0.005 inch (0.13 mm), the component may not seat properly for soldering.

J-form leads are designed for better protection against damage in handling. Other advantages of the J-form leads are that there is better clearance between the PWB and the package, which makes for easier cleaning of solder flux from under the package, and that the package takes up less area on the PWB.

The primary disadvantage of the J-form lead is that visual inspection of the solder joint is much more difficult. Also, the forming of J-form leads is more difficult, which means the process is more expensive and quality levels are lower.

The I-lead provides many of the benefits of the J-form lead, and it is more simple to manufacture. It is also less susceptible to handling damage than either the J-form or gull-wing lead. The one difficulty is that all leads must be sheared simultaneously to provide coplanarity of lead ends.

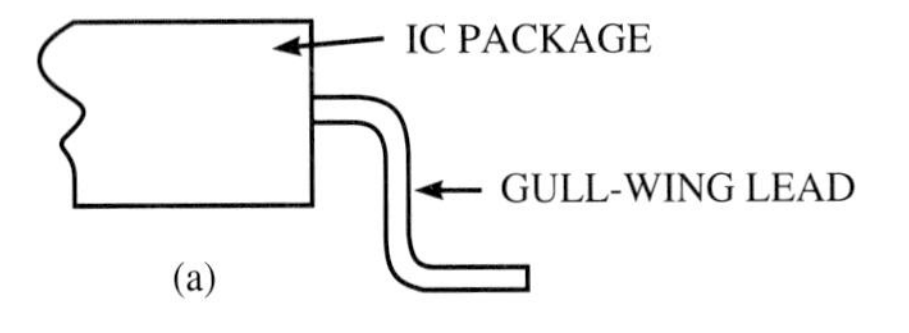

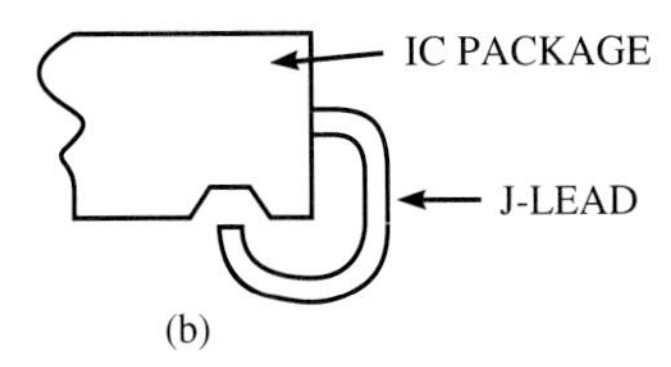

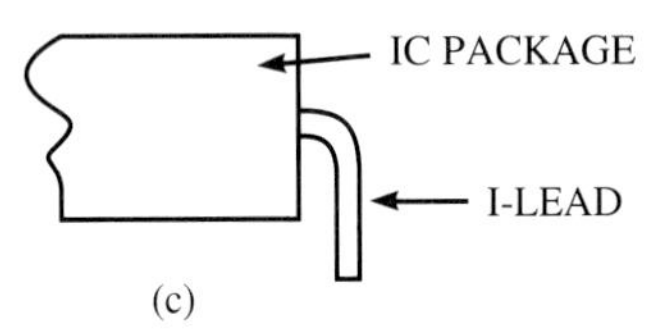

Figure 3-53 IC lead configurations: (a) gull-wing; (b) J-lead; (c) I-lead.

3.9.10 Leadless Packages

Leadless packages come in cylindrical and rectangular varieties. The rectangular packages have metallized pads or edges to permit soldering of the package to a PWB, and the cylindrical types have metallized ends for the same purpose. Leadless packages will be discussed more fully when specific leadless components are addressed.

3.9.11 Package Types

Figure 3-54 shows the structure of a **flat pack.** The flat pack comes in variations with up to 200 leads, usually mounted on 0.050 inch (1.27 mm) centers. It is one of the smallest multiple-lead component types; it may be as small as ⅛ inch (3.2 mm) wide, ¼ inch (6.4 mm) long, and 1⁄32 inch (1.6 mm) thick (see Fig. 3-55). The flat pack also has the distinction of being the first SMC package type, introduced in the 1960s.

The **quadpack** is one of the earliest SM plastic IC packages (see Fig. 3-56). Quadpacks have gull-wing leads on all four sides and are available in lead pitches from 0.005 inch (0.13 mm) down to 0.002 inch (0.05 mm). These fine lead pitches make this package especially suited for lead counts of 84 pins or more. To achieve lead pitches this fine, the leads are made from very thin metal stock. Thus, the leads are extremely fragile and susceptible to damage during handling, tending to discourage widespread use of quadpacks.

Chip carriers (CCs) come in two main classes (plastic and ceramic) and are also available in leadless and leaded versions (see Fig. 3-57). The Joint Electronic Device

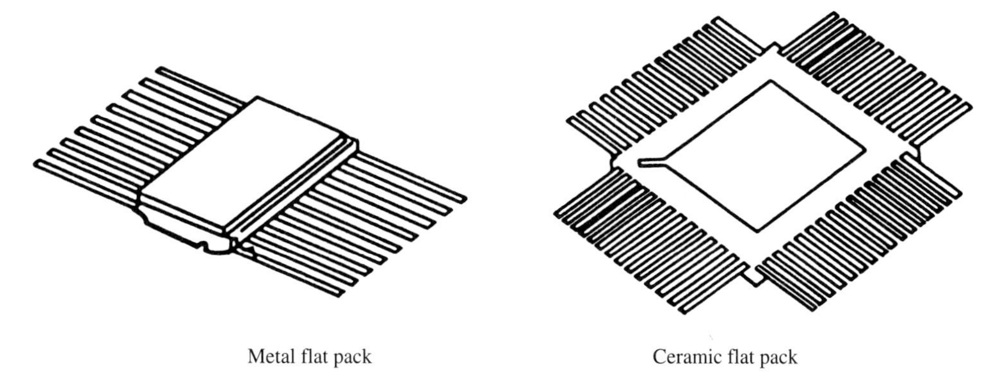

Figure 3-54 Flat-pack multiple-lead package.

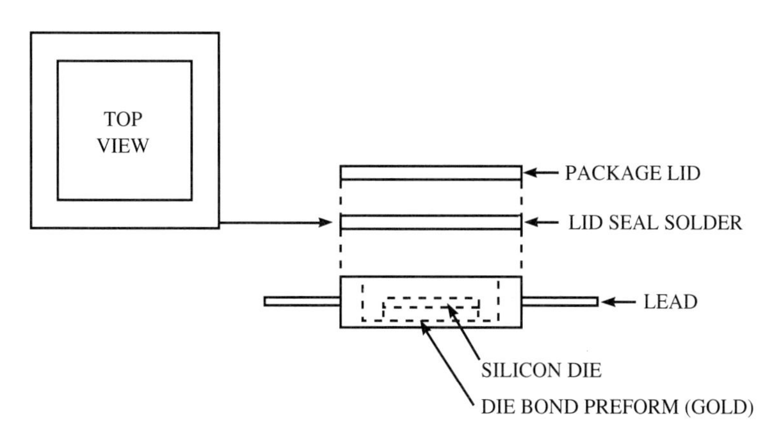

Figure 3-55 Exploded view of a ceramic flat pack.

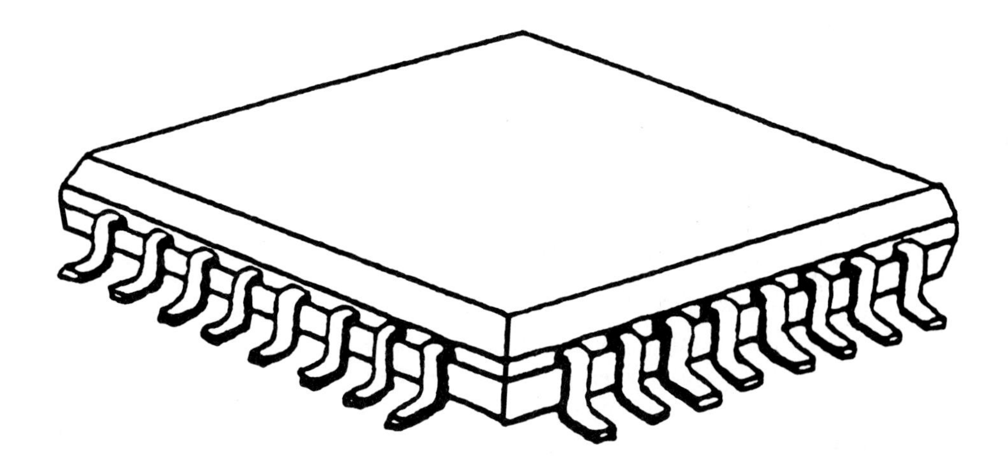

Figure 3-56 Quadpack.

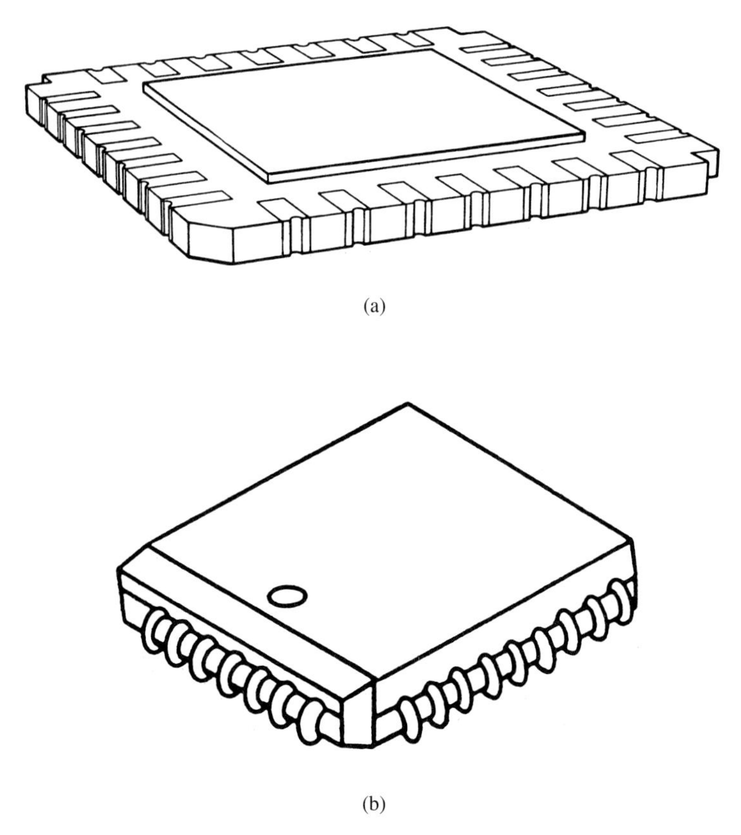

(a)

(b)

Figure 3-57 Chip carriers: (a) leadless; (b) leaded.

Engineering Council (JEDEC) has established standards for CCs that are flexible, in that they allow for a multiple-design approach.

JEDEC has standardized two basic package styles (on two lead-center spacings):

- 0.050 inch (1.3 mm) centers
- 0.040 inch (1.0 mm) centers.

The 50-mil package is the most common package used in industry, and the 40-mil package is used mainly in military applications, for which high packaging densities are desired. Of the seven 50-mil package variations, four are leadless ceramic chip carriers that have hermetically sealed ceramic or metal lids, two are **plastic leaded chip carriers,** and one is a ceramic leaded chip carrier. Output pin counts range from 16 to more than 150.

The class distinctions between the leadless-type CCs are based on internal features, thermal performance, and mounting orientation with respect to the substrate. The leadless CCs all have metallized pads around the perimeter of the package that allow the package to be soldered to a PWB (see Fig. 3-58).

Ceramic leaded chip carriers were developed because of the problems caused by

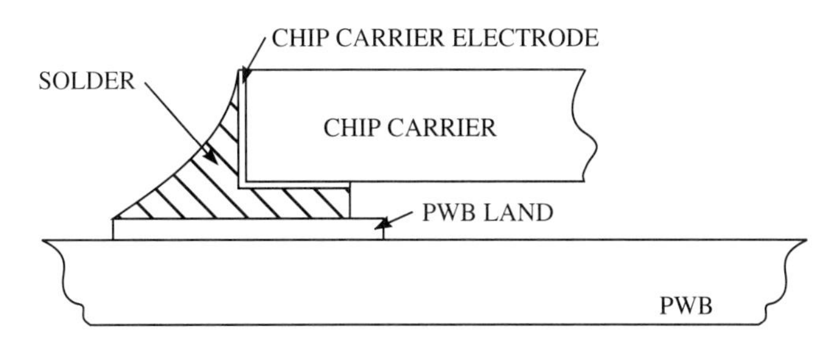

Figure 3-58 Solder joint cross-section.

mounting LCCCs to epoxy-based PWBs. The coefficient of thermal expansion (CTE) for ceramic materials is much less than for epoxy-glass. Consequently, the solder joint is subject to severe stresses over the full operating temperature range of the components, which can cause cracking and subsequent failure of the joint.

Methods have been tried to prevent thermal cracks, such as constraining expansion of the PWB by laminating it to material with a CTE similar to ceramic. Another method is to add leads to the ceramic chip carrier. The leads then act as compliant members between the chip carrier and the PWB, but this approach makes the chip carrier more expensive than its leadless counterparts because of the extra step needed to add the leads. The added leads may also degrade the high-frequency electrical performance of the component, due to the additional lead length.

Small outline integrated circuits (SOICs) were originally designed for digital wristwatches. They are similar to DIPs, except smaller (refer back to Fig. 3-50). SOICs have gull-wing leads on two sides, with a lead pitch of 0.050 inch (1.3 mm).

SOICs are available in several body widths, and in the United States, the EIA has adopted two widths for standardization:

- 0.150 inch (3.8 mm) for packages with 8 to 16 leads
- 0.300 inch (1.6 mm) for packages with 16 to 28 leads

In Asia, the following body widths are common:

- 4.4 mm (0.173 inch)
- 5.4 mm (0.213 inch)
- 7.5 mm (0.295 inch)
- 9.4 mm (0.370 inch)

3.9.12 Comparing SOICs and PLCCs

Both the SOIC and PLCC packages, shown in Figs. 3-50 and 3-59, are widely used, and sometimes either can be used for the same application. The choice of which should be used is usually determined by the specifics of the application.

The smallest practical PLCCs are 20-pin square packages for digital and li

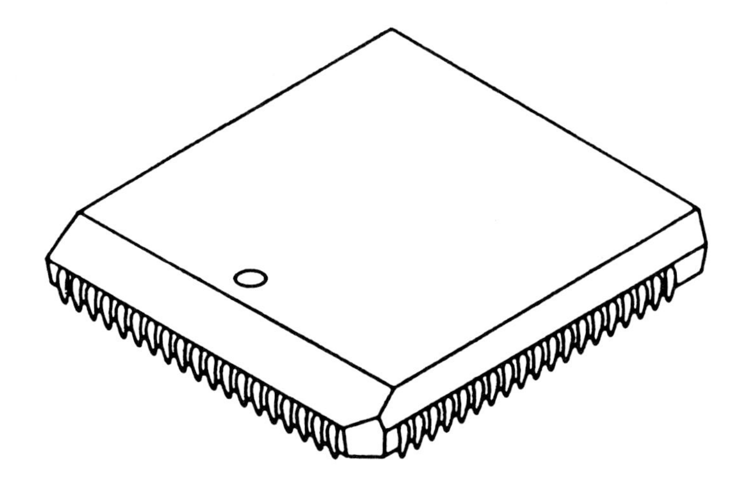

Figure 3-59 Plastic leaded chip carrier (PLCC).

circuits, and 18-pin rectangular packages for memory products. The SOIC package is more space-efficient for lead counts below these limits. The PLCC package is more space-efficient for lead counts of 28 pins or more.

In the range of 20–24 pins, the packages are about equal in efficiency, but other factors may need to be considered. For example, for all pin counts the SOIC has a lower profile than the PLCC. Therefore, if vertical clearance is of concern, the SOIC is preferred.

For the assembly process, the gull-wing configuration of a SOIC is especially suited for wave-soldering. Small SOICs may be mounted on the underside of a PWB and soldered in this manner. The J-form leads of PLCCs are not suited for wave-soldering; they must be reflow soldered.

Taking PWB layout into consideration, it is usually easier to make a layout for SOICs than for PLCCs, since SOIC traces can be routed under the package and out either end. PLCCs, on the other hand, have leads on all four sides, which makes routing more difficult. The traces must be routed between the leads of the package, or traces must be brought down to inner layers of a multilayer board.

3.10 SUMMARY

Electronic components used with printed wiring boards are designated as either through-hole ... surface-mount. These components can be further categorized as either active or ... devices. Passive discrete devices are available in a large variety of values and ...eir parameter values are designated using several marking schemes. Active devices ...mplexity from discrete diodes and transistors to large integrated circuits. Two ...minant methods used in component manufacturing are film technology and ...sion technology. Film technology is generally divided into thick film, where ...onents are formed using screening and firing techniques, and thin film, which ...acuum deposition and etching. Planar-diffusion technology is used to create electrical components beneath the surface of a silicon wafer. For all electrical components, power dissipation and input/output lead count determine the package size and shape. Consequently, there are a large variety of package types available commercially.

KEY TERMS

Axial-lead components
Capacitors
Ceramic leaded chip carriers (CLCCs)
Chip carriers
Depletion region
Dielectric
Diffusion
Diode
Discrete chip components
Discrete semiconductors
Doping
Epitaxial method
Film technology
Flat pack
Holes
Hybrid circuit
Integrated circuits
Lead configurations
Leadless packages
Monolithic
Monolithic circuits
Multiple-lead components
Planar
Planar-diffusion technology
Plastic leaded chip carrier (PLCC)
P-N junctions
Quadpack
Radial-lead components
Resistors
Semiconductor material
Small-outline integrated circuits (SOICs)
Solid-state devices
Space-charge region
Surface-mount (SM) components
Thick-film technology
Thin-film technology
Through-hole (TH) components
Triode
Transistors

EXERCISES

1. Distinguish between through-hole (TH) and surface-mount (SM) components. Which technique has been in existence for the longest period of time?
2. Discuss the reasons one might choose SM components over TH components.
3. A resistor is known to dissipate 250 milliwatts of power when a current of 5 milliamperes flows through it. What is the voltage across the resistor, and what is the value of the resistor?
4. Describe the physical construction of the three basic types of resistors.
5. Distinguish between the performance characteristics of the three basic types of resistors.
6. What are the classifications and the corresponding ranges of discrete resistor tolerances?
7. What is the resistance and tolerance of each standard resistor type, with color bands as follows:
   a. yellow, violet, red, silver
   b. red, yellow, yellow, gold
   c. orange, white, black, no color
   d. gray, red, blue, gold
   e. brown, brown, brown, gold

8. MIL-SPEC resistors have a prefix code which identifies something about them. What do the following prefixes indicate?
   a. RC
   b. RL
   c. RN
   d. RCR
   e. RLR
   f. RNR
9. What factors affect the capacitance value of a capacitor? Give an equation that relates capacitance to these factors.
10. Discuss what happens if the dielectric material of a capacitor is made too thin for the applied voltage or the applied voltage is made too high for the thickness of the dielectric material.
11. What are the two most common semiconductor materials used to produce diodes and transistors? Which one presently dominates the market?
12. Briefly discuss the process called "doping" as it relates to silicon.
13. How are donor atoms used to "dope" silicon related to each other in reference to the periodic table of the elements? Answer the same question for acceptor atoms.
14. What is a hole as it relates to doped silicon?
15. In your own words, describe how a "space-charge-layer" is formed in a p-n junction.
16. Draw a circuit containing a voltage power supply, a resistor, and a diode that has the diode in reverse bias.
17. Can an NPN transistor be fabricated by externally connecting two p-n diodes together? If not, why not?
18. Distinguish between the two dominant methods of component manufacturing.
19. Distinguish between thick- and thin-film technologies.
20. Determine the resistance value of the structure given in Figure 3-60.

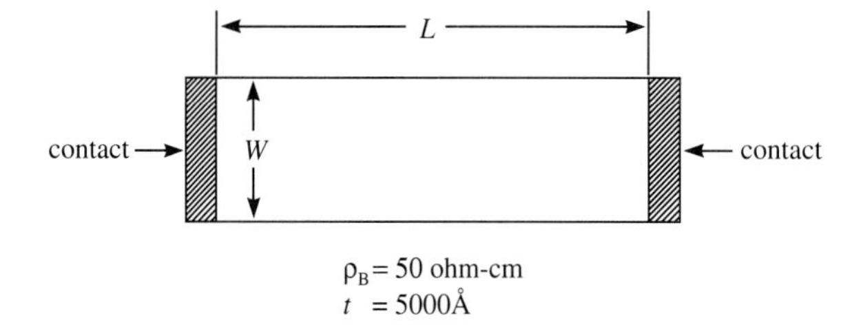

Figure 3-60 Thin-film or diffused-resistor structure.

21. Briefly discuss the three methods used to trim resistors. What are the advantages and disadvantages of each?
22. Distinguish between monolithic and hybrid integrated circuits.
23. Distinguish between the three basic types of through-hole components.
24. List and give a short description of five types of integrated circuit packages.
25. List and sketch the three lead configurations available for surface-mount component packages. What are the advantages and disadvantages of each?
26. What are the three general classifications of chip carriers in terms of their construction?

BIBLIOGRAPHY

1. JONES, T. H., *Electronic Components Handbook.* Reston, Va.: Reston Publishing Company, 1978.
2. FINK, DONALD G., AND DONALD CHRISTIANSEN, eds., *Electronics Engineers' Handbook* (3rd ed.). New York: McGraw-Hill Book Company, 1989.
3. COOMBS, CLYDE F., JR., *Printed Circuits Handbook* (3rd ed.). New York: McGraw-Hill Book Company, 1988.
4. *Reference Data for Radio Engineers* (6th ed). Indianapolis, Ind.: Howard W. Sams & Company, Inc., 1977.
5. MILLMAN, JACOB, *Microelectronics: Digital and Analog Circuits and Systems.* New York: McGraw-Hill Book Company, 1979.
6. HALLMARK, CLAYTON L., *Microelectronics.* Blue Ridge Summit, Pa.: TAB Books, 1976.
7. TOWERS, T. D., *Hybrid Microcircuits.* New York: Crane, Russak & Company, Inc., 1977.
8. TRAISTER, JOHN E., *Design Guidelines for Surface Mount Technology.* San Diego, Calif.: Academic Press, Inc., 1990.
9. *Electronic Materials Handbook, Vol. 1: Packaging.* ASM International, 1989.
10. TUMMALA, RAO R., AND EUGENE J. RYMASZEWSKI, eds., *Microelectronics Packaging Handbook.* New York: Van Nostrand Reinhold, 1989.
11. MAIWALD, WERNER, ed., "Soldering in SMD Technology" (brochure in series of publications entitled: *SMD Technology*), Siemens, Aktiengensellschaft, Munich, Germany, 1990.

4 Interconnections

4.1 INTRODUCTION

The previous chapter on electronic components discussed three general categories of integrated circuits (ICs). These categories included:

- *Thick/Thin Film ICs:* Conductive paths and resistive or dielectric film layers are deposited on a substrate to create passive electronic components.
- *Monolithic ICs:* The entire circuit is formed within the monolithic body of a silicon chip (also known as a die) and can include both passive and active circuit components.
- *Hybrid ICs:* These circuits incorporate at least one component from both the film and monolithic IC categories.

Figure 4-1 shows an example of each of the IC types noted above.

If an IC chip is to be useful in an electronic circuit, the chip has input/output (I/O) lines that must be accessed by that circuit. The interface between the chip and the surrounding circuit occurs at the chip I/O pads. Typically, the I/O pads are located around the outside edge, or periphery, of the chip, but sometimes they are in an array configuration on one face of the chip itself.

There are three primary **chip-level interconnection** methods currently being used: wire bonding (Sec. 4.2), controlled collapse chip connect (Sec. 4.3), and tape automated bonding (Sec. 4.4). This chapter also introduces multichip packaging (Sec. 4.5) and interconnection devices, including cables and connectors (Sec. 4.6–4.9).

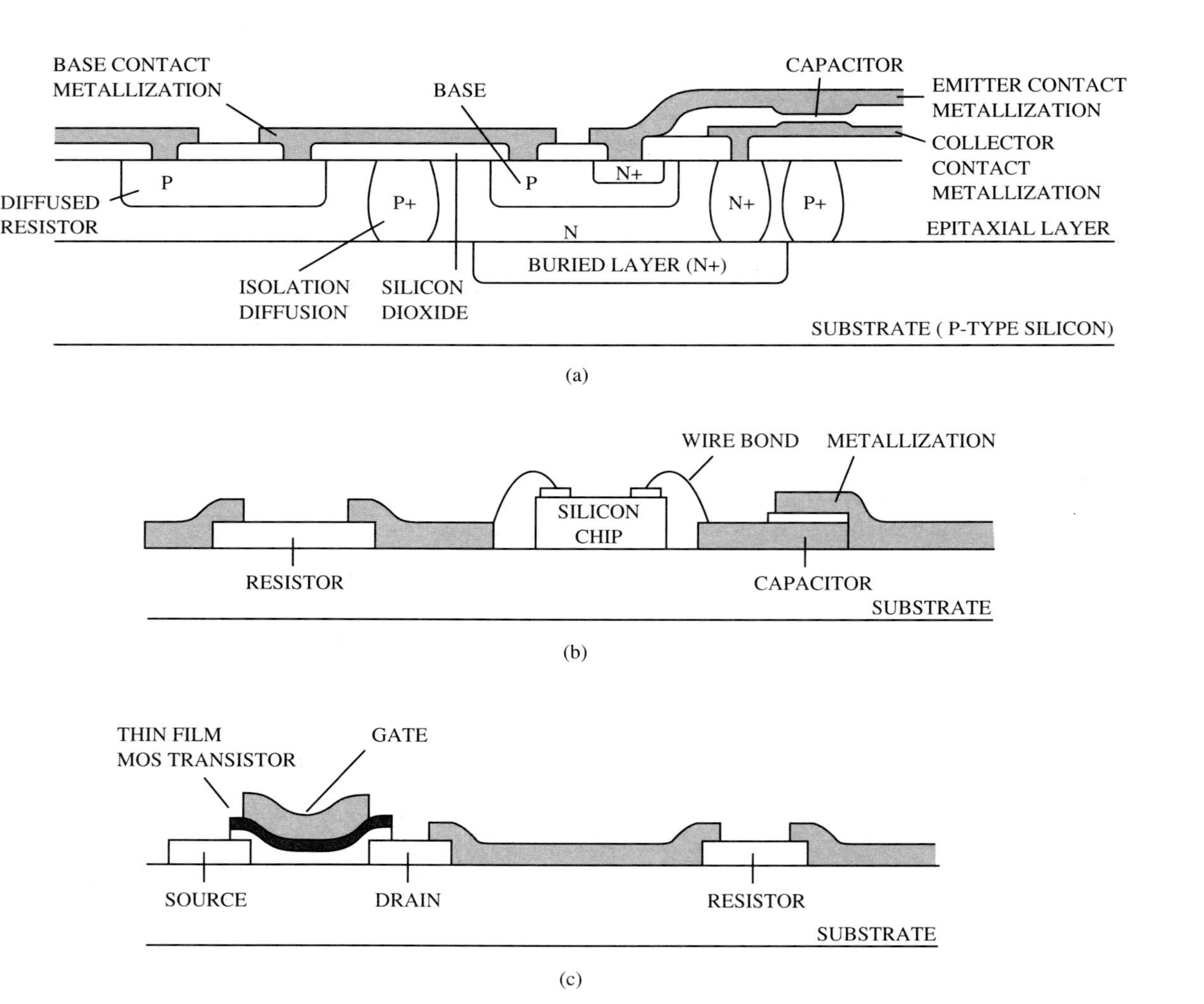

Figure 4-1 Cross-sectional diagrams of the three types of microelectronic circuits that use thin films: (a) monolithic silicon integrated circuit; (b) hybrid thin-film circuit; (c) all thin-film integrated circuit (note that drawings are not to scale).

4.2 WIRE BONDING

Wire bonding was introduced in the 1950s and is presently the most commonly used method of chip-level interconnection. It was originally done entirely by hand but has progressed to a fully automated process. The typical rate for this process is around five bonds per second, but it may be performed up to eight or ten times per second. It has been used for all types of chips and packaging styles.

In the wire bonding process, the chip must first be firmly attached to a proper substrate (usually ceramic or metallic) or to a lead frame (see Figs. 4-2 and 4-3 for some examples of lead frames). The chip is bonded to the central pad of the lead frame by silver-filled epoxy, silver-filled glass, or a metal-alloy reflow process. Figure 4-4 shows chips attached to a package substrate.

After the chip is attached, the interconnect wires are then bonded, or welded, one at a time. Figure 4-5 shows chips that have been wire bonded into a circuit. Figure 4-6 illustrates the assembly sequence for a single-chip, ceramic DIP package. Figure 4-7 shows end views of wire-bonded chips in other single-chip packages.

Figure 4-8 gives a profile view of wires after they have been bonded to the chip

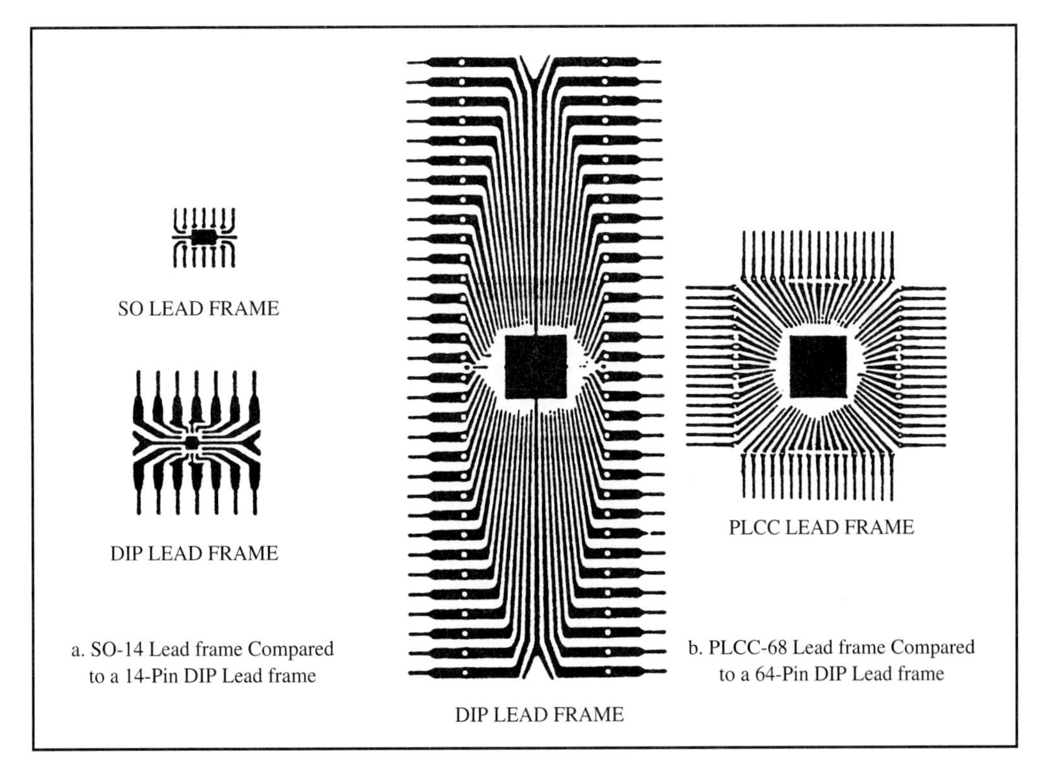

Figure 4-2 Comparison of SO, SO-14, and PLCC-68 lead frames to a DIP lead frame.

(*Source:* John E. Traister, *Design Guidelines for Surface Mount Technology* (San Diego, Ca.: Academic Press, 1990), p. 58.)

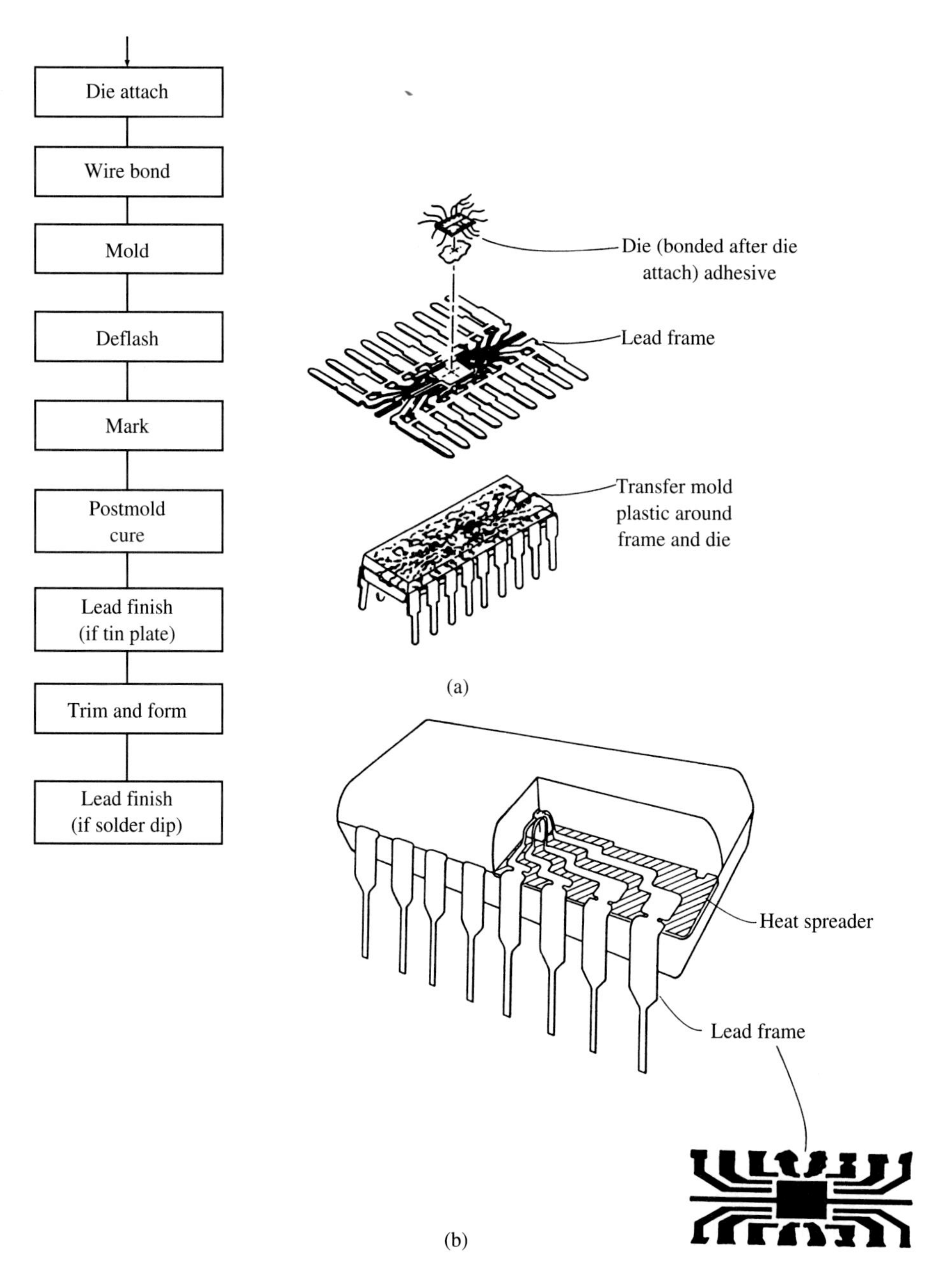

Figure 4-3 (a) Assembly sequence for plastic postmolded dual-in-line package. The lead frame serves as the chip carrier after die bond and wire bond. (b) View of lead frame positioning in postmolded nonhermetic package showing wire interconnects from chip to inner leads of lead frame.

(*Source: Electronic Materials Handbook, Vol. 1: Packaging* (Materials Park, Ohio: ASM International, 1989), pp. 487, 489.)

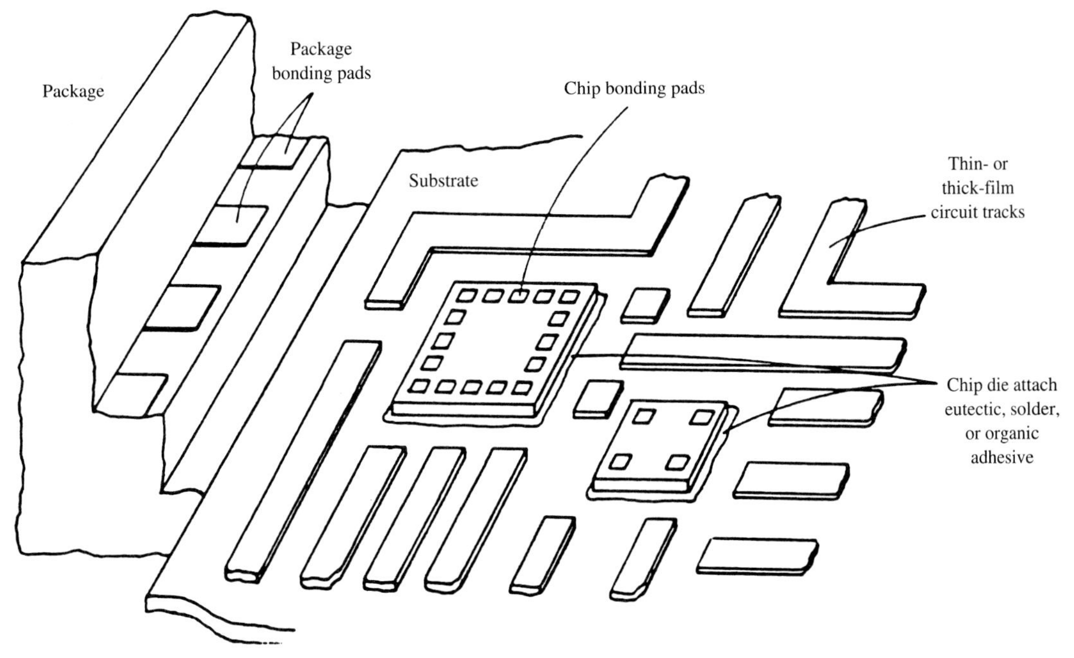

Figure 4-4 Multiple chips attached to a substrate.

(Source: Electronic Materials Handbook, p. 224.)

and its surrounding circuit. Note that the upper profile is the ball-wedge type of bond and the bottom profile is known as the wedge-wedge type of bond.

Wire bonding is performed in one of three ways: thermocompression bonding, thermosonic bonding, or ultrasonic bonding. These will be discussed in the following sections.

4.2.1 Thermocompression Bonding (Ball-Wedge Profile)

Gold wire is generally used in the **thermocompression bonding** technique, and wire diameters may be as small as 18 μm (0.74 mils). Gold is chosen because it deforms easily under heat and pressure, and because of its resistance to oxide formation, which can inhibit joining.

Figure 4-9 gives the thermocompression and thermosonic bonding sequences. These two wire bonding methods are known as the ball-wedge type of bonding and are performed by a tool called a capillary.

The steps involved in thermocompression bonding are as follows:

- Gold wire is threaded through the capillary, and the capillary is heated to approximately 350°C.
- Hydrogen flame or electric arc is used to cut wire and form a ball on the end of the wire.

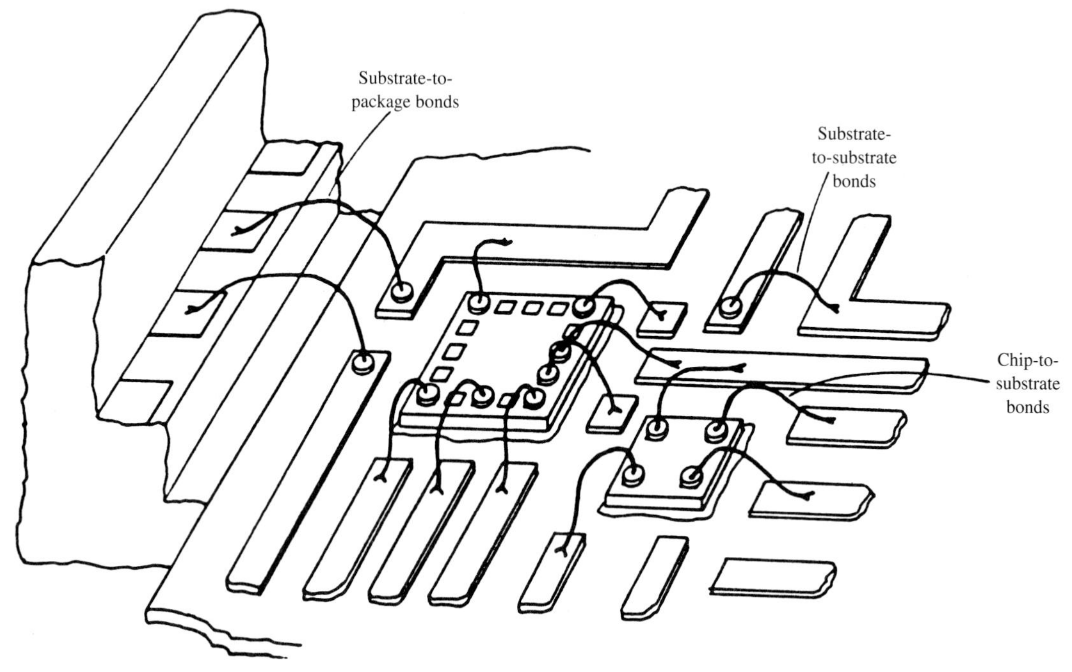

Figure 4-5 Circuit illustrated in Figure 4-4 with wire bonds in place.

(*Source: Electronic Materials Handbook,* p. 225.)

- The still-plastic ball is brought into contact with the gold or aluminum bonding pad, which is also maintained at a temperature of 150–250°C.
- Pressure is applied with the capillary, creating the weld by atomic interdiffusion between the wire and bonding pad.
- The capillary is raised, allowing wire to play out from the spool and be repositioned over the second bonding site.
- The second bond is made with capillary pressure. The wire clamp is closed, and the capillary is raised, which breaks the wire at the heel of the second bond.
- The wire is then cut by flame or arc, which creates a new ball on the end to continue the process.

4.2.2 Thermosonic Bonding (Ball-Wedge Profile)

Gold wire is principally employed for the **thermosonic bonding** technique also, but aluminum, copper, and palladium are also used. The same process is used for thermosonic bonding as was used for thermocompression, with the following exceptions:

- The capillary is not heated.
- The substrate is only heated to between 100°C and 150°C.
- Ultrasonic bursts of energy are applied for 20 to 50 ms to develop the bond.

4.2.3 Ultrasonic Bonding (Wedge-Wedge Profile)

Figure 4-10 illustrates the steps involved in the **ultrasonic bonding** process. This process differs from the thermocompression and thermosonic bonding techniques in that a low temperature process and aluminum or aluminum-alloy wire are normally used. The energy source for welding the wire is a transducer that vibrates the bonding tool (wedge), parallel to the bonding pad, in frequency ranges of 20 to 60 kHz. The wire is threaded through the wedge and hangs below the bonding tip. The wedge is lowered to the first bonding pad (gold or aluminum) and a burst of ultrasonic energy is applied while the wedge presses the wire against the pad. This combination of pressure and vibrational energy creates a metallurgical (cold) weld at the wire-pad interface. The wedge is lifted and moved into position over the second bonding site, then lowered, and another burst of ultrasonic energy effects the second bond. The wedge is then rocked slightly to weaken the wire at the heel of the second bond, the clamp is closed, and the wedge is raised, breaking the wire.

4.2.4 Limitations of Wire Bonding

IC chips with up to a few hundred I/O pads have been interconnected with wire bonding. I/O of this density usually requires two rows of bonding pads with pad sizes as small as

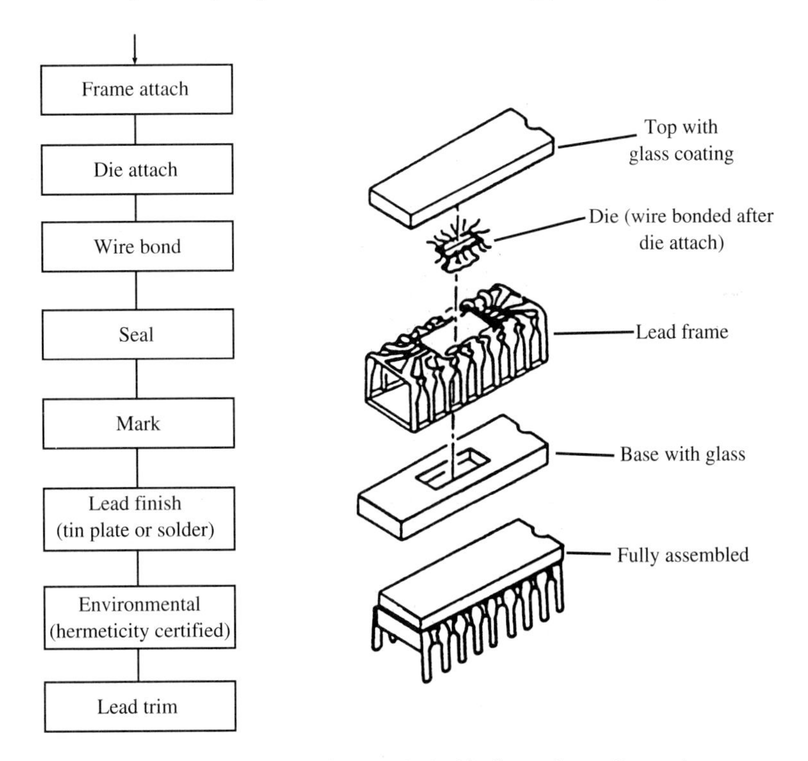

Figure 4-6 Assembly sequence for ceramic dual-in-line packages. Base and top components come already glaze-coated with glass for lead frame sink and seal.

(*Source: Electronic Materials Handbook,* p. 487.)

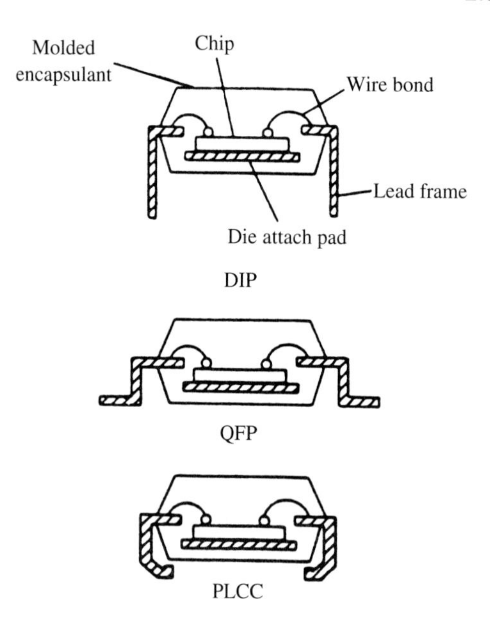

Figure 4-7 Chips wire-bonded to single-chip packages.

(Source: Electronic Materials Handbook, p. 471.)

2 × 2 mils (50 × 50 μm), with 4 mils (100 μm) between on-row pad centers. Figure 4-11 shows an example of staggered bonding pads. Capillary size and ball-pad size are the main density-limiting factors in ball bonding, as can be seen in Fig. 4-12.

4.3 CONTROLLED COLLAPSE CHIP CONNECT

The **controlled collapse chip connect** (C4) technique was introduced in 1964 for use in the hybrid circuit modules for IBM's System/360, to overcome the expense, unreliability, and low productivity of manual wire bonding. However, high-density chips that require smaller perimeters than can be accomplished with wire bonding have helped maintain the popularity of this method, which allows I/O pad layouts in full-population area arrays, as can be seen in Fig. 4-13.

In wire bonding, chips must be bonded to the substrate with the active side of the chip facing up in order to gain access to the chip I/O pads. In the C4 technique, the chip is flipped upside-down and is directly connected to the circuit substrate via rows or an array of solder bumps. These solder bumps are deposited on the wettable metal I/O terminals on the chip in a pattern that matches the solder wettable terminals on the substrate. All solder joints are then made simultaneously by reflowing the solder. The solder bump is restrained from overflowing the terminal pad by using thick-film glass dams which limit the solder flow. This process is also known as flip-chip joining and controlled collapse bonding (CCB). Figure 4-14 gives a closer look at the solder connection. Figure 4-15(a) is a photomicrograph of an IBM logic flip chip with a full array of solder bumps. Figure 4-15(b) is a 15K scanning electron micrograph (SEM) of the C4 pads on a chip.

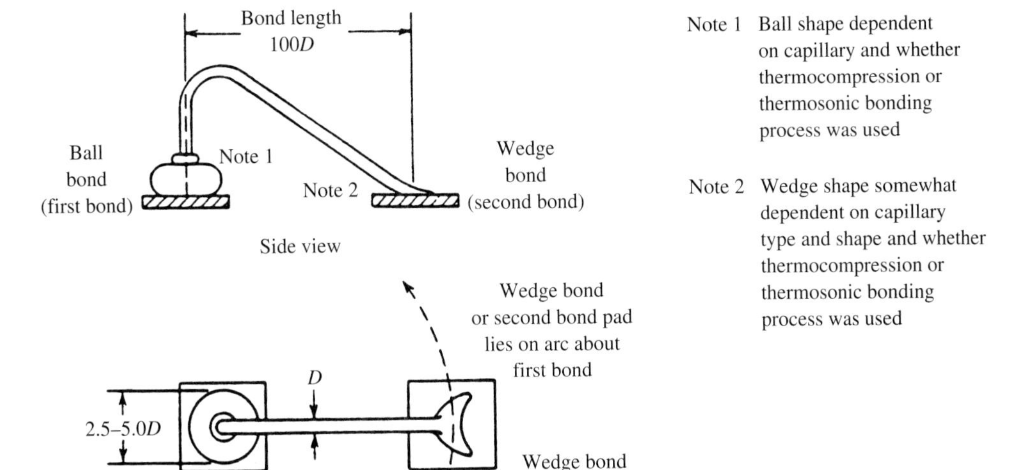

Top and side view of ball-wedge bonds. Generalized pictures applicable to both thermocompression and thermosonic wire bonding processes.

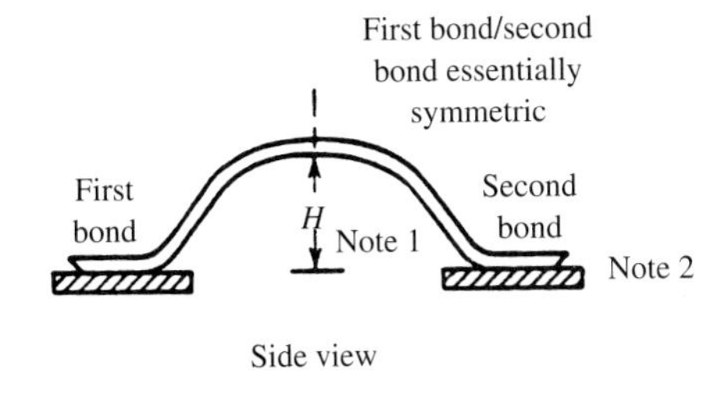

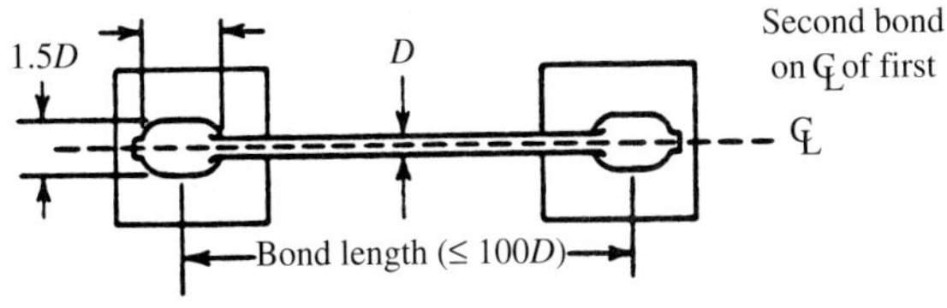

Figure 4-8 Top and side views of ball-wedge and wedge-wedge wire bonds.

(*Source: Electronic Materials Handbook,* p. 225.)

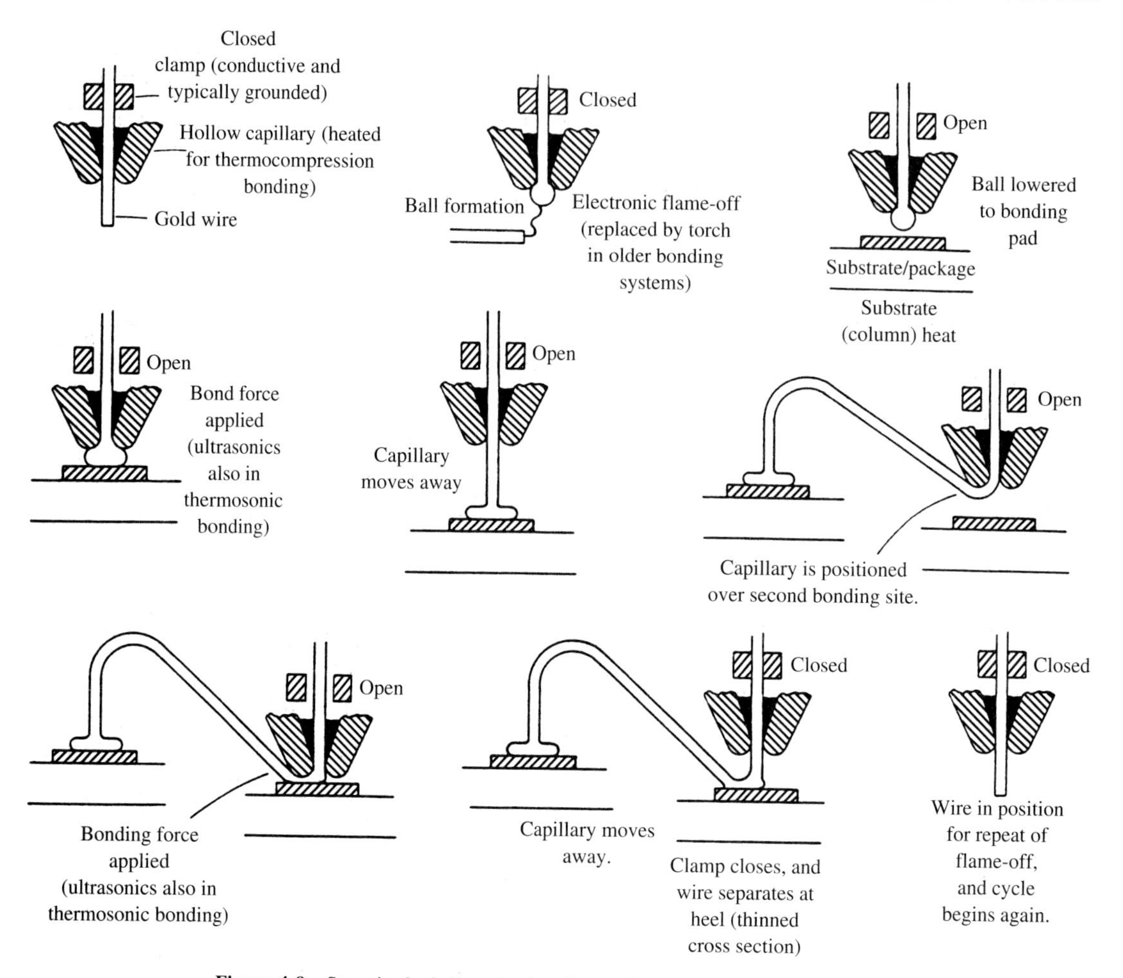

Figure 4-9 Steps in the ball-wedge bonding cycle.

(*Source: Electronic Materials Handbook*, p. 226.)

4.4 TAPE AUTOMATED BONDING

Tape automated bonding (TAB) was initiated in the 1960s by General Electric. It was originally developed as a highly automatable technique for large-volume low-I/O devices, but it has more recently been found to be useful for high-I/O devices up to several hundred chip interconnections. Practically every major computer and semiconductor manufacturer is now involved in some aspect of TAB.

4.4.1 The TAB Process

The TAB process involves bonding silicon chips to patterned metal-on-polymer tape (for example, copper on polyamide) using thermocompression bonding. The chip then resides on a roll of tape. Subsequent processes can be carried out in strip form such as testing, encapsulation, and excising of the individual packages from the tape, and attachment to the PWB or some other substrate. Figures 4-16 and 4-17 are pictures of carrier tapes for high-I/O and low-I/O TAB, respectively.

Bonding of the leads to the chip is by the inner lead bonding (ILB) process and is carried out as shown in Figs. 4-18 and 4-19.

Figure 4-20 shows the chip as it resides on the tape after the ILB process and after it has been excised and outer lead bonded (OLB) to the substrate using a reflow process.

4.4.2 TAB Tapes

There are three major divisions of **TAB tapes,** whose cross sections can be seen in Fig. 4-21.

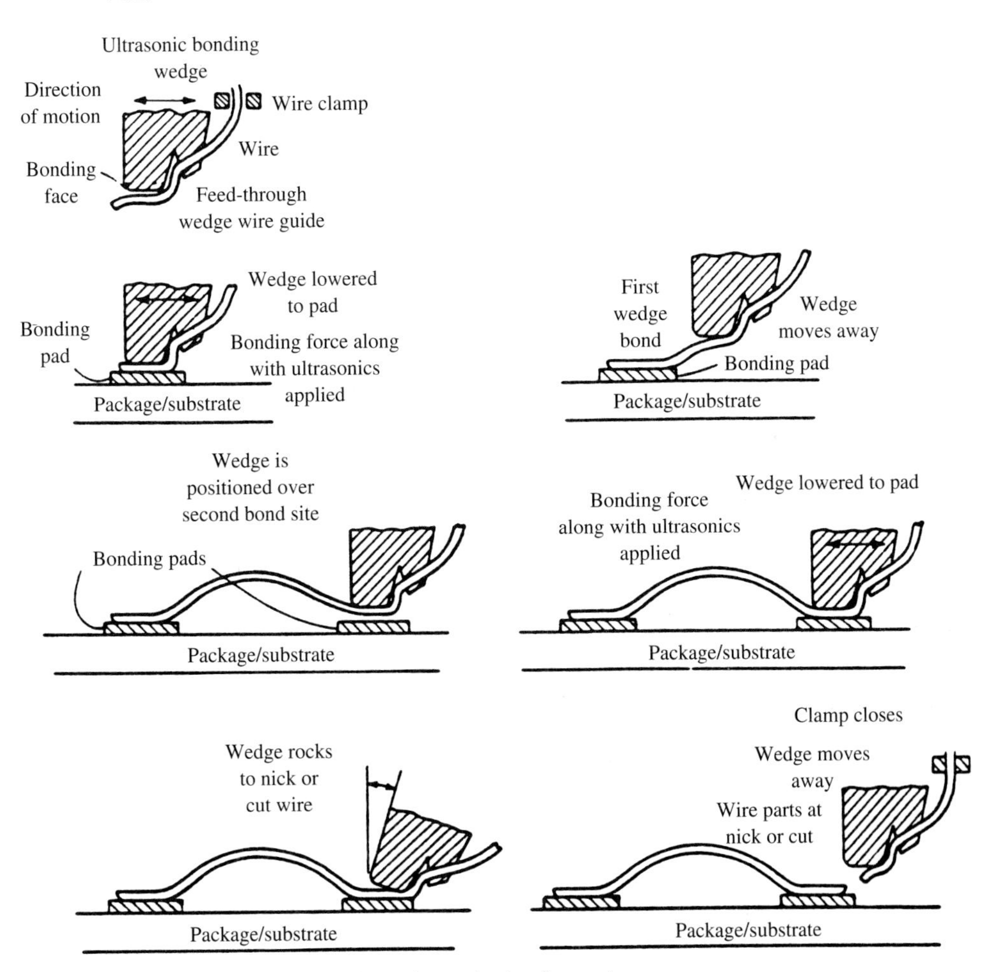

Figure 4-10 Steps in the wedge-wedge bonding cycle.

(*Source: Electronic Materials Handbook,* p. 227.)

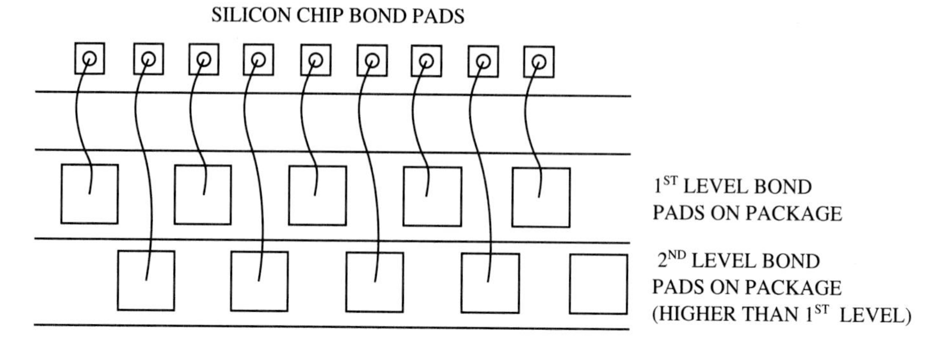

Figure 4-11 Chip and package bonding pad configuration showing bond pads staggered on different levels of the package.

- *One-Layer (Single-Level) Tape:* This is made of etched metal, normally copper, approximately 70 μm thick. Since the leads are all shorted together, the bonded chip must be excised from the tape before testing can be carried out.
- *Two-Layer (Double-Level) Tape:* Consists of polymer film onto which a metal lead frame is deposited and etched or pattern plated. Typical metal thickness is 20 to 40 μm. These tapes support isolated leads, thus allowing device testing on tape.
- *Three-Layer (Triple-Level) Tape:* A polymer film is again used, and copper foil is applied to both sides of the polymer, much in the same way PWB substrates are made. The lead patterns are then photolithographically defined, and the polymer is punched with metal dies instead of being etched as in the double-level structure.

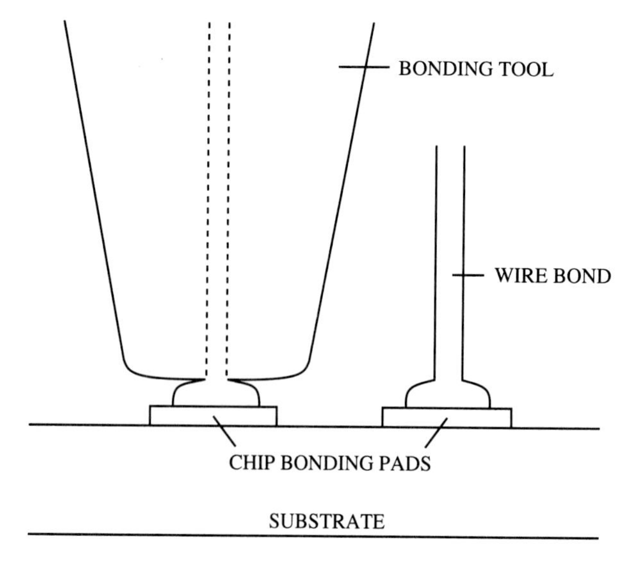

Figure 4-12 Ball bond spacing. Limiting factors are capillary to next ball separation and capillary to next wire separation.

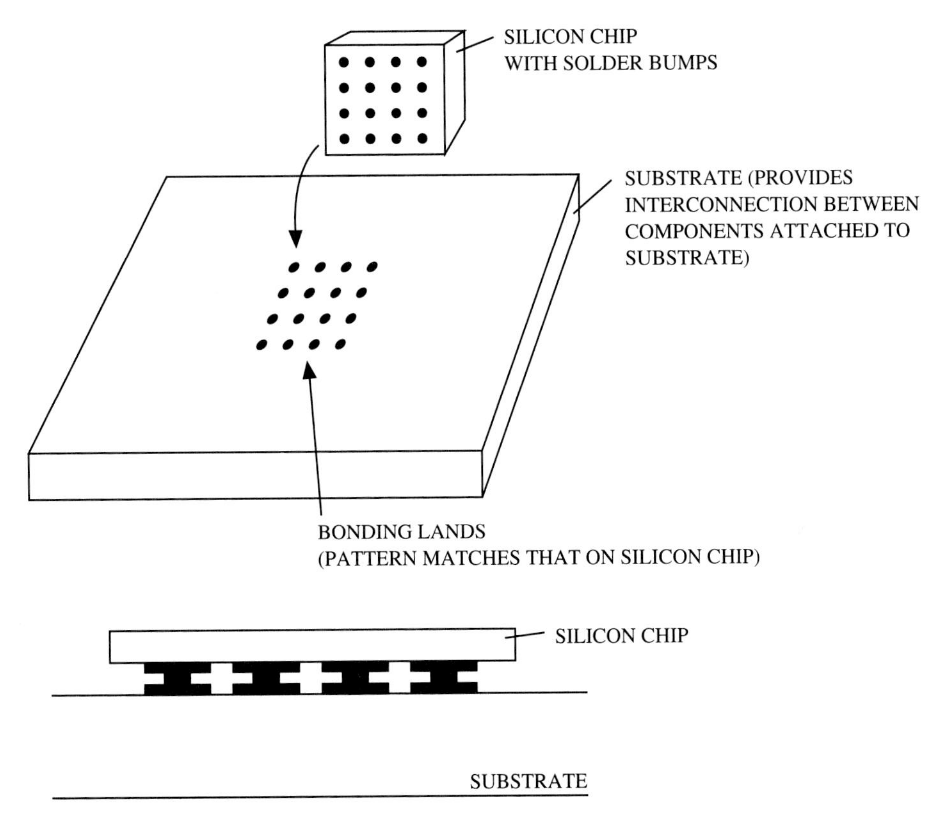

Figure 4-13 Controlled collapse chip connection (C4): The chip is mounted upside down (flip-chip), and all bonds are made at the same time by reflow soldering.

4.4.3 TAB Types

There are at least three popular kinds of chip-level interconnections currently in use in TAB technology: bumped chip, bumped tape, and area (array). A characteristic of all TAB technology is that a bonding projection (**bump**) is required between the chip and the beam lead. This bump provides the necessary bonding metallurgy for inner lead bonding, and it also provides a physical standoff to prevent the leads from shorting on the edge of the chip. The conventional **bumped chip TAB** technique is to use specially processed bumped wafers. The established procedure is to plate metal (gold) bumps on each pad site to a height of about 1 mil. All die on an entire wafer are bumped in a single operation. Figure 4-22 gives magnified views of the bumped chip and copper lead.

The **bumped tape TAB** technique was developed in the late 1970s. The bumps are put on the inner end of the tape rather than on the chip. This makes it possible for organizations that have no wafer processing capability to use TAB processes. Bumped tape is illustrated in Fig. 4-23.

Bumped chip and bumped tape TABs are known as periphery TABs. That is, the

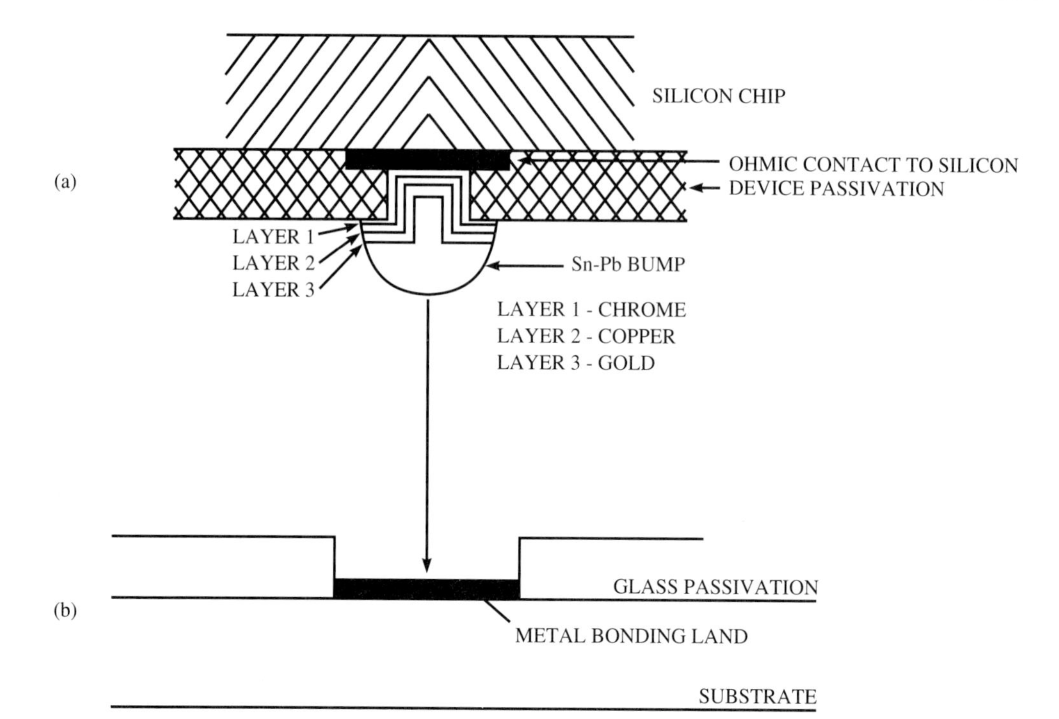

Figure 4-14 C4 configuration: (a) side view of a solder bump; (b) side view of a bonding land on a substrate (dam method).

Figure 4-15 Area array C4 configuration: (a) a logic flip-chip; (b) 15K SEM view of the C4 pads on a chip.

(Courtesy of IBM Corporation.)

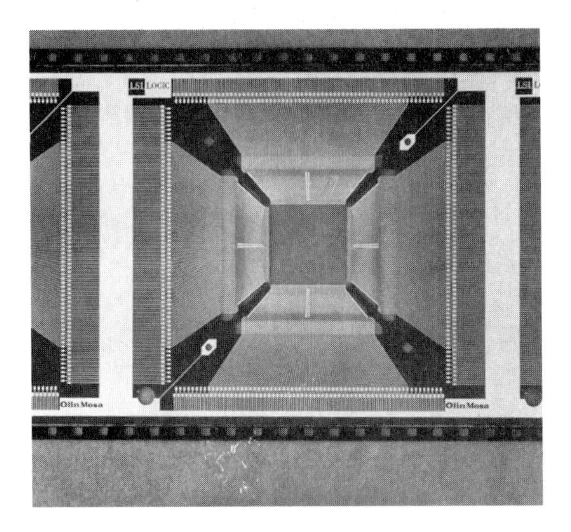

Figure 4-16 High I/O TAB: (a) 328-lead two-layer tape; (b) 204-lead three-layer tape; (c) 308-lead three-layer tape.

(Courtesy of Olin Interconnect Technologies.)

leads are attached along the periphery of the chip. **Array TAB** is designed to contact bumps within the internal area of the chip as well as the periphery. The array TAB technique uses multilayer tape to provide interconnection between the grid of solder bumps of a VLSI chip and the bonding pads of a ceramic package. Figure 4-24 shows a side view of an array TAB.

4.4.4 Inner Lead Bonding

Inner lead bonding is the process whereby the inner leads on the tape are attached to the IC chip. There are two main inner lead bonding processes: single-point and gang (mass).

Single-point bonding may be carried out using thermocompression, thermosonic,

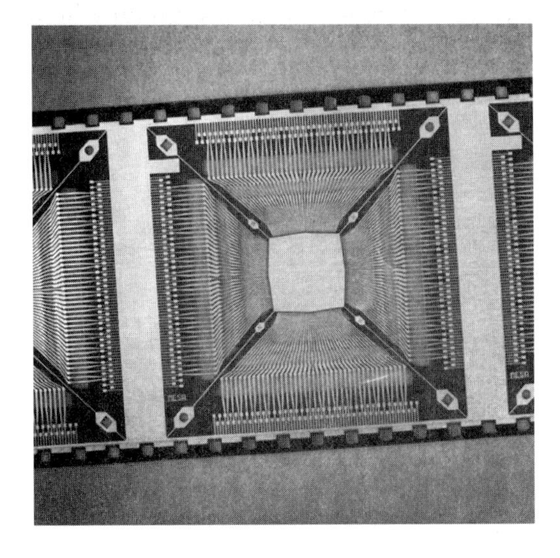

Figure 4-17 Low I/O TAB: (a) 28-lead three-layer tape; (b) 16-lead one-layer tape; (c) 40-lead three-layer tape.

(Courtesy of Olin Interconnect Technologies.)

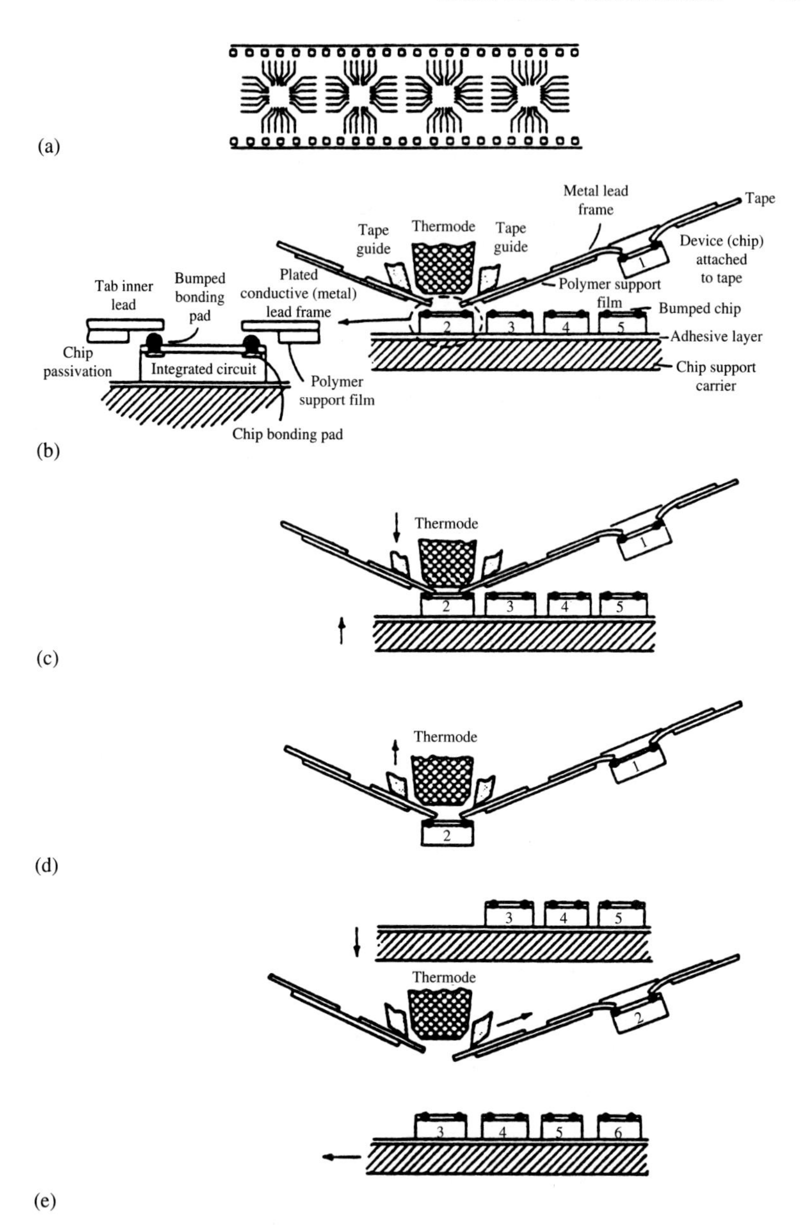

Figure 4-18 Inner lead bonding process. (a) Film tape with sprocket holes. (b) Overall process and enlarged view of tape leads and bonds. (c) Inner lead bonding with thermode down, support carrier up. (d) Chip attached to tape with thermode up, support carrier down. (e) Indexing to next chip-bonding position.

(*Source: Electronic Materials Handbook, Vol. 1: Packaging,* p. 232.)

and ultrasonic bonding, as well as solder reflow and laser bonding. The disadvantage of single-point bonding is speed. The advantages include:

- No chip planarity problem
- Minimal set-up time
- Consistent bonding pressure on each lead
- Each lead can be serviced individually
- Very similar to wire bonding, which is old technology

Unlike wire bonding technology, TAB beam leads can be bonded simultaneously using thermocompression bonding, dynamic alloy formation, and solder reflow.

Gang (mass) bonding increases throughput, but on chips with high-density I/O, planarity problems can lead to cracked chips, missed bonds, or inconsistent bond pressure, all of which contribute to poor bond quality.

4.4.5 Encapsulation

Beam leads of a tape are fully bonded to the chip and the chip is tested and burned-in (powered for a time under specified conditions to detect failures). The next operation is **encapsulation** (Fig. 4-25), which consists of depositing a protective coating on the top and/or bottom surface of the chip. The purpose of this coating is to protect the chip, bumps, and leads from mechanical loads, shock and vibration, and edge shorts. The encapsulant is usually epoxy or silicone.

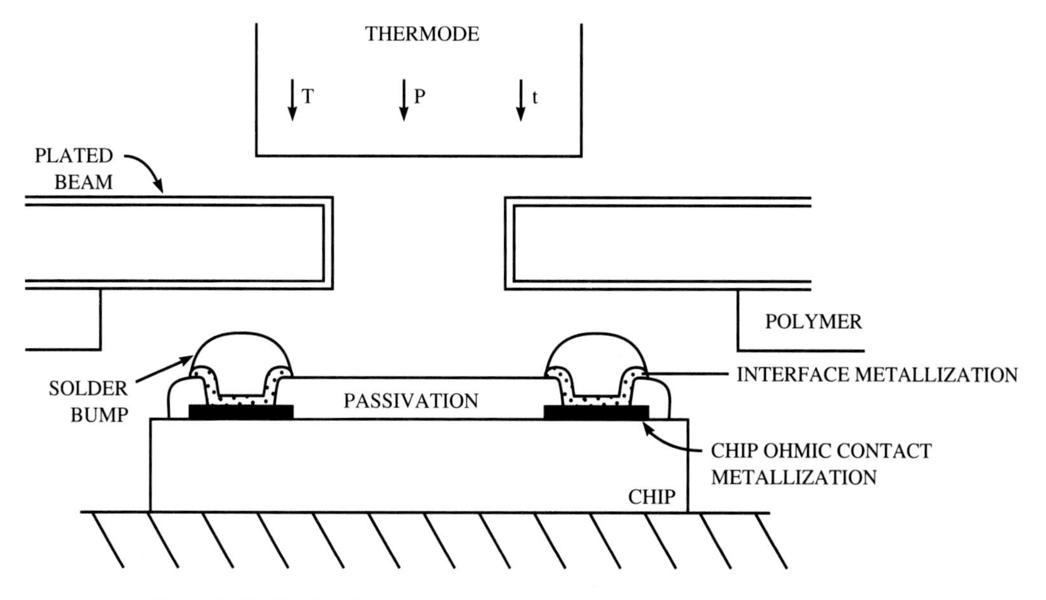

Figure 4-19 Details of chip in position for ILB.

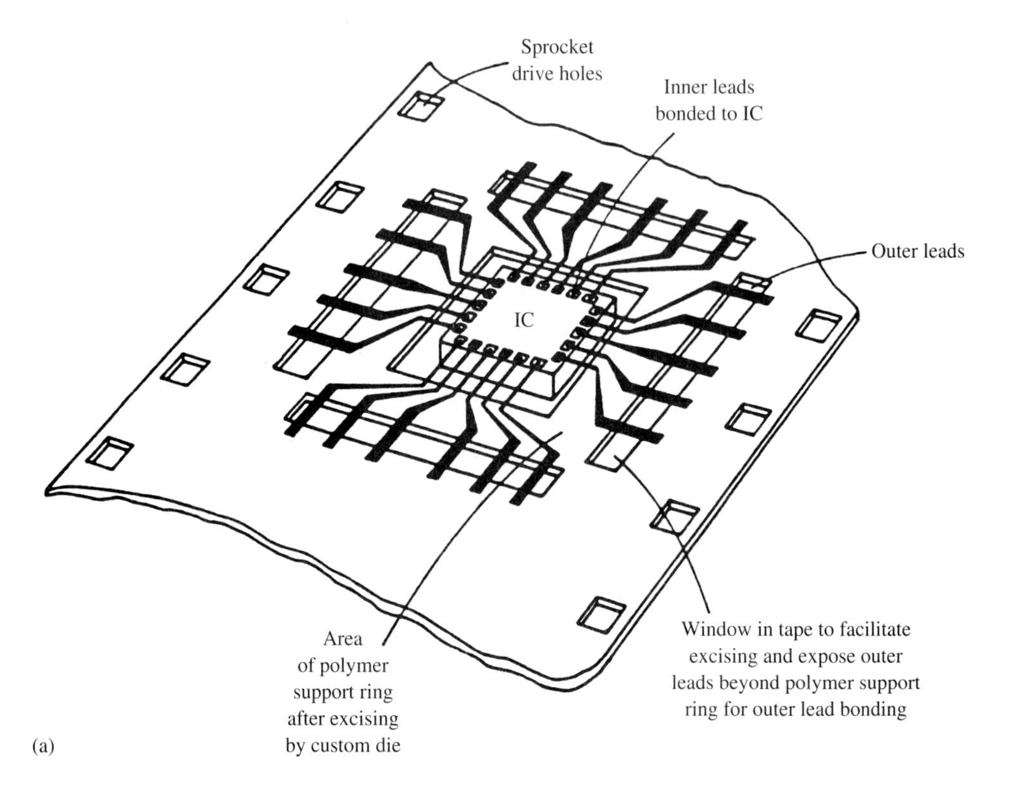

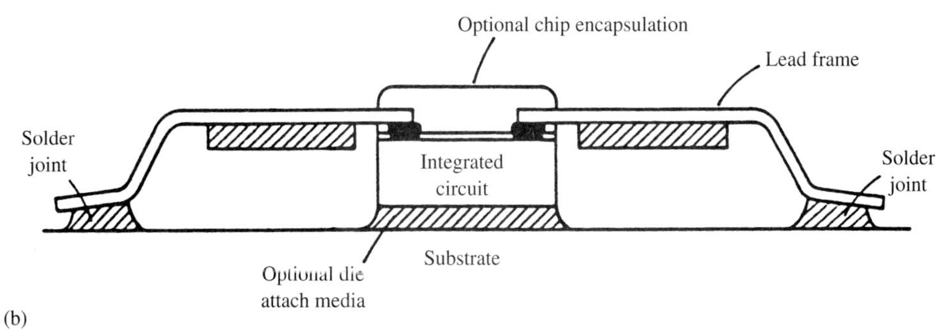

Figure 4-20 Outer lead bonding process. (a) Chip in position on tape after ILB, ready for excising prior to outer lead bonding. (b) Excised assembly outer lead bonded to substrate using reflow.

(*Source: Electronic Materials Handbook,* p. 233.)

4.4.6 Outer Lead Bonding

Outer lead bonding is the final step of the TAB assembly process and involves transferral of the chip, with leads in place, to the PWB. There the leads are attached to the land patterns on the substrate. Outer lead bonding requires a four-step process (usually performed on the same piece of equipment), as follows:

- Excising of leads from carrier tape
- Lead forming
- Transporting to the PWB (or other substrate) location
- Welding or soldering in place

Figures 4-26 and 4-27 illustrate the steps of this process. The last step attaches the outer end of the leads to the lands of the PWB by either single-point bonding or by gang (mass) bonding. Like the ILB single-point bonding procedure, the OLB process may use ultrasonic, thermosonic, and laser bonding.

Unlike the ILB gang bonding, OLB gang bonding rarely uses thermocompression bonding because the high temperatures and pressures needed for this technique may deform the PWB laminate. Instead, a technique called pulse-thermode reflow soldering is used. This method uses much lower temperatures and pressures than the thermocompres-

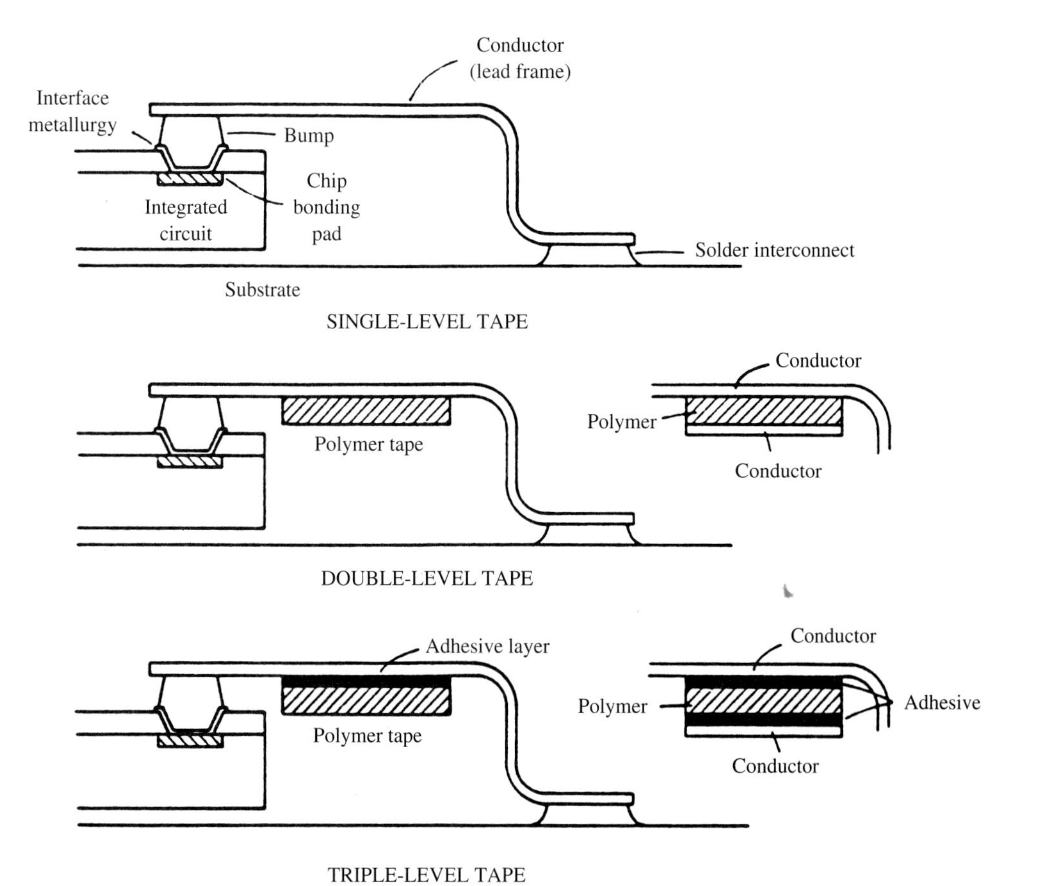

Figure 4-21 Examples of single-level, double-level, and triple-level tape assembly cross sections. (*Source: Electronic Materials Handbook,* p. 234.)

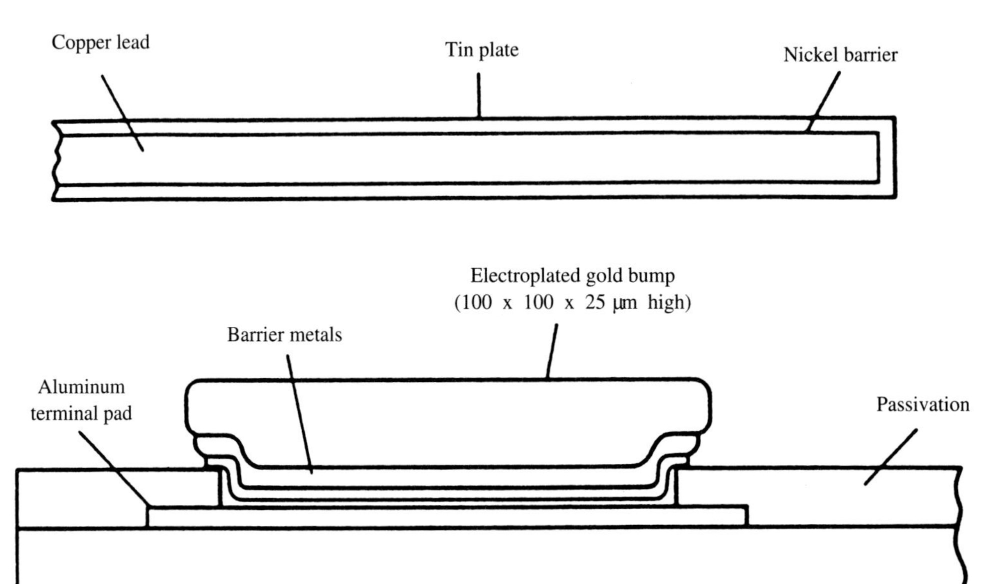

Figure 4-22 Bumped chip (planar lead) TAB.

(*Source: Electronic Materials Handbook,* p. 275.)

sion method. Solder at the bond site is heated by conduction until it melts, and the bond is held in place, by slight pressure, until the solder resolidifies.

4.4.7 TAB Advantages

One of the primary motivations for using TAB is that wire bonding has reached its limit at bonding wires to pads on 6-mil (152.4 μm) centers along the periphery of the die. This means that a 300-I/O die (75 pads per edge) requires a die of about 0.450-inch square (2.9 cm^2). For the same die using TAB, where 4-mil (102 μm) contact centers are now common (2 mils (51 μm) of metal and 2 mils (51 μm) of space), the required die size is reduced to about 0.300-inch square (1.9 cm^2). In this case, the TAB chip requires less than half the surface area of the wire-bonded chip, and yield from a 5-inch (12.7 cm) silicon wafer is increased by approximately 120 die. Figure 4-28 is a picture of gang-bonded 4-mil (102 μm) ILBs on gold bumps; note that a second row of gold bumps is provided for test probes.

Other TAB advantages include:

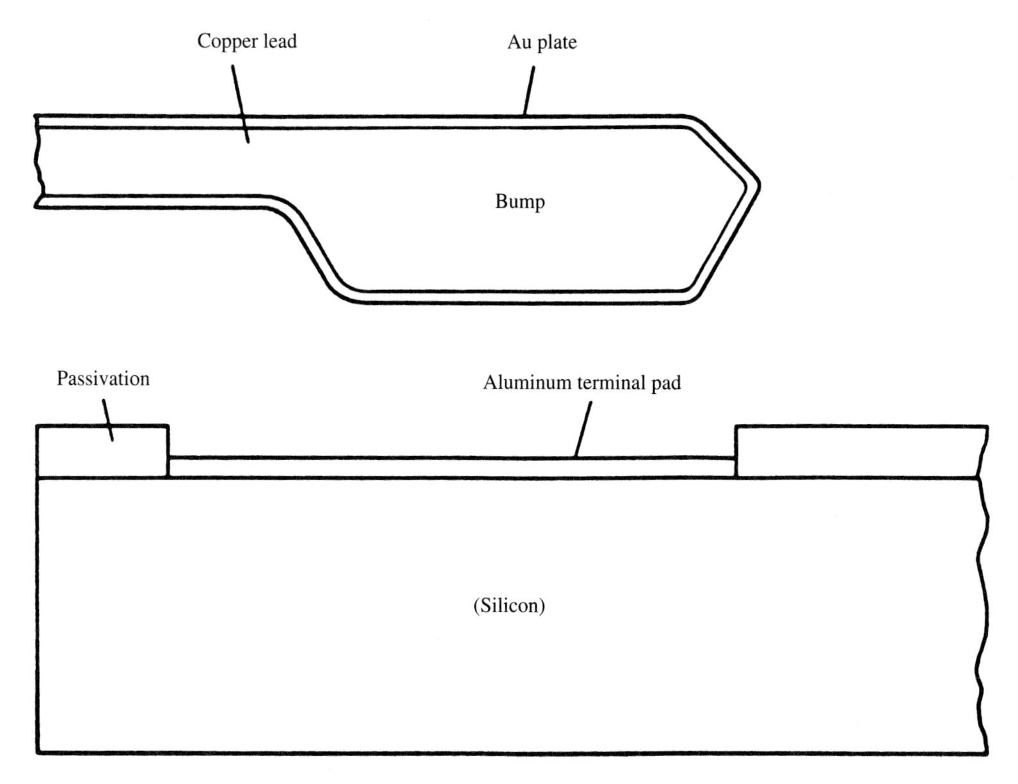

Figure 4-23 Bumped tape (planar chip) TAB.

(*Source: Electronic Materials Handbook,* p. 275.)

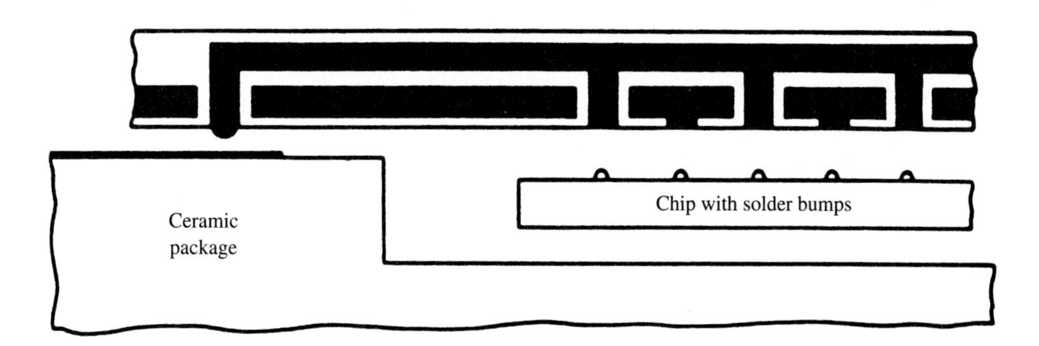

Figure 4-24 Area (array) TAB.

(*Source: Electronic Materials Handbook,* p. 276.)

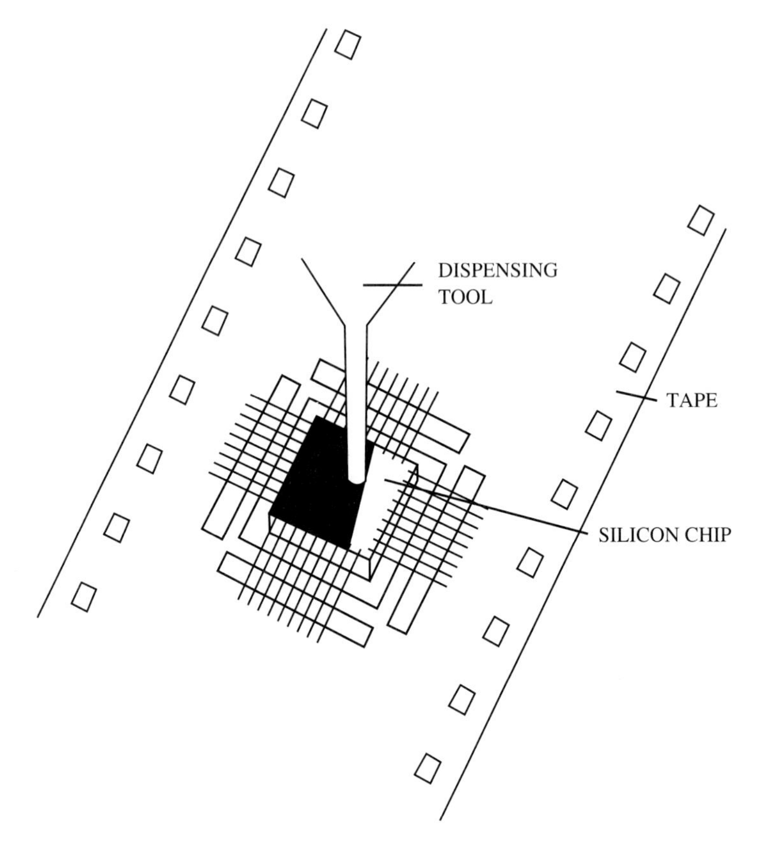

Figure 4-25 Encapsulation of a chip bonded to a TAB tape by using a nozzle to dispense encapsulant.

- Lower profile (makes TAB applicable for smart cards, watches, credit-card calculators, read/write head circuity, and the like)
- More precise geometry
- Better electrical performance
- Provision for testing before insertion into circuits to ensure that only working devices are outer lead bonded
- Improved heat transfer (copper beam leads provide means for efficient heat removal; the heat generated at the surface of a chip has a highly conductive thermal path through the 2-mil wide by 1-mil thick copper leads)

Some TAB disadvantages are as follows:

- Inflexible process, in that TAB requires a special tape design and retooling for each different chip design

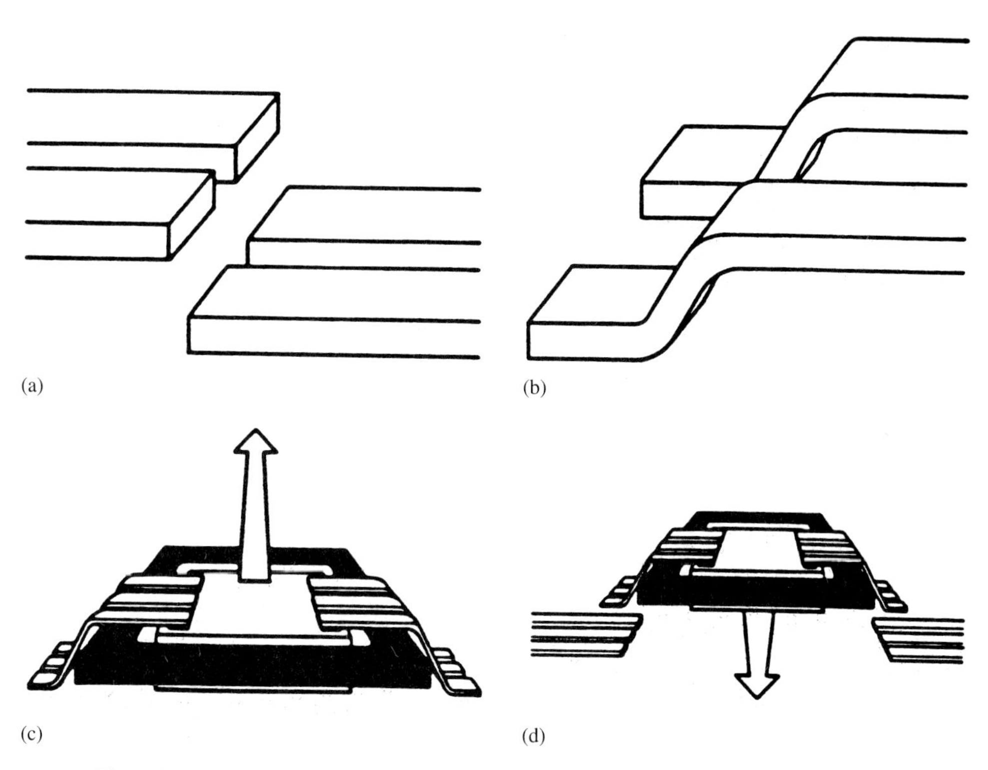

Figure 4-26 Steps of outer lead bonding: (a) excising; (b) lead forming; (c) transporting; (d) welding or soldering.

(*Source: Electronic Materials Handbook,* p. 284.)

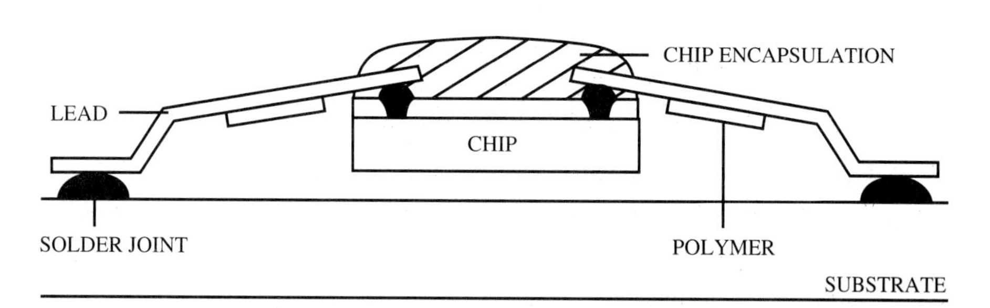

Figure 4-27 TAB outer lead bonding (OLB). After the TAB package is excised from the tape and placed on the substrate or board, it is normally solder bonded.

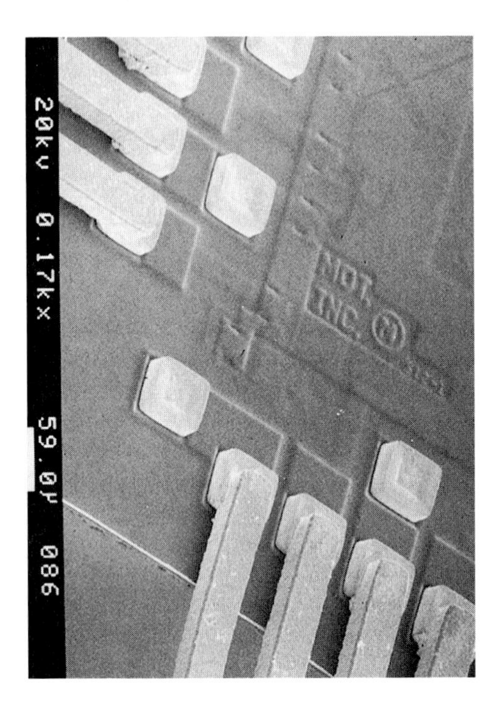

Figure 4-28 These inner-lead bonds were formed by gang bonding. The second row of gold bumps is for test probes.

(Courtesy of Motorola, Inc.)

- Large capital investment (ILB and OLB equipment is very expensive; there is limited availability of bumped die, so bumped-chip TAB is mainly available to large manufacturers)

4.5 MULTIPLE CHIP PACKAGING

4.5.1 Introduction

To this point, the discussion has been directed at two types of applications for electronic components: PWBs (which may have only a few dozen, or at most a few hundred, circuit components per square foot of board surface area) and silicon chips (which may have many thousands or millions of circuit components on a device less than a half inch square). The logical question is, are there any applications of circuit components that bridge the gap between these two techniques? The answer to this question is yes. There are at least two other methods of **multiple chip packaging** that fall between these two classes: wafer-scale integration (WSI) and multichip modules (MCMs). The latter category includes hybrid circuits.

4.5.2 Wafer-Scale Integration

In **wafer-scale integration** (WSI), the area of the entire silicon wafer is used to create an electronic circuit. That is, the wafer is not designed to be separated into numerous silicon die but is designed to perform as one system.

WSI has not achieved wide application because of several inherent problems. When silicon wafers are used to create multiple individual chips, it is not uncommon for less than 50 percent of the chips to be defect-free per wafer. This means that, in order to ensure that a wafer-scale circuit will work, a lot of redundancy must be designed into the wafer.

The redundant subsystems on the wafer are known as cells. Once the wafer has been fabricated, it must be tested to determine which cells are good and which contain faults. This testing of the wafer is a difficult process in itself, because conducting simple static tests is not sufficient. The wafer must be tested dynamically (simulated operation) to ensure that there are no problems in phasing, cross-talk, and timing delay. Effective fault detection and isolation is possible only if a very high degree of testability is built into the wafer. If enough good cells are located on the wafer, they are routed together through an irreversible process such as laser cutting.

The optimum degree of cell redundancy, for acceptable quality/cost considerations, has been determined to be 100 percent. Considering that, in IC technology, 60 percent of the wafer surface is committed to the aluminum interconnections, this leaves only 40 percent of the silicon area that is actually used for active circuits, half of which are primary circuits and half of which are redundant circuits. Thus, only 20 percent of wafer surface is utilized for primary functionality; 80 percent is consumed by overhead functions. More space is taken up in the next level of packaging and interconnection to incorporate the wafer into a circuit on a PWB. In total, WSI provides an effective utilization of space on a PWB that is generally less than 5 percent. Consequently, WSI is too expensive and will not be competitive until further breakthroughs are made.

4.5.3 Multichip Modules

The second approach to increasing the density of electronic circuits is through the use of **multichip modules** (MCMs). It can be noted in Fig. 4-29 that packaging densities available to MCMs fall between those available to PWBs and ICs.

Before silicon chips were packaged individually, they were mounted directly onto circuit substrates. It was decided then that chips were easier to handle and were better protected if they were encased in plastic or ceramic packages. Even with the packages taken into consideration, the miniaturization that ICs made possible in the 1970s was spectacular compared to previous eras. The 1980s brought about a need for further reduction in scale and greater packaging densities and chips on board (COB), that is, the use of chips on board the substrate without their protective packages became popular again. Examples of COB were shown in previous sections when C4 and TAB mounting techniques were discussed.

For several years, ceramic (an alumina/silica combination) substrates have been used as a platform to mount bare chips in combination with thick/thin film techniques. These circuits have been known as hybrid circuits. Up to 35 layers of cofired ceramic have been used to provide chip-to-chip interconnections, much in the same way that multilayer PWBs are designed. In recent years, due to the low speed and high cost of these devices, they have given way to other materials and techniques.

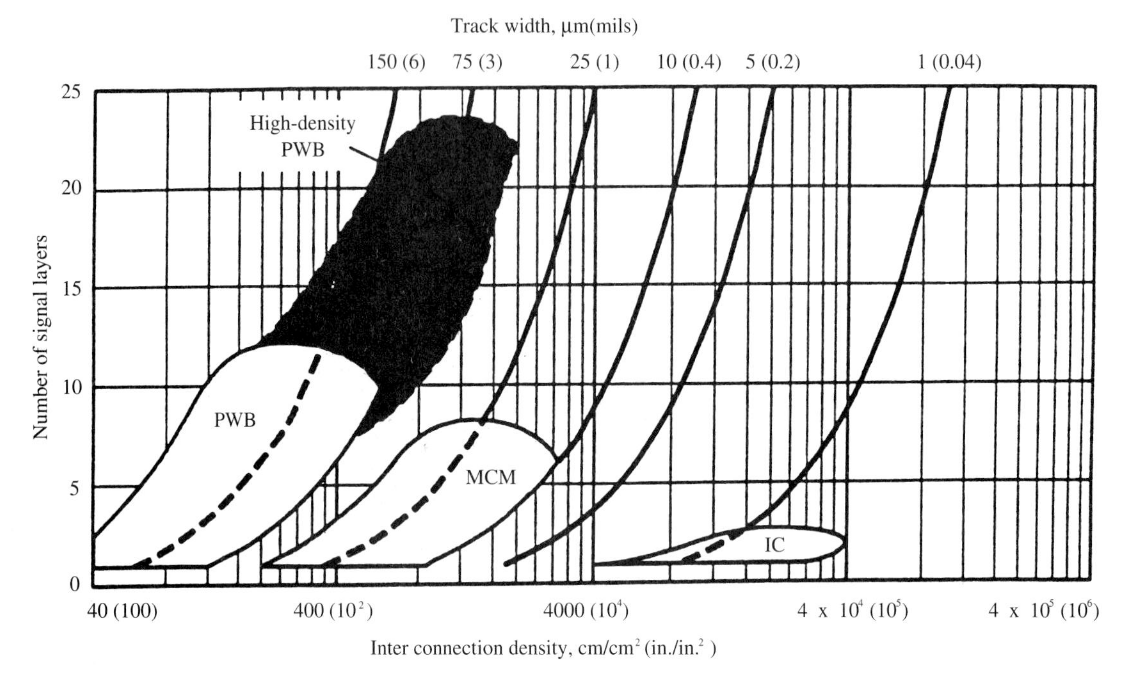

Figure 4-29 Interconnection density versus line technology.

(*Source: Electronic Materials Handbook,* p. 297.)

One factor that contributes to the high cost of ceramic hybrid technology is the use of thick-film screening techniques. Recall that thick-film technology is a screen printing process that deposits film that is typically 10 to 50 μm (0.4 to 2 mils) in thickness. This requires the use of path widths of at least 100 μm (4 mils). These sizes decrease the available circuit density, thus creating a need for more layers to maintain a relatively small package footprint. More layers increase the cost and decrease the yield.

The latest developments in hybrid circuits have become known as multichip modules. Instead of using a substrate made up of multiple layers of ceramic materials, most of the new multichip modules are made up of layers of organic or inorganic dielectric material.

The dominant dielectric medium being used is polyamide, and the layers are 10 to 25 μm (0.4 to 1.0 mil) thick. The dielectric layers may be metallized with thin circuits of aluminum or copper and placed on a silicon substrate for stability and heat removal. The important feature of the new technology is its use of many thin dielectrics and metallization layers for interconnection purposes.

The improved density and circuit interconnection made possible by this new technology can, with as little as three dielectric layers and four metallization layers, accomplish what required 35 layers in the ceramic technology. The use of four dielectric and five metal layers is typical.

In a sense, PWBs can also be defined as MCMs, since PWBs may also have multiple, unpackaged chips on a single substrate. For this reason, the interconnection industry has proposed the following three categories of multichip modules:

- MCM-L (high-density, laminated PWBs)
- MCM-C (ceramic substrates, either cofired or low-dielectric-constant ceramics)
- MCM-D (deposited wiring on organic, silicon, or ceramic and metal substrates)

Figure 4-30 gives some comparisons of these classifications. The MCM-D group has been the most popular in recent years. Figure 4-31 shows several multichip modules.

4.6 INTERCONNECTION DEVICES

Typical electronic systems are made up of several layers or levels of packaging, and each level of packaging has distinctive types of **interconnection devices** associated with it. One way in which this hierarchy of interconnection levels can be divided is as follows:

- *Level 0:* Gate-to-gate interconnections on the monolithic silicon chip.
- *Level 1:* Packaging of silicon chips into DIPs, SOICs, chip carriers, and so on, and the chip-level interconnects that join the chip to the lead frames. Occasionally this level is skipped when TAB and COB technologies are utilized.
- *Level 2:* PWB level of interconnection. Printed conductor paths connect the device leads of components to PWBs and to the electrical edge-connectors for off-the-board interconnection.
- *Level 3:* Connections between PWBs. This may include PWB-to-PWB interconnections or card-to-motherboard interconnections, as illustrated in Figs. 4-32 and 4-33.
- *Level 4:* Connections between two subassemblies. For example, a rack or frame may hold several shelves of subassemblies that must be connected together to make up a complete system. This situation is illustrated in Fig. 4-34.
- *Level 5:* Connections between physically separate systems such as host computer to terminals, computer to printer, and so on.

The following sections focus primarily on levels 3, 4, and 5 and the interconnection devices that are required at each level. These interconnection devices can be separated into two general divisions: cables and connectors.

4.7 CABLES

Cables are generally required at levels 4 and 5. A cable may be terminated at one or both ends by connector(s) or solder joint(s). Common choices of cable include:

- Discrete wire
- Twisted pair (see Fig. 4-35)

Parameter	MCM-C		MCM-D		MCM-L*
	Cofired ceramic	Low-K ceramic	Silicon-on-silicon	Low-K dielectrics on ceramic, solicon, metal substrates	Printed-circuit boards
Line density (centimeters per square centimeter per layer)	20	40	400	200	30
Line width/separation, micrometers	125/125-375	125/125-375	10/10-30	15-25/35-75	750/2250
Maximum substrate size, cm	10-15	15	10	10	66
Dielectric coefficient	9	5	4	2.4-4	4.5-5
Pinout: per cm^2 peripheral	15-60 1600-6400	15-60 1600-6400	8-30 800-3200	8-30 800-3200	15-30 1600-3200
Structural integrity	All military specifi-cations	Mil Spec 202	Young's modulus of mild steel	Can be mounted on other substrate material	All military specifications
Turnaround time	1 month	1 month	1-10 days	1 month	9-13 weeks
Substrate costs, dollars per square centimeter prototype production	$300 $80	--- $10	$6-$10 $2-$4	$6-$10 $2-$4	$0.03-$0.06/layer/cm^2
Number of years commercially available	>10	>5	5	5	>50
Electrical properties: transmission line type terminating resistors	Telegraph Built-in acceptable	Ideal (50-Ω) Built-in possible	Ideal (50-Ω)** Difficult**	Ideal (50-Ω) Built-in possible, or surface-mounted	Ideal (50-Ω) Built-in possible, or surface-mounted
Power distribution capacitors	Surface-mounted	Surface-mounted	Can have depletion layer, dual dielectric layer, or high-capacitance layers***	High-capacitance layer or surface-mounted	Surface-mounted

* Comparable with low-K ceramic, assuming separate heat removal.
** Mosaic Systems Inc.'s user-programmable silicon-on-silicon Optimum transmission line provides 50-ohm dynamic driver load without requiring terminating resistors.
*** Mosaic's silicon-on-silicon user-programmable substrates do not require separate power-decoupling capacitors.

Figure 4-30 Classifications of modern multichip modules.

(*Source:* Robert R. Johnson, "Multichip modules: next-generation packages," *IEEE Spectrum,* 5 (March 1990), p. 34.)

- Trilead (see Fig. 4-36)
- Ribbon cable (see Fig. 4-37)
- Coaxial cable (see Fig. 4-38)
- Optical fiber (see Fig. 4-39)

When choosing cable for a specific application, the designer should consider the following:

Figure 4-31 Three different multichip modules.

(Courtesy of nCHIP of San Jose, Ca.)

- Capability to propagate signals over the range of the required distance (may be centimeters to kilometers) with acceptable
  - — Delay
  - — Noise
  - — Power loss
  - — Distortion/dispersion
- Flexibility of the cable that is needed to
  - — Connect various planes of packaging
  - — Insert/remove the connected device
  - — Accommodate relative movement of the devices that are to be connected
- Capability of the cable to fulfill the required connection density

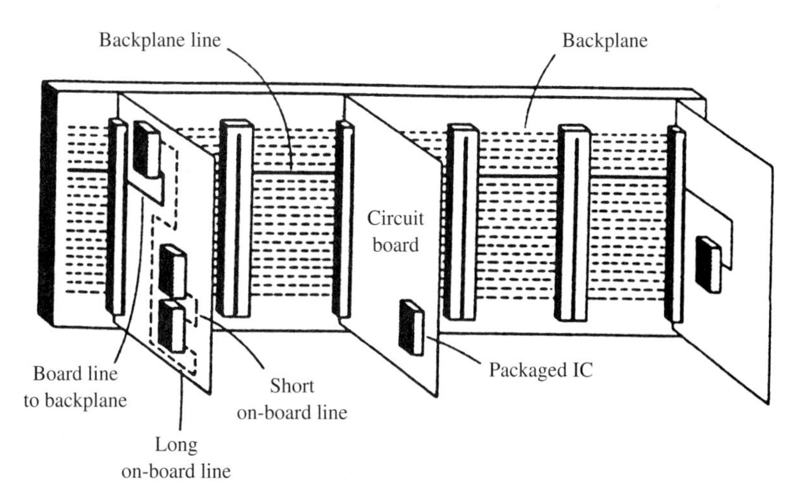

Figure 4-32 Level 3, card-to-motherboard connections.

(*Source: Electronic Materials Handbook,* p. 3.)

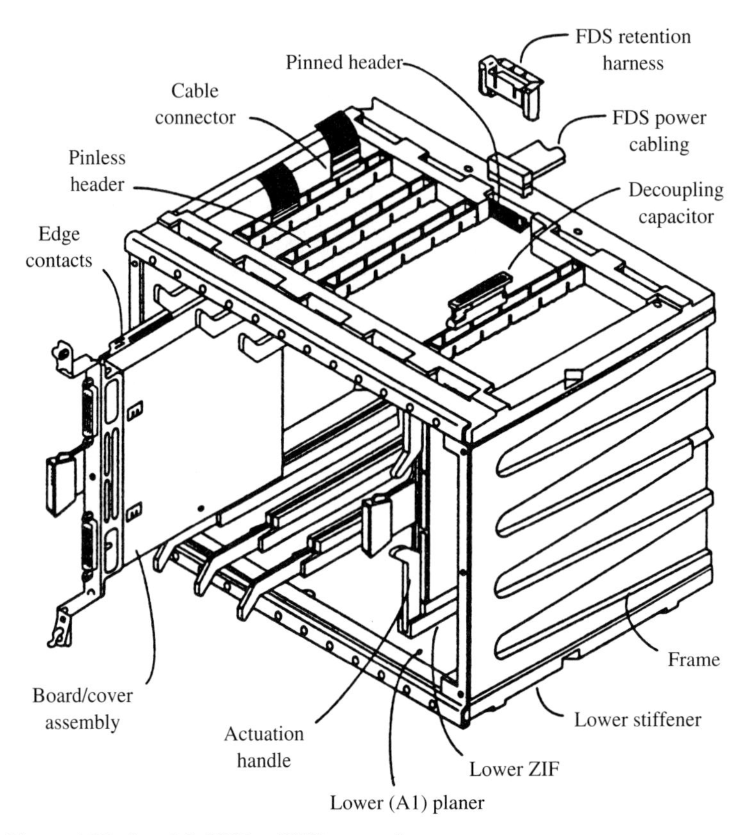

Figure 4-33 Level 3, PWB-to-PWB connections.

(*Source: Electronic Materials Handbook,* p. 22.)

4.8 CONNECTORS

4.8.1 Introduction

Connectors have sometimes been defined as sources of trouble between two pieces of electronic equipment. A more technical definition will be used here: a **connector** is a device used to repeatedly separate and reconnect pathways, for both electrical and optical signals, in an electronic system.

In some applications, connectors serve a mechanical role as well as an electrical role. As electrical devices, connectors provide a means of transmitting signals between components and packaging levels. As mechanical devices, connectors provide mechanical linkage and structure to the system.

The variety of connectors available today fill volumes of vendors' catalogs, thus a comprehensive discussion of connectors is not possible. This discussion will be limited

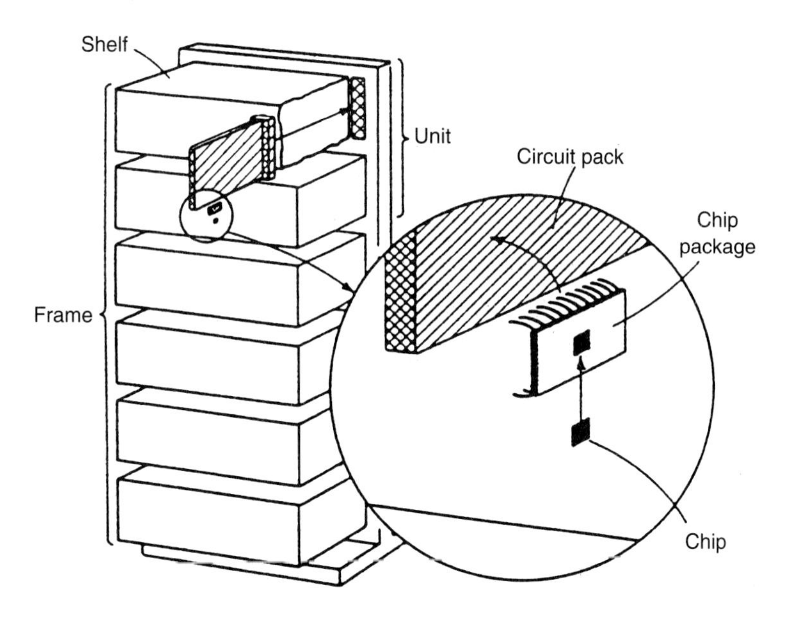

Figure 4-34 Interconnections of level 1 (chip), 2 (printed wiring board), 3 (shelf), 4 (unit), and 5 (frame).

(*Source: Electronic Materials Handbook,* p. 13.)

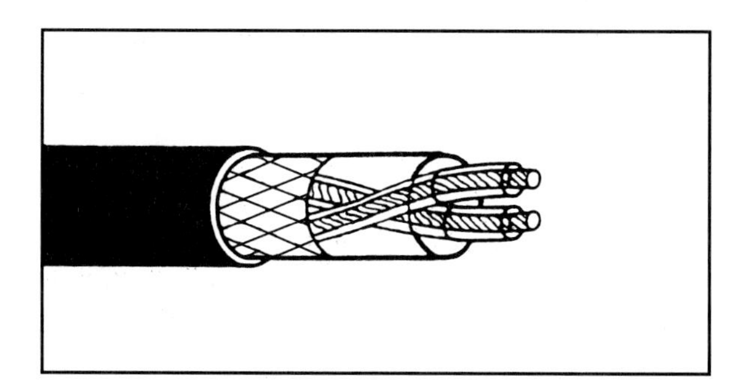

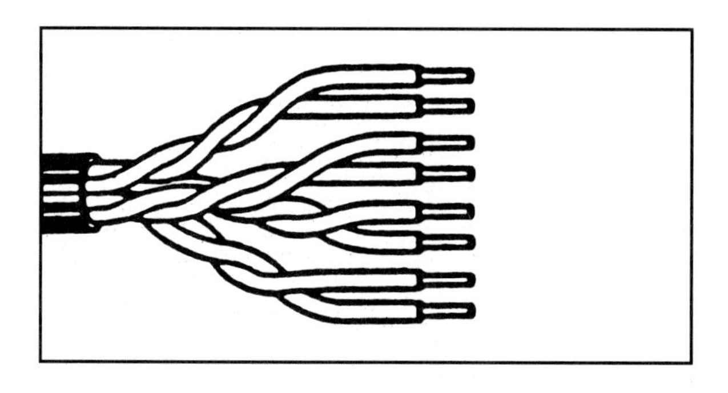

Figure 4-35 Twisted-pair cable.

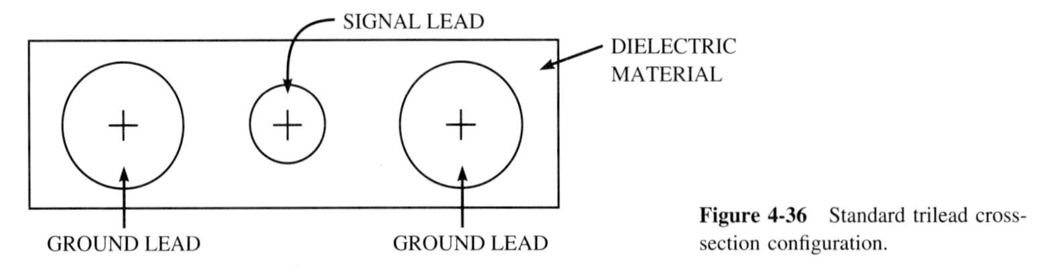

Figure 4-36 Standard trilead cross-section configuration.

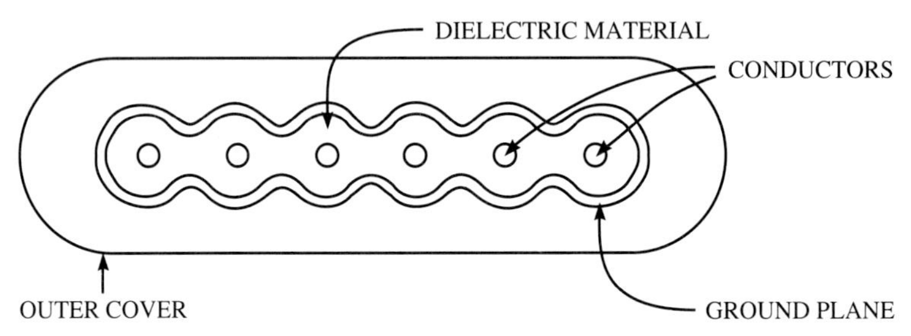

Figure 4-37 Shielded flat (or ribbon) cable.

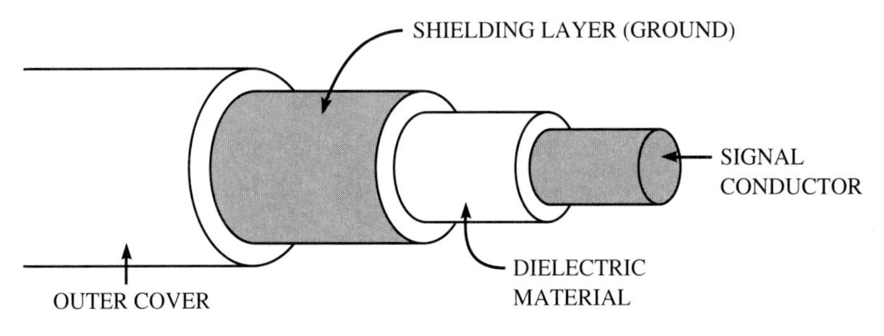

Figure 4-38 Coaxial cable.

to the fundamental properties of good connector design and will then present some examples of families of connectors. Some connector symbols that are often used in the industry can be seen in Fig. 4-40.

4.8.2 Connector Reliability

Connector reliability depends on connector design and operating environment. There are two major measures of connector reliability: time-dependent reliability and plug-dependent reliability.

Time-dependent reliability is the measure of the failure rate of a connector relative to the length of time the connector operates. This is the primary cause of connector failure and the most difficult for systems designers to overcome.

In an office environment, a well-designed connector is expected to demonstrate less than one failure in 10,000 contacts over a machine operating life of 10 years. To achieve this degree of reliability, the connector must be able to avoid fretting corrosion by maintaining a gas-tight, noncorrosive contact interface over its operating life. Fretting action is a cyclic motion of two contacts relative to each other. The motion may be caused by vibration or a differential thermal expansion. This slight motion traps small particles or oxide at the contact interface and, over a period of time, increases the contact resistance to an unacceptable degree. Fretting corrosion is illustrated in Fig. 4-41.

Plug-dependent reliability is a function of the number of actuations/deactuations of a particular plug and is the second-largest cause of connector failures. Plug-dependent reliability problems can be further divided into two classes:

- Those that recover with subsequent connector insertions
- Those that never recover, therefore requiring repair or replacement of the package

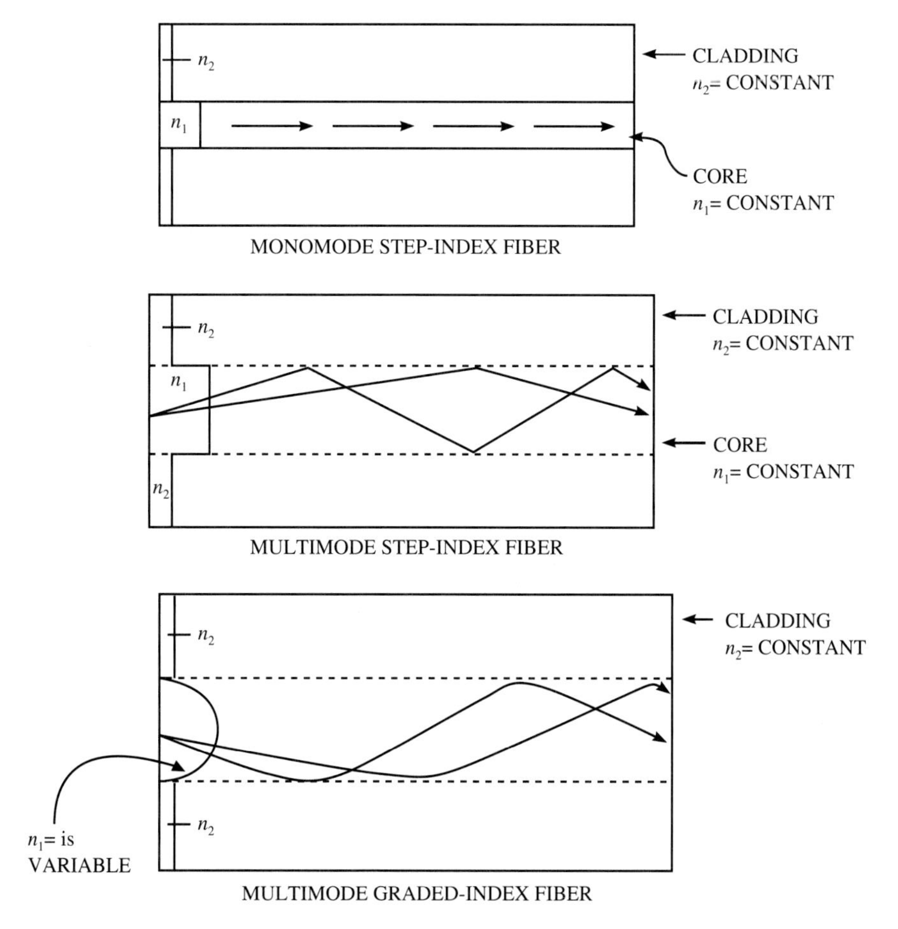

Figure 4-39 Schematics of three common optical fiber types: single-mode step-index, multimode step-index, and multimode graded-index structures.

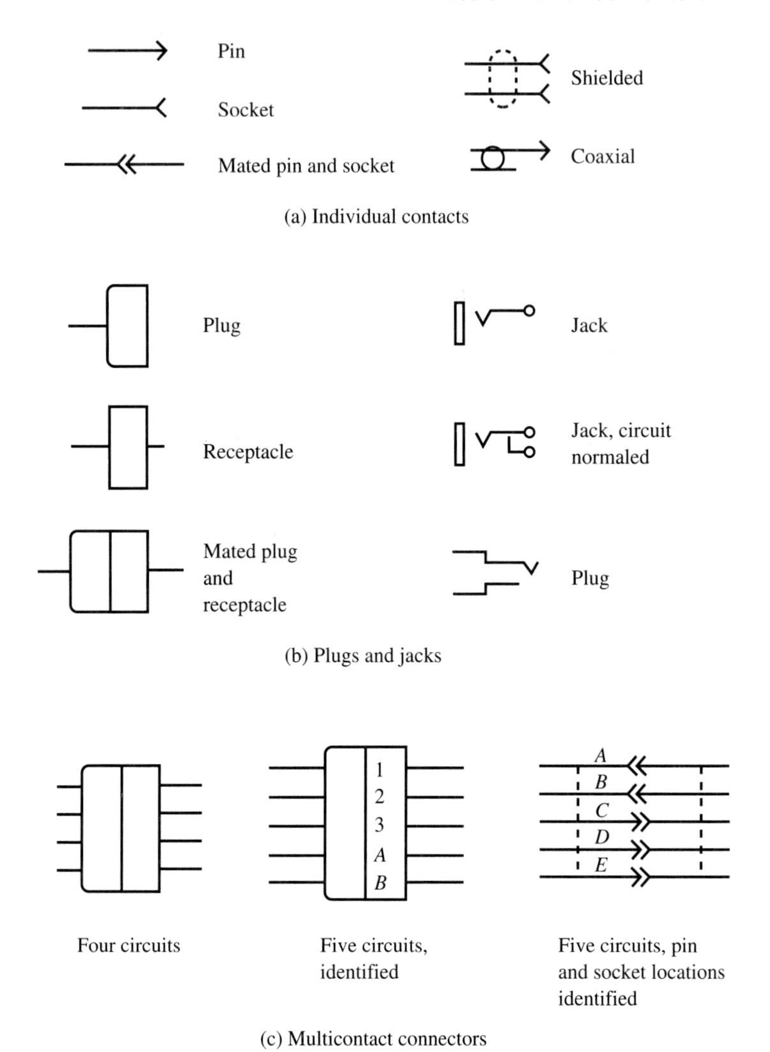

Figure 4-40 Circuit symbols of commonly used connectors.

(*Source: Electronic Materials Handbook,* p. 313.)

Airborne particulates are the major cause of plug reliability problems. A reliable contact system is expected to demonstrate on the order of ten failures per million contact-insertions in a normal use environment.

4.8.3 Connector Design

Connector design factors that must be considered are the connector material, the contact design, and the operating environment. Connector metallurgy is the single most important factor in designing for a reliable connector. High-quality connectors use noble metals, such as gold or palladium, plated over a diffusion barrier of nickel. The diffusion barrier serves to prevent the connector base metal (usually spring material) from diffusing through the noble metal plating and corrupting the contact interface.

The noble metal is plated over the diffusion barrier and must be both noncorrosive and thick enough to resist wear-through and corrosion. Gold or palladium is usually plated to a thickness of 0.75 to 5.0 μm on high-quality contact systems. High-quality connectors exhibit resistance changes of less than 150 milliohms over their operating lives.

Connector systems must be designed to provide sufficient contact force and wipe. Contact force is the radial force applied by the receptacle toward the center of the individual pins of the plug. Typical values of 0.29 to $>$0.98 Newtons per contact are required to ensure a gas-tight joint at the contact interface and to provide displacement of particles and oxide during mating.

Contact wipe, which is the amount of relative sliding motion between mating pairs, must be sufficient to expose new metallurgy during mating. This helps to guarantee a contact interface free of particles and oxides.

The number of contact interfaces is also an important parameter in assessing reliability. The addition of multiple contact interfaces greatly reduces the probability of contact failures.

To ensure long life for a connector, the operating environment must be taken into consideration during design. Attention must be given to factors such as:

- Expected amount and size of airborne particles
- Presence of corrosive materials
- Presence of moisture
- Forced-air-cooling

Some plugs may house the contacts in a shield for protection from particulate matter, while others may be hermetically sealed. The effectiveness with which the contact area is protected from a hostile environment can be a major contributor to the plug-dependent performance.

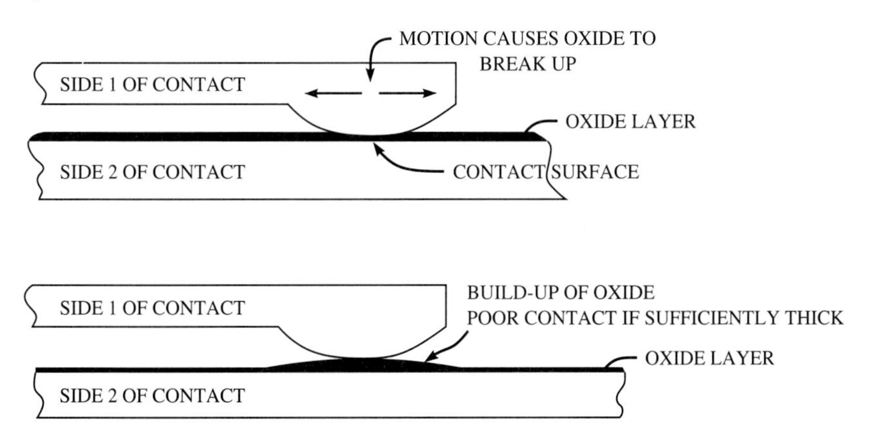

Figure 4-41 Fretting corrosion of a contact surface.

4.9 CONNECTOR DESCRIPTIONS

4.9.1 Contact Types

Connectors come in many varieties. The first distinction to be made is whether the connector is of the one-part or two-part type. Examples of one-part varieties include PWB component sockets and edge-connectors for PWBs. The edge-connector assemblies use one edge of the PWB as the plug on which conductor contacts have been plated. The other half of the connector is usually an assembly of mating contacts in a chassis-mounted receptacle. Figure 4-42 shows several examples of multiple-lead component sockets. Card edge-connectors (both loose and installed in a PC) are shown in Fig. 4-43.

The following common one-part connector contacts can be seen in Fig. 4-44: bellows, tuning fork, cantilever, and ribbon.

Figure 4-42 Multiple-lead sockets.

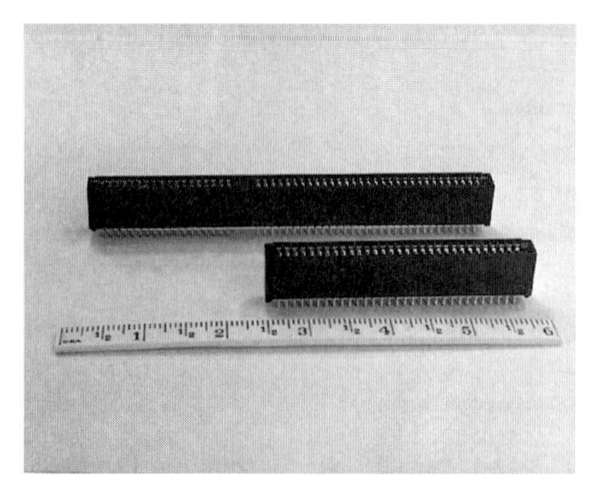

Figure 4-43 Card edge-connectors (top); installed in a PC shown supporting a PWB (bottom).

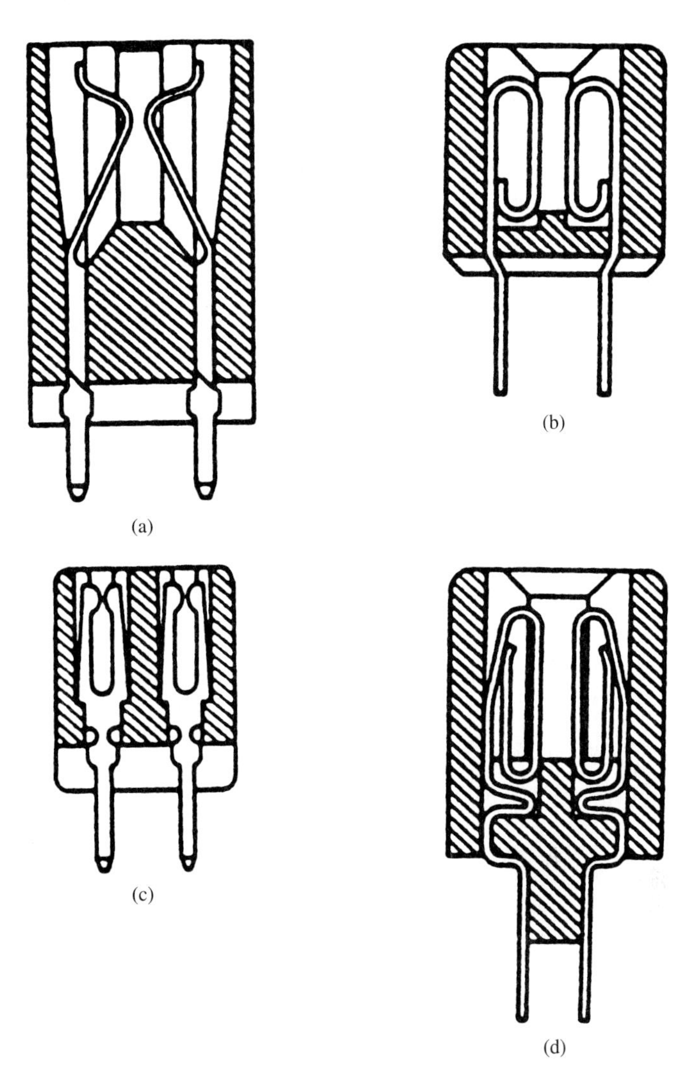

Figure 4-44 One-part connectors: (a) cantilever contact; (b) bellows contact; (c) tuning fork contact; (d) ribbon contact.

Two-part connectors consist of self-contained multiple-contact plug-and-receptacle assemblies. With few exceptions, the plug contact assembly mounts to the plug-in assembly as shown in Fig. 4-45.

The receptacle (plug-in) half of the connector mounts to an interconnection wiring panel (motherboard) or plate. The receptacle half may be soldered to the wiring panel, or it may be attached by the wire-wrap method illustrated in Fig. 4-46.

The most common types of two-part contacts are the pin-and-socket contact (Fig. 4-47(a)) and the hermaphroditic contact (Fig. 4-47(b)). Another variety of the pin-and-socket connector can be seen in detail in Fig. 4-48.

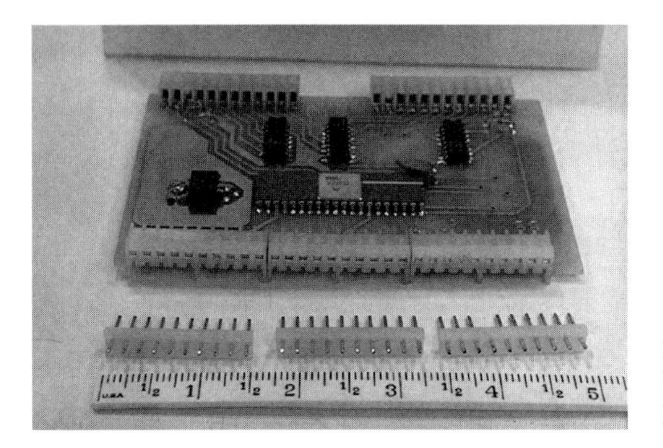

Figure 4-45 Two-part connector: printed board assembly with receptacle (top) and plug half (bottom).

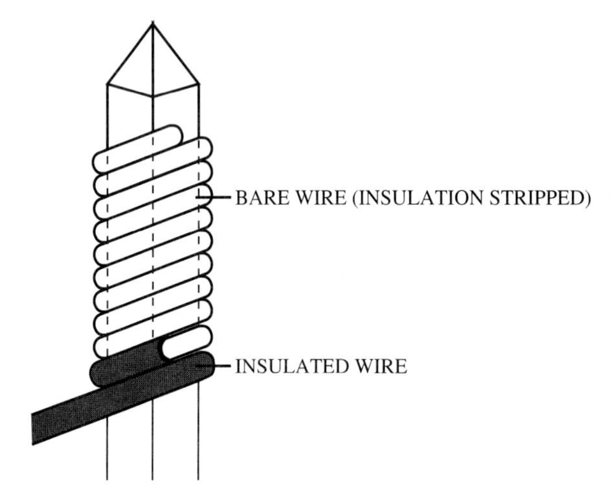

Figure 4-46 Solderless wire-wrap termination.

4.9.2 Connector Families

Electronic components are often identified generically—for example, wirewound resistors, FETs, and DIPs—but connectors are usually identified by trade name or military specification. Often, families of connectors have been developed to serve a specific need or to fulfill the requirements of military specifications.

Of the families of connectors presented here, some will be referred to by trade names, some by mil specs, and others by application. This is not meant to be a comprehensive view, but rather a sampling of some of the many varieties that are available.

Audio connectors (Figs. 4-49 and 4-50) are designed for use in audio and other low-frequency applications (for example, test instruments, computers, and medical instrumentation) and are supplied in two grades, as follows:

- Commercial
  High-fidelity

Home tape recording
Amateur use

- Broadcast
  Interconnecting broadcasting
  Recording studio
  Television
  Computer
  Medical electronic equipment (excluding life-support systems)

These connectors are designed to be used in an environment that may include rough handling (for example, the normal hazards of being stepped on, slammed around, and dropped) but usually do not include the possibility of extreme temperatures, vibration, and shock.

(a)

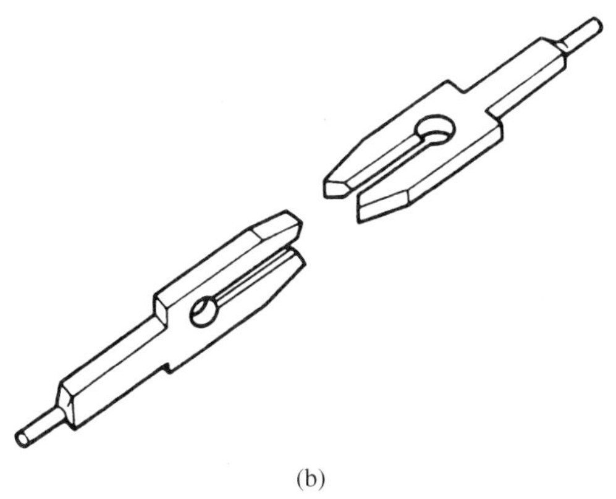

(b)

Figure 4-47 Two-part connectors: (a) pin-and-socket contact; (b) hermaphroditic contact.

(*Source:* Jones, *Electronic Components Handbook,* p. 327. Courtesy of Elco Corp.)

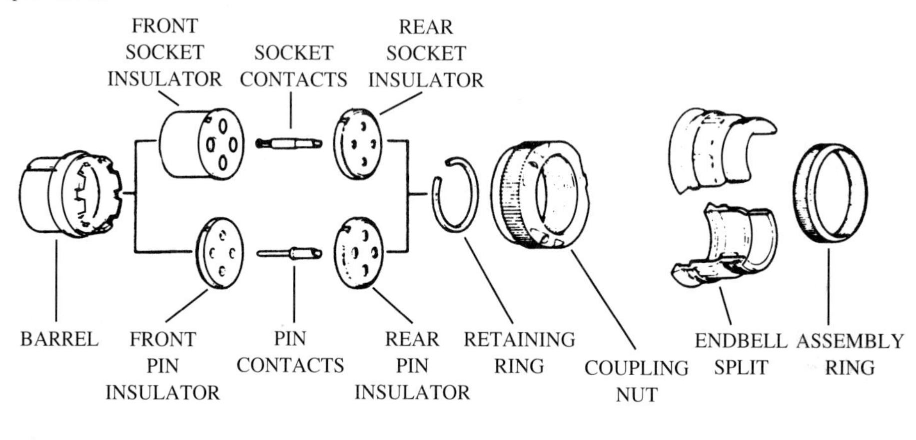

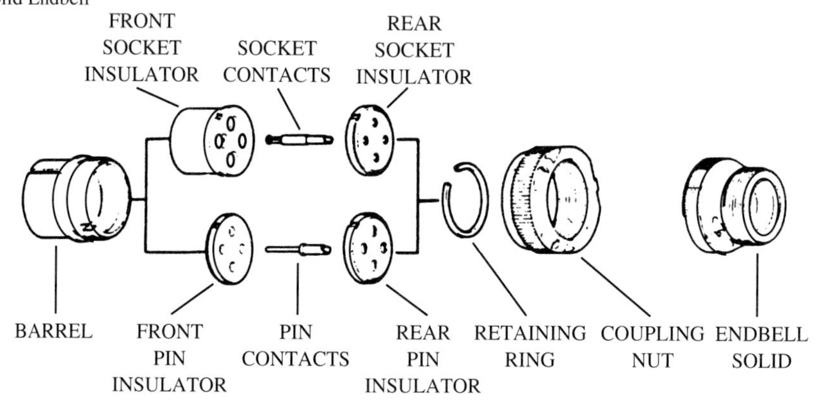

Figure 4-48 The parts of a typical connector. The MIL-C-5015 connector and its descendants are probably the most widely used electronic equipment connectors in both military and industrial systems.

(*Source:* Jones, *Electronic Components Handbook,* p. 314. Courtesy of ITT Cannon Electric.)

Radio frequency (RF) connectors are used for higher than audio frequency applications and so must be designed more carefully to prevent problems such as unwanted signal attenuation and reflection. Some common types of RF connectors are shown in Fig. 4-51.

Factors to consider when selecting RF connectors for a design include:

- For optimum electrical performance, connector size should approximate the cable size.
- Voltage and power-handling capability should be considered for the particular application.
- RF connectors may be either bayonet or threaded type.

Table 4-1 gives descriptions of some common RF connectors.

The BNC is a bayonet type of RF connector with the advantage of quick connect or disconnect. BNCs are small, weather resistent, and lightweight. They are rated for 500 Vrms at sea level and, when used properly, offer excellent electrical performance for DC and for AC up to 4 GHz. BNC connectors are used with a coaxial-type cable, and the outer shield of the cable is terminated by either crimping or clamping.

The TNC connector is similar to the BNC, except it is threaded instead of having the bayonet coupling. TNCs are small, lightweight, and weatherproof, and they have a nominal impedance of 50 ohms. The threaded feature enables this connector to withstand high vibration levels without introducing excessive signal noise, and it also extends the frequency range of application up to 11 GHz.

The UHF connector is a low-cost RF connector, of the threaded variety, used in the megahertz range and at lower RF frequencies where impedance mismatch is not a

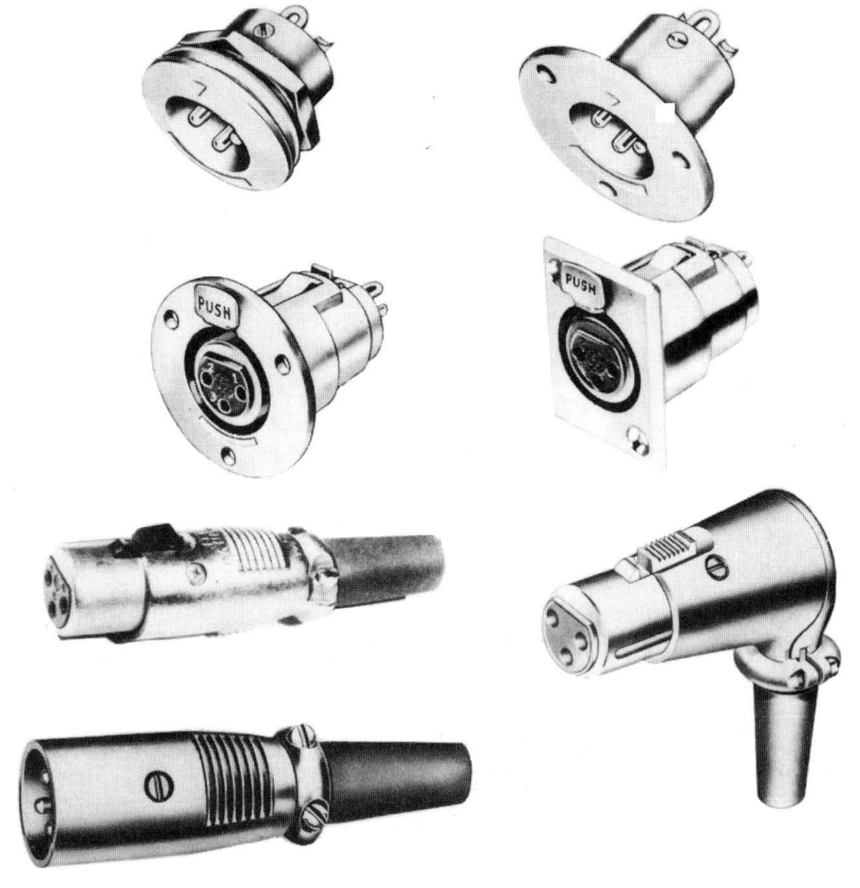

Figure 4-49 Audio connectors.

(*Source:* Jones, *Electronic Components Handbook,* p. 318.)

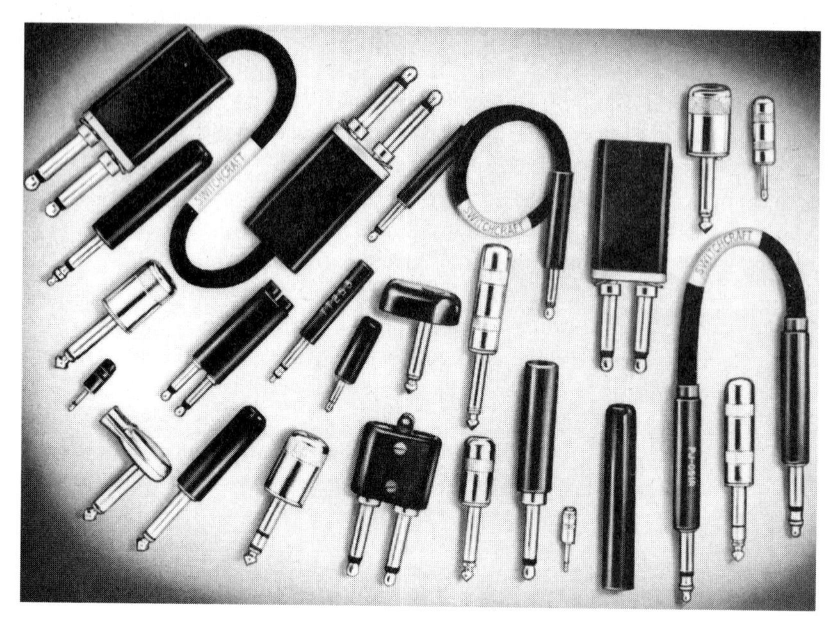

Figure 4-50 Phone-plug type of audio connectors.

(*Source:* Jones, *Electronic Components Handbook,* p. 319. Courtesy of Switchcraft, Inc.)

Figure 4-51 Common types of RF connectors.

(*Source:* Jones, *Electronic Components Handbook,* p. 321. Courtesy of Bunker Ramo Amphenol.)

problem. These connectors operate satisfactorily up to 300 MHz and have a peak voltage rating of 500 V. UHF connectors find extensive use in video applications.

PWB connectors are available in two types:

- One-piece receptacles that mate directly with the edge of the PWB or card
- Two-piece connector assemblies in which one of the mating connectors is mounted on and hard-wired to the PWB

Edge connectors are one-piece connectors and are available in a variety of sizes, configurations, contact spacings, and grades. They are used in military, commercial, and industrial applications, though they are normally not used in airborne or life-support equipment. Examples are presented in Fig. 4-52.

Edge connector contact may be made with one or both sides of the PWB. The most commonly used contact spacing is 0.156 inches (0.4 cm), but 0.100 (0.25 cm), 0.125 (0.32 cm), 0.150 (0.38 cm), and 0.200 inch (0.5 cm) spacings are also standard. Current ratings for these contacts range from 3 A to 7.5 A per contact.

Several forms of contacts are used in card-edge connectors, as can be seen in Fig. 4-53. These include:

- *Bellows:* These contacts have a folded-ribbon surface that maintains contact well with PWBs of uneven or off-spec thickness. The wire leads must be soldered to this type of edge-connector.

TABLE 4-1 DESCRIPTIONS OF COMMON RF CONNECTORS[39]

	Connector type			
	BNC	TNC	N	UHF
Coupling	Bayonet	Threaded	Threaded	Threaded
Impedence	50	50	50	50
Operating Frequency (max.)	4 GHz	11 GHz	11 GHz	300 MHz
VSWR	1.3 to 1	1.3 to 1	1.3 to 1	—
Working Voltage	500 Vrms	500 Vrms	1000 Vrms	500 Vrms
RG8/U			x	
RG9/U			x	
RG55/U	x	x		
RG58/U	x	x		any
RG87/U			x	coaxial
RG141/U	x	x		cable
RG142/U	x	x		that
RG213/U			x	will
RG214/U			x	physically
RG223/U	x	x		fit
RG225/U			x	
RG303/U	x	x		

Source: T. H. Jones, *Electronic Components Handbook,* p. 323.

- *Ribbon:* These contacts have characteristics similar to the bellows, except that they cost less to make and can be terminated by soldering, taper tab, or wire wrap.
- *Cantilever:* These contacts are less compliant than either the bellows or ribbon contacts, and also cost less to make.
- *Tuning Fork:* These contacts must contact both sides of the card. These contacts are the cheapest to make and are incompatible with off-thickness boards.

Two-piece connectors are available in two styles: enclosed and open. Examples of the enclosed-contact variety are the D-subminiature contacts shown in Fig. 4-54. Enclosed-contact two-piece connectors cost more than either the card-edge or open contacts, but they also have a longer life. Other advantages include a higher contact density, better reliability than card-edge connectors, and a lower engagement and withdrawal force.

Among open-contact connectors, the best-known type is the Elco Varicon™ with

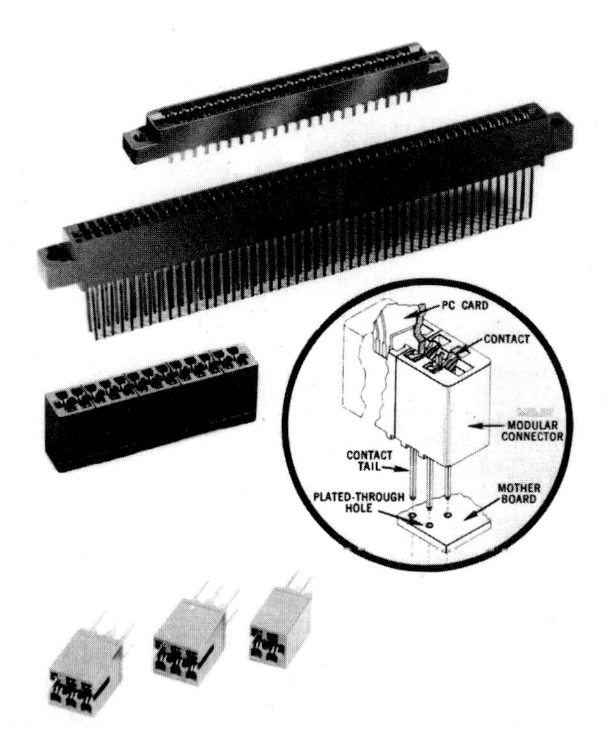

Figure 4-52 PC board card-edge connectors. Ribbon contacts, solder tab terminals with 5 A rating (top). Bifurcated bellows contacts with wire-wrapable terminals (center). Connector made up of four and six contact modules, which can be ganged for custom designs (bottom). Contact rating is 5 A.

(*Source:* Jones, *Electronic Components Handbook,* p. 325. Courtesy of Elco Corp.)

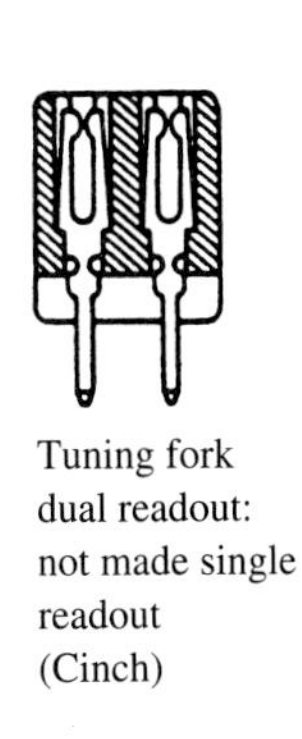

Tuning fork
dual readout:
not made single
readout
(Cinch)

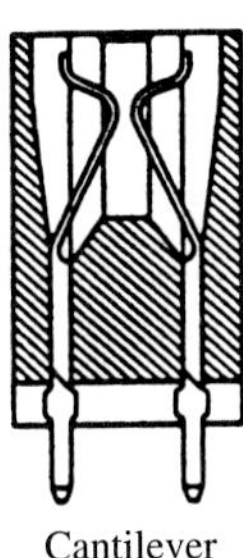

Cantilever
dual readout
(Cinch)

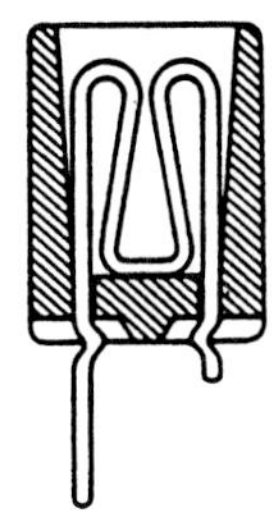

Ribbon
single readout
(ELCO)

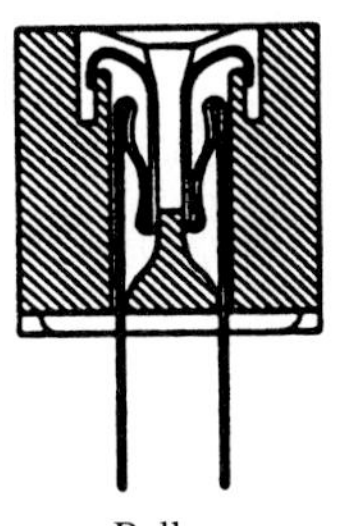

Bellows
dual readout
(Cannon)

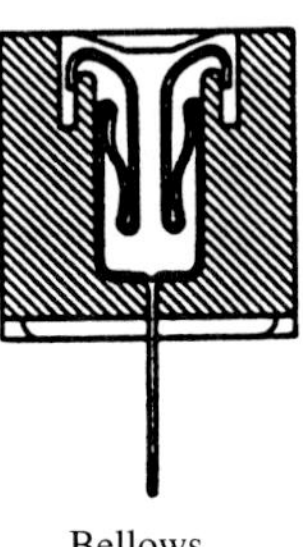

Bellows
single readout
(Cannon)

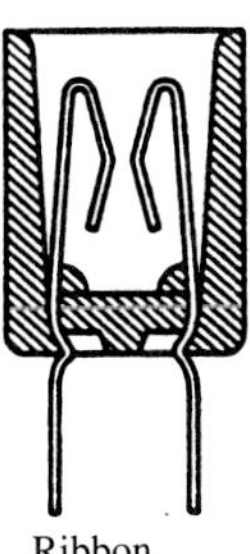

Ribbon
dual readout
(ELCO)

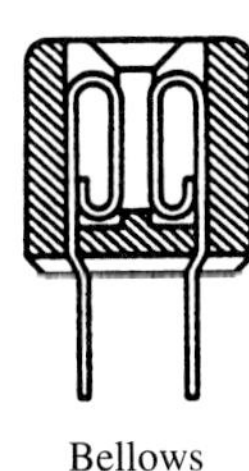

Bellows
dual readout
(Cinch)

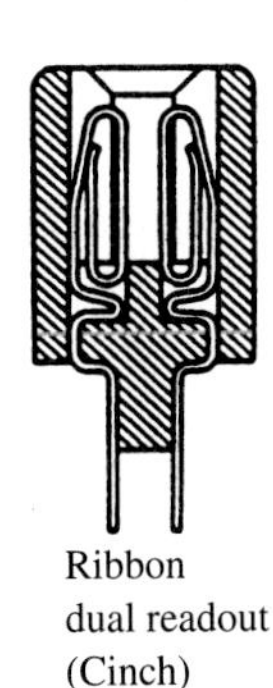

Ribbon
dual readout
(Cinch)

Figure 4-53 PC board card-edge connector contacts.

(*Source:* Jones, *Electronic Components Handbook,* p. 326.)

hermaphrodite-type contacts, pictured in Fig. 4-55. The forked contacts mate along four beveled surfaces that are coined for hardness and smoothness. The individual contacts are attached to the PWB by soldering. These contacts allow for considerable flexibility in contact layout, but they cannot be used in areas where high contact-density is required.

Miniature and subminiature rack/panel connectors are designed to save space and weight and are especially suited for use in aircraft, missile, satellite, and ground support systems. Although designed for rack and panel applications, these connectors are also used as cable connectors.

D-subminiature connectors have contact arrangements with up to 50 contacts at 5 A per contact for #20 wire with normal spacing. High-density arrangements with 78 size-22 contacts and double-density arrangements with up to 100 contacts are also available. The polarization of this connector is provided by the keystone shape of the shell. Coupling of the connector is by friction or by locking mechanisms that are available. Figure 4-56 shows some examples of typical contact arrangements of D-subminiature connectors.

The standard threaded circular connector family began as AN (Army-Navy) connectors originally designed for aircraft use and are now used widely in military, aerospace, and industrial applications. AN connectors are popular because of their

- Low cost
- Uniform high quality

Figure 4-54 D-subminiature connectors.

(*Source:* Jones, *Electronic Components Handbook,* p. 328. Courtesy of ITT Cannon Electric.)

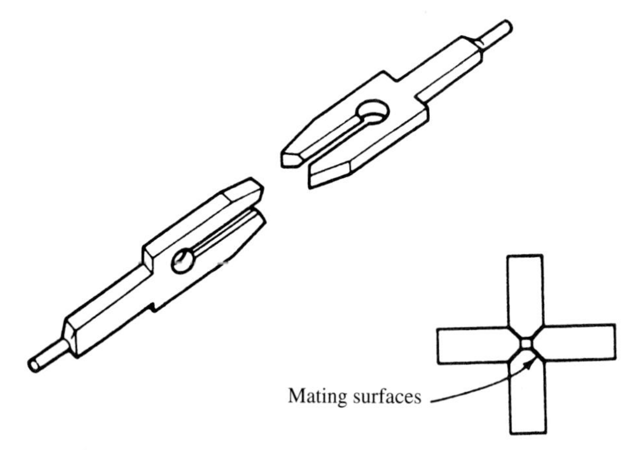

Figure 4-55 Elco Varicon hermaphrodite contact. Contacts are used in multicontact connectors and individually staked to PC boards.

(*Source:* Jones, *Electronic Components Handbook,* p. 327. Courtesy of Elco Corp.)

- Broad range of sizes, styles, variations, and pin arrangements
- Ready availability from many vendors

Figure 4-57 shows two of the many varieties of this connector. All vendor models of this type are interconnectable, intermountable, and interchangeable, as long as the proper match is made between shell size, pin configuration, and pin size.

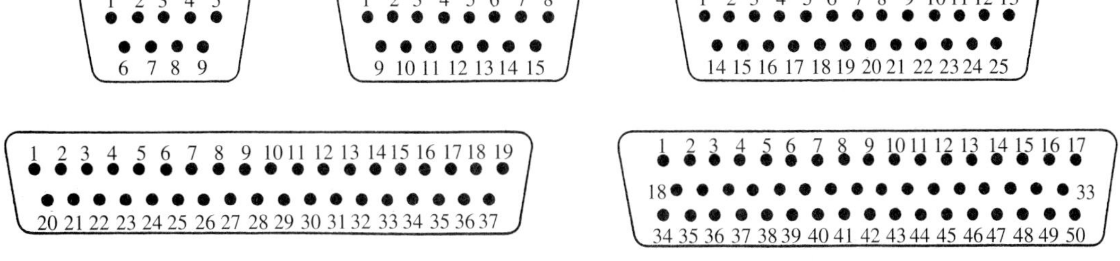

Basic *D*-Subminiature (ITT-Cannon, Matrix Science, TRW-Cinch)
or MIN-RAC (Bunker Ramo-Amphenol) connectors

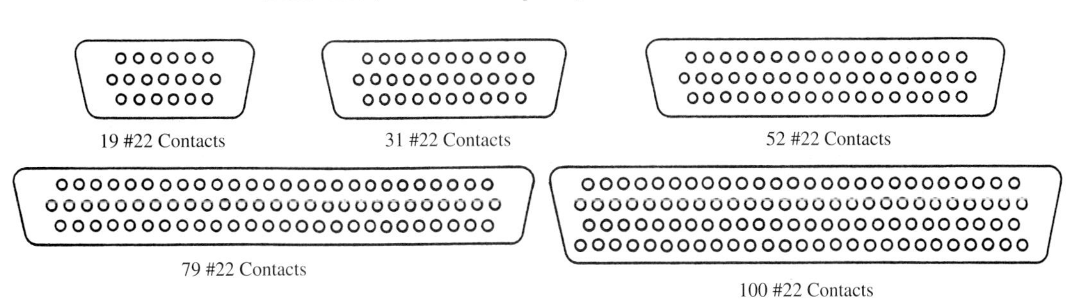

Figure 4-56 Typical contact arrangements of D-subminiature connectors.

(*Source:* Jones, *Electronic Components Handbook,* p. 330.)

The shell sizes of these connectors are based on the diameter of the coupling threads in sixteenths of an inch. For example, a size-18 shell has a diameter of 1⅛ inch (2.86 cm) and a size-24 shell has a diameter of 1½ inch (3.81 cm).

The type and number of contact pins in a connector are defined by a combination of the shell size and a contact arrangement number (for example, for an 18-22 or 32-8 connector, the left-hand number indicates shell sizes in sixteenths of an inch and the right-hand number indicates the contact arrangement number). The number does not indicate the quantity or size of contact pins. For such information, the vendor catalog must be consulted.

Figure 4-57 Standard threaded circular connector.

(*Source:* Jones, *Electronic Components Handbook,* p. 337. Courtesy of ITT Cannon Electric.)

There is a variety of threaded circular connectors to meet a wide range of applications. Some examples are as follows:

- *Mil-C-5015:* This connector is for extreme environments and high altitude, offering protection from condensation, vibration, corona discharge, and flashover.
- *Fireproof (Mil-C-5015):* These connectors (Fig. 4-58) are used for military and commercial aircraft firewall applications. They provide a means of penetrating the engine firewall of aircraft with electrical circuits while maintaining flame-barrier integrity of the firewall. They are resistant to fuel, cleaning solvents, coolants, and hydraulic fluid.
- *Hermetically Sealed (Mil-C-5015):* These connectors are used where a partial vacuum, inert gas, or constant controlled pressure is used inside a piece of equipment to eliminate adverse effects created by atmospheric changes. Applications include aircraft instruments, tachometers, and direction finders.
- *Waterproof (Mil-C-5015):* These connectors can withstand mud, ice, and water encountered around ground support equipment, radar installations, heavy construc-

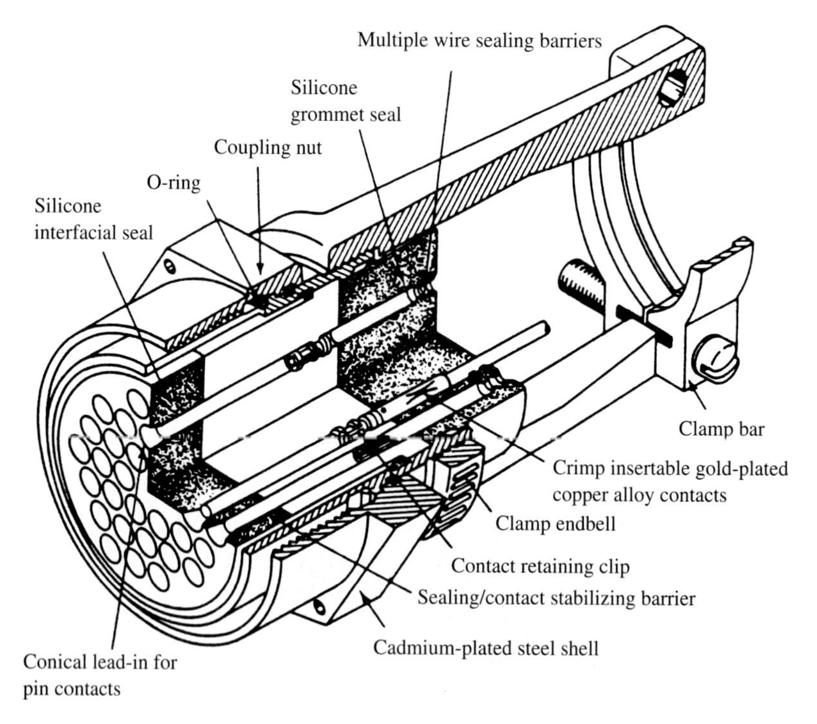

Figure 4-58 Fireproof MIL-C-5015 connector construction.

(*Source:* Jones, *Electronic Components Handbook,* p. 339. (Courtesy of ITT Cannon Electric.)

tion sites, outdoor rapid transit, radio/tv stations, and marine equipment. When used properly, MIL-C-5015 connectors can be completely submersed.

Microminiature connectors were developed due to needed miniaturization and the increased complexity of aerospace vehicles, aircraft, and electronic data processing equipment. The contacts in these connectors are usually on 0.05 inch (0.127 cm) centers instead of the larger 0.10 inch (0.254 cm) center. Figure 4-59 gives several examples.

Hydrospace connectors were developed to meet the need for connectors that would be able to withstand environments encountered by submarines and other undersea vehicles. They are used for both military and commercial applications at pressures up to 10,000 psi (680 atmospheres) and for long periods of saltwater immersion. The construction of a hydrospace connector is shown in Fig. 4-60.

Fiber-optic connectors are a recent addition to interconnection technology. The most critical specifications of a fiber-optic connector are insertion loss and reflected noise.

Figure 4-61 is a picture of a permanent optical splice with a rated average loss of less than 0.25 dB. A typical optical power budget allows for about 1.0 to 1.5 dB loss per connector. It should be remembered that a 3 dB loss represents about ½ the signal power. Losses can be minimized with proper shaping, cleaning, and polishing of the optical interface before making the connection. Special tools are required to make these splices

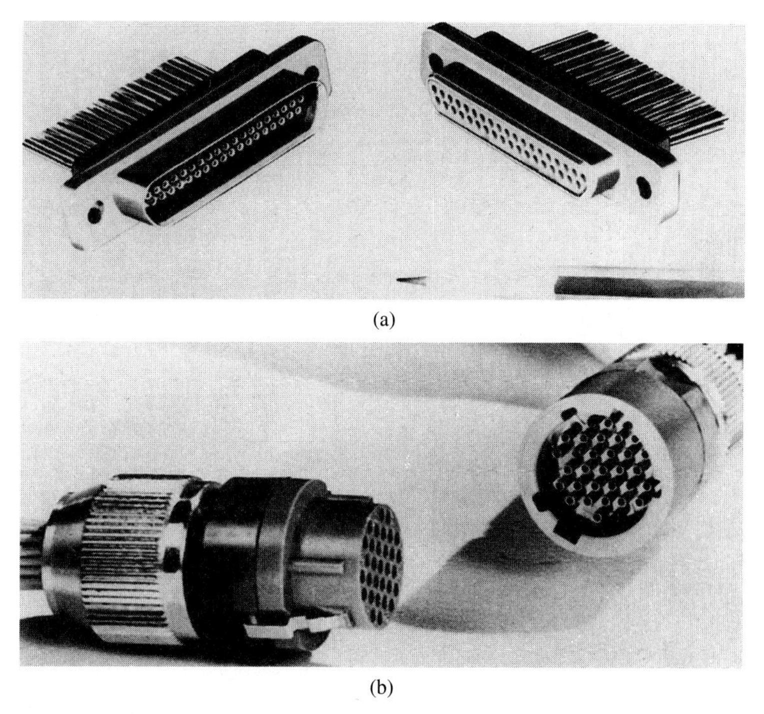

(a)

(b)

Figure 4-59 Microminiature connectors. (a) Amphenol Mighty-Mite has 7 to 61 contacts on 0.085 in. centers. (b) Cannon MDM series offers 9 to 51 contacts on 0.050 in. centers.

(*Source:* Jones, *Electronic Components Handbook,* p. 345.)

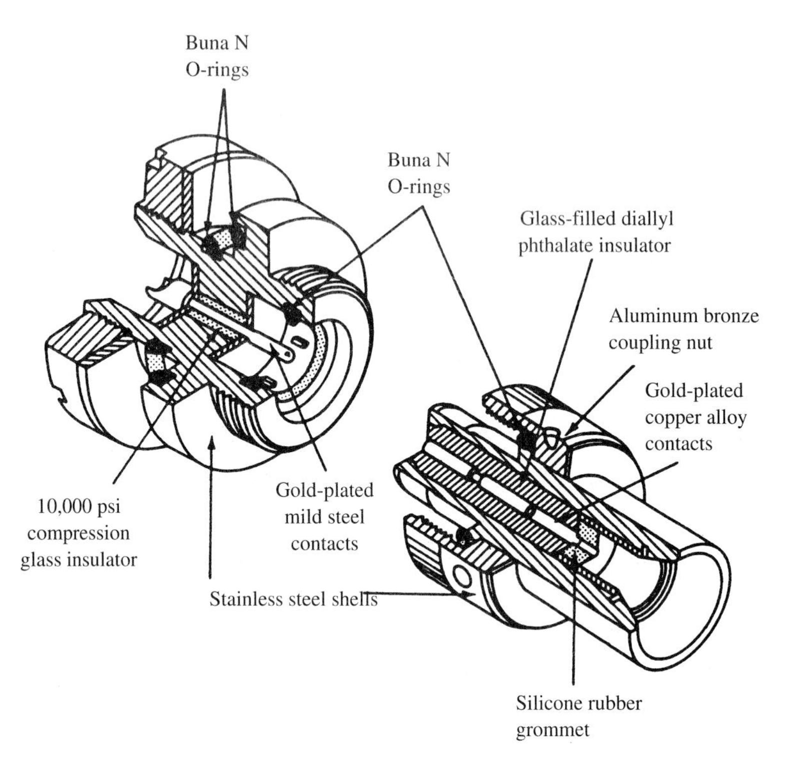

Figure 4-60 Hydrospace connectors. These MIL-C-24217 connectors are capable of withstanding pressures up to 10,000 psi.

(*Source:* Jones, *Electronic Components Handbook,* p. 348. Courtesy of ITT Cannon Electric.)

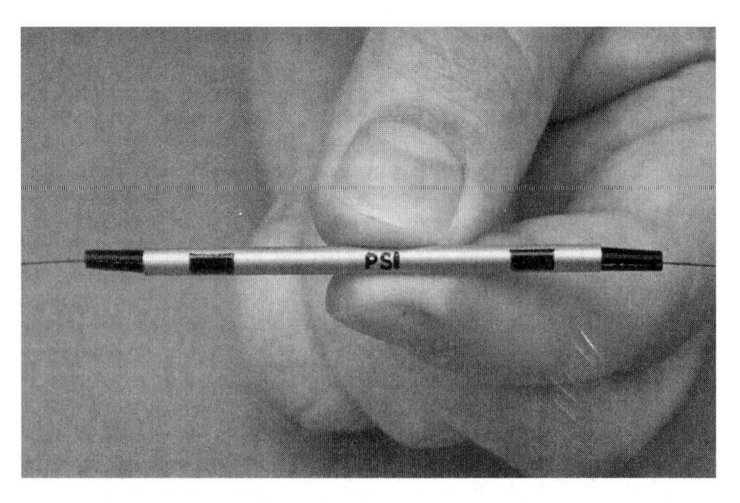

Figure 4-61 Mechanical splice for optical fibers.

(Courtesy of Aurora Optics.)

and are available from connector vendors. Low noise is easier to achieve with permanent optical splices than with separable connections.

Separable optical connectors come in several varieties, including the biconic connector used by the telephone industry, the BNC connector used by the communications industry, and the subminiature SMA connector which has been adopted for general use by the Electronic Industries Association. Figure 4-62 includes a photograph and explosion drawing of miniature plastic connectors.

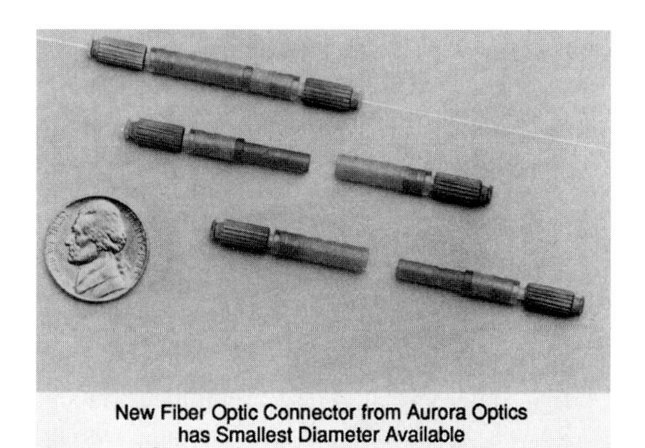

(a)

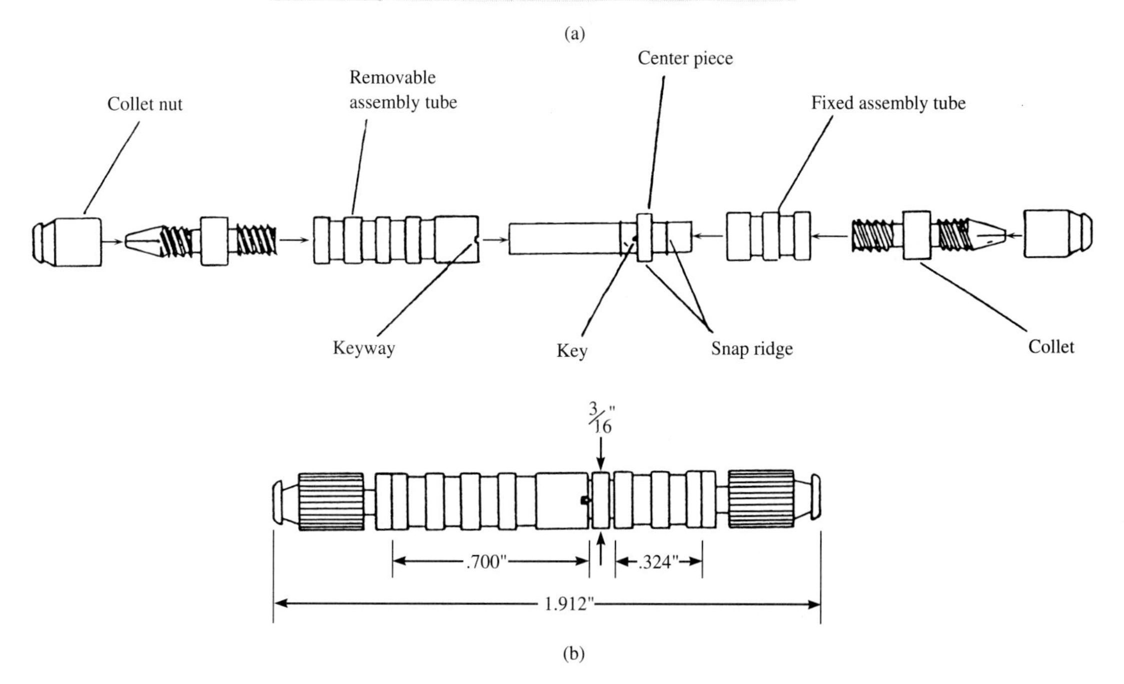

(b)

Figure 4-62 Plastic connectors for optical fibers: (a) photograph of several connectors; (b) construction details.

(Courtesy of Aurora Optics.)

4.10 SUMMARY

In general, electronic systems are composed of multiple levels of packaging which must be interconnected. The hierarchy of interconnection varies from gate-to-gate interconnections on silicon chips to cables used to interconnect subsystems. At the chip level, vacuum-deposited thin-film metal is used to interconnect individual devices. Chip I/O is accomplished using a variety of techniques such as wire bonding, controlled collapse chip connect, and tape automated bonding. For higher levels of integration, electronic systems utilize cables for signal propagation between subsystems. The type and size of cable chosen for a specific application is based on its ability to propagate signals over a specified distance without significant degradation of signal quality. The use of cables necessitates the use of some type of connector to provide mechanical and electrical linkage between packaging levels in a system. Since connector reliability is a function of both operation time and the number of contacts, connector metallurgy is the single most important factor in connector design.

KEY TERMS

Area (array) TAB
Bump
Bumped chip TAB
Bumped tape TAB
Cables
Chip-level interconnections
Connector
Controlled collapse chip connect (C4)
Encapsulation
Gang (mass) bonding
Inner lead bonding (ILB)
Interconnection devices
Multichip module (MCM)
Multiple chip packaging
Outer lead bonding (OLB)
Single-point bonding
TAB tapes
Tape automated bonding
Thermocompression bonding
Thermosonic bonding
Ultrasonic bonding
Wafer-scale integration (WSI)
Wire bonding

EXERCISES

1. Distinguish between thermocompression, thermosonic, and ultrasonic wire bonding.
2. Distinguish between a ball bond and a wedge bond. Which of these two types of bonds is used in each of the wire bonding processes of exercise 1?
3. List and discuss the three major divisions of tape automated bonding (TAB) tapes, in terms of tape cross sections.
4. List and distinguish between the three popular kinds of chip-level interconnection in TAB technology.
5. Discuss the advantages and disadvantages of the use of single-point and gang bonding in TAB.
6. List the advantages and disadvantages of TAB.
7. Briefly discuss the two categories of multichip systems.
8. Define the six levels in the hierarchy of interconnection in electronics.

9. What factors must be considered when choosing a cable for a specific application?
10. What is fretting? How and why does it degrade a connector?
11. What are the four categories of one-part connectors?
12. Describe a wire-wrap connector.
13. List and distinguish between the three types of RF connectors.
14. Distinguish between one-piece and two-piece PWB connectors.
15. Why have AN connectors been well received?
16. What are the distinguishing features of hydrospace connectors?
17. What are the three primary chip-level interconnects?
18. When choosing a cable, what considerations must be given in regard to a signal traveling over a given distance?
19. What are the primary and secondary causes of connector failure?
20. For a good connector design, what three things must be considered?

BIBLIOGRAPHY

1. FINK, DONALD G., AND DONALD CHRISTIANSEN, eds., *Electronics Engineers' Handbook.* New York: McGraw-Hill Book Company, 1989.
2. COOMBS, CLYDE F., JR., *Printed Circuits Handbook,* 3rd Edition. New York: McGraw-Hill Book Company, 1988.
3. *Reference Data For Radio Engineers,* 6th Edition. Indianapolis, Ind: Howard W. Sams & Company, Inc., 1977.
4. MILLMAN, JACOB, *Microelectronics: Digital and Analog Circuits and Systems.* New York: McGraw-Hill Book Company, 1979.
5. HALLMARK, CLAYTON L., *Microelectronics.* Blue Ridge Summit, Pa.: Tab Books, 1976.
6. TOWERS, T. D., *Hybrid Microcircuits.* New York: Crane, Russak & Company, Inc., 1977.
7. TRAISTER, JOHN E., *Design Guidelines for Surface Mount Technology.* San Diego, Calif.: Academic Press Inc., 1990.
8. *Electronic Materials Handbook, Vol. 1: Packaging,* ASM International, 1989.
9. TUMMALA, TAO R., AND EUGENE J. RYMASZEWSKI, eds., *Microelectronics Packaging Handbook.* New York: Van Nostrand Reinhold, 1989.
10. JONES, T. H., *Electronic Components Handbook.* Reston, Va.: Reston Publishing Company, 1988.

5 Printed Wiring Boards

5.1 INTRODUCTION

A **printed wiring board** (PWB) is a type of substrate, consisting of one or more layers of metal conductors and insulating material that allow for electronic components to be electrically interconnected and mechanically supported. Electrical connection between the various layers of metal circuitry is achieved by punching (drilling) holes (**vias**) through the substrate and then plating. PWBs are the most commonly used platform for manufacturing electronic systems. Electronic devices used in these systems range from passive discretes to very large scale integrated (VLSI) circuits.

This chapter provides an introduction to types of PWBs (Sec. 5.2) and base materials (Sec. 5.3), and the processes of fabrication (Sec. 5.4), image transfer (Sec. 5.5), plating (Sec. 5.6), etching (Sec. 5.7), and bare-board testing (Sec. 5.8).

5.2 TYPES OF PRINTED WIRING BOARDS

The majority of PWBs are in one of three categories: single-sided, double-sided, or multilayer. These categories will be discussed in the following sections.

5.2.1 Single-Sided Boards

The **single-sided board** is a PWB having circuits on only one side. The base material of the board has copper foil laminated to one side. The circuit image is applied foil by means of photoresist, then etched to remove all the copper foil ex conductive path.

5.2.2 Double-Sided Boards

Double-sided boards have circuits on both sides. The circuits on the opposing side of the board may be interconnected by drilling a hole through the board and depositing metal on the inner surface of the hole. This is known as **through-hole metallization.** If the connecting hole is completely filled with conductive silver ink it is known as silver-through hole (STH). If the inner surface of the connecting hole is plated (instead of filled) it is known as **plated-through hole** (PTH).

5.2.3 Multilayer Boards

As electronic components became more miniaturized, it became obvious that the wiring connecting the components was causing severe restriction on circuit density. The wiring also contributed greatly to quality problems and to the volume and weight of the circuitry. **Multilayer boards** were the solution to this problem. Multilayer PWBs are made up of three or more boards that have been pre-etched, then laminated together before being drilled. The inner layers are etched with the conductive paths, and components are mounted on the outer layers.

5.2.4 Basic Process

A printed wiring board undergoes the following basic processes:

- Fabricating of a paper-based or epoxy-glass substrate that has a copper overlay on one or both surfaces
- Hole drilling to mechanical print specifications
- Electroless copper plating operation
- Image application
- Electroplating
- Etching to leave the printed wiring exposed
- Machining to mechanical print specifications

5.3 BASE MATERIALS

5.3.1 Types of Base Materials

There are many types of copper-clad laminates used as **base materials** in the fabrication of PWBs. Some of the most common types are: FR-2, FR-4, FR-5, CEM-1, CEM-3, and GI. The FR prefix indicates that the base material is impregnated with a flame-retardant so that it is self-extinguishing. The laminates have the following features:

- *FR-2:* Laminated paper impregnated with phenolic resin for a flame retardant. The advantage of the paper laminates is that they are the least expensive, and their disadvantage is that they are also the least durable. They are used for low-cost consumer electronics such as radios, calculators, and toys.

- *FR-3:* Laminated paper impregnated with an epoxy-resin binder. This is a stronger board than FR-2 and is used in computers, televisions, and communication equipment.
- *FR-4:* Laminated layers of woven glass cloth that have been impregnated with epoxy resin. This is a very durable material and is widely used because of its excellent physical and electrical properties. FR-4 is used in control systems, the aerospace industry, and communications.
- *FR-5:* Laminated layers of woven glass cloth that have been impregnated with polyfunctional epoxy resin. FR-5 is used in applications much the same as FR-4, except that it has better heat-resistant qualities.
- *CEM-1:* A composite material made up of paper impregnated with epoxy resin and coated on both sides with woven glass cloth that has been impregnated with the same resin. CEM-1 is used in smoke detectors, television sets, and industrial electronics.
- *CEM-3:* A composite material, made up of a nonwoven fiberglass core and coated on both surfaces with woven glass cloth that has been impregnated with epoxy resin. CEM-3 is more durable and more expensive than CEM-1, and its uses include computers, stereos, video cassette recorders, and automobiles.

5.3.2 Laminating Process

In the **laminating process** base materials are joined together and are treated and cured, then stored in an area where temperature and humidity are strictly controlled. This prepared material is known as **prepreg.**

The next step is to take the prepreg to a clean room so that copper foil can be laminated to one or both sides. Copper foil is placed on a large polished stainless steel press plate, and the proper number of layers of prepreg are placed on top of the foil. The number of layers of prepreg used will depend on the final thickness of substrate desired. If the material is to have copper foil deposited on both sides, a final layer of foil is placed on top. If only one side is to have copper foil, a release film such as DuPont Tedlar™ is placed on top instead.

5.3.3 Pressing

The stack of laminate materials is moved from the buildup room to the next process called **pressing,** in which a hydraulic press bonds the layers together with heat and pressure. Steam is the usual heat source, and it is applied to the press platens while pressure in excess of 1000 psi is applied.

Once the desired stage of cure is achieved, cold water is pumped through the press platens until the material reaches a temperature of about 80°F. The laminate is then removed from the press, and the edges are trimmed. The finished sheets of substrate are typically $\frac{1}{16}$ inch (0.16 cm) thick and are in 36 × 48 inch (91 × 122 cm) or 48 × 144 inch (122 × 366 cm) sheets.

5.3.4 Quality Control

Product design requires that the copper-clad sheets pass standard specifications and evaluation if they are to be acceptable. The substrates that pass tests with varying degrees of success are graded and priced accordingly.

Standard quality control tests to which the PWB substrates are subjected include:

- *Copper Surface:* Designates the acceptable number and size of pits and dents per unit area of surface.
- *Punchability and Machinability:* Defines the ease with which holes can be punched through the PWB and the PWB can be machined into a specified shape.
- *Peel Strength:* Designates copper-bond strength for various thicknesses of copper-foil laminate.
- *Bow and Twist:* Designates the amount of bow and twist per unit length that is acceptable.
- *Solder Resistance:* Determines how the substrate will react to the soldering process. Specimens are floated horizontally on a solder bath for a specified time, then evaluated for blistering, measling, delamination, and weave exposure.
- *Autoclave:* Uses high temperature and pressure to determine moisture resistance.
- *Degree of Cure:* Determines whether or not the substrate was pressed and cured properly.
- *Insulation Resistance:* Determines the resistance between two conductors on the copper-clad laminate.
- *Fungus Resistance:* Determines the base material's resistance to attack by various molds and fungi.
- *Dielectric Strength:* Determines the ability of the base material to resist the passage of a disruptive discharge (voltage spike) produced by an electrical stress. This is measured by applying a 60 Hz voltage through the thickness of the laminate, in varying amplitudes, over a period of time.

Boards that pass the appropriate standards are then used for product fabrication.

5.4 FABRICATION

Before the image transfer or the plating and etching of a PWB takes place, the boards are cut to size, holes are drilled or punched, and edges and holes are deburred in preparation for the chemical processes. The quality of the finished board can only be as good as the quality of these initial steps.

5.4.1 Sizing Boards

Boards that are cut to rectangular shapes may be sheared to proper size. To prevent cracks and edge deformation of the copper-clad lamination during **shearing,** the shears should be set with about 0.001 inch (0.025 mm) to 0.002 inch (0.05 mm) clearance between the blades, as shown in Fig. 5-1.

When sawing paper-base laminates, the best results are usually obtained with circular saws. The blades should be hollow-ground, have carbide teeth, usually have 10 to 12 teeth per inch (4 to 5 teeth per cm) of diameter, and should be operated at 7,500–10,000 ft/min (38–51 m/sec).

For glass-based laminates, saws with carbide teeth are acceptable, but diamond-steel-bonded saws will have a much longer lifetime. Manufacturers of these blades usually recommend a cutting speed of about 15,000 ft/min (76 m/sec) at the perimeter of the blade.

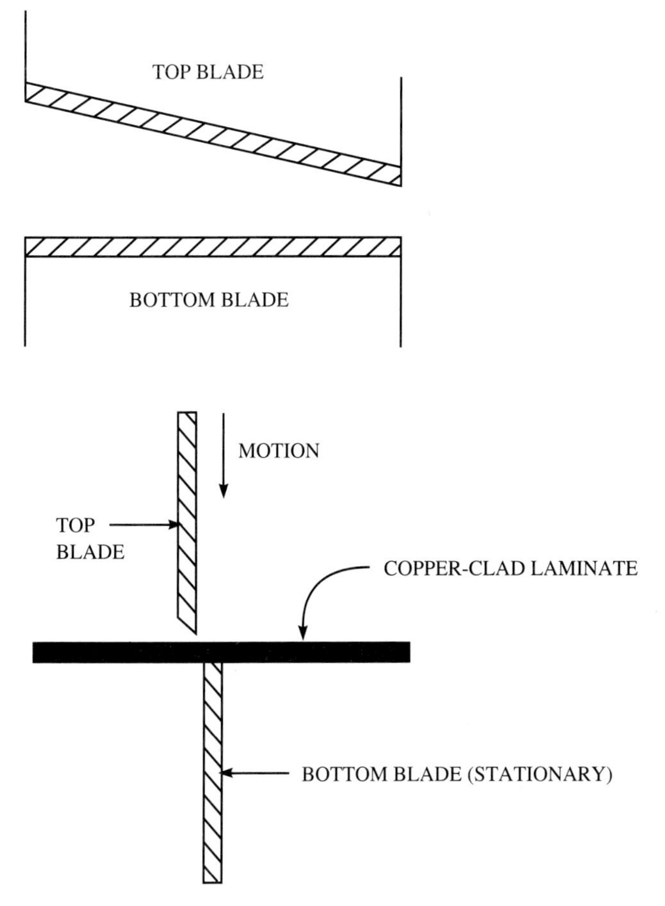

Figure 5-1 Construction of typical adjustable shear blades for copper-clad laminates.

Some boards need to be produced in shapes that are not rectangular. One method for cutting these boards is with the use of a blanking die, which functions similarly to a cookie-cutter. This process is only used when a large quality of a specific shape of PWBs is to be produced because of the time and expense that is involved in retooling for a new job.

An alternative to the use of blanking-dies for odd-shaped boards is routing. The use of multiple-spindle routing machines precludes the long lead-time and expense of making new blanking dies. Routing generally produces better-quality edges than does blanking (refer to the case study in Sec. 11.4.6). However, the production cycle-time per board is longer for routing than for blanking.

Router bits are small cylinders made of solid carbide that travel at a spindle rate of 12,000 to 24,000 rpm. They must be driven with sufficient power to make a cut without a measurable drop in speed. Both the cutter speeds and feed rates are much higher than can be achieved with the sawing operations. These higher speeds yield the advantages of cleaner edge finishes and closer tolerances than can be accomplished with blanking or sawing.

5.4.2 Punching Holes

One of the major problems in hole **punching** is the tendency for the board to crack. Punch-dies are designed so that many holes are punched in a single stroke. If the boards are stripped from the perforators unevenly, cracks will almost always occur. Also, the boards must be held securely in place before the perforators enter the board. It is important, therefore, for the stripping mechanism to compress the board an instant before the perforators make contact and to strip the board evenly after the holes are punched.

Even when all aspects of the punching operation work perfectly, there are some physical limitations that must be respected if board failures are to be prevented. The minimum distance that is allowed from hole to hole and from hole to edge is a function of the thickness of the base laminate, as is shown in Fig. 5-2.

If the boards being punched are of paper-based materials, there will be some shrinkage of the punched holes due to elasticity (the tendency of the material to spring back). The amount of shrinkage will depend upon the thickness and temperature of the board being punched. In order for the punched holes to remain within tolerance, the perforators must be larger than the finished hole size. Table 5-1 gives guidelines for oversizing of the perforator.

5.4.3 Drilling Holes

Another method of creating holes for through-board connections or component mounting is by **drilling.** The quality of the drilled hole will affect follow-up processes such as plating and soldering, and it is important for the formation of a highly reliable electrical/ mechanical connection. When proper procedures are followed, corrective procedures such as deburring, desmearing, and etchback are minimized. This results in more efficient PC board production.

The drill bits are usually made of tungsten carbide. However, all manufactured drill bits are not of the same quality, nor do they have the same physical profile. Desirable

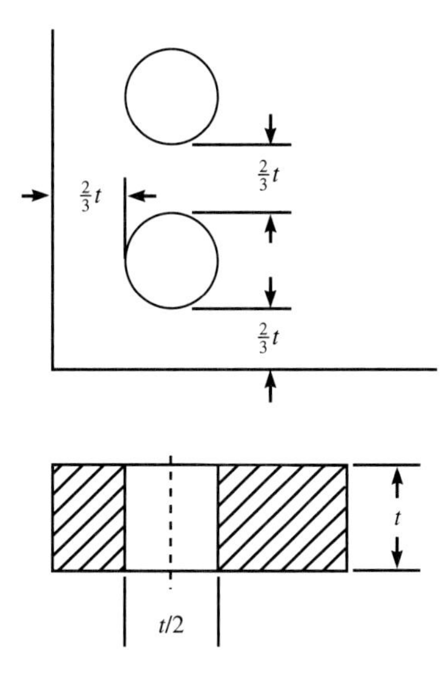

Figure 5-2 Illustration of the proper sizing and locating of punched holes with respect to one another and to the edge of paper laminates. Minimum dimensions are given as multiples of laminate thickness *t*.

drill bit geometries are determined by the specific base material of the laminate that is to be drilled. Drill characteristics that must be decided upon for specific jobs are shown in Fig. 5-3. In general, the best drill designs are those that promote the lowest drilling temperatures, remove chips efficiently, and retain a sharp cutting edge.

During the drilling operation, a stack of PWBs is sandwiched between an entry-sheet and a drill backup. The entry-sheet is used to prevent damage to the top copper laminate surface from the pressure foot, reduce entry burrs, and help minimize drill wandering.

Drill backup material is placed under the bottom laminate. The backup material is used to prevent an exit burr from forming when the boards are drilled. At the bottom of its stroke, the drill should penetrate the backing material to a drill lip penetration depth of 0.005 to 0.015 inch (0.127 to 0.38 mm), as can be seen in Fig. 5-4.

The composition of the entry-sheet and backup material should not contribute to

TABLE 5-1 SHRINKAGE IN PUNCHED HOLE DIAMETERS FOR PAPER-BASE LAMINATES

Material thickness		Shrinkage at room temperature	
Inches	mm	Inches	mm
1/64	0.4	0.001	0.025
1/32	0.8	0.002	0.05
3/64	1.2	0.003	0.075
1/16	1.6	0.004	0.100
3/32	2.4	0.006	0.15
1/8	3.2	0.010	0.25

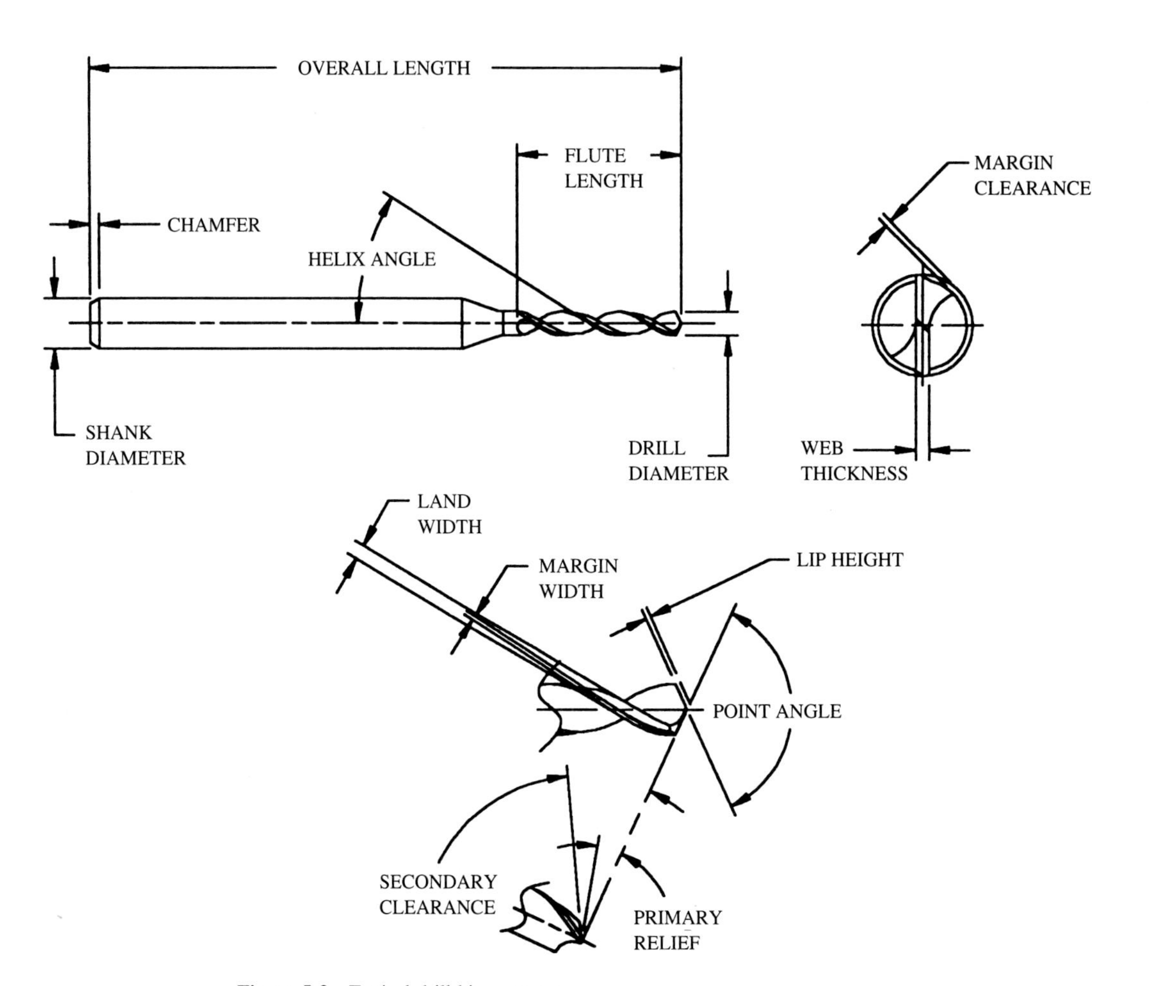

Figure 5-3 Typical drill bit geometry.

(Courtesy of Rogers Tool Works, Inc.)

increased drilling temperatures or contamination of the hole due to resin-smear. Typical entry-sheet materials include:

- *Solid aluminum:* 5–30 mils (0.127–0.76 mm) thick.
- *Aluminum-clad:* Consists of two thin Al alloy skins bonded to a noncontaminating core.

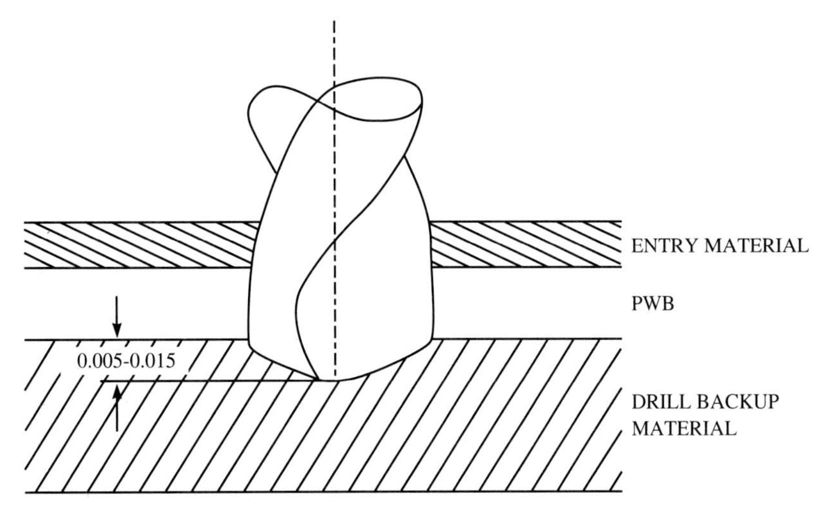

Figure 5-4 Setting drill termination depth in the backup.

- *Phenolic, resin-based:* 12–22 mils (0.3–0.56 mm) thick, 60% + paper with the rest being resin.
- *Paper:* 5-10 mils (0.127 - 0.25 mm) thickness.

The following are some common backup materials:

- *Paper-clad or phenolic-clad:* Two thin paper or phenolic skins bonded to a wood core.
- *Phenolic:* Consists of 60% + phenolic resin with the remaining material being paper.
- *Hardboard:* Pressed wood fibers with resins and oils.
- *Vented aluminum:* A structure permitting the drill to terminate in air and reduce burring at the exit.
- *Aluminum-clad:* Consists of two thin aluminum skins bonded to a noncontaminating core.

Types of drilling machines vary from single-spindle manually operated types to automatic types that drill stacks of laminate and that change tools automatically.

Laminate sizes of up to 24 × 24 inches (61 × 61 cm) are accepted by some machines, and costs may range from $10,000 to over $500,000.

5.5 IMAGE TRANSFER

There are two primary types of **image transfer** used in the PWB industry: screen printing and photoprinting. Screen printing involves preparation of the screen, setting up the screen, screening the image, oven curing the paste, inspecting, retouching if necessary, either etching or plating, and then stripping the resist.

The photoprinting process can be further divided into wet-film and dry-film categories. A typical dry-film process requires preparation of the laminator, lamination of the film, exposure of the film, development of the film, inspection of the image, etching or plating, and stripping of the resist. The wet-film or liquid resist process involves preparation of the applicator; coating by spraying, dipping, or spinning; oven drying; exposing and developing the film; postbaking; etching or plating; and stripping the film.

5.5.1 Screen Printing

In **screen printing,** the circuit design is applied to a screen fabric that has been stretched on an aluminum frame. This image of the circuit acts as a stencil. Liquid resist is deposited on the top surface of the screen, and a PWB blank is placed under the screen frame. The liquid resist is forced through the mesh, except for the area protected by the stencil, by pushing a squeegee along the top of the screen as shown in Fig. 5-5.

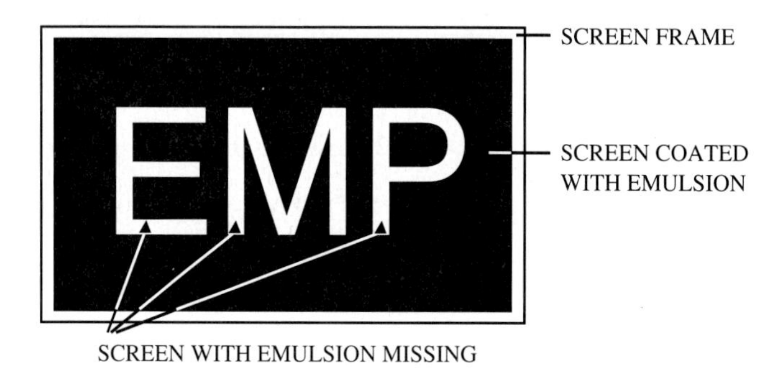

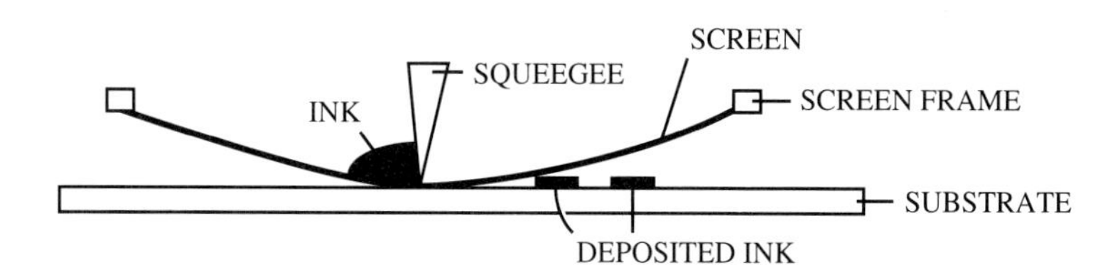

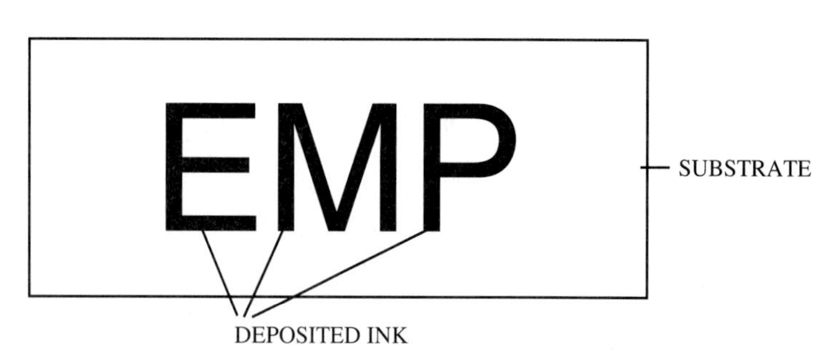

Figure 5-5 Screen-printing process.

5.5.2 Photoprinting

Two methods of **photoprinting** are currently used as an alternative to screen printing. Wet-film photoresist has been in use since the 1940s, and dry-film resist was introduced by DuPont in 1968.

Wet-film photoresist was introduced as an alternative to screen printing and became popular because higher resolution enabled finer lines. The disadvantages of wet-film photoresist are that it is difficult to control, the coating thickness is generally difficult to apply evenly, and it is easily contaminated by dust which causes pinholes in the image.

Dry-film photoresist is a sheet of photopolymer that is 0.0007 to 0.003 inch (0.018 to 0.076 mm) thick, sandwiched between DuPont Mylar™ and polyolefin film, as shown in Fig. 5-6. This method is two to three times more expensive than screen printing or liquid resists, but it offers the advantages of higher production rates and less sensitivity to dust and handling.

The photopolymer film is pressed against the PWB substrate by rollers, as shown in Fig. 5-7. A positive or negative phototool of the circuit image is placed on top of the board, then exposed to UV light. The protective sheet of Mylar is then removed, and the photo-sensitive film is developed in a special process. The unexposed areas of photoresist are removed in this developing process.

If the circuit is to be applied by panel plating or pattern plating, a positive phototool is used. If the circuit is to be created by the print-and-etch process, a negative phototool is used.

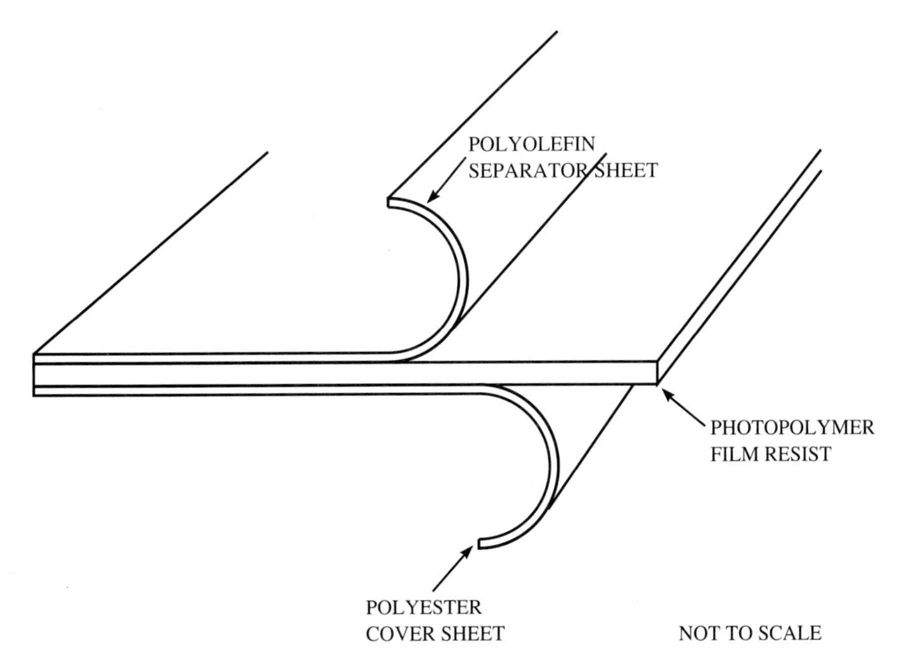

Figure 5-6 Dry-film three-layer photoresist structure.

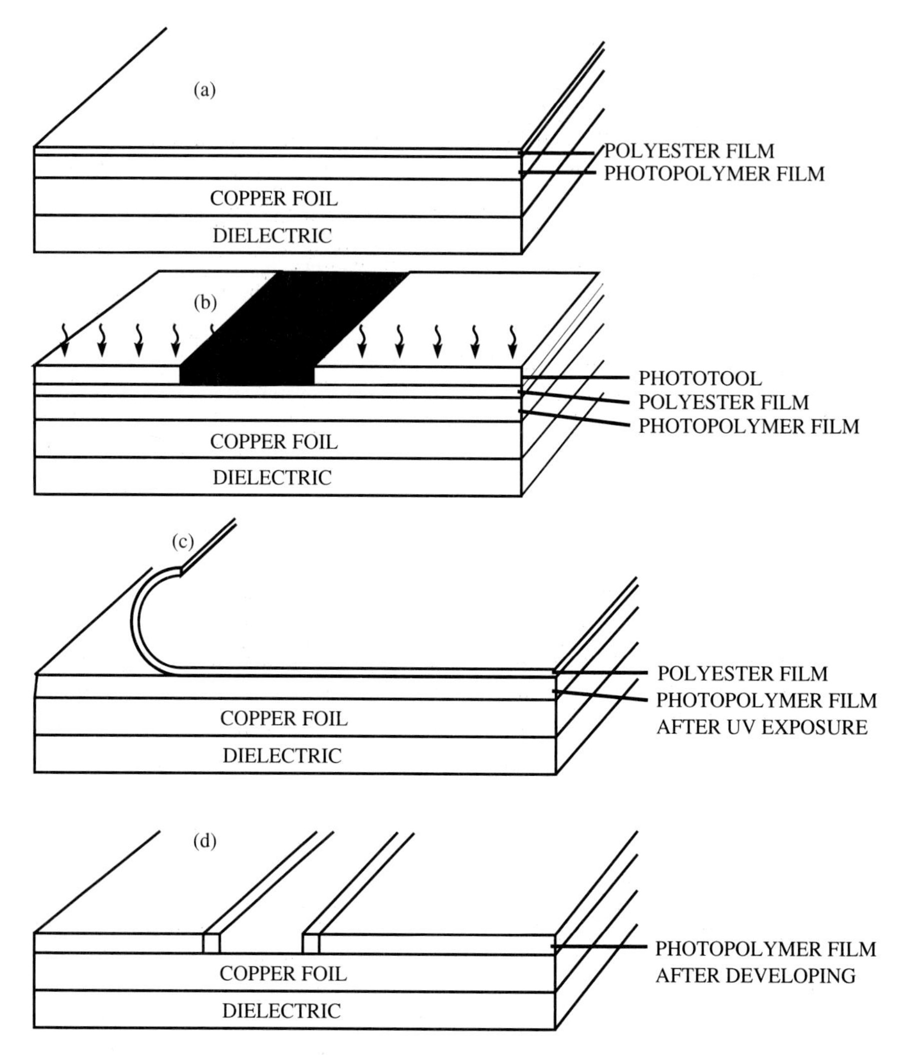

Figure 5-7 The dry-film process: (a) remove polyolefin film and apply to copper surface with a roller; (b) expose with UV using either a positive or negative phototool (negative for print and etch or positive for plating); (c) remove protective polyester film; (d) develop photopolymer film.

After the board has gone through its appropriate plating or etching processes, the remaining photoresist film is removed by using a chemical stripping agent such as methylene chloride or methyl alcohol.

5.6 PLATING

Plating is the coverage of a material with a thin layer of metal. Before going to the plating process, single-sided, double-sided, and multilayer boards must have all their holes deburred. After deburring, multilayer boards must go through the additional steps of smear removal and etchback. Figure 5-8 is a flowchart showing the **plating processes.**

5.6.1 Smear Removal

When holes are drilled through multilayer boards, it is likely that some or all of the holes will suffer from drill smear. **Drill smear** occurs, during the drilling process, when epoxy-resin in the base material of the board is heated up by the drill bit and the resin coats the inner surface of the drilled hole. This layer of resin must be removed before the plating-through-hole process to ensure electrical continuity from the inner layers to the PTH. When the resin is removed, the inner layer connection will be flush with the drilled hole, as can be seen in Figs. 5-9 and 5-10.

5.6.2 Etchback

Sometimes it is desirable to have the inner layer connections extend 0.5 mil (12.7 μm) beyond the inner surface of the drilled hole. **Etchback** is done by extending the smear-removal process long enough to remove some of the inner layer of dielectric material, as illustrated in Fig. 5-11. Smear removal is required to meet some military specifications.

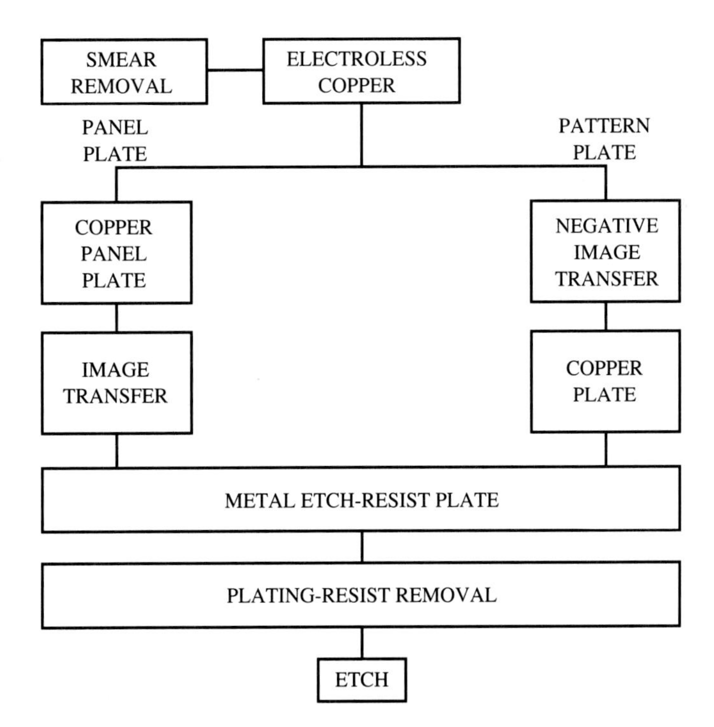

Figure 5-8 Partial plating flowchart for PWB.

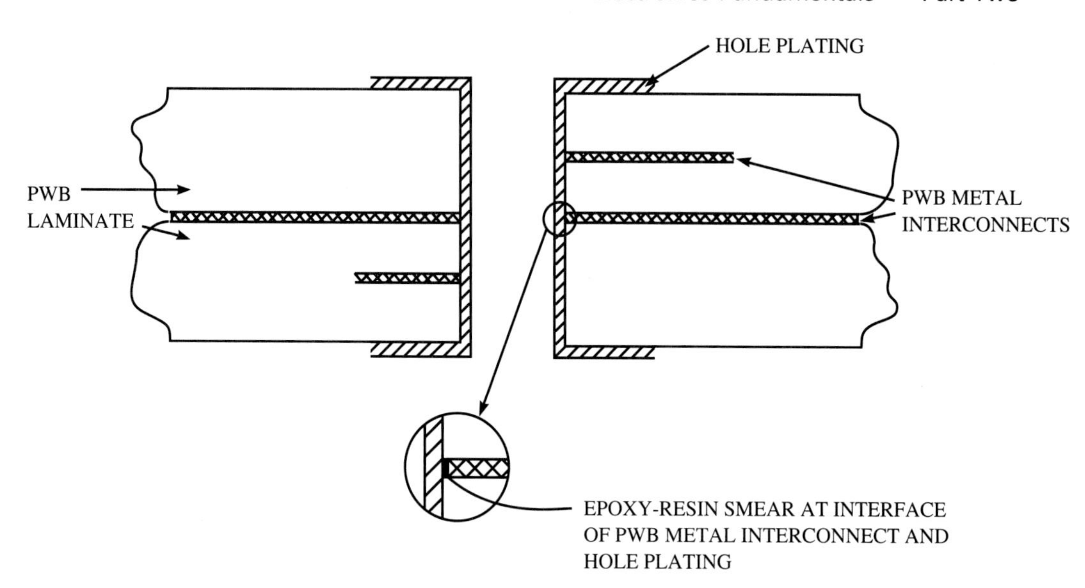

Figure 5-9 Cross-section of PWB showing PTH and illustrating inner layer smear.

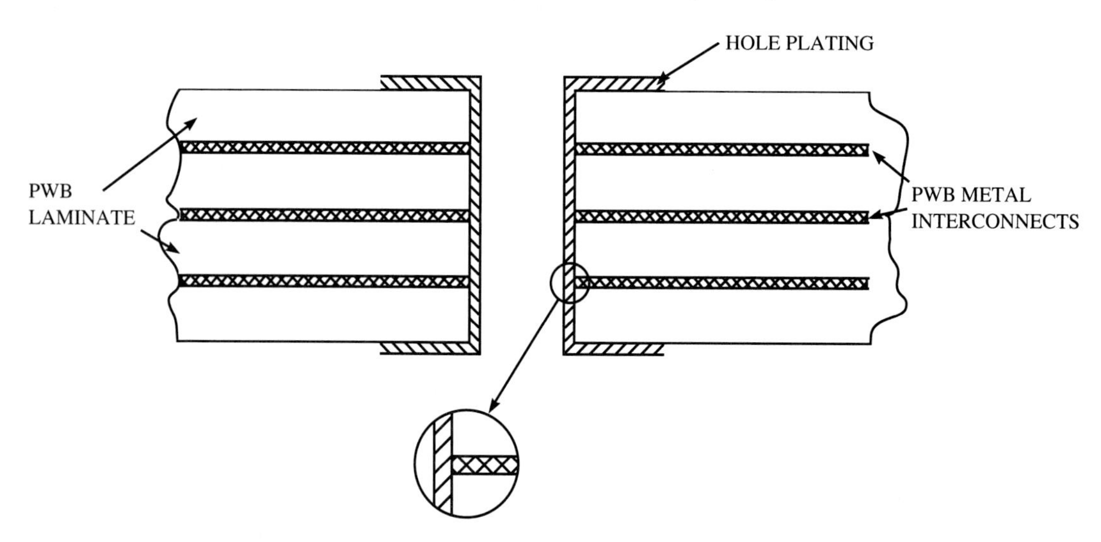

Figure 5-10 Cross-section of PWB showing PTH and illustrating proper smear removal and ideal PWB metal interconnect to hole plating.

5.6.3 Plating Processes

PWBs that are produced using an image-transfer process can be categorized into two classes: subtractive and additive. Since the additive process has not passed military specifications, only the subtractive process will be discussed here. The subtractive processes that will be discussed are printing and etching, pattern plating, and panel plating.

After smear removal and etchback, the board goes through a chemical process that makes panel side-to-side and inner-layer (PTH) connections by metallizing the board with copper. An electroless copper process is used after the boards are drilled and before the boards go through the image transfer process.

The print-and-etch process uses a positive photoresist image to protect the copper beneath the resist. This is accomplished by applying a negative phototool and exposing the photoresist to light, then developing and washing away the unexposed photoresist. The unprotected copper can then be etched or subtracted away. This process is illustrated in Fig. 5-12.

Pattern plating (selective plating) is the process where only the desired circuit pattern and holes receive the copper build-up and etch-resist metal plate. In this process, a positive phototool is used to define a negative photoresist image of the circuit. The negative image area is then electroplated with an etch-resistant alloy such as tin-lead, tin-nickel, or nickel-gold.

After the electroplating process, the copper that is unprotected by the etch-resist is etched away and the remaining dry-film is stripped off the PWB. This process is illustrated in Figs. 5-13 and 5-14.

Panel plating is the process whereby the entire surface area and the drilled holes are copper-plated, stopped-off with resist on the unwanted copper surfaces, then plated with the etch-resist metal. The panel plating process is best used on PWBs with circuit features greater than 0.015 inch (0.38 mm) due to the high degree of **undercut** that accompanies etching away the relatively thick copper. Compare Fig. 5-15 with Fig. 5-16 to note the thickness of panel plating versus pattern plating.

If tin-lead is used as the etch-resist, the **overhang** caused by undercutting can be removed by melting the tin-lead by a hot-oil or infrared (IR) fusing process. The results of the fusing process can be seen in Fig. 5-17.

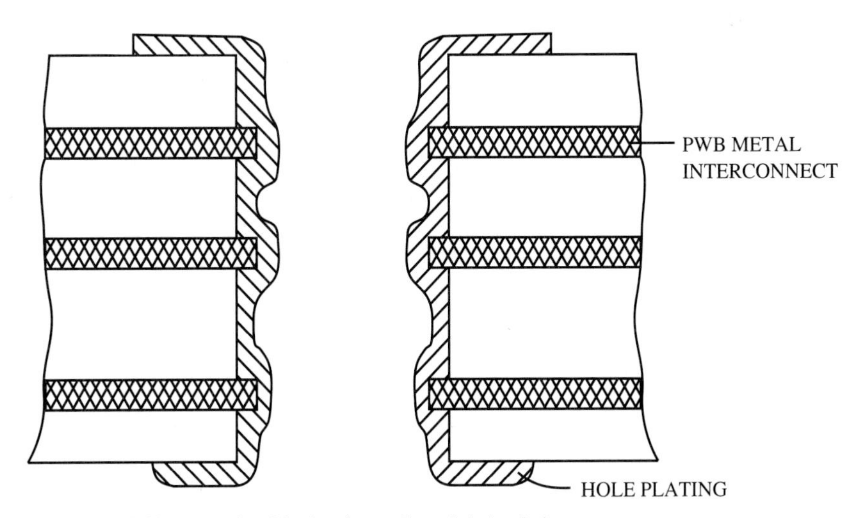

Figure 5-11 Use of etchback prior to through-hole plating.

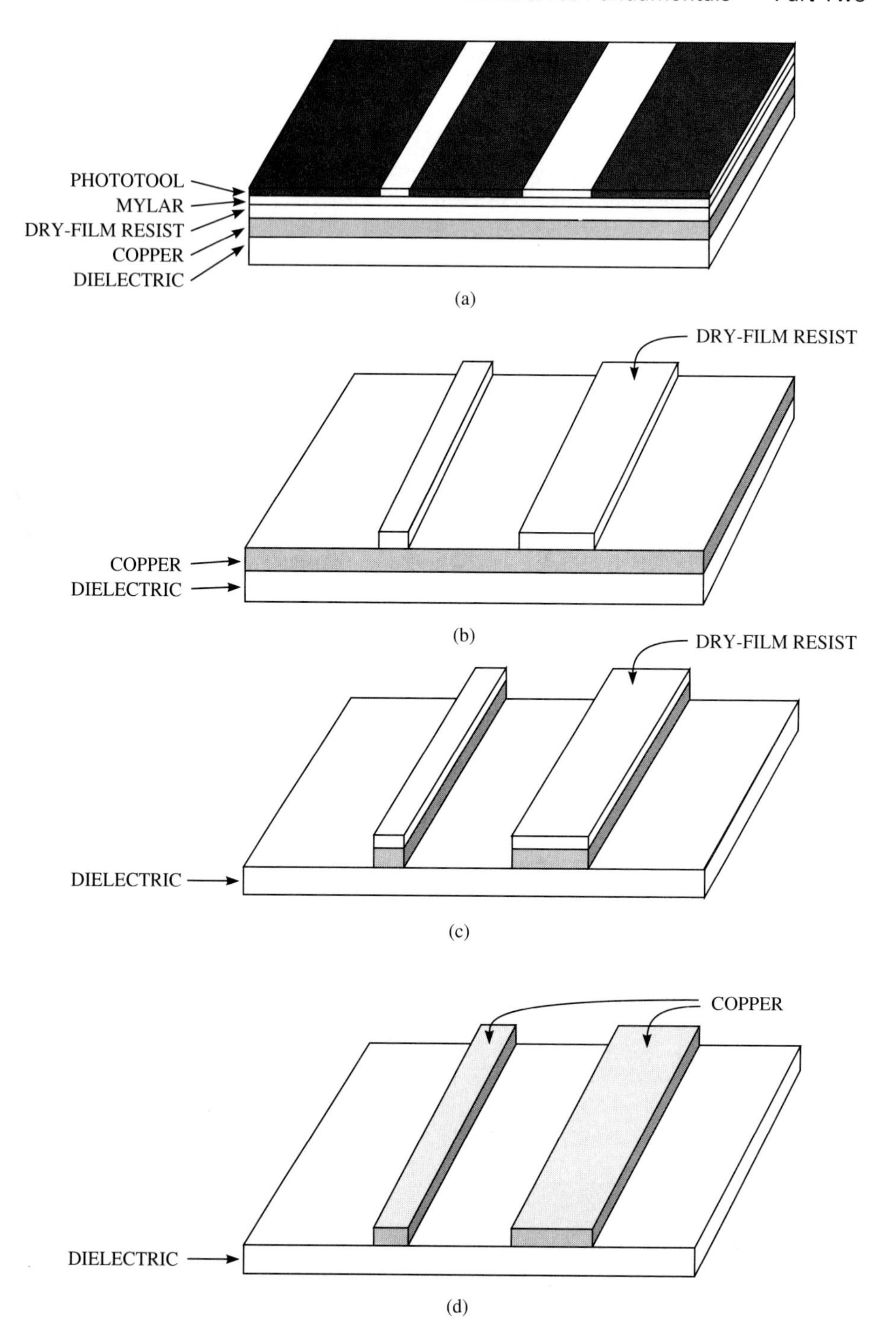

Figure 5-12 Print-and-etch: (a) PWB coated with dry-film resist and in contact with a negative phototool; (b) unexposed resist removed; (c) copper etched away; (d) remaining resist removed.

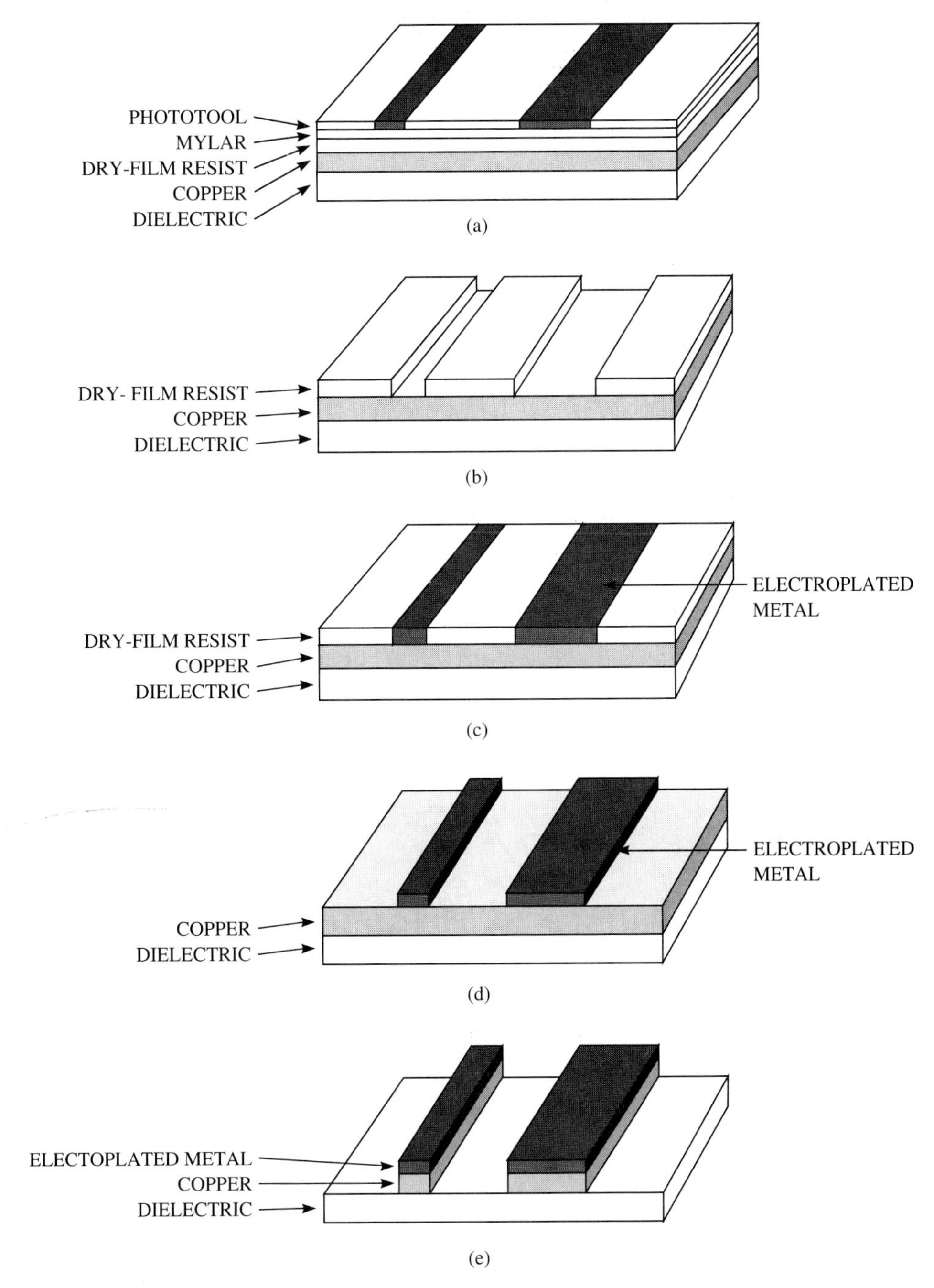

Figure 5-13 Print, plate, and etch process: (a) PWB coated with dry-film resist and in contact with a positive phototool; (b) unexposed resist removed; (c) pattern plate resist metal; (d) dry-film resist removed; (e) copper etched using plated resist metal as a mask.

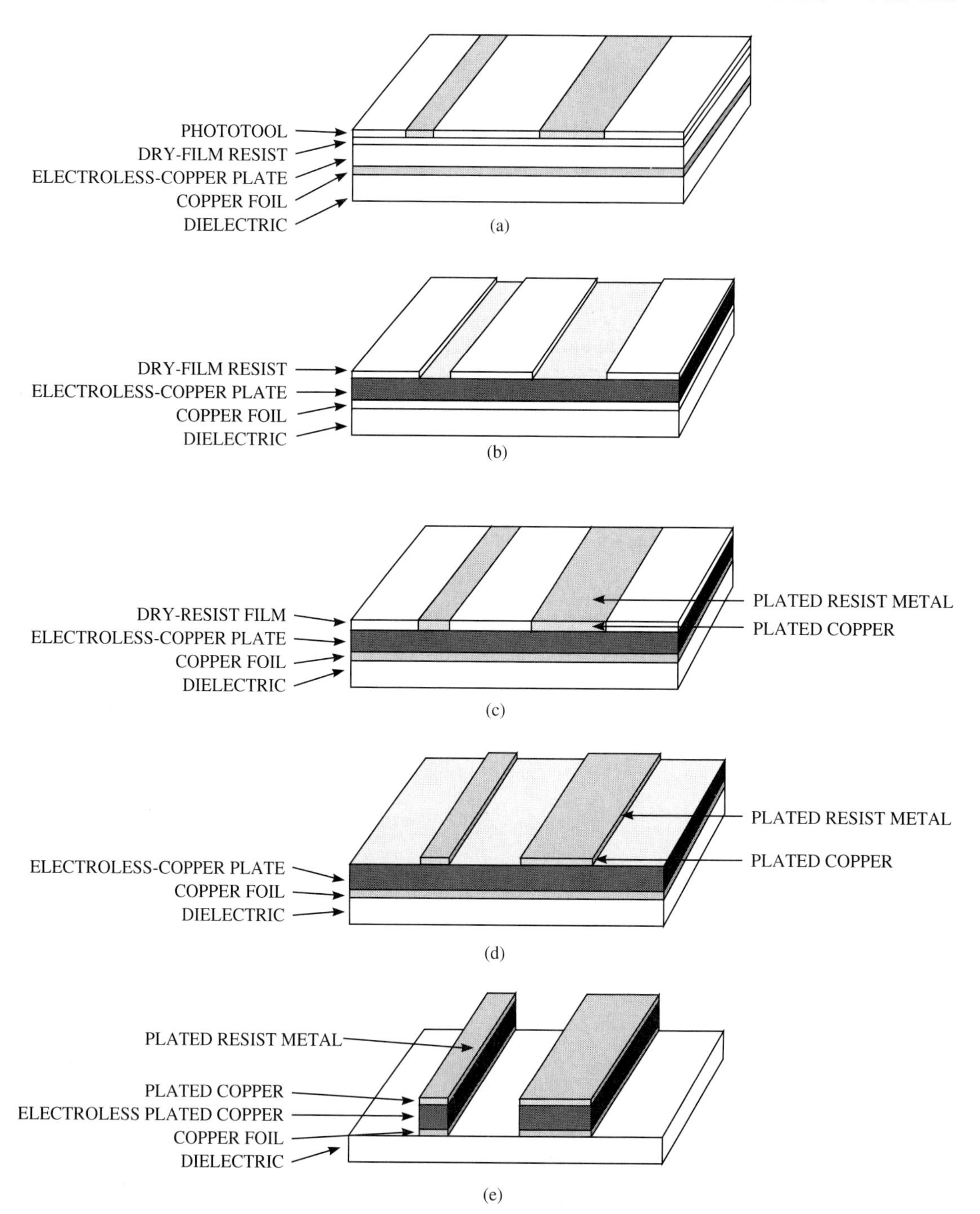

Figure 5-14 Pattern-plating process: (a) PWB coated with dry-film resist and in contact with a positive phototool; (b) unexposed resist removed; (c) pattern plate resist metal; (d) remove dry-film resist; (e) remove unwanted copper.

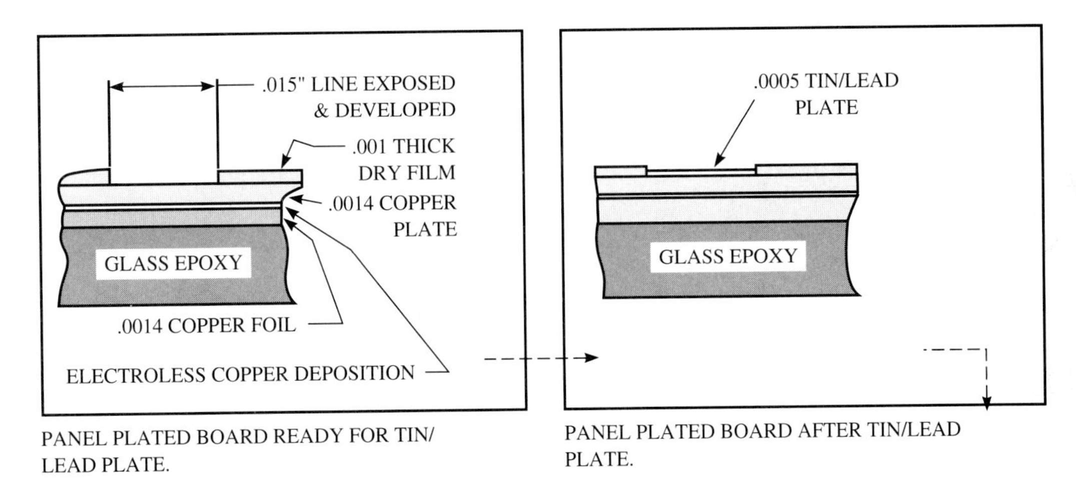

PANEL PLATED BOARD READY FOR TIN/LEAD PLATE.

PANEL PLATED BOARD AFTER TIN/LEAD PLATE.

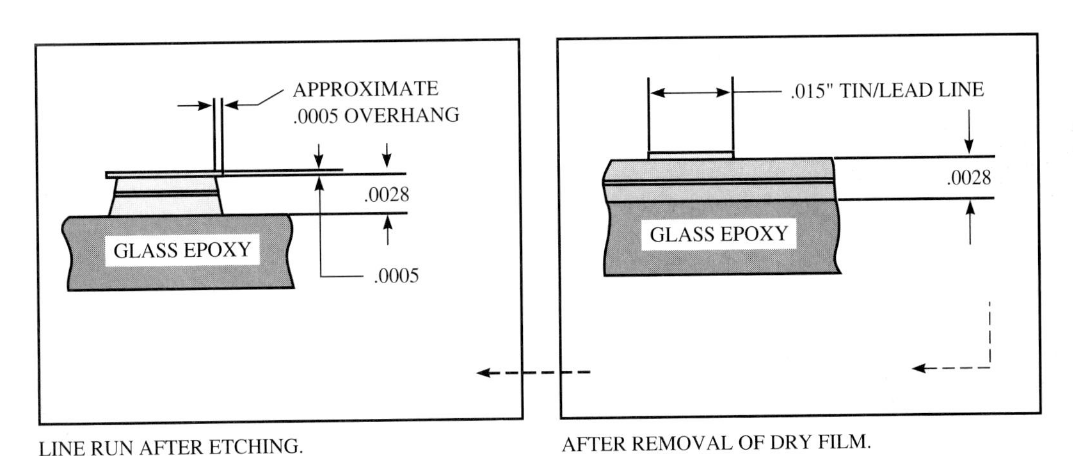

LINE RUN AFTER ETCHING.

AFTER REMOVAL OF DRY FILM.

Figure 5-15 Example of panel plating.

(*Source:* Norman S. Einarson, *Printed Circuit Technology* (Burlington, Mass.: Printed Circuit Technology, 1977), p. 89.)

5.7 ETCHING

Etching is a subtractive process which removes unwanted copper from copper-clad substrates for the purpose of creating a circuit on a PWB. Very fine-lined conducting paths of 0.004 to 0.006 inch (0.1 to 0.15 mm) can be created using this process. An etch-resistant image of the circuit is transferred to a copper-clad PWB substrate, and the board is placed in an etching solution. This solution removes all metal from the surfaces of the board except for the areas defined by the etch-resist. After all unnecessary metal is etched away, the board goes through a process which washes away residual etching solution and then strips the remaining etch-resist from the board.

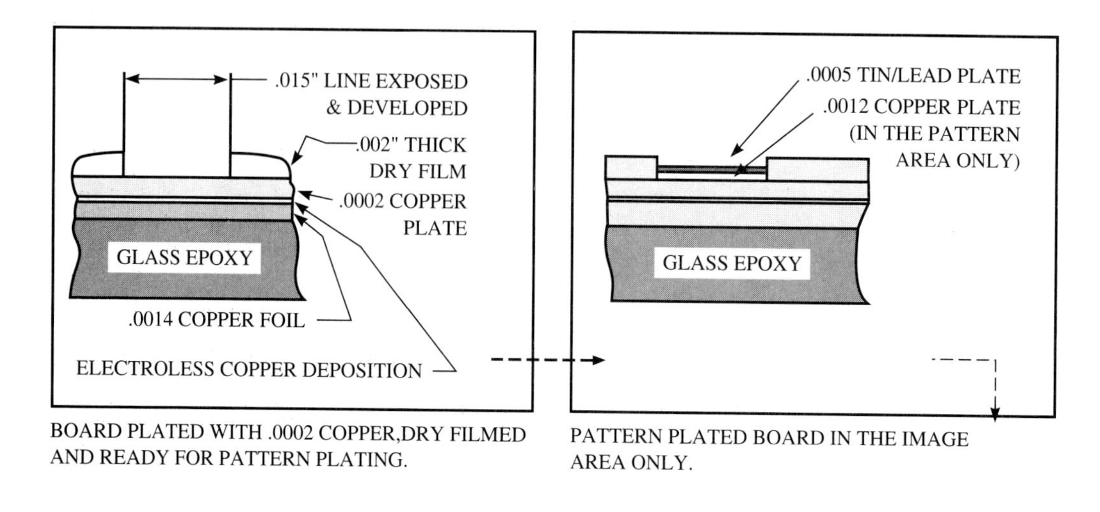

BOARD PLATED WITH .0002 COPPER,DRY FILMED AND READY FOR PATTERN PLATING.

PATTERN PLATED BOARD IN THE IMAGE AREA ONLY.

ETCHED LINE RUN

GLASS EPOXY

TIN/LEAD AND COPPER PLATE

GLASS EPOXY

.0016

LINE RUN AFTER ETCHING.

AFTER REMOVAL OF DRY FILM.

Figure 5-16 Example of pattern plating.

(*Source:* Einarson, *Printed Circuit Technology,* p. 90.)

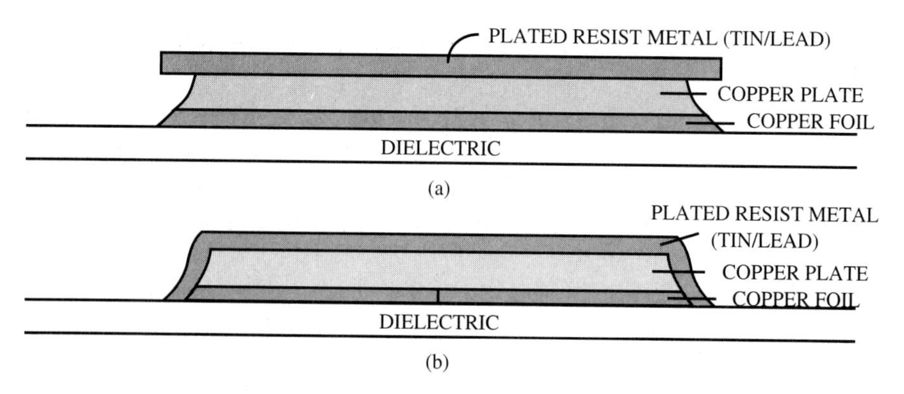

Figure 5-17 Solder slivering and fusing.

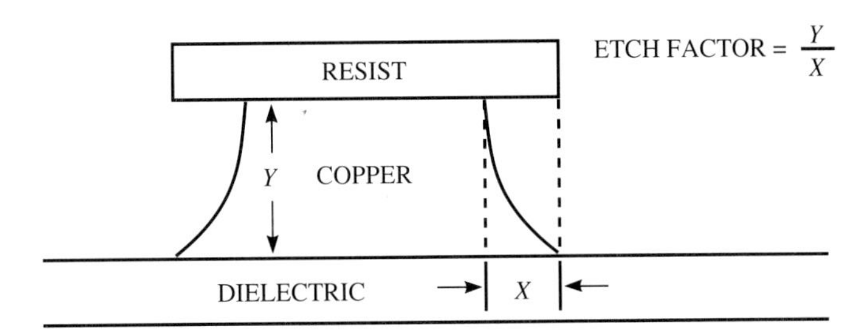

Figure 5-18 Etch factor in PWB etching.

While the board is in the etching solution, etch occurs on all surfaces that are in contact with the solution. This creates an undercutting action, because as the etch moves down through the copper plate, it also moves laterally under the etch-resist, as shown in Fig. 5-18. The ratio of the vertical etch depth to the lateral etch depth is called the **etch factor.** This etch factor can be controlled to some degree through the choice of etch application. Spray etching tends to yield a high etch factor, which is desirable, while immersion etching usually results in a low etch factor.

5.8 BARE-BOARD TESTING

The trend toward highly complex circuits has created the necessity for **bare-board testing.** With some multilayer boards being made up of 10 or more layers having a trace separation of 0.01 inch (0.25 mm) or less, the defect rate may be as high as 30 percent. The cost of finding faults on PWBs increases rapidly at higher levels of the product structure (that is, at progressively later stages in the manufacturing process). The repair cost can increase by a factor of ten or more from one level (for example, repair of a part) to the next level (repair of a PWB). Refer to Sec. 13.4 and Table 13-2.

Faults which occur on single-sided or simple double-sided boards may be easy to correct at any stage of production. For example, a sliver of metal that shorts two circuit paths may be removed with a knife, or a discontinuity in a circuit trace may be bridged with a bit of solder. These same faults occurring on one of the inner layers of a multilayer board would be difficult, if not impossible, to repair.

An isolation test is used to check for short circuits or leakage between networks on a PWB. The network being tested is checked for isolation from all other networks simultaneously with an applied voltage of 100–250 volts. An isolation resistance of less than 100 Mohms will usually indicate leakage current between networks.

Breakdown tests check for current leakage between networks as a function of the amount of time the voltage is applied. A breakdown test is usually applied for 10 ms or more. If the typical operating voltage of a board is low, as is the case with, for example, logic boards, this test may not be applied.

Continuity tests are usually applied as an initial step for the previous two tests. In order for the high voltage tests to be valid, it must first be determined that the test probes are indeed connected to the network that is being tested. This is determined by making continuity measurements.

5.9 SUMMARY

Printed wiring boards, consisting of one or more layers of conductors and insulating material, are the substrates most commonly used to provide mechanical support and electrical connectivity for electronic components comprising an electronic system. Materials used to form the base laminates (prepreg) include paper, epoxy-resin, woven glass cloth, and fiberglass. The layers of metal and insulating material of multilevel boards are laminated using heat and pressure after the interconnection circuitry is defined. Circuits on the various layers of the board are interconnected by drilling or punching holes through the board and depositing metal on the inner surface of the hole. Sizing of boards is accomplished by shearing, sawing, routing, or using blanking dies. Processes used to transfer circuit images to the board are screen printing and photoprinting. Photoprinting can be further divided into wet-film and dry-film processes. Two other heavily used chemical processes are plating and etching. Prior to insertion of components, electrical testing of the board is performed.

KEY TERMS

Bare-board testing
Base materials
Double-sided boards
Drill smear
Drilling
Etch factor
Etchback
Etching
Image transfer
Laminating process
Multilayer boards
Overhang
Photoprinting
Plated-through hole (PTH)
Plating
Plating processes
Prepreg
Pressing
Printed wiring board (PWB)
Punching
Screen printing
Single-sided board
Shearing
Through-hole metallization
Undercut
Vias

EXERCISES

1. The majority of printed wiring boards are of what three types?
2. What are three types of base material used in printed wiring boards?
3. List four ways of sizing a printed wiring board.
4. What are two ways to transfer an image onto a printed wiring board?
5. What are the four ways of plating?
6. What are the major considerations for printed wiring board design specifically concerning solderability?
7. List two methods of photoprinting, and discuss their advantages and disadvantages.
8. What is the best design for a drill bit to produce holes in a printed wiring board?
9. What are two problems that can occur when punching holes into a printed wiring board?

10. Why does a manufacturer use bare-board testing rather than other testing methods later? What are the three types of bare-board tests?

BIBLIOGRAPHY

1. COOMBS, CLYDE F., *Printed Circuits Handbook* (3rd ed.), New York: McGraw-Hill Book Company, 1988.
2. FINK, DONALD G., and DONALD CHRISTIANSEN, eds., *Electronics Engineers' Handbook* (3rd ed.), New York: McGraw-Hill Book Company, 1989.
3. NOBEL, P. J. W., *Printed Circuit Board Assembly*. New York: Halsted Press, 1989.
4. EINARSON, NORMAN S., *Printed Circuit Technology*. Burlington, Mass., 1977.
5. EINARSON, NORMAN S., *PWB Commercial Workmanship Standards*. Burlington, Mass., 1986.

6 Soldering and Solderability

6.1 INTRODUCTION

To be considered reliable, a solder joint must perform two functions: bond a component to a PWB (or other substrate) and provide electrical continuity between the component and substrate. Factors that affect these functions are the connection materials and the process. Connection materials are the base metals (for example, component leads, PWB traces), soldering alloys, and solder fluxes. The process is the method whereby these materials and thermal energy are applied to effect the bond.

Soldering is a metallurgical joining technique involving a molten filler metal which wets the surface of both metals to be joined and, upon solidification, forms the bond. At soldering temperatures, the physical state of the base material is solid, the solder alloy is liquid, and the flux is either liquid or vapor. The metals to be soldered do not become molten, and the bond occurs at the interface of the two metals. Although the base metal does not become molten, some alloying (mixing of metals, forming an intermetallic compound at the interface) may take place if the base metal is soluble in the filler metal.

The bonding is strongly dependent on the **wettability** (**solderability**) of the base metal by the molten alloy. Wettability can be thought of as a measure of affinity between two materials. Consider a drop of molten solder resting on a solderable surface. The forces that must be considered are the cohesive forces (surface tension) in the drop of solder and the adhesive forces that occur between the solder and the base metal.

The cohesive forces tend to pull the molten solder into a sphere in an effort to reduce the surface area of the drop. The adhesive forces tend to spread the liquid over the surface of (or wet) the metal. The difference between these two forces can be noticed at the point where the drop of solder contacts the metal surface. The angle between the surface and the line that is tangent to the curve at this point is called the dihedral angle. Figure 6-1 demonstrates several examples of the dihedral angle. As the wettability, or solderability, of the surface increases, the dihedral angle decreases. Total **wetting** occurs as $\theta \rightarrow 0°$.

The wettability of a base metal is improved if the surface is kept free of dirt, grease, and other contaminants. Surface oxides also reduce wettability, and solder flux is used to react with the oxides and move them away from the area to be soldered. Flux also influences the surface tension equilibrium and encourages solder to spread by reducing the dihedral angle.

Soldering processes have changed a great deal since the earliest applications. During the time from 1920 to the 1940s, all solder interconnections were made using point-to-

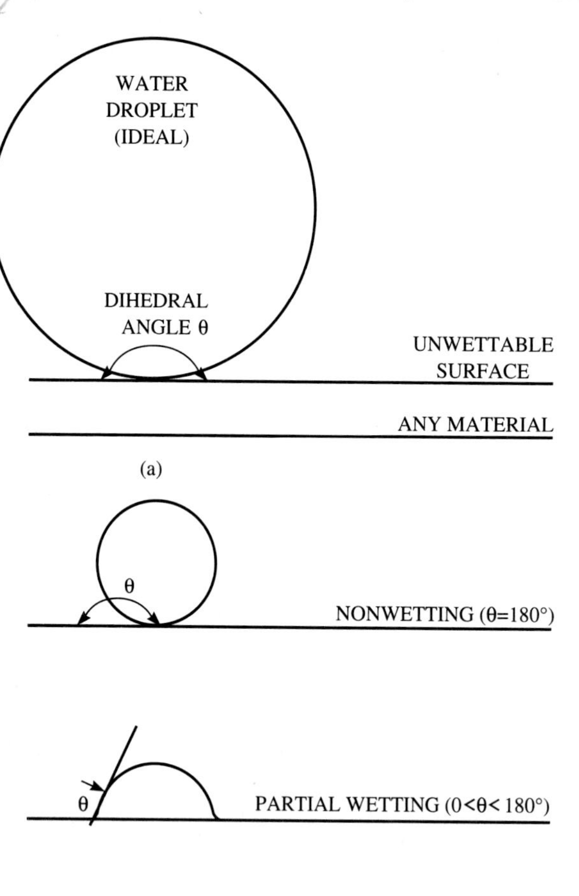

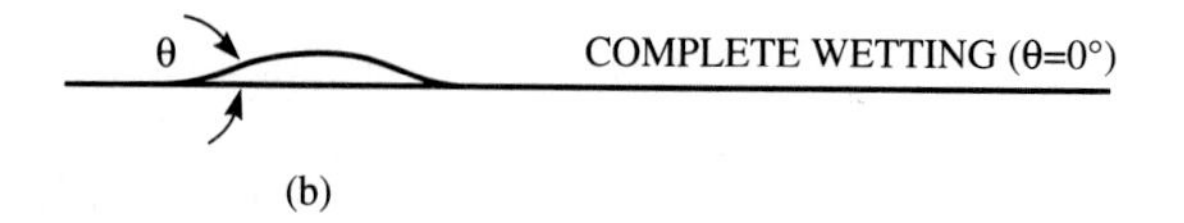

Figure 6-1 a) Complete nonwetting of an ideal surface. b) Relation between dihedral angle and the degree of wetting.

point wiring methods and all soldering was done by hand with soldering irons. After World War II the demand for consumer electronics began to increase, and by the 1950s a mass market had developed. Dip soldering was introduced to help meet this demand for products.

By the late 1950s wave soldering had been developed in England and was soon introduced into the United States. This was necessitated by the increased use of PWBs. Wave soldering became the method of choice through the 1980s and is still widely used.

The 1980s also saw the development of surface-mount devices (SMDs). With the appearance of SMDs, new techniques in making solder joints were needed. These new processes are known as reflow soldering techniques. The reflow soldering methods include vapor-phase soldering (VPS), infrared (IR) soldering, and convection.

Section 6.2 provides an overview of soldering materials. Solderability and PWB soldering processes are described further in Sec. 6.3. Section 6.4 surveys the automatic soldering techniques, and Sec. 6.5 surveys the increasingly important topic of postsolder cleaning operations.

6.2 SOLDERING MATERIALS

Solder is available in a variety of compositions and forms. Fluxes serve several vital functions and are applied either by foam, spray, or wave.

6.2.1 Solder Composition

In the earliest applications, for point-to-point hand soldering, the **solder composition** was 50 percent tin (Sn) and 50 percent lead (Pb). This is the same composition that was used during the time of the Roman Empire for plumbing joints. The introduction of machine soldering brought about a need for solder with a faster solidifying time. The compositions of 63Sn-37Pb and 60Sn-40Pb became standards.

The appearance of SMDs required more changes in composition due to a wider range of requirements. In SMT, the solder joint must support the entire weight of the component and must absorb varying stresses caused by shock, vibration, and the differential coefficients of thermal expansion (CTE) during heating/cooling cycles. Other factors that must be considered are adhesive qualities, gas dispersion, and paste requirements, which will be discussed later.

The most commonly used compositions are covered by federal specification QQ-S-571 and ASTM specification B-32 given as Tables 6-1 and 6-2, respectively. Note that most solders in these tables are still primarily binary mixtures of tin and lead. Additive metals can be introduced into tin/lead solder alloys to tailor them for certain processing requirements or to enhance desirable properties.

6.2.2 Available Forms

Solder comes in a variety of physical forms. Some of these include:

- *Solder bars:* Rectangular shaped, weigh up to 1 kg (2.2 lb), used to replenish baths in wave-soldering and solder-leveling machines.

TABLE 6-1 ALLOY COMPOSITIONS FROM FEDERAL SPECIFICATION QQ-S-571

Composition	Tin, %	Lead, %	Antimony, %	Bismuth, max %	Silver, max %	Copper, max %	Iron, max %	Zinc, max%	Aluminum, max %	Arsenic, max %	Cadmium, max %	Total of all others, max %	Approximate melting range Solidus °C	Solidus °F	Liquidus °C	Liquidus °F
Sn96	rem	0.10, max	. . .	. . .	3.6–4.4	0.20	. . .	0.005	. . .	0.05	0.005	. . .	221	430	221	430
Sn70	69.5–71.5	rem	0.20–0.50	0.25	0.015	0.08	0.02	0.005	0.005	0.03	0.001	0.08	183	361	193	379
Sn63	62.5–63.5	rem	0.20–0.50	0.25	0.015	0.08	0.02	0.005	0.005	0.03	0.001	0.08	183	361	183	361
Sn62	61.5–62.5	rem	0.20–0.50	0.25	1.75–2.25	0.08	0.02	0.005	0.005	0.03	0.001	0.08	179	354	179	354
Sn60	59.5–61.5	rem	0.20–0.50	0.25	0.015	0.08	0.02	0.005	0.005	0.03	0.001	0.08	183	361	191	376
Sn50	49.5–51.5	rem	0.20–0.50	0.25	0.015	0.08	0.02	0.005	0.005	0.025	0.001	0.08	183	361	216	421
Sn40	39.5–41.5	rem	0.20–0.50	0.25	0.015	0.08	0.02	0.005	0.005	0.02	0.001	0.08	183	361	238	460
Sn35	34.5–36.5	rem	1.6–2.0	0.25	0.015	0.08	0.02	0.005	0.005	0.02	0.001	0.08	185	365	243	469
Sn30	29.5–31.5	rem	1.4–1.8	0.25	0.015	0.08	0.02	0.005	0.005	0.02	0.001	0.08	185	365	250	482
Sn20	19.5–21.5	rem	0.80–1.2	0.25	0.015	0.08	0.02	0.005	0.005	0.02	0.001	0.08	184	363	270	518
Sn10	9.0–11.0	rem	0.20, max	0.03	1.7–2.4	0.08	. . .	0.005	0.005	0.02	0.001	0.10	268	514	290	554
Sn5	4.5–5.5	rem	0.50, max	0.25	0.015	0.08	0.02	0.005	0.005	0.02	0.001	0.08	308	586	312	594
Sb5	94.0, min	0.20, max	4.0–6.0	. . .	0.015	0.08	0.08	0.03	0.03	0.05	0.03	0.03	235	455	240	464
Pb80	rem	78.5–80.5	0.20–0.50	0.25	0.015	0.08	0.02	0.005	0.005	0.02	0.001	0.08	183	361	277	531
Pb70	rem	68.5–70.5	0.20–0.50	0.25	0.015	0.08	0.02	0.005	0.005	0.02	0.001	0.08	183	361	254	489
Pb65	rem	63.5–65.5	0.20–0.50	0.25	0.015	0.08	0.02	0.005	0.005	0.02	0.001	0.08	183	361	246	475
Ag1.5	0.75–1.25	rem	0.40, max	0.25	1.3–1.7	0.30	0.02	0.005	0.005	0.02	0.001	0.08	309	588	309	588
Ag2.5	0.25, max	rem	0.40, max	0.25	2.3–2.7	0.30	0.02	0.005	0.005	0.02	0.001	0.03	304	579	304	579
Ag5.5	0.25, max	rem	0.40, max	0.25	5.0–6.0	0.30	0.002	0.005	0.005	0.02	0.001	0.03	304	579	380	716

Note: rem, remainder

Source: Electronic Materials Handbook, Vol. 1, p. 633.

TABLE 6-2 SOLDER COMPOSITIONS PER ASTM B-32

Alloy grade	Composition, % (a)(b) Sn 1	Pb 2	Sb 3	Ag 4	Cu 5	Cd 6	Al 7	Bi 8	As 9	Fe 10	Zn 11	Approximate melting range Solidus °C	Solidus °F	Liquidus °C	Liquidus °F
Sn96	rem	0.10	0.12, max	3.4–3.8	0.08	0.005	0.005	0.15	0.01, max	0.02	0.005	221	430	221	430
Sn95	rem	0.10	0.12	4.4–4.8	0.08	0.005	0.005	0.15	0.01	0.02	0.005	221	430	245	473
Sn94	rem	0.10	0.12	5.4–5.8	0.08	0.005	0.005	0.15	0.01	0.02	0.005	221	430	280	536
Sn70	69.5–71.5	rem	0.50	0.015	0.08	0.001	0.005	0.25	0.03	0.02	0.005	183	361	193	377
Sn63	62.5–63.5	rem	0.50	0.015	0.08	0.001	0.005	0.25	0.03	0.02	0.005	183	361	183	361
Sn62	61.5–62.5	rem	0.50	1.75–2.25	0.08	0.001	0.005	0.25	0.03	0.02	0.005	179	354	189	372
Sn60	59.5–61.5	rem	0.50	0.015	0.08	0.001	0.005	0.25	0.03	0.02	0.005	183	361	190	374
Sn50	49.5–51.5	rem	0.50	0.015	0.08	0.001	0.005	0.25	0.025	0.02	0.005	183	361	216	421
Sn45	44.5–46.5	rem	0.50	0.015	0.08	0.001	0.005	0.25	0.025	0.02	0.005	183	361	227	441
Sn40A	39.5–41.5	rem	0.50	0.015	0.08	0.001	0.005	0.25	0.02	0.02	0.005	183	361	238	460
Sn40B	39.5–41.5	rem	1.8–2.4	0.015	0.08	0.001	0.005	0.25	0.02	0.02	0.005	185	365	231	448
Sn35A	34.5–36.5	rem	0.50	0.015	0.08	0.001	0.005	0.25	0.02	0.02	0.005	183	361	247	447
Sn35B	34.5–36.5	rem	1.6–2.0	0.015	0.08	0.001	0.005	0.25	0.02	0.02	0.005	185	365	243	470
Sn30A	29.5–31.5	rem	0.50	0.015	0.08	0.001	0.005	0.25	0.02	0.02	0.005	183	361	255	491
Sn30B	29.5–31.5	rem	1.4–1.8	0.015	0.08	0.001	0.005	0.25	0.02	0.02	0.005	185	365	250	482
Sn25A	24.5–26.5	rem	0.50	0.015	0.08	0.001	0.005	0.25	0.02	0.02	0.005	183	361	266	511
Sn25B	24.5–26.5	rem	1.1–1.5	0.015	0.08	0.001	0.005	0.25	0.02	0.02	0.005	185	365	263	504
Sn20A	19.5–21.5	rem	0.50	0.015	0.08	0.001	0.005	0.25	0.02	0.02	0.005	183	361	277	531
Sn20B	19.5–21.5	rem	0.8–1.2	0.015	0.08	0.001	0.005	0.25	0.02	0.02	0.005	184	363	270	517
Sn15	14.5–16.5	rem	0.50	0.015	0.08	0.001	0.005	0.25	0.02	0.02	0.005	225	437	290	554
Sn10A	9.0–11.0	rem	0.50	0.015	0.08	0.001	0.005	0.25	0.02	0.02	0.005	268	514	302	576
Sn10B	9.0–11.0	rem	0.20	1.7–2.4	0.08	0.001	0.005	0.03	0.02	0.02	0.005	268	514	299	570
Sn5	4.5–5.5	rem	0.50	0.015	0.08	0.001	0.005	0.25	0.02	0.02	0.005	308	586	312	594
Sn2	1.5–2.5	rem	0.50	0.015	0.08	0.001	0.005	0.25	0.02	0.02	0.005	316	601	322	611
Sb5	94.0, min	0.20	4.5–5.5	0.015	0.08	0.03	0.005	0.15	0.05	0.04	0.005	233	450	240	464
Agl.5	0.75–1.25	rem	0.40	1.3–1.7	0.30	0.001	0.005	0.25	0.02	0.02	0.005	309	588	309	588
Ag2.5	0.25	rem	0.40	2.3–2.7	0.30	0.001	0.005	0.25	0.02	0.02	0.005	304	580	304	580
Ag5.5	0.25	rem	0.40	5.0–6.0	0.30	0.001	0.005	0.25	0.02	0.02	0.005	304	580	380	716

(a) Limits are maximum percent unless shown as a range or stated otherwise. (b) For purposes of determining conformance to these limits, an observed value or calculated value obtained from analysis shall be rounded to the nearest unit in the last right-hand place of figures used in expressing the specified limit, in accordance with the rounding method of Recommended Practice E 29.

Source: Electronic Materials Handbook, Vol. 1, p. 634.

- *Solder ingots:* Similar to solder bars but weighing 0.9 to 4.5 kg (2 to 10 lb).
- *Solder pigs:* Similar to solder bars but weighing 4.5 to 45 kg (10 to 100 lb).
- *Solder wires:* Wire wrapped into spools and used in hand soldering or for some automatic feeding in machine-soldering operations.
- *Flux-core wire:* Solder wire containing internal channels that are filled with solder flux.
- *Other forms:* Sheet, foil, ribbon, wire segments, castings, balls, pastes, and powders.

6.2.3 Soldering Fluxes

The strength and dependability of a solder joint is reduced by the presence of impurities. These impurities may be chemically bound to the base metal—such as layers of oxide, sulfide, and carbonate—or they may be due to absorptive forces which attract residue such as water, grease, gases, and dust. The function of **solder flux** is to remove these contaminants.

The chemical function of solder flux is to react with surface tarnish films, such as oxides and sulfides, by dissolving some of the substrate molecules, thus initiating the formation of intermetallic compounds. In essence, the flux chemically etches the surface. The reaction products are then displaced by the molten solder alloy. During the soldering process, the flux also acts as a chemical blanket to prevent reoxidation during the heating period.

The thermal function of solder flux is to assist in the transfer of heat from molten solder to the joint area so that base metals reach sufficiently high temperature to be wetted by the solder. The physical function of solder flux is to reduce the surface tension between the solder and the base metal, enabling the solder to flow over and metallurgically wet the solderable surface.

Flux activity is the degree of efficiency with which a given flux promotes the wetting of surfaces by molten solder. Flux corrosivity is defined as a chemical attack by the flux or its residues on the base metals. Generally, the greater the activity of the flux, the more corrosive it and the residues are likely to be. Flux cleanability refers to the ease with which flux residues can be removed.

Active fluxes may contain carboxylic acids or halides (for example, amino hydrochlorides). Active fluxes also promote wettability more than nonactivated or mildly activated fluxes because they tend to etch the substrate more aggressively. Using active fluxes allows for a broader range of substrate cleanliness during the manufacturing process, because the flux will etch away most oxides and contaminants. More stringent postsolder cleaning is required to prevent the corrosive effects of active flux residue from harming components, substrate, and/or solder joints.

Mildly active or nonactivated fluxes do not etch the surface and thus require the substrate to have more inherent solderability. The advantage of these fluxes is that postsolder cleaning is less critical and, in some cases, totally unnecessary. This is a very important feature due to the increasing costs of PWB cleaning. Cleaning usually requires the use of highly restricted chlorinated and fluorinated hydrocarbons, which are subject to stringent environmental regulations. These no-clean fluxes are less likely to be toxic, thereby minimizing environmental and disposal problems.

The nonactivated fluxes find widespread use with SMDs. The close proximity of SMD components to the substrate reduces the probability of complete removal of flux residues. If left in place, these residues are likely to lead to the degradation of surface insulation resistance (SIR) of the substrate, especially under adverse climatic conditions.

When applying flux, the object is to apply a thin, even coating of flux so that the board surface to be soldered is completely covered. There are three primary methods of flux application: foam, spray, and wave.

Foam fluxing is one of the simplest and most common methods used. A foam wave is generated by pumping compressed air through a porous cylinder located under the surface of the liquid flux. The aeration is contained in a chimneylike structure to concentrate the mass of air bubbles as they rise. The height of the foam wave is usually about 10 mm above the lip of the chimney. The PWB passes over the fluxer on a conveyor and flux is deposited evenly across the underside, as shown in Fig. 6-2.

Foam fluxing is very effective in its ability to penetrate plated through-holes, as the foam bubbles burst against the underside of the PWB. It also works well with a high

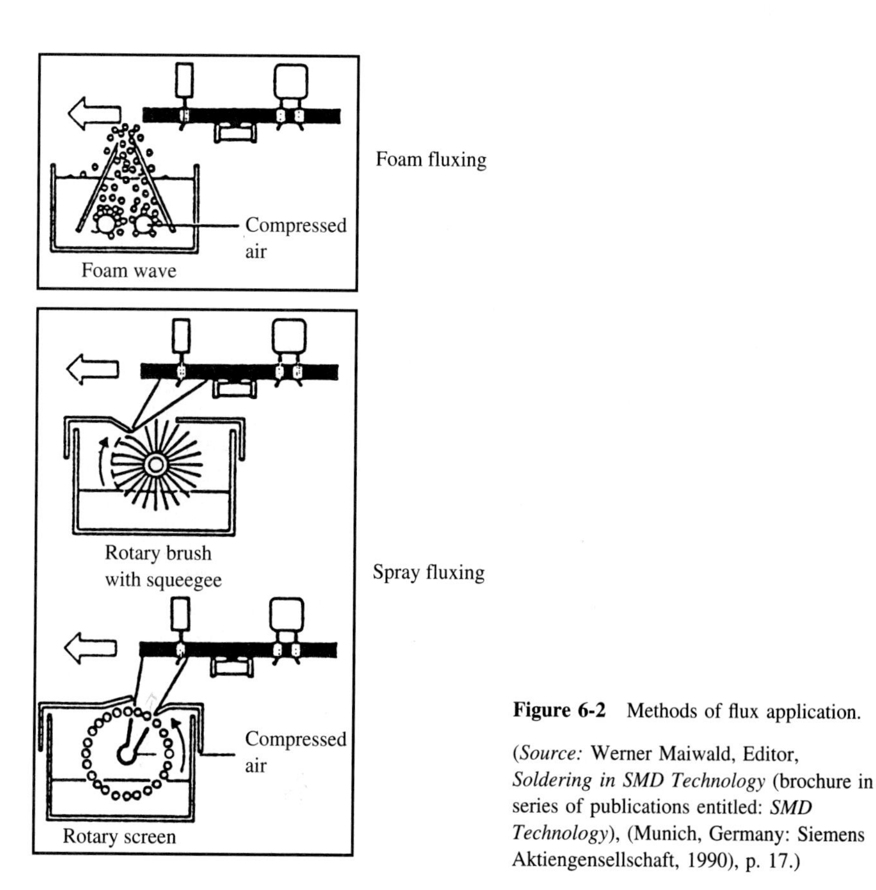

Figure 6-2 Methods of flux application.

(*Source:* Werner Maiwald, Editor, *Soldering in SMD Technology* (brochure in series of publications entitled: *SMD Technology*), (Munich, Germany: Siemens Aktiengensellschaft, 1990), p. 17.)

population density of SMDs mounted on the underside of **mixed-technology** (MT) boards utilizing both through-hole and surface-mount components.

With spray fluxing, very thin, even coatings of flux are attainable. However, this is a somewhat messy technique, and precautions must be taken to control overspray. There are two types of spray fluxing, as seen in Fig 6-2, as follows:

- *Rotary Brush with Squeegee:* The brush is rotated in the flux container, and bristles take up the flux from the reservoir. The bristles are wiped against the squeegee, and flux is catapulted onto the underside of the PWB.
- *Rotary Screen:* Flux is taken up from the reservoir by a rotating screen. A jet of compressed air is ejected through a nozzle that is located in the center of the rotary screen. This propels the flux droplets onto the underside of the PWB.

Wave fluxing is similar to foam fluxing except that it does not employ aeration. A pump forces flux up through a chimney-type chute in contact with the bottom of the board. Wave height is controlled by pump speed. An air knife squeezes excess flux from the board surface at the exit of the flux tank. Its advantage is complete flooding of components and holes with flux, although care must be exercised to minimize deposition on the top surface of the board. However, fluxing tends to be messy.

6.3 SOLDERING AND PRINTED WIRING BOARDS

Effective soldering of PWBs depends upon solder resist, solderability properties, wetting, and choice of solder alloy.

6.3.1 Solder Resist

After a PWB has undergone bare-board testing and before components are assembled on its surfaces, a solder mask or **solder resist** must be applied to the board, as shown in Fig. 6-3. The solder resist is a coating material which prevents solder from sticking to any part of the board surface where solder is unwanted. Other desirable characteristics have also been designed into the solder resist material to

- Reduce solder **bridging** (formation of unwanted conductive paths) and electrical shorts
- Reduce solder pot contamination
- Reduce the volume of solder pickup (saving on cost and weight)
- Protect PWB circuitry from handling damage
- Provide an insulation barrier between electrical components and conductor lines if components are mounted on top of conductor lines

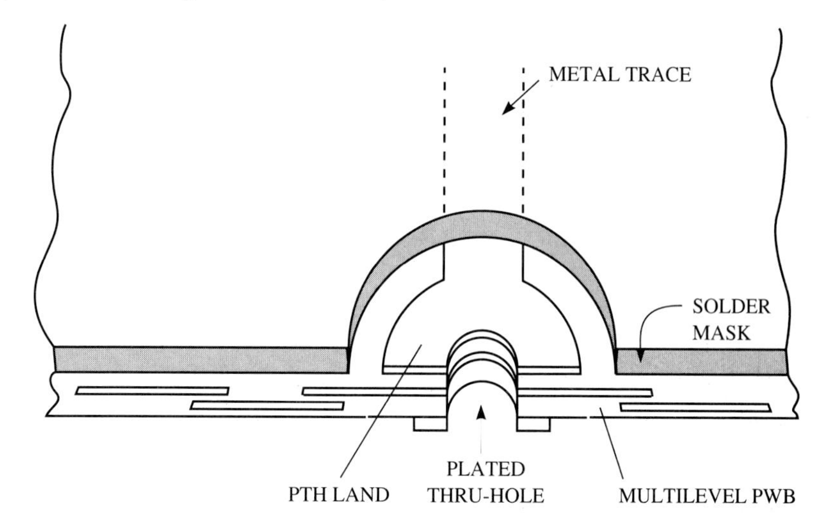

Figure 6-3 Solder resist on a printed wiring board. The mask should not contact the plated-through hole land or pad, the trace should be covered, the laminate area should be completely covered, and the adjacent conductors should not be exposed.

Maximum performance of the solder mask can only be achieved if the surface is completely clean, so that there is total contact between the mask and the PWB. To ensure this, the PWB surface may be mechanically scrubbed and/or degreased with solvents such as 1,1,1-trichloroethane, and then oven-dried to remove surface moisture. Application of the solder mask should immediately follow the drying process to minimize absorption of moisture.

The same three methods that are available for the application of the image on the PWB are used for the application of the solder mask. They are as follows:

- *Screen Printing:* Applied in the same way as for the circuit image transfer.
- *Liquid Photoresist:* May be applied by roller application, curtain coating, or by using a printing screen with no stencil (image) on the screen. The actual mask will then be defined with a phototool the same as is used to apply the circuit image on a PWB.
- *Dry-Film:* The dry film cannot be applied with rollers the same as it is for image transfer on a PWB, due to the three-dimensional nature of the PTH and conducting paths. The rollers can only be used effectively on a smooth, flat surface. A vacuum laminator is used to apply the dry-film solder mask. The film is held away from the PWB surface, air is removed from the surface, and atmospheric pressure forces the film onto the PWB. This method ensures total contact between the mask and the PWB. A phototool is used to define the mask, the PWB is exposed to UV light, and the unexposed solder resist areas that are defined by the phototool are washed away during the development step.

6.3.2 Solderability

A board must be designed with solderability in mind. Major considerations are

- Wire-to-hole ratio
- Size and shape of component pads (lands)
- Direction of parallel trace runs with respect to the direction of wave flow soldering
- Population density of solder joints

The rule of thumb on wire-to-hole ratio is that the hole diameter should be no less than the lead diameter plus 0.004 inch (10 mm). The maximum hole diameter should be no more than 2.5 times the lead diameter. For multilayer boards, the hole should be less than 2.5 times the diameter to encourage capillary action to the solder during the soldering process. For size and shape of the land area, the rule of thumb is to use circular land shapes that are not more than three times the diameter of the hole in the board. Larger land areas will cause excessive amounts of solder to be picked up, causing bridging and **webbing** (excess solder that collects during peel-back of the solder wave from the PWB).

When boards have large groups of conducting paths that run parallel and are close together, the direction of run should be parallel to the direction of flow in the solder wave to help prevent bridging and webbing.

A large number of solder joints in a small area may suffer from the effects of excessive heat buildup. It may also contribute to bridging, webbing, and **icicling** (excess solder left hanging icicle-like after wave soldering).

6.3.3 Wetting

Soldering is a metallurgical joining technique. The base metal to which solder is applied does not become molten, so there is no chemical reaction that covalently or ionically bonds the solder to the metal surface. In order for solder to form a good bond between two metals, it must wet the surface of both metals that are to be joined.

The property of wettability or solderability can be visualized by using a drop of water in contact with a surface, as discussed in Sec. 6.1 and shown in Fig. 6-1(a). When the drop of water does not have affinity for the surface it rests upon, the surface tension of the water tries to pull the drop up into a perfect sphere so that it is in contact with the surface, ideally, at only one point.

If the drop of water does have affinity for the surface, it tries to spread out and come into intimate contact with it. The degree of wettability of the surface, then, will be related to the degree to which the drop tries to spread out. The angle between the drop and the surface at the point of contact is called the dihedral angle. Refer to the examples of different wetting states and their dihedral angles discussed in Sec. 6.1 and shown in Fig. 6-1(b).

Wetting of a surface is improved if the metal is free of grease and dirt and if there is no oxide layer on the surface of the metal. The first step then, to ensure wettability, is

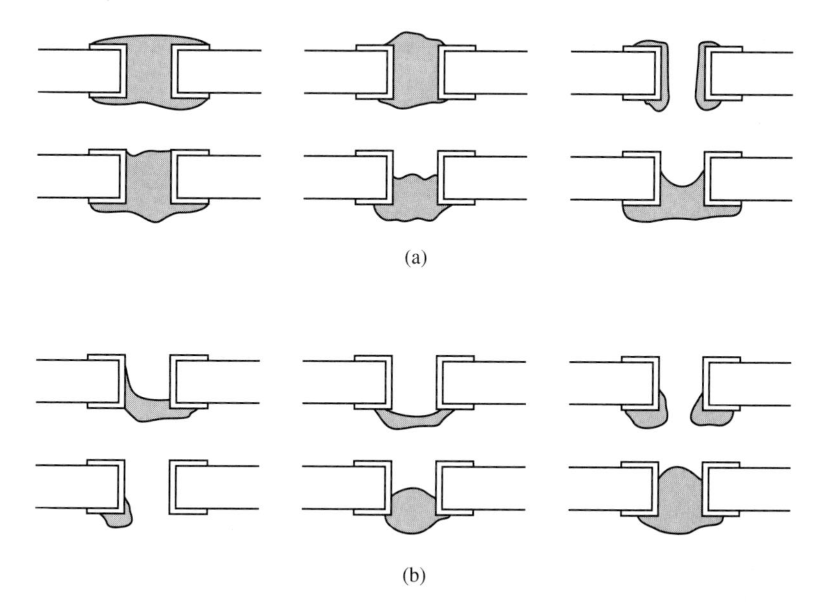

Figure 6-4 Solder wetting of plated-through holes: (a) acceptable; (b) unacceptable because solder has not wetted hole wall surfaces.

to make sure the surface is clean. Examples of PTHs with acceptable and unacceptable wetting conditions can be seen in Fig. 6-4.

Fluxes are used to clean the surface as noted previously, and rosin-based fluxes are the most common type used in the electronics industry. They may be applied by dipping, spraying, or roller coating and will provide solderability protection for six weeks to four months.

6.3.4 Solder Alloys

A desirable characteristic of solder is that it have a single melting point rather than a range of temperatures over which it melts. This is accomplished by developing **solder alloys** with a eutectic composition. A eutectic composition is that alloy composition with minimal melting point, called the eutectic temperature.

A eutectic solder alloy is composed of 62 percent tin and 38 percent lead. The melting point of this alloy is approximately 354°F. There are solders made of other alloys that cover a range of 200 to 600°F, but the 63/37 composition is the one most often used for automated soldering processes (see Sec. 6.2.1).

6.4 AUTOMATIC SOLDERING TECHNIQUES

The competitive pressures for high-quality, low-cost electronic products necessitate automated soldering processes. There are some instances where it is necessary to install and solder components of a PWB by hand, but the majority of solder joints are formed via an automatic soldering operation. There are two major divisions of soldering techniques, bath or wave soldering and reflow soldering, and each of these has several variations.

The techniques used in bath or wave soldering (Fig. 6-5) are known as the conventional wave, dual-wave, drag, and dip techniques. The conventional and dual-wave soldering techniques are the most commonly used methods in bath soldering and will be the only types discussed in Sec. 6.4.1.

Condensation and radiation and/or convection are methods used to perform reflow soldering. Both of these techniques will be discussed in Sec. 6.4.2.

6.4.1 Wave Soldering

In the **wave soldering** method, the solder alloy is kept in the molten state in a heated vat called a solder pot. The surface of the molten solder is maintained as a standing wave by a pumping action and nozzles inside the solder pot.

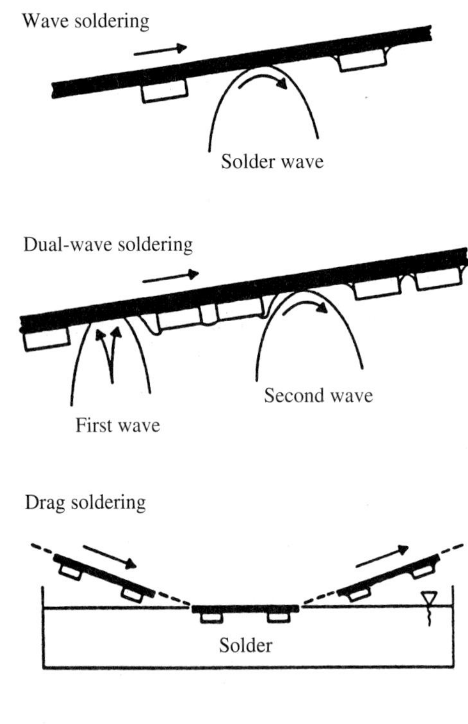

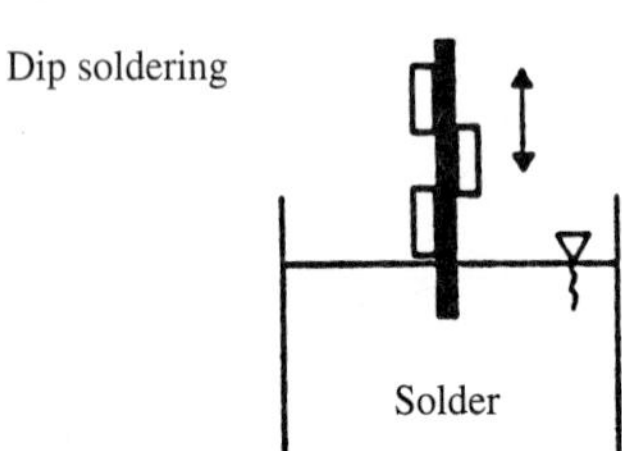

Figure 6-5 Illustration of bath soldering methods.

(*Source:* Maiwald, *Soldering in SMD Technology,* p. 16.)

Due to air coming in contact with the molten solder, oxides of tin and lead form a layer of impurities on the surface of the solder pot. This layer is called "dross" and is harmful to the soldering process and to the definition of the solder wave. It should be removed from the pot at least once per shift, and new solder should be added to make up for the dross that was removed.

The formation of dross can be minimized by protecting the surface of molten solder from the air. A blanket of rosin or oil is an effective method for preventing oxidation. However, when solder blankets are used they must be changed regularly. Oil blankets should be changed every 8 to 16 hours, and rosin-based blankets should be changed after 4 hours of use.

Proper formation of the solder wave is an important aspect in the process of obtaining smooth, shiny solder joints that are free of icicles and webbing. A cross-section view of the wave-forming components of a solder pot is shown in Fig. 6-6. The wave must be even and smooth so that it contacts the entire surface of the board as the board passes over the wave. An uneven wave could cause molten solder to flow over the board and ruin the assembly. Dross should be minimized, because it contributes to an uneven wave form by causing excessive wear in the pumping system.

The relationship between the speed of board travel and width of wave crest is also of importance. Due to the high temperature of the molten solder, the board must travel across the wave fast enough to prevent damage to the assembly, yet slow enough to allow wicking of the solder up and around component leads. Also, to minimize icicling, the relative velocity of the board to the solder must be zero when it leaves the wave.

Bidirectional solder wave flow, as shown in Fig. 6-7, is an effective way to help minimize icicling. Since there is forward and backward flow of solder in the wave, there must be an area of zero velocity on the surface of molten solder. The velocity profile of the flowing solder near the zero-velocity area is critical to the formation of icicle-free joints.

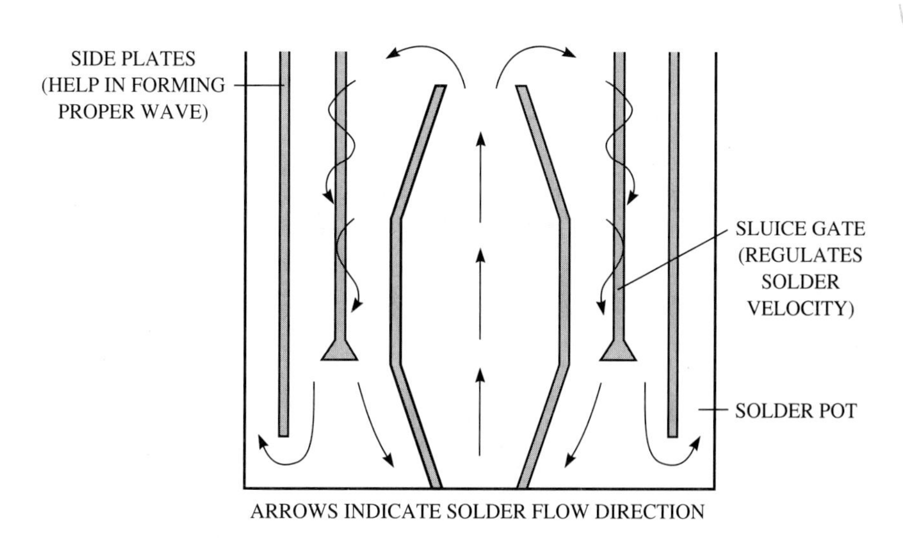

Figure 6-6 Solder nozzle used to form solder wave.

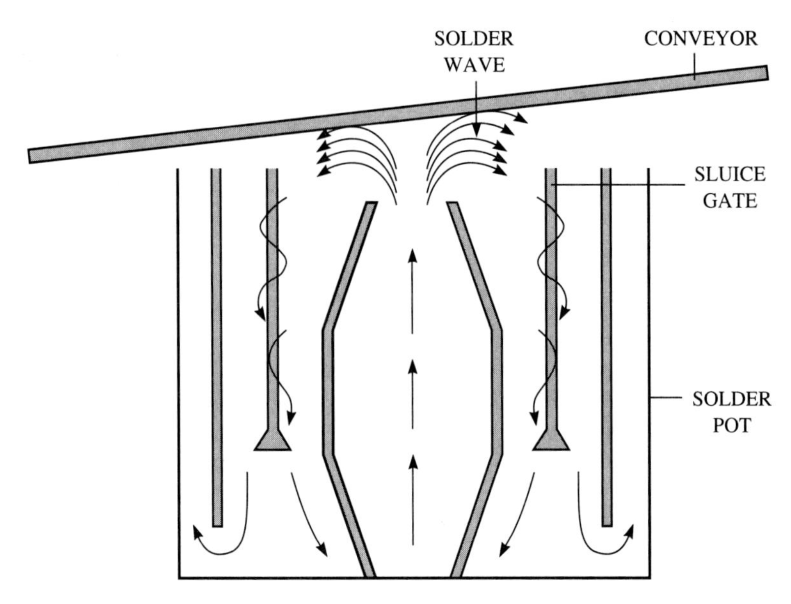

Figure 6-7 Solder nozzle and its relationship to PWB.

To understand the relationship between the velocity profile and icicling, we have to recall the discussion on surface tension. Surface tension causes a drop of liquid such as water or solder to try to pull itself into the shape of a sphere when it is in contact with a surface that it is incapable of wetting. In Fig. 6-8, we see an assembled PWB passing over, and contacting, a bidirectional wave of solder.

Figure 6-8 shows the tendency of the board to pull a **web** of solder from the wave as it passes over. The size of this web is influenced by the surface tension of the solder, the velocity profile of the wave at the point of contact with the web, and the weight of the molten solder in the web. As the size of the web increases, it becomes more difficult

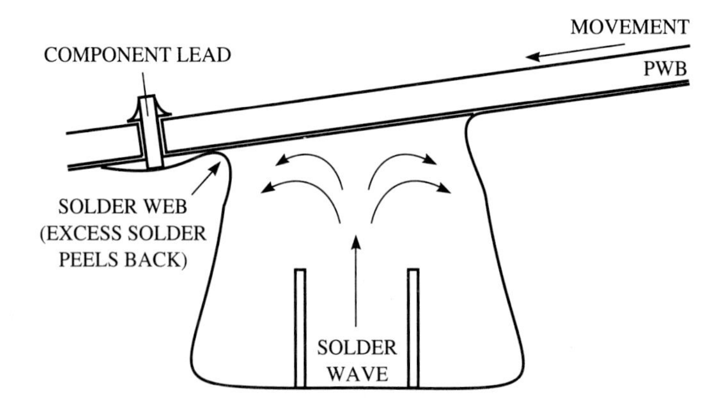

Figure 6-8 PWB passing over a solder wave.

for the surface tension forces to pull the solder back into the wave. A critical size will be reached where surface tension forces will cause separation of the web from the wave. When this occurs, the excess solder that has not been pulled back into the wave will form an icicle. To minimize icicle formation, web formation must be minimized. This can be done by altering the surface tension of the solder or the velocity profile of the wave at the point where the web is formed.

Surface tension of the solder can be altered by increasing the temperature of the molten solder or by injecting oil into the solder wave. Increasing temperature has very minimal effect on the surface tension and can cause thermal damage to components.

The disadvantages of injecting oil into the solder wave are that it makes boards more difficult to clean after the soldering process, it may adversely affect mechanical and electrical properties of the joint, and it could be a source of corrosive residues in the joints. Methods of altering the velocity profile of the wave/web contact point are to

- Incline the board conveyer 4 to 9 degrees so that the solder web peels back more quickly
- Form a much wider wave, as in Fig. 6-9, so that the relative velocity of the board to the wave is near zero as the board leaves the wave

The wave soldering line (for dual and single waves) is usually comprised of the following stations:

- *Loading station:* PWBs mounted onto the conveyor.
- *Fluxing station:* Flux applied to underside of PWB.
- *Preheating station:* Helps prevent thermal shock of bath.
- *Soldering station:* Solder applied to the PWB.
- *Unloading station:* PWBs removed from conveyor.

Wave soldering techniques are useful for both PTH and SMT wiring boards. Foam fluxing is normally used to promote penetration of flux through PTHs and small via holes in multilayer boards. After passing through the foam, the PWB passes over an air-knife to wipe off excess flux, returning it to the flux tank. This helps to minimize board cleanup and provides a safer process if flammable fluxes are being used.

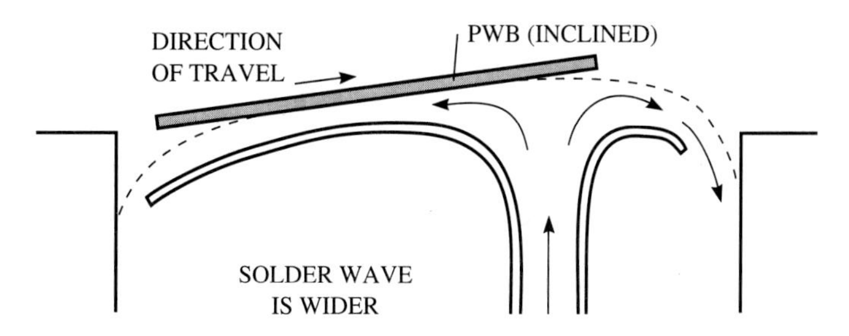

Figure 6-9 Lambda soldering wave.

The preheating station has become increasingly important as SMT has increased in popularity. The station serves the following functions:

- Raises flux temperature, making flux become more active, thus increasing its cleaning and wetting action.
- Improves capillary action of flux up through PTHs and multilayer vias.
- Evaporates flux solvents on the substrate surface. Preheat elevates substrate temperature uniformly to approximately 125°C. This controlled temperature rise helps avoid excessive thermal shock when the board enters the solder wave.
- Heats multilayer boards on both sides simultaneously to avoid excessive side-to-side thermal gradient. This helps to prevent twisting, bowing, or warping of the board as it enters the solder wave.

Figure 6-10 includes two illustrations of single-wave soldering. This single-wave soldering is limited to PTH boards and some SMT boards if the population of SMDs on the bottom of the board is not very dense. A high density of SMDs on the bottom side leads to problems known as shadowing and capillary action.

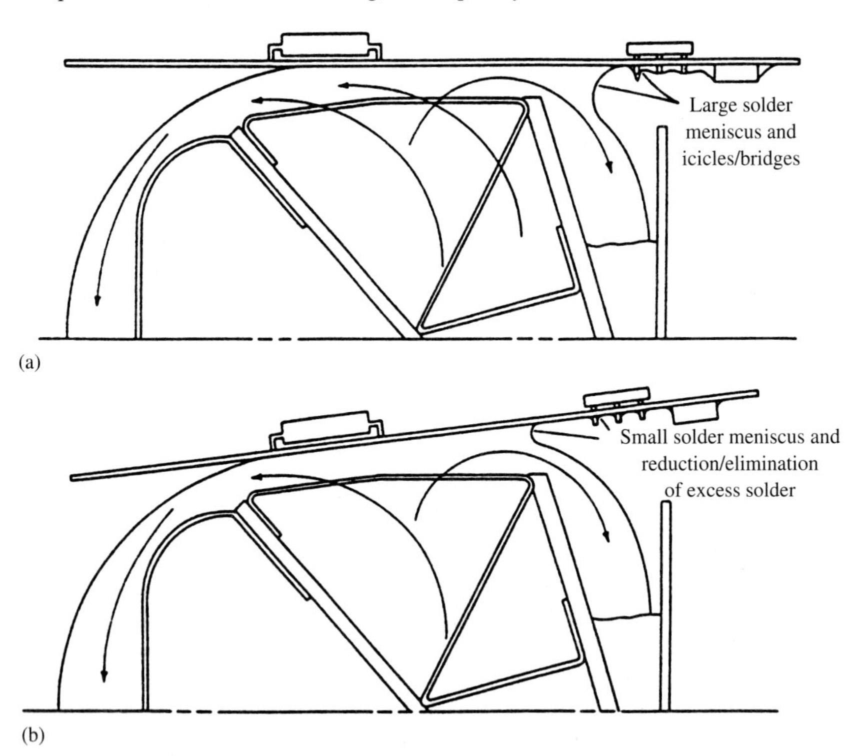

Figure 6-10 Single-wave flow soldering: (a) horizontal; (b) inclined.

(*Source: Electronic Materials Handbook, Vol. 1: Packaging* (Materials Park, Ohio: ASM International, 1989), p. 688.)

The leading edge of the solder wave moves in a direction opposite to the direction of the PWB travel. This opposing motion not only presents solder to the PWB but also provides a mechanical scrubbing action. As the board passes the crest of the wave, the laminar flow of the wave is in the same direction and, preferably, at the same velocity as the movement of the board, permitting total contact between the pads, components, PTHs in the board, and solder wave.

Normal practice is to immerse the board in the wave at a depth equal to one-third to one-half the thickness of the board. At this depth, the wave applies a dynamic pressure to the underside of the board, forcing solder up through the PTHs and vias beyond that which normal capillary action would accomplish. Thick multilayer boards may require immersion depths of up to three-quarters the substrate thickness to ensure adequate heat transfer and wetting.

Figures 6-11 and 6-12 give illustrations of dual-wave soldering, which was developed to solve the problems of shadowing and capillary effects. The two waves divide the soldering process into two steps: primary wetting and final soldering.

The first wave is a high-pressure wave and is very turbulent. This turbulence offers greater scrubbing action around tightly packed SMDs and ensures good wetting of all areas that would not be reached by a normal wave. The turbulent wave cannot be controlled as to how much solder is left on the PTH leads, so there is no control over bridging and icicling.

The second (bidirectional) wave is needed to correct the irregularities of the first wave. It is of much lower pressure than the first wave and has a laminar flow profile. This laminar wave removes excess solder that may have been applied by the turbulent wave, which reverses the bridging generated by the first wave.

The angle of inclination and speed of the conveyor can be adjusted such that the formation of icicles is minimized. Figure 6-13 shows an example of the correct peel-back area that helps minimize icicling. Increasing the angle of inclination encourages more solder to be pulled (peeled) back into the wave, due to surface tension, before the through-

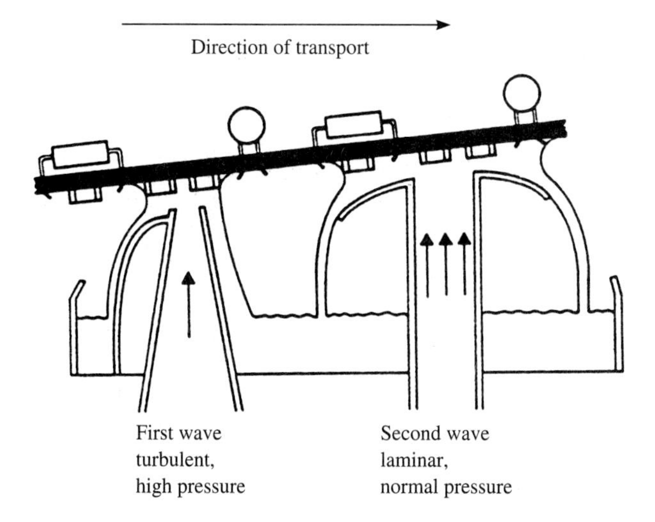

Figure 6-11 Dual-wave soldering showing division of the solder stream into a first and second wave.

(*Source:* Maiwald, *Soldering in SMD Technology,* p. 20.)

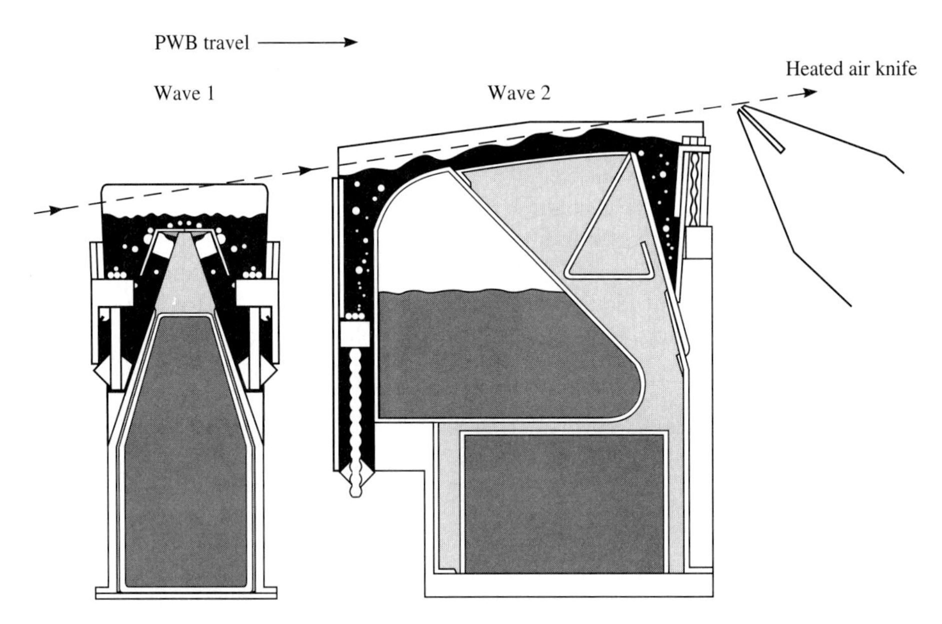

Figure 6-12 Dual waves: turbulent wave 1, laminar wave 2.

(*Source: Electronic Materials Handbook,* p. 689.)

hole leads leave the wave. Thus, the increased angle of inclination reduces the formation of icicles. However, an angle which is too steep will cause the solder fillets to be too lean. The optimal peel-back is usually achieved when the relative velocity of the board leaving the wave is zero and the incline of the conveyor is 3° to 9°.

Wave soldering is not without its problems. Some of the more notable ones are

- Shadowing effect
- Capillary effect

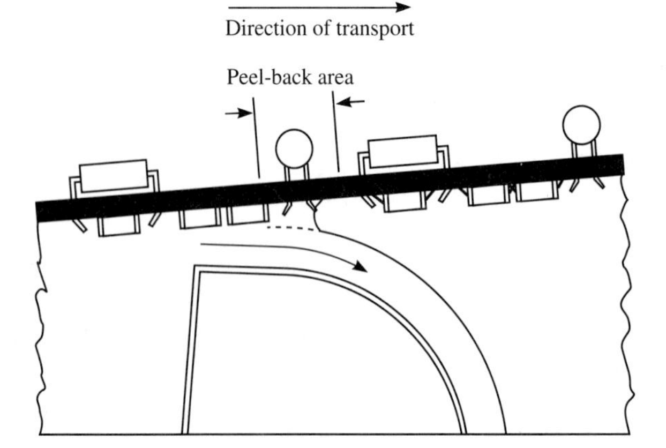

Figure 6-13 Solder peel-back of laminar main wave.

(*Source:* Maiwald, *Soldering in SMD Technology,* p. 21.)

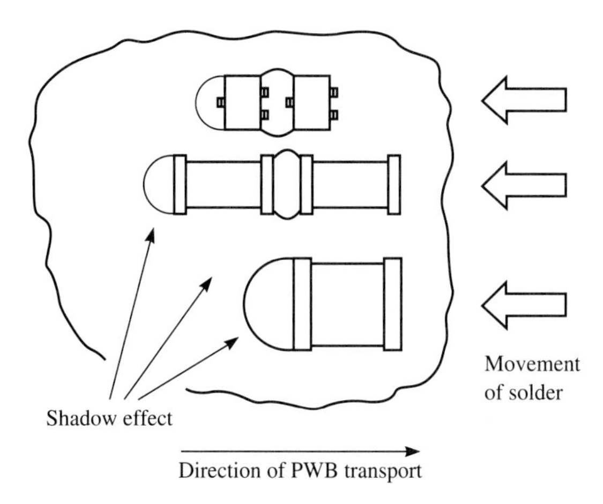

Figure 6-14 Shadow effects in wave soldering.

(*Source:* Maiwald, *Soldering in SMD Technology,* p. 18.)

- Contaminated pads
- Outgassing
- Misaligned components

The **shadowing effect** is due to the SMDs mounted on the underside of the board, as shown in Fig. 6-14. The board and wave move in opposite directions toward each other. Liquid solder flows unimpeded to the pads lying in front of the SMD. The component body then prevents the solder from reaching the pads at the rear of the SMD. Reliable wetting at the rear of the SMD can be achieved by extending the pads.

Capillary effects are due to dense population of SMDs. When a capillary tube is immersed in a nonwetting liquid, the liquid will be depressed (capillary depression). In contrast, if the tube is immersed in a wetting liquid, it will cause an elevation of the liquid (capillary attraction). The same effects can be observed on boards densely populated with SMDs. Figure 6-15 gives an example of capillary depression due to dense population of SMDs. Liquid solder cannot reach the pads lying between the component bodies. Figure

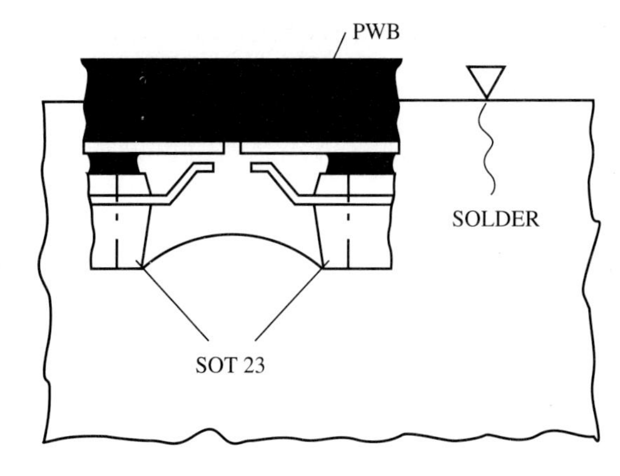

Figure 6-15 Capillary depression of solder between two SOT 23 components (causes open circuit).

(*Source:* Maiwald, *Soldering in SMD Technology,* p. 18.)

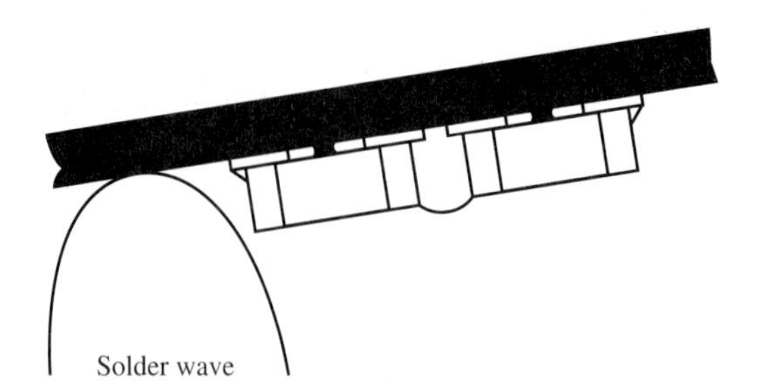

Figure 6-16 Capillary attraction of solder between two SMDs (causes short circuit).

(*Source:* Maiwald, *Soldering in SMD Technology,* p. 19.)

6-16 shows an example of capillary attraction. A short circuit is the result of capillary attraction on a densely populated board.

Figure 6-17 shows an example of pad contamination by glue. Inaccurate or excessive application of glue may prevent proper wetting of pads. Bleeding or overflow of the solder resist mask may have a similar effect in case of screen printing in conjunction with extremely small pads.

When solvents in the flux are not adequately driven off (especially during the preheating process), gases may become entrapped in the solder and cause voids or blow holes, as illustrated in Fig. 6-18. This is referred to as outgassing and is primarily observed with single-wave soldering. It is usually not a problem with dual-wave soldering.

Using adhesives on the underside of a board for SMDs which are not temperature-stable may lead to component misalignment when the board passes the solder wave.

6.4.2 Reflow Soldering Techniques

Figure 6-19 shows the basic layout of a reflow soldering line. **Reflow soldering** is used for surface-mount components only. It involves lower soldering temperatures than wave soldering, which allows heat-sensitive components to be soldered. As a rule of thumb, the cost per reflow soldered joint is twice to four times higher than for a wave soldered joint. In reflow soldering, the application of the solder and the heating of the solder to make it fluid are two separate processes. Figure 6-20 compares heat applications in wave soldering and reflow soldering.

The most common method of solder application is by the use of solder paste. Solder

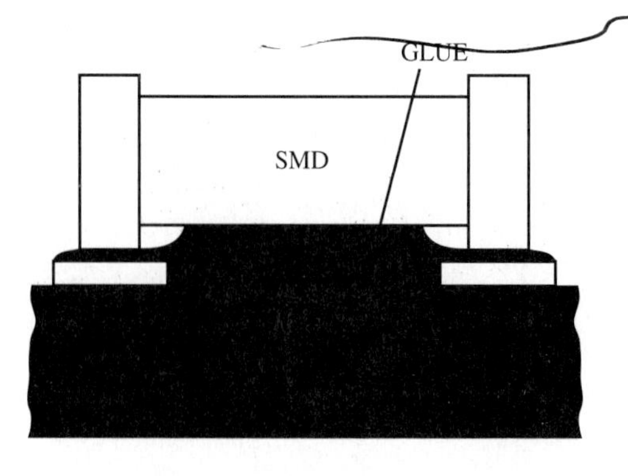

Figure 6-17 Contamination of pads by glue impairs the soldering process.

(*Source:* Maiwald, *Soldering in SMD Technology,* p. 19.)

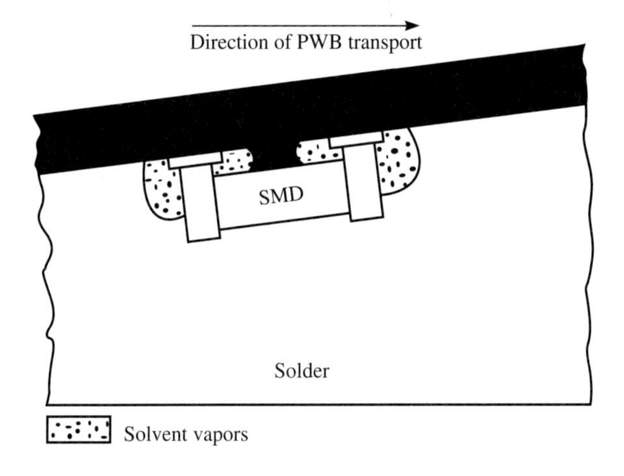

Figure 6-18 Soldering defect (outgassing) due to insufficiently driven off solvents.

(*Source:* Maiwald, *Soldering in SMD Technology,* p. 19.)

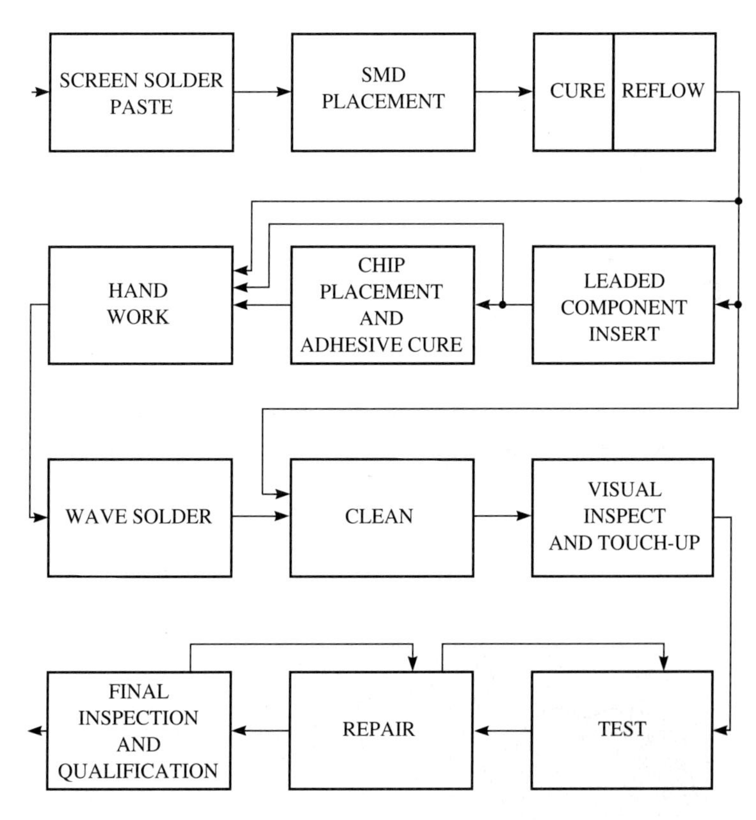

Figure 6-19 Flowchart for a mixed technology assembly line.

Wave soldering

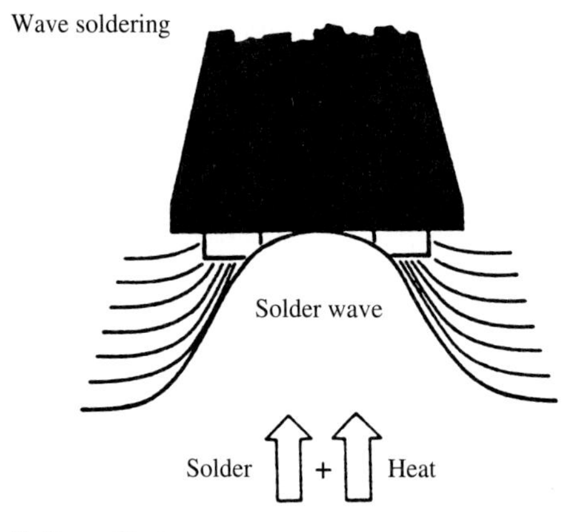

Reflow soldering

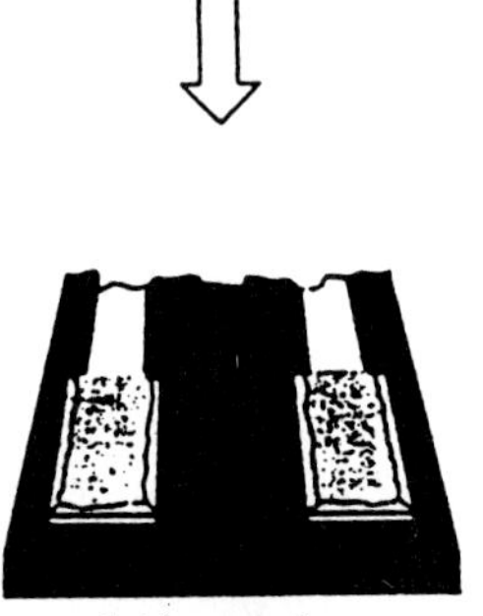

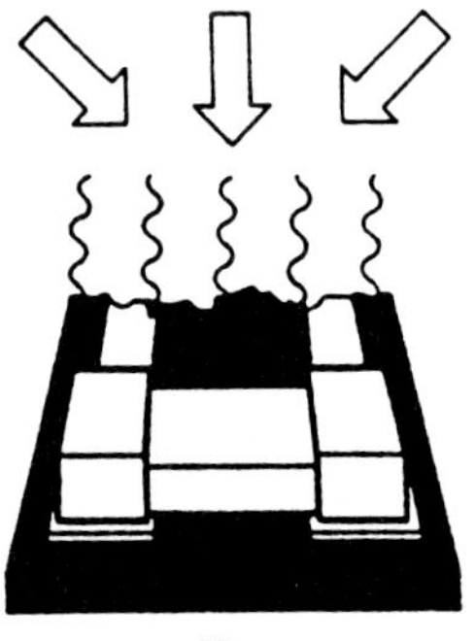

Figure 6-20 Differences in heat application between the wave soldering and reflow soldering processes.

(*Source:* Maiwald, *Soldering in SMD Technology,* p. 22.)

paste is a complex mixture of very fine particles of solder, solder flux, and adhesives. The size of solder particles in the paste will vary from 20 to 150 μm (0.8 to 6 mils), and they may be combined with as many as fifteen chemical components. Table 6-3 gives specifications for paste types and corresponding particle sizes.

Solder paste is normally applied by screen printing, stencil printing, or dot placement. The screen printing used in solder paste application is the same type of screen printing process that is used for PWBs and thick-film circuit fabrication. The finer the screen mesh, the finer the size and spacing of paste deposit. Usually about 80 to 105 mesh steel gauze is used for the screen, onto which a negative of the pad placement pattern has been applied. A squeegee is then used to push paste through the screen and onto the PWB lands. The minimum wet layer thickness of paste should be about 150 μm. The major drawback in this process is the difficulty of achieving uniform paste deposition across the entire width of screen.

Stencil printing is the most widely used method of applying solder paste for reflow soldering. The use of metal-foil stencils provides increased control of both solder volume deposited and its uniformity. To create the stencil, a metal foil has the land patterns applied using photolithography, and these are chemically etched, making very fine patterns

TABLE 6-3 SPECIFICATIONS OF SOLDER PASTE INGREDIENTS

Solder paste distribution
Based on IIW/1SO specification

	Percent of sample by weight—nominal sizes					
	Less than 1%, larger than:		80% minimum, between:		10% maximum, less than:	
	μm	mil	μm	mil	μm	mil
Type 1	150	6	75–150	3–6	20	0.8
Type 2	75	3	45–75	2–3	20	0.8
Type 3	45	1.8	20–45	0.8–1.8	20	0.8
Type 4	38	1.5	20–38	0.8–1.5	20	0.8

Powder size classification
Based on IIW/1SO specification

	1% W/W maximum, larger than:		90% W/W minimum, between:		10% W/W maximum, less than:	
Class	μm	mil	μm	mil	μm	mil
1	150	6	75–150	3–6	75	3
2	75	3	45–75	2–3	45	1.8
3	45	1.8	20–45	0.8–1.8	20	0.8
4	36	1.4	20–36	0.8–1.4	20	0.8

Screen print application

	mm	in.
Powder size, largest particle		
Type 2	0.075	0.003
Type 3	0.045	0.0018
Type 4	0.038	0.0015
Screen printing opening		
80 mesh	0.22	0.0088
105 mesh	0.17	0.0065
120 mesh	0.14	0.0057
180 mesh	0.09	0.0034

Source: Electronic Materials Handbook, Vol. 1, p. 652.

possible. This method has been used effectively for depositing solder paste with land spacings on 25 mil (635 mm) centers.

In dot placement (or pressure dispensing), the paste is dispensed from a supply cartridge by pneumatic or mechanical pressure. The paste is forced through small diameter tubes or orifices to place small discrete deposits of paste at interconnection sites. Both single-point and multipoint dispensers are used.

The single-point dispensers are best suited for local repair of individual joints and low-volume production runs. Multipoint dispensers can be used to dispense paste at a complete package site. Pressure dispensing requires the use of solder paste that has a finer particle size than is used for paste printing in the same applications.

In all paste applications, the paste is allowed some set-up time to become tacky before components are placed. This time may vary from seconds (the conveyer time between the screen printer and the placement machine) to hours (a set-aside time), depending on the type of paste that is used. The SMD is then pressed into the solder paste and held in place by the inherent tackiness of the paste. During the reflow process the SMD literally floats on the liquid solder, so slight variations in the thickness or symmetry of the solder dot may cause component displacement during the reflow process.

The application of heat is the second step in the reflow soldering process. Table 6-4 summarizes the methods of heat application used in reflow soldering.

Infrared (IR) soldering became the preferred SMT mass soldering technique in the 1980s although **forced-air convection** has recently become the preferred method for surface-mount technology. The IR is carried out in-line, in a chamber, and is a noncontact process in that no condensing fluid or solder bath contacts the workpiece. IR allows the temperature profile of the system to be changed rather quickly. In contrast to other soldering methods, not only the surface but also the component body is heated by the absorption of IR. There are two wavelengths of IR that are used most often for reflow soldering: short-wave and long-wave.

The source of heat for short-wave IR is usually quartz lamps. This 1.2 to 2.5 μm IR radiation heats the solder paste from the inside out. There is minimal convection heat involved in this process.

The source of heat for long-wave (4 to 6.5 μm) IR is usually heated panels. The long-wavelength IR tends to heat the surrounding air due to the high absorption capacity of air. Convection heat transfer may make up 60 percent of the total heat transferred in this process.

In the condensation process, a vaporized fluid condenses on a PWB that is immersed in the vapors over the boiling liquid. The latent heat of vapor is given up to the PWB, raising the temperature of the entire workpiece to the temperature of the vapor. This method is also known as **vapor phase soldering** and may be a batch or an in-line process.

TABLE 6-4 METHODS OF HEAT APPLICATION IN REFLOW SOLDER PROCESSES

Heat transfer/ Heat input	Heat input from below over surface, simultaneous	Heat input from above		Heat input from all directions simultaneous
		local, sequential	over surface, simultaneous	
Radiation	IR radiation	Laser beam	IR radiation	IR radiation
Condensation	—	—	—	Vapor phase
Conduction	Hot plate Hot belt	Thermode (stamp)	Thermode (bracket)	—
Convection	—	Gas nozzle	Hot air Hot gas	Hot air, circulating air Hot gas

Source: Werner Maiwald, *Soldering in SMD Technology,* p. 15.

Figure 6-21 illustrates the vertical batch process. Note the primary and secondary vapor levels. The vertical batch process requires the use of two fluids known as the primary fluid and secondary fluid.

The primary fluid is often a synthetic inert fluorocarbon fluid with boiling points in the range that provides sufficient temperatures to reflow solder alloys used in SMT. The secondary fluid is used as a blanket to prevent loss of the primary fluid.

The properties of the secondary fluid are as follows:

- Vapor density of less than primary fluid
- Boiling point lower than primary fluid
- Much lower cost than the primary fluid
- Chemical compatibility with primary fluid and vapor
- Nontoxic, nonflammable, stable, and noncorrosive

Figure 6-22 gives an illustration of the in-line process in which the PWB is horizontally conveyed into the reflow environment. The input and output throat is above the working zone to decrease the loss of primary vapor; thus, there is no need for a blanket of secondary fluid. One advantage of the in-line system is that it is a continuous system and is easily integrated into existing production lines.

Figure 6-23 is a chart that compares reflow solder techniques in terms of component/substrate temperature profiles.

The reader is referred to Section 6.4.3 and to Fig. 6-24 for detailed illustrations of solder joint connections including defective solder joints. The following are commonly encountered solder defects:

- *Tombstoning:* **Tombstoning** (component stands on end) of two-terminal SMDs may be due to

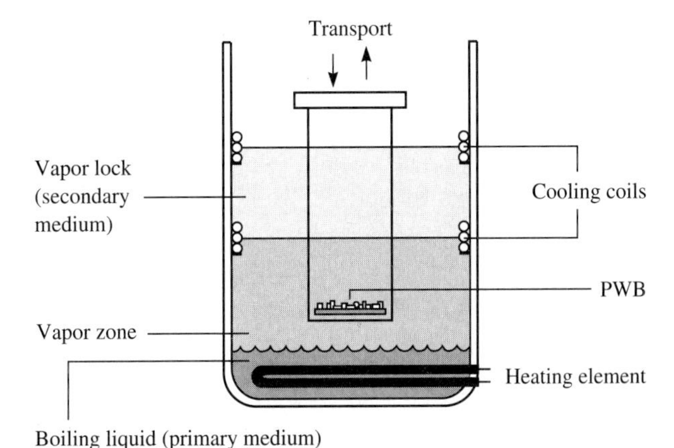

Figure 6-21 Vapor phase batch system.

(*Source:* Maiwald, *Soldering in SMD Technology,* p. 26.)

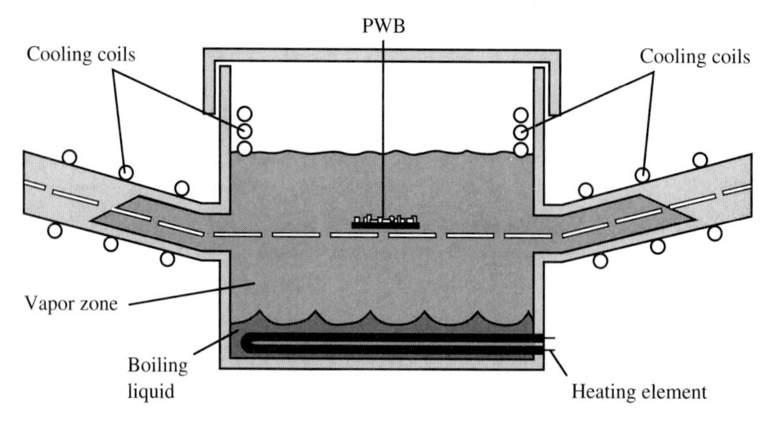

Figure 6-22 Vapor phase in-line system.

(*Source:* Maiwald, *Soldering in SMD Technology,* p. 27.)

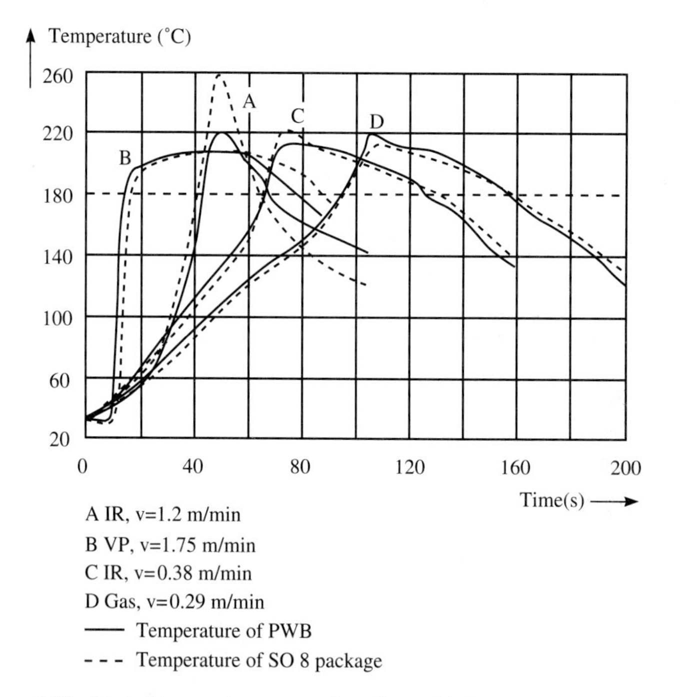

Figure 6-23 Typical temperature curves for reflow soldering.

(*Source:* Maiwald, *Soldering in SMD Technology,* p. 29.)

— Too large pads and too large separation of pads
— Different quantities of solder on the pads
— Nonsimultaneous melting of solder on the pads
— Different solderability of the terminals or pads
— Insufficiently dried solder paste

- *Solder ball formation:* Solder ball formation may be due to
— Unsuitable solder paste
— Too fast heating
— Insufficient or excessive drying of solder paste
- *Formation of voids:* Formation of voids in the solder joint is usually due to insufficient soldering time, which does not allow complete outgassing.
- *Dissolution of metallization:* Dissolution usually results because of excessive soldering time.
- *Lean solder fillet:* Lean solder fillet may be due to
— Insufficient quantity of solder paste
— Solder drain-off
- *Short circuits:* These are usually caused by solder bridging between the pins of packages with close pin spacings.
- *Microcracks in solder joints:* These cracks result in accelerated fatigue failure.
- *Nonwetting (dewetting) of metal surfaces:* This problem is usually due to poor solderability caused by contamination of the metal surfaces or inadequate fluxing of the solder joints.
- *Wicking:* This process is observed mainly on PLCC packages. The solder is drawn from the pads and climbs up the PLCC leads. This occurs because the PLCC leads have reached solder temperature faster than the PWB, and the solder moves to the location of highest heat.

6.4.3 Illustrations of Solder Connections

Illustrations of solder connections are shown in Fig. 6-24. The illustrations show both good and poor connections with relevant comments about each.

6.5 POSTSOLDER CLEANING

The environment in which printed wiring boards are assembled, the equipment which handles the boards during assembly (for example, the solder pot and conveyor), and the postsolder PWB cleaning are extremely critical factors in determining the quality and reliability of these electronic assemblies. The philosophy and process of **postsolder cleaning** begin with designing the PWB to facilitate cleaning. Elements that should be considered include:

- *Component orientation:* Every effort should be made to orient components on the PWB so that cleaning solvents will drain off the board.

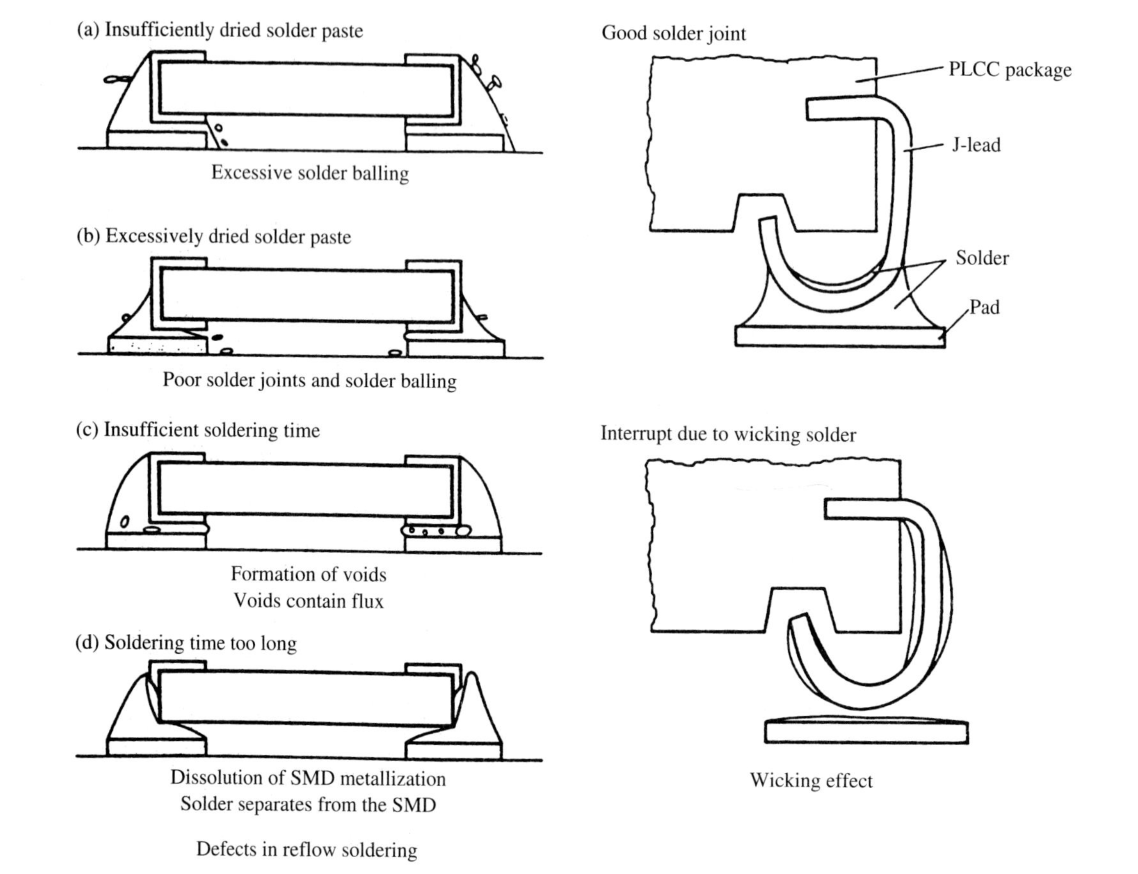

Figure 6-24 Illustrations of solder connections.

(*Source:* Maiwald, *Soldering in SMD Technology,* pp. 31, 36–49.)

- *Solder mask:* A solder mask should be used over all areas where solder is not required.
- *Process control:* A cleanliness control program should be a routine part of the overall program, including quantitative testing for cleanliness.

The following subsections discuss the sources and effects of contamination; cleaning methods, equipment and tests; and the emergence of processes eliminating the need for cleaning operations.

6.5.1 Sources and Types of Contamination

Contaminants of PWBs are of various types and come from a variety of sources. The general terminologies associated with contaminants are as follows:

- *Organics (also referred to as nonpolars or nonionics):* This category includes such materials as rosin, oil, grease, wax, make-up, and hand lotion. These materials

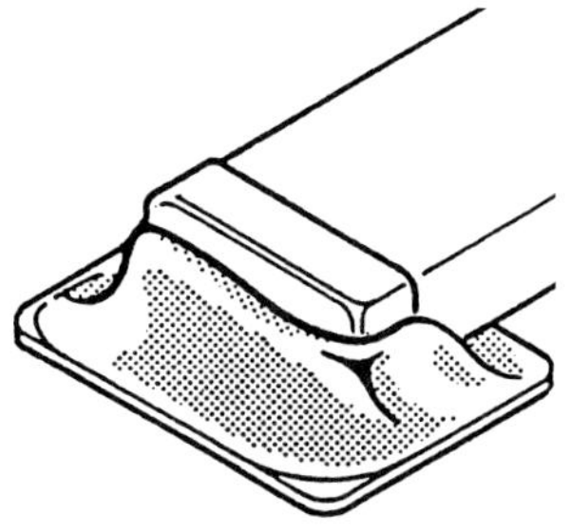

Cubic SMDs with inadequate solder joints after wave soldering: SMD terminal displays poor wetting.

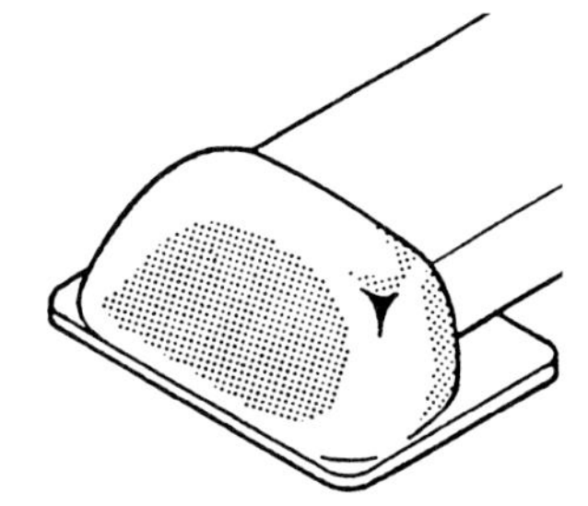

Cubic SMDs with good solder joints after wave soldering: Convex meniscus, good wetting recognizable.

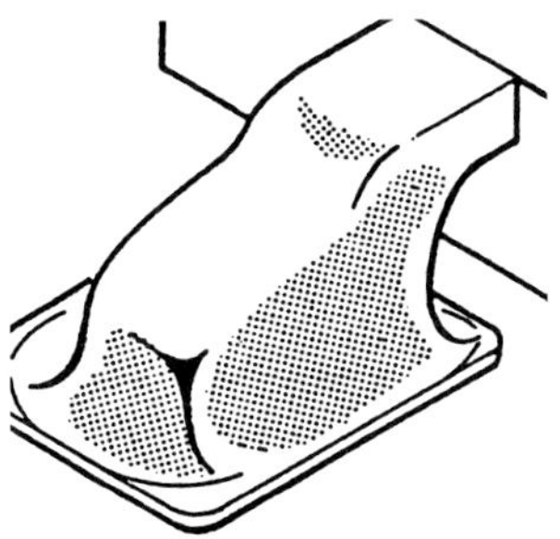

Multiterminal SMDs with good solder joints after wave soldering: The SMD terminal is completely wetted. The contour of the terminal is visible.

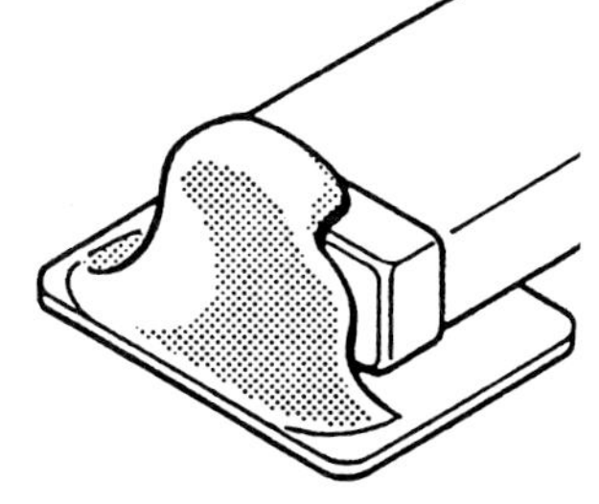

Cubic SMDs with inadequate solder joints after wave soldering: Solder does not completely (<75%) cover the edges of the SMD terminal that rest on the pad.

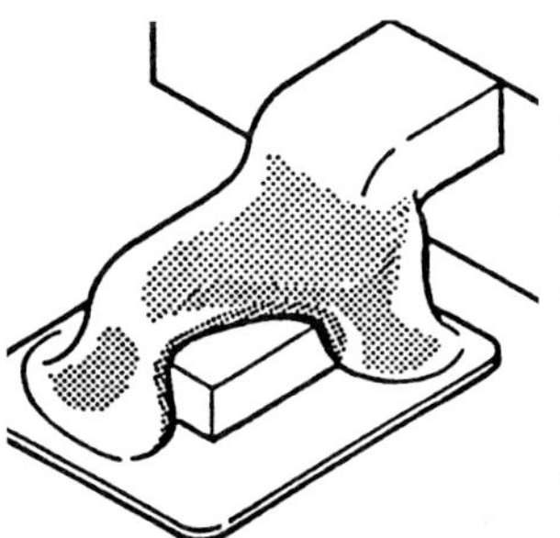

Multiterminal SMDs with inadequate solder joints after wave soldering: The solder does not completely (<75%) cover the edges of the SMD terminal that rest on the pad.

Cylindrical SMD with inadequate solder joint after wave soldering: SMD terminal displays poor wetting.

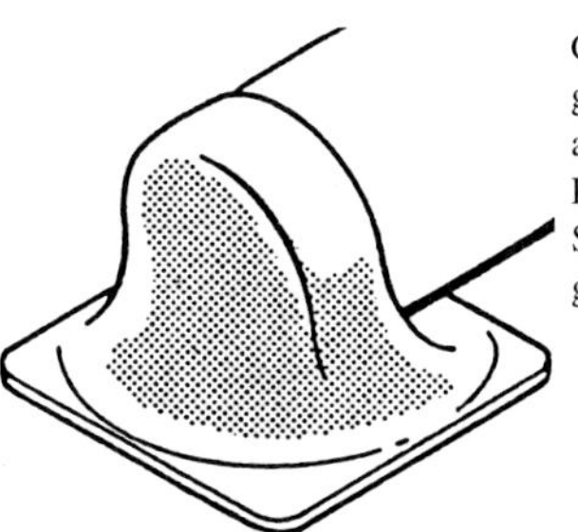

Cylindrical SMD with good solder joint after wave soldering: Proper solder meniscus. SMD terminal displays good wetting.

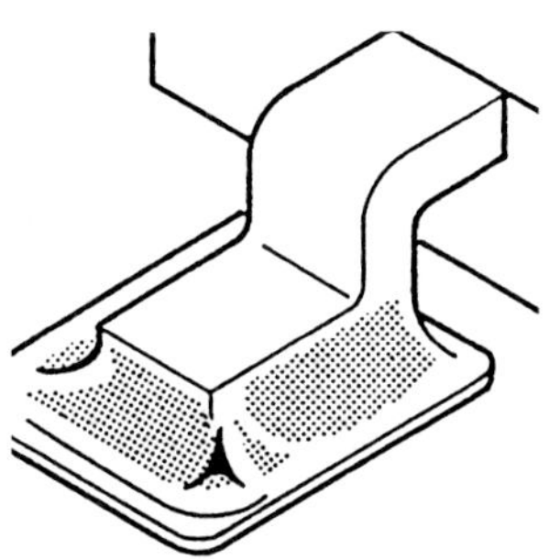

Multiterminal SMDs with good solder joints after reflow soldering: Proper solder fillet formed at the edges of the SMD terminal that rest on the pad.

Figure 6-24 *(continued)*

Cubic SMDs with good solder joints after reflow soldering: Solder covers the edges of the SMD terminal that rest on the pad (≥75%). Proper solder fillet.

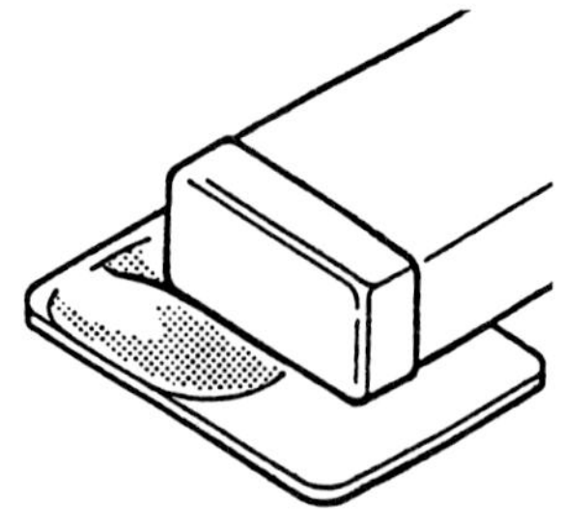

Cubic SMDs with inadequate solder joints after reflow soldering: Solder does not completely (<75%) cover the edges of the SMD terminal that rest on the pad. Solder fillet not recognizable.

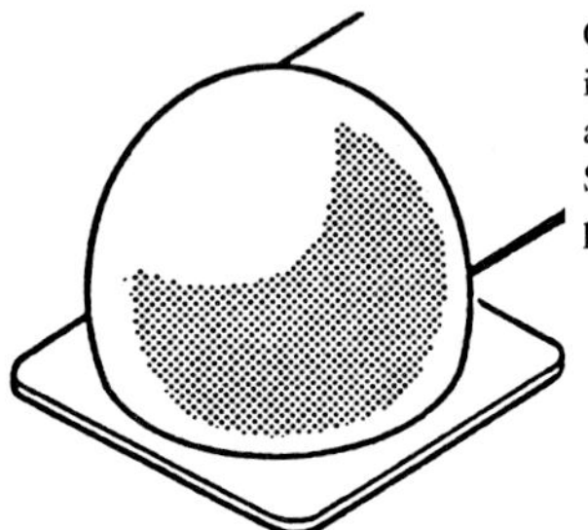

Cylindrical SMD with inadequate solder joint after wave soldering: Solder accumulation, poor wetting of pad.

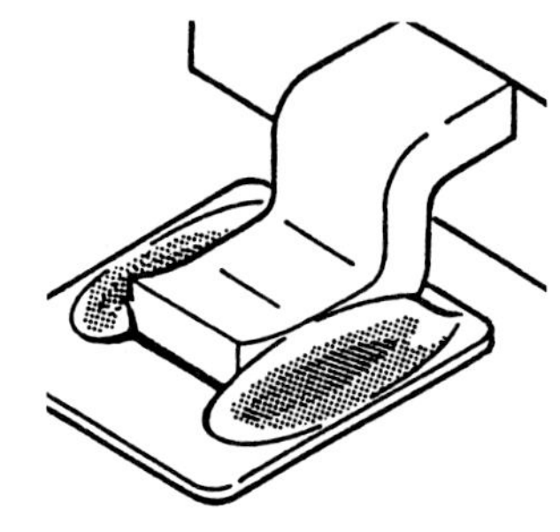

Multiterminal SMDs with good solder joints after reflow soldering: With nonplanar SMD terminals or untinned edges, a solder fillet is visible on both sides: solder covers (≥75%) of the terminal edges resting on the pad.

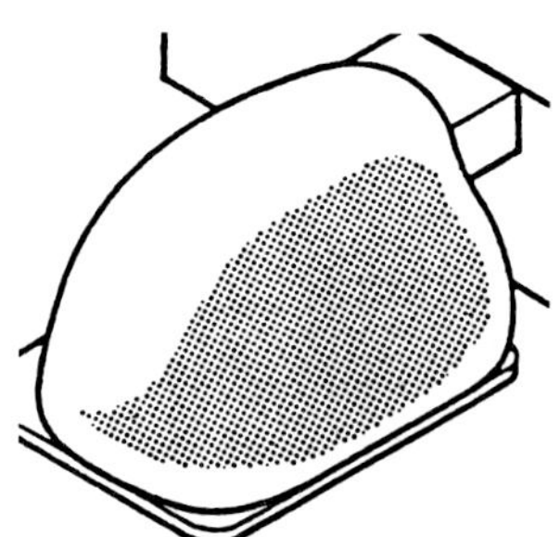

Multiterminal SMD with inadequate solder joints after reflow soldering: solder accumulation.

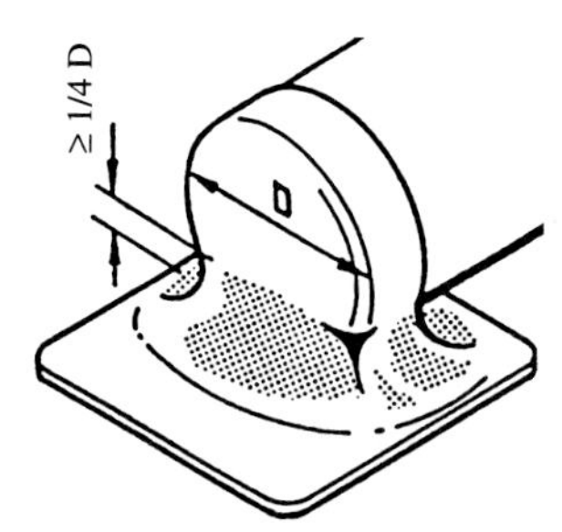

Cylindrical SMD with good solder joint after reflow soldering: SMD terminal is wetted along end face and sides. Proper solder fillet.

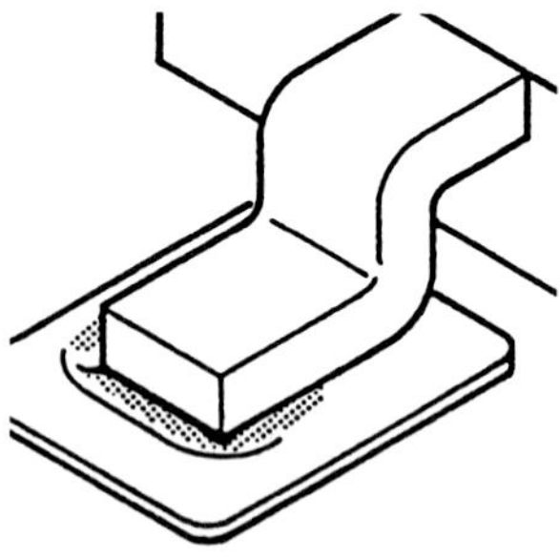

Multiterminal SMDs with inadequate solder joints after reflow soldering: Amount of solder too small for nonplanar SMD terminals. Solder seam recognizable, solder covers <75% of the terminal edges resting on the pad.

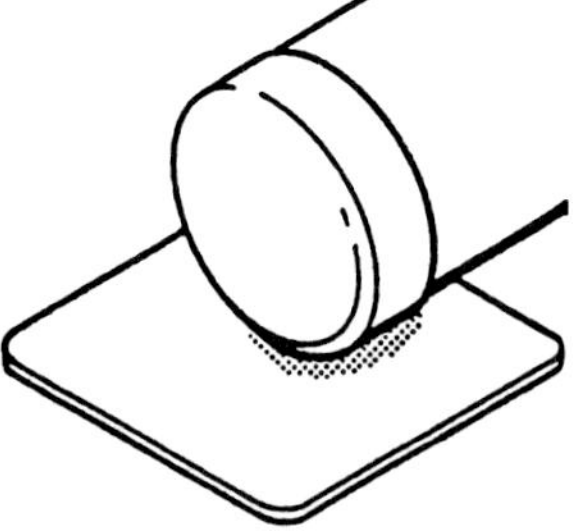

Cylindrical SMD with inadequate solder joint after reflow soldering: No solder fillet visible at end face and sides.

Figure 6-24 *(continued)*

contain carbon and generally are good insulators. They come from handling of the boards and rosin from flux residues.

- *Inorganics (also referred to as polars or ionics):* This category includes soldering fluxes, flux activators, halides, acids, salts, and the like. The sources of these contaminants include plating solutions (salts), sweaty hands (salt), and fluxing operations. Ionic molecules form ion pairs in solution. In film form, the molecules of polar material align with an electric field, since they possess an electric dipole moment.

6.5.2 Effects of Contamination

Organic contaminants (nonionic) generally form insulating films which can cause intermittent opens at connector contacts or probe points, retention of dust or dirt particles, and interference with solder mask adhesion. Although these type contaminants can cause circuit performance problems, they are not as detrimental as inorganic contaminants.

Inorganics (ionic) can cause chemical reactions associated with corrosion of conductor lines and, because they dissociate into ions in the presence of moisture, can carry an electric current, leading to dielectric breakdown or electrical shorts due to electromigration.

6.5.3 Cleaning Methods

When selecting a cleaning process, there are several factors that must be considered, including:

- *Performance:* Cleaning solution should perform its intended function.
- *Component and substrate compatibility:* Process must not have any adverse effects on cleaning equipment, component packages, printed wiring boards, or marking inks.
- *Economics of usage (including energy efficiency and operating costs):* The length of cleaning cycle and the costs of solvents, equipment, and waste disposal should be considered.
- *Toxicity and environmental factors:* Ability to comply with environmental, health, and safety regulations must be considered.
- *Surface tension:* Cleaning solution must be able to penetrate the tightest spaces on the PWB (for example, between a surface-mount device and the PWB) and dissolve the solder flux residue and other contaminants.
- *Cleaning solution residue:* Cleaning solution should leave a minimum amount of residue upon evaporation from the PWB.

Of all of the factors listed, the primary consideration is the selection of a cleaning process that thoroughly removes all of the contamination for which it is being used but does not damage the assembly.

The fundamental rule to be followed when selecting a solvent is that polar solvents dissolve polar contaminants and nonpolar solvents dissolve nonpolar contaminants. If both type contaminants are present, then a mixture of these two type solvents (called an azeotrope solvent) should be selected. There are two basic categories of cleaning solutions, as follows:

- *Aqueous:* These solutions use water as the primary cleaning fluid and are recommended for organic acid fluxes. Chemical additives are used to convert the nonpolar contamination to a form which will dissolve in water.
- *Solvent:* These solutions use chlorinated or fluorinated hydrocarbon liquids and are recommended for rosin-based fluxes. Chlorinated solvents are more aggressive and are better at removing contaminants from tight spots, but they are not compatible with plastic and are more hazardous to humans. Fluorinated solvents are less aggressive but easier and safer to use.

6.5.4 Cleaning Operations

There are two broad approaches used in a cleaning operation:

- *Batch processing:* Operator-intensive and generally restricted to small volumes of PWBs including reworked assemblies.
- *In-line processing:* Essentially operator-independent and normally used for high-volume production to achieve repeatability and consistency.

Cleaning operations can be further divided into solvent and aqueous processes. Solvent cleaning operations take a number of forms. These include:

- *Batch:* Commonly referred to as vapor degreasers. The process is based on the physical property that hot vapors of a solvent will condense on the PWB if it is at a lower temperature than the boiling temperature of the solvent. The solvent that condenses on the PWB dissolves the contamination and drains off of the PWB into a sump if the PWB is oriented properly. The system consists of two sumps, a condensate chamber, and cooling coils to minimize escape of the solvent vapor and to supply the clean sump with redistilled solvent. A diagram of this system is shown in Fig. 6-25. The solvent in the first sump is boiling and contains the bulk of the flux residues. The second sump is a cooler solvent reservoir which contains continuously redistilled solvent because the clean solvent drips back into it and dirty solvent flows back into the first sump. Spray wands and ultrasonic cleaners for use in mechanically dislodging particulates may also be part of the system.
- *In-line:* Boards are carried by a conveyer through a series of cleaning operations or zones. The number of zones in a cleaning system varies, but they are variations of three basic processes referred to as spray, immersion, and vapor.

The various zones of an in-line cleaning process (as many as eight) generally fall into one of the following categories:

- *Horizontal spray:* Board is subjected to a solvent spray with the board in a horizontal position. Spray pressures can be relatively high.

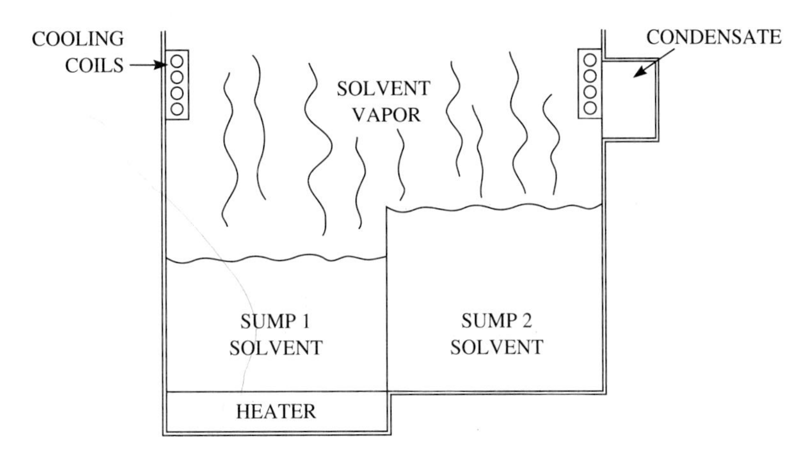

Figure 6-25 A vapor degreaser used for batch solvent cleaning.

- *Inclined spray:* Similar to horizontal spray except that the board is carried at an angle to enhance solvent drainage.
- *Immersion:* Board is immersed into a boiling solvent.
- *Spray and immersion:* Board is subjected to both immersion and spray in a single zone.
- *Vapor:* Similar to a batch degreaser.

In-line cleaning systems are usually closed systems to minimize the loss of vapor. They are highly automated for high-volume production and allow for high-pressure spraying.

Aqueous cleaning systems are used to remove water-soluble and rosin flux residues, although the viscosity and surface tension of water make it less efficient in removing contamination from tight spaces. Aqueous cleaning systems are available in both batch and in-line configurations, with a batch system being nothing more than an industrial dishwasher. In-line systems convey boards through a prewash zone, a recirculating detergent wash zone, several rinse zones, and a drying zone.

Aqueous cleaning solutions are not purely water but contain chemical additives to enhance their cleaning effectiveness. Some of the additives are:

- *Neutralizers:* Convert insoluble compounds formed during soldering to more soluble forms.
- *Antifoaming agents:* Limit foam in cleaning solutions caused by spraying operations.
- *Dispersants:* Assist in the removal of insoluble particulates.
- *Saponifiers:* Convert rosin flux residues to water-soluble soaps.
- *Surfactants:* Reduce the surface tension of the water, which improves its wetting characteristics.

Because of these additives, there is a concern about the environment if the solutions are dumped into the sewer system. Also, some of the additives in concentrated form are dangerous to humans. Therefore, they should only be handled by trained personnel, in strict compliance with environmental regulations.

6.5.5 Cleanliness Test Methods

No matter what process is used to assemble PWBs, a program must be in place to continuously monitor the cleanliness of the product using both qualitative and quantitative methods. This monitoring is necessary to ensure that the process remains in control and to prevent the manufacture of large quantities of defective PWBs. Common defects related to improper cleaning are white residues (polymerized rosin) and mealing (ionic contamination). The most commonly used methods for monitoring assembly line quality are:

- *Visual inspection:* A qualitative inspection accomplished by viewing the PWB under an optical microscope for gross contamination levels.
- *Surface insulation resistance (SIR):* A nondestructive quantitative technique used to detect flux residues using a special circuit built on the PWB.
- *Solvent extraction resistivity:* PWBs are soaked in an ionic solvent and the conductivity of the resulting solution is measured. This method relies on all contamination being removed from the assembly.

6.5.6 No-clean Processing

In October 1987, a group of atmospheric scientists, referred to as the Ozone Trends Panel, gave a disturbing report on the depletion of the ozone layer (approximately 2 percent since 1976). They noted high levels of stratospheric chlorine and suggested that this came from the use of fully halogenated chlorofluorocarbons (CFCs), carbon tetrachloride, and methyl chloroform. CFC manufacturers and users of these materials supported a call for their regulation. In 1990 a plan was developed to totally phase out the use of CFCs, carbon tetrachloride, and methyl chloroform by the year 2000.

Because of the impending phaseout of many solvents and difficulty in disposing of aqueous cleaning waste and wastewater used in PWB assembly, there are ongoing research and development efforts directed at **no-clean processing** (eliminating postsolder reflow cleaning). For this purpose, fluxless and no-clean flux soldering have been developed. However, postsolder cleaning continues to be an important process in the assembly of PWBs out of concern for long-term reliability of the final product.

6.6 SUMMARY

Soldering, a metallurgical technique used to join metals, must provide electrical continuity between the metals. In some cases, solder must also provide for mechanical support. Although solder is available in several compositions, the more common ones are binary compositions of tin and lead. The soldering process also uses solder fluxes, applied using the foam, spray, or wave technique, to improve the wettability of the surfaces being joined by chemically removing contamination and assisting in the transfer of heat from the molten solder to the joint area. Generally, electronic components are attached (soldered) to printed wiring boards using either the wave or reflow soldering technique. Wave soldering requires the use of solder resist (that is, a solder mask), applied by screen printing or photoprinting, to define the area on the PWB where a solder joint is to be formed. The placement of solder paste in reflow soldering is accomplished using either screen printing, stencil printing, or dot placement. Postsolder cleaning is a critical factor in determining the quality and reliability of PWBs. However, due to environmental concerns, the industry is moving toward use of no-clean processing.

KEY TERMS

Bridging
Capillary effects
Contaminants
Forced-air convection
Icicling
Mixed-technology (MT)
No-clean processing
Postsolder cleaning
Reflow soldering
Shadowing effect
Solder alloys
Solder composition
Solder resist
Solderability
Solder flux
Tombstoning
Vapor phase soldering (VPS)
Wave soldering
Web
Webbing
Wettability
Wetting

EXERCISES

1. Define the following terms:
   a. alloying b. wettability c. soldering
   d. cohesive force e. adhesive force
2. What are the requirements for a good soldering process?
3. To be considered reliable, what must a solder joint provide?
4. List the chemical, thermal, and physical functions of a flux.
5. Discuss the following:
   a. flux activity
   b. flux corrosivity
   c. flux cleanability
6. What are the three primary methods of applying flux to a PWB?
7. What is the purpose of a solder resist, and what desirable characteristics should it have?
8. What three methods are available for applying a solder mask?
9. What are the major considerations for designing a PWB for solderability?
10. What causes icicling to occur in wave soldering?
11. What procedures can be used to minimize icicling?

12. List five problems encountered in wave soldering.

13. What are the three methods used to apply solder paste for reflow soldering?

14. What are the four methods used to apply heat in reflow soldering?

15. List nine commonly encountered solder connection defects encountered in wave soldering.

16. What are the two broad terminologies associated with PWB contaminants? What are their sources?

17. What are the two types of automated soldering techniques?

18. What are the factors that should be considered when selecting a PWB cleaning process?

19. Discuss the two basic categories into which cleaning solutions fall.

20. What are the chemical additives commonly found in aqueous cleaning solutions for PWBs, and why are they used?

21. Discuss the three commonly used methods for monitoring assembly line quality.

22. What are three design elements that should be considered to facilitate PWB cleaning?

BIBLIOGRAPHY

1. JONES, T. H., *Electronic Components Handbook.* Reston, Va.: Reston Publishing Company, 1978.
2. COOMBS, CLYDE F., JR., *Printed Circuits Handbook* (3rd ed.). New York: McGraw-Hill Book Company, 1988.
3. *Electronic Materials Handbook, Vol. 1: Packaging.* Materials Park, Ohio: ASM International, 1989.
4. FREDERICKSON, MICHAEL D., *Conductive Point-to-Point Soldering.* Indianapolis, Ind: U.S. Navy Electronics Manufacturing Productivity Facility, 1989.
5. CLARK, RAYMOND, *Handbook of Printed Circuit Manufacturing.* New York: Van Nostrand Reinhold Company, 1985.
6. CLARK, RAYMOND, *Printed Circuit Engineering, Optimizing for Manufacturability.* New York: Van Nostrand Reinhold, 1989.
7. HINCH, STEPHEN W., *Handbook of Surface Mount Technology.* New York: Longman Scientific and Technical (copublished with John Wiley & Sons, Inc.), 1988.
8. PRASAD, RAY P., *Surface Mount Technology, Principles and Practice.* New York: Van Nostrand Reinhold, 1989.
9. WOODGATE, RALPH W., *The Handbook of Machine Soldering.* New York: John Wiley & Sons, Inc., 1988.
10. United States Department of Defense, MIL-STD-2000: Standard Requirements for Soldered Electrical and Electronic Assemblies. Washington, D.C.: U.S. Government Printing Office, 1989.

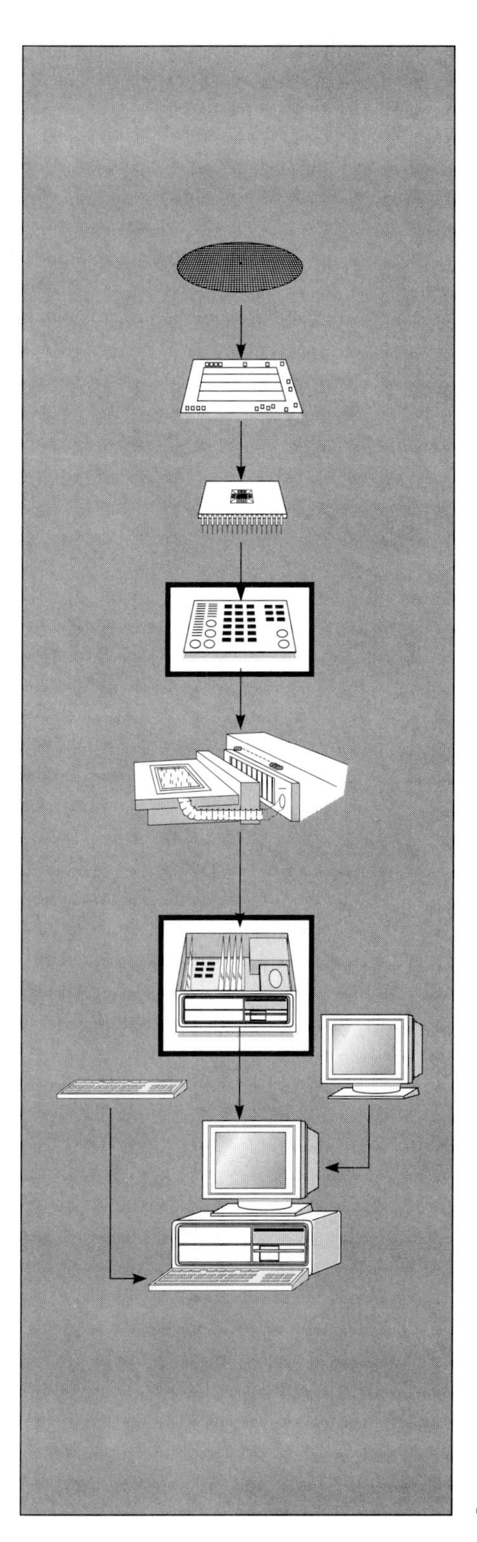

PART THREE
AUTOMATIC ASSEMBLY

7 Principles of Automation

7.1 INTRODUCTION

There are many reasons for automation of manufacturing processes, including

- *Quality:* Automation produces parts and assemblies with greater consistency and conformity to quality specifications than do manual processes.
- *Productivity:* Automation can result in greater output per hour of direct labor input and higher production rates.
- *Lead Time:* Automation provides flexibility to rapidly change over production among products and thus reduce the time between customer order and product delivery.

In this chapter, we discuss the principles of automation as applied to electronics manufacturing processes. Many of these automation principles are utilized in automatic insertion and placement equipment for PWB assembly. There are a variety of processes which must be considered: robotics and component feeding and handling (Sec. 7.2), integrated machine vision systems for assembly and inspection operations (Sec. 7.3), and other automated inspection technologies (Sec. 7.4).

7.2 ROBOTS

Before we describe the **robotics** applications in electronics manufacturing (Chapters 8 and 9), the technical features and construction of robots and the terminology of robotics must be defined and illustrated. We will discuss general principles of robotics which are applied throughout industry for **pick-and-place** tasks and in the electronics assembly industry in robotic placement/insertion mechanisms.

7.2.1 Physical Construction

The physical construction of a robot involves a body, an arm, and a wrist. Most robots are mounted on a base which is fastened to the floor. The body is attached to the base and the arm assembly is attached to the body. At the end of the arm is a wrist. The wrist is composed of a number of components that allow it to be oriented in various positions. Relative movements among the wrist, arm, and body are provided by a series of joints. These movements of joints may either involve sliding or rotating motions. The term **manipulator** is sometimes used to describe the body, arm, and wrist assembly.

Attached to the wrist of the robot is a hand or **end effector.** The end effector is positioned by the arm and body joints of the manipulator, and the orientation of the end effector is controlled by the wrist joints of the manipulator.

There is a wide variety of shapes, sizes, and physical configurations of industrial robots, but the following five configurations (illustrated in Fig. 7-1) are most common:

- *Cartesian coordinate:* Cartesian (*XYZ,* gantry, or rectilinear) robots use three perpendicular slides to construct the *X, Y,* and *Z* axes. The robot operates within a rectangular work space by moving its three slides relative to one another.
- *Cylindrical:* The cylindrical configuration uses a vertical column and a slide that moves up and down the column. The arm of the robot is attached to the slide and moves radially with respect to the column. Because of the rotation about the column, the robot work space is approximately a cylinder.
- *Jointed-arm:* This configuration consists of two straight components mounted on a vertical pedestal. A rotary joint connects one of the straight components to the pedestal, while another joins the two straight components. A wrist is attached to the end of the second straight component and provides several additional joints.
- *Polar:* The polar (spherical coordinate) robot uses a telescoping arm that can be lowered or raised about a horizontal pivot, which is mounted on a rotating base.
- *SCARA:* The **selective compliance assembly robot arm** (SCARA) configuration is a special type of jointed-arm robot in which the two straight components rotate about the vertical axes. Because this configuration provides compliance in the horizontal plane and substantial rigidity in the vertical direction, this robot is ideal for many light-duty assembly tasks.

Industrial robots are designed to perform work by the movement of the body, arm, and wrist through a series of positions and motions. A specific work task is done by the end effector which is attached to the wrist. There are two general categories of robot movements: arm and body motions and wrist motions. The motions of individual joints are called the degrees of freedom. Most robots in industry possess four to six degrees of freedom. Figure 7-2 shows the degrees of freedom for a polar robot.

The arm and body joints for robots of cylindrical, polar, or jointed-arm configuration have three degrees of freedom:

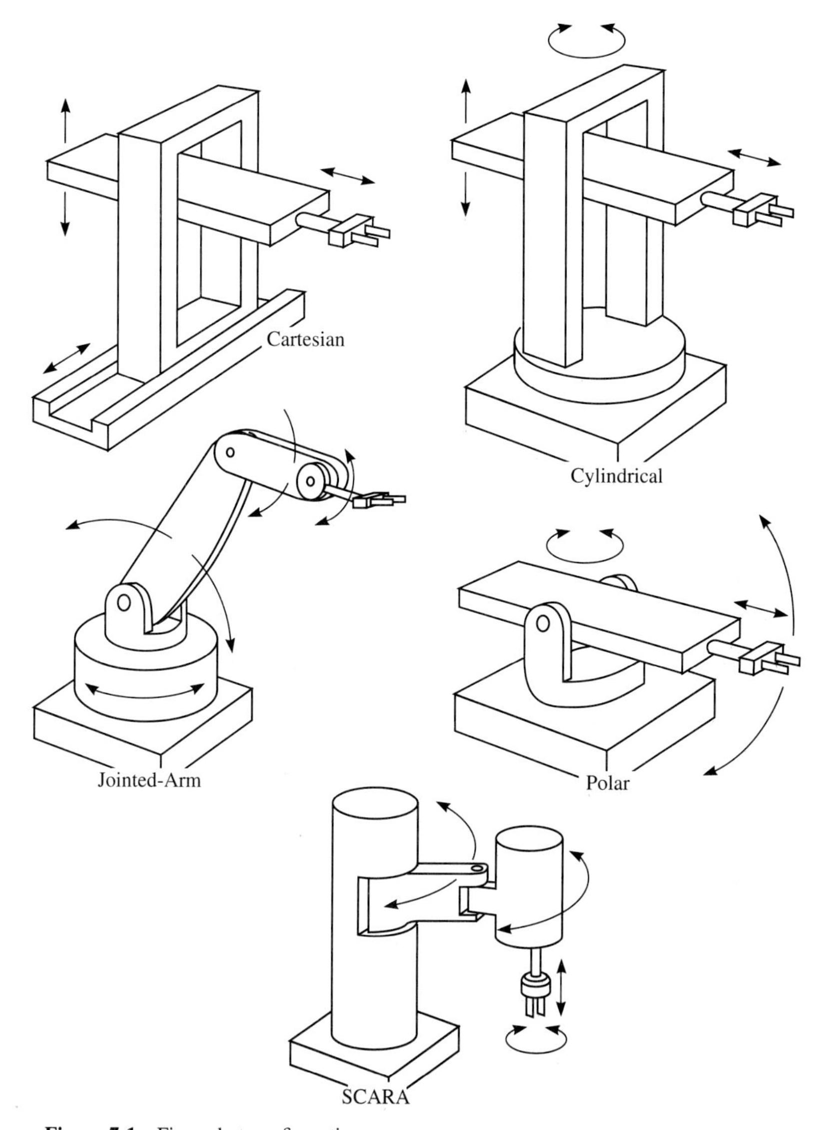

Figure 7-1 Five robot configurations.

- *Rotational traverse:* The rotation of the arm about the vertical axis.
- *Radial traverse:* The telescoping motion of the arm from the vertical center of the robot.
- *Vertical traverse:* The movement of the wrist up and down.

The wrist motion enables a robot to properly orient the end effector with respect to the task to be accomplished. A wrist is typically given up to three degrees of freedom in order to solve an orientation problem. The following terms describe wrist motion:

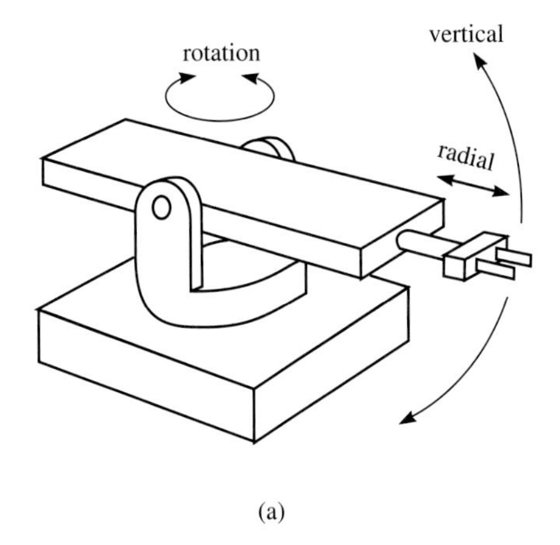

(a)

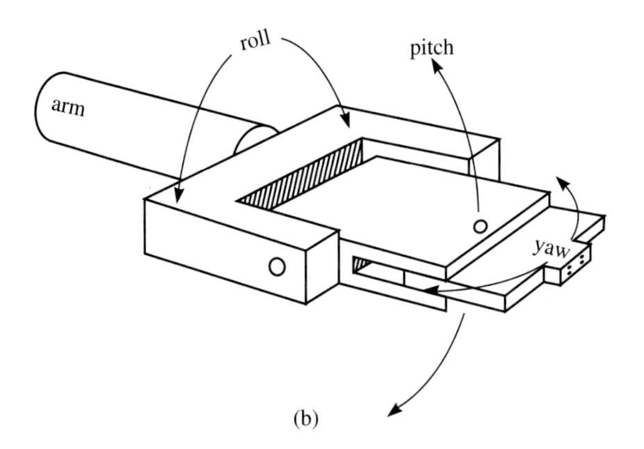

(b)

Figure 7-2 Robot motions: (a) arm and body motions; (b) wrist motions.

- *Wrist roll:* Involves the rotation of the wrist about the arm axis.
- *Wrist yaw:* Involves the left and right rotation of the wrist when the wrist roll is in the center position of its range.
- *Wrist pitch:* Involves the up and down rotation of the wrist when the wrist roll is in its center position.

7.2.2 Work Envelope

The work envelope is the space within which the robot can use its wrist. The reason for defining the work envelope in this manner is to avoid the complication of different sizes and shapes of end effectors which might be attached to the wrist. The work envelope for a robot is determined by its physical configuration (cartesian, cylindrical, SCARA), the sizes of the three components of the manipulator, and the limits of the joint movements for the manipulator. See Fig. 7-3 for the work envelopes of cartesian, cylindrical, and polar robots.

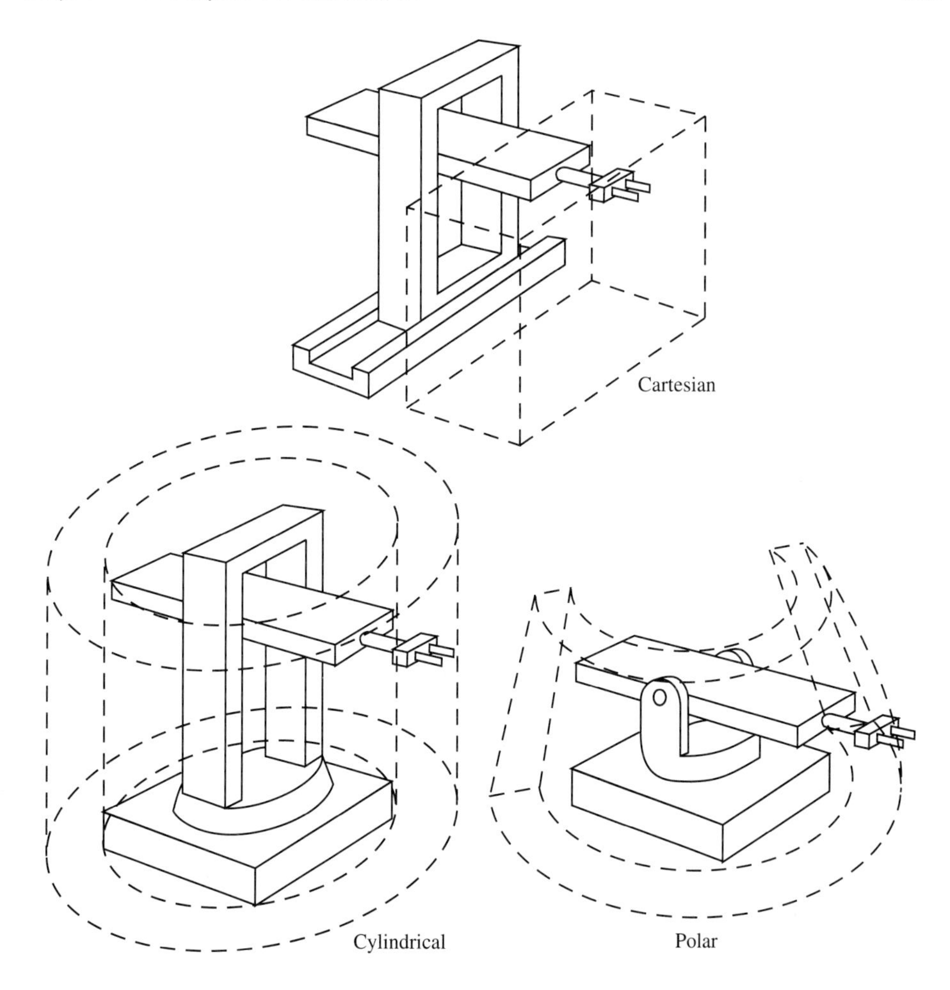

Figure 7-3 Work envelope for various robot configurations.

7.2.3 Drive Systems

The power required to move the robot is provided by the drive system, which determines the speed and strength of the arm movements and, thus, the feasible applications of the robot. Most industrial robots are powered by one of the following three types of drive systems:

- *Pneumatic:* Pneumatic drive is usually reserved for the smaller robots, which are limited to simple, fast-cycled, pick-and-place operations. Robots with pneumatic drive usually have only two to four degrees of freedom.
- *Hydraulic:* Hydraulic drive is used in larger robots. Hydraulic robots are generally heavy and require both large floor space and heavy floor loadings.

- *Electric:* Electric drive is good for robots in light-duty, precision applications but does not offer the speed and strength of the hydraulic drive. The electric-driven robots are smaller in size and are used in applications such as electronics assembly, where precision is required.

7.2.4 Motion Control Systems

A robot must control its drive system in order to regulate its motions. Therefore, a robot must use one of four types of motion control systems:

- Axis-limited
- Point-to-point playback
- Contour-path playback
- Intelligent.[1]

Robots with axis-limited control do not use feedback to indicate the relative positions of the joints; they use limit switches or mechanical stops to define the end points of travel for each joint. The positions and the sequences of stops involve a mechanical setup rather than a software approach for programming a robot. Any one of the three types of drives can be used, but pneumatic is most commonly employed.

Point-to-point playback robots perform motion cycles that consist of a series of point locations and related actions, which are taught and recorded into a control unit. During playback, the robot moves from point to point in the proper sequence, but the control unit does not control the path between points. This type of control is adequate for spot welding and for loading and unloading of machines.

Contour-path playback robots perform motion cycles in which the control unit directs the path of the robot from point to point. The robot moves through a series of closely spaced points defined by the control unit rather than a programmer. A control unit capable of storing a large number of point locations is used for defining a smooth, curved-path motion. Arc welding and spray painting are two applications for this type of control system.

Intelligent robots include the most sophisticated controls and possess the playback capability, but they also can alter their programmed cycle by making logical decisions based on sensor data obtained from the environment. Communication during the work cycle with human operators or other computer-based systems, such as machine vision and host computer, is possible. The programming for intelligent robots is performed using Englishlike, symbolic language similar to general-purpose computer programming languages. Typical applications for this type of control systems are arc welding and assembly.

[1]M. P. Groover et al., *Industrial Robotics: Technology, Programming, and Applications* (New York: McGraw-Hill Book Company, 1986), pp. 35–37.

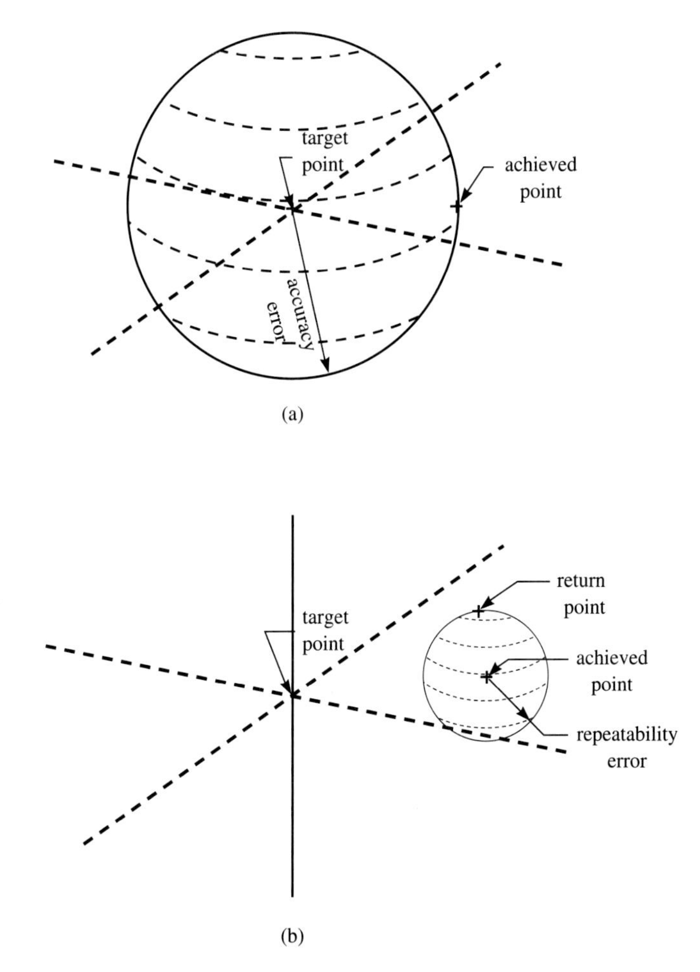

Figure 7-4 Precision of movement: (a) accuracy; (b) repeatability.

7.2.5 Precision

An important measure of robot performance is the **precision** with which a robot is capable of targeting a given position in its work envelope. Precision is commonly defined as a function of three features: accuracy, repeatability, and compliance.[2] The definitions for these terms are based on Fig. 7-4 and the following assumptions:

[2]C. Ray Asfahl, *Robots and Manufacturing Automation,* 2nd ed. (New York: John Wiley & Sons, Inc., 1992), pp. 164–65.

- There is no end effector attached to the wrist.
- Only worst-case conditions will be considered. Generally, this means that the robot arm is fully extended.
- Definitions are developed in the context of a point-to-point robot.

The ability of a robot to position its wrist at a designated target point within the work envelope is referred to as **accuracy**. Accuracy is expressed as a range (plus and minus) about a specified value. **Local accuracy** measures the ability to reach a designated target point within a segment of the total work envelope. When the entire work envelope is considered, then the term **global accuracy** is used.

The following three factors influence accuracy:

- Within the work envelope, accuracy is at a maximum where the arm of the robot is close to its base and becomes worse as the arm is extended to the outer range of the work envelope.
- If the motion cycle is restricted to a limited work space, then accuracy of the robot can be improved.
- The load carried by the robot influences accuracy because heavier loads cause greater deflection of the mechanical components, resulting in lower accuracy.

The ability of a robot to reposition its wrist or end effector at a point in the work space that has been programmed or previously taught to the robot is known as **repeatability.** Accuracy and repeatability refer to two quite different aspects of robot precision. Accuracy is the ability to be programmed to achieve a designated target point. Repeatability is the ability to return to a programmed or taught point.

Repeatability errors, when mapped in three-dimensional space with the designated programmed target point as center, can be conceptualized as a spherical region. Manufacturers of robots typically quote repeatability as the radius of the idealized sphere. Repeatability will tend to degrade in the areas of the work envelope that are further away from the center of the manipulator.

Compliance is the displacement of a robot wrist due to a force being exerted upon it. The displacement can either be caused by the weight of a load or as the reaction to a force being exerted on a workpiece by the robot. If the wrist is displaced a large amount by a small force, the compliance is high; conversely, if the wrist is displaced a small amount, there is low compliance. Robot performance will be degraded when operating under loaded conditions if the compliance has not been considered.

7.2.6 Mechatronics

Recall from Chapter 2 that **mechatronics** is a term used to describe the combined technologies of electrical, mechanical, and computer engineering applied in the design and development of high-precision machines, including assembly robots. The mechatronics concept is employed in the design of robot components (for example, stepper motors, bearings, and end effectors) and of the robot as a complete system with controlling computers.

To achieve the accuracy and repeatability required in electronics assembly applications, there is a need for miniature robots to execute precise positioning. There have been two types of these miniature robots developed, either of which attach as end effectors to a larger robot wrist.[3] The master robot manipulator executes large motion commands, and the miniature robot executes the precision motions to pick up a workpiece and perform the appropriate assembly tasks.

In one approach, the miniature robot is attached to a mounting plate by a compliant coupling, which allows limited rotation and three-dimensional translation. Using three locating legs with conical tips and a set of springs, the miniature robot can be positioned using a fixturing principle commonly used to locate a part for accurate machining.

Another miniature robot is attached to the master robot wrist by electromagnetic coupling across an air bearing (layer of compressed air). The precision of this approach is ten times finer than that of the plate-mounted miniature robot and one hundred times finer than that of a typical industrial robot.

7.2.7 End Effectors

The end effector is the gripper or tool attached to the robot wrist, including the sensor system which allows the robot to interact with its work environment. Grippers are used to grasp and hold a workpiece during the work cycle. There are mechanical grippers, vacuum cups, magnetic grippers, adhesive grippers, and inflatable diaphragms. When an application requires a robot to perform some operation, other than grasping and holding a workpiece, the end effector is usually some type of tool (for example, for arc welding, spot welding, spray painting, drilling, grinding, and adhesive application).

7.2.8 Robotic Sensors

Sensors are used in industrial applications to allow the robot to operate with other pieces of equipment in a workcell. Some of the more common sensor devices used in robotic workcells are as follows[4]:

- *Electrical contact switch:* Used to indicate presence or absence of a conductive object by establishing an electrical potential between two objects.
- *Limit switch:* An electrical on-off switch actuated by mechanical means for measuring the presence or absence of an object.
- *Microswitch:* A small electrical limit switch used to measure the presence or absence of an object.
- *Strain gauge:* Used to measure force, torque, or pressure applied to grasp an object.
- *Eddy current detectors:* Used to indicate presence or absence of a conductive object by inducing eddy currents in any conductive object.

[3]Stephen Derby, "Mechatronics for Robots," *Mechanical Engineering,* July 1990, pp. 40–42.

[4]Groover et al., *Industrial Robotics,* p. 146.

- *Infrared sensor:* Used to indicate presence or absence of a hot object by using the infrared light emitted from the surface of an object.
- *Optical pyrometer:* Used to indicate presence or absence of a hot object by detecting the brightness of the object's surface.
- *Photometric sensors:* Used to indicate presence or absence of an object by using light.
- *Radiation pyrometer:* Used to indicate presence or absence of a hot object by using the thermal radiation emitted from an object's surface.
- *Vacuum switches:* Used with vacuum grippers to indicate presence or absence of an object by using negative air pressure.
- *Ammeter:* Used to measure electrical current.
- *Piezoelectric accelerometer:* Used to measure vibration.
- *Potentiometer:* Used to measure voltage.
- *Pressure transducers:* Used to indicate fluid pressure and air pressure.
- *Thermistor:* Used to measure temperatures based on electrical resistance.
- *Ohmmeter:* Used to measure electrical resistance.
- *Thermocouple:* Used to measure temperatures based on the electromotive force emitted at the junction of two dissimilar metals.
- *Machine vision:* Used in inspection, part identification, location, and orientation.

These sensors are used in four types of industrial applications, as follows:

- *Workcell control:* Coordination of the individual pieces of automated equipment by regulating the work cycle sequence and verifying that certain elements of the work cycle have been completed satisfactorily.
- *Positional data collection:* Part identification and random part position and orientation.
- *Quality control:* Automated inspection of all parts for a limited range of part characteristics and defects.
- *Safety monitoring:* Protection of human operators working near robots or other automated equipment.

7.2.9 Robot Applications

There is a wide assortment of robot applications in industry. Most of the industrial applications can be assigned to one of the following four categories:

- Material handling (Chapter 14)
- Processing (Chapters 3–6)
- Assembly (Chapters 8 and 9)
- Inspection (Chapter 11).

There is interest in robotic assembly because of the high manual labor content of most assembly operations. Inspection applications are important because of the emphasis placed on product quality.

An assembly process is a sequence of operations where discrete components are fitted together to form a new, more complex unit. A considerable amount of part handling and orientation is involved in assembly operations. Therefore, in robotic assembly, accuracy, repeatability, variety of motion, complex gripper devices, and multiple gripper mechanisms are required.

In industry there are a variety of assembly processes to join parts together using fasteners (screws, nuts, bolts, rivets) or by other joining (swaging, crimping, press fit, snap fit, welding, brazing, soldering, and adhesives). Each assembly process requires specialized grippers or tools.

A successful robotic assembly usually requires a variety of tactile (touch) and visual feedbacks to the robot controller to monitor the progress of the assembly and permit corrective action(s) if necessary. Tactile force sensors mounted on the gripper provide the controller with data so that the robot can detect when a part is misaligned and realignment is required. Feedback to a robot controller provides information as to part identification, part location and orientation, and for determining if parts are missing or lost.

If an assembly robot has tactile sensory and visual capabilities, inspection of products for quality control is feasible. **Inspection** checks the product in relation to design specifications, such as dimensional accuracy, surface finish, and completeness/correctness of assembly.

Generally, other pieces of equipment are used in conjunction with robots in order to perform inspection. For example, a robot presents a large PWB for inspection of both sides. The robot will reposition the PWB as needed so that the inspection system can check the dimensions of the PWB, locate and count the mounted components, and identify components by type.

7.2.10 Placement/Insertion of Electronic Components

Robotic assembly of PWBs reduces product cost and increases product quality.[5] In electronics assembly, two types of robot configurations are common for assembly of large or irregularly shaped parts onto PWBs: the cartesian and the SCARA. These robots were designed for assembly operations and provide the accuracy and tight tolerances of component placement. Whether a cartesian or a SCARA robot is used in a particular application is determined by the constraints of the actual work space.

With the proper component delivery system, end effectors, and tools, an assembly robot is capable of assembling a wide variety of irregular-shaped electronic components, such as connectors, transformers, potentiometers, radial devices, crystals, light-emitting diodes, and large DIPs (for example, microprocessors). Lead configuration (shape, length, thickness, and spacing) significantly affects the assembly process.

Other applications of robots in the electronics industry are the preparation of compo-

[5]J. Storjohann, "Reducing Labor Costs Using Robotics in Electronic Assembly Manufacturing Operations," *Industrial Engineering,* 18, no. 2 (February 1986), pp. 52–56.

nent leads for insertion (such as lead forming, trimming, and tinning), the material handling functions in a clean room, parts kitting where parts are grouped into an assembly kit, dispensing solder mask onto PWBs, and application of adhesives and other types of viscous fluids.[6]

Before a robot can perform any assembly, the parts that compose the assembly must be presented to the robot in a known position and orientation. There are several approaches that can be taken to obtain a known position and orientation for parts, including:

- Vibratory
- Centrifugal (rotary)
- Magazine
- Matrix trays[7]

There are two types of vibratory feeders: bowl and straight-line. In bowl feeders, a track attached in a spiral arrangement to inside of the bowl creates a circular flow of parts from the bottom of the bowl up an inclined pathway to an outlet point. Straight-line feeders use linear part movement to obtain required part orientation.[8]

Centrifugal feeders, also known as rotary feeders, employ a spinning drum to separate parts. The centrifugal feeders are limited in application when compared to the vibratory feeders, since they rely on feeding simple part configurations. In certain applications, the centrifugal feeders offer higher feed rates, gentler parts handling, and reduced noise levels.

Once the parts approach the outlet point of a feeder, proper part orientation begins with the use of two methods[9]:

- *Selection:* Removing the parts from the track that are not properly oriented and allowing properly oriented parts to pass to the outlet point.
- *Orientation:* Physically moving parts into the desired orientation.

Both selection and orientation use a series of obstacles to allow only those parts to pass which meet certain specified orientation criteria. Another approach which does not use mechanical obstacles is photo-optic pattern recognition. This approach allows verification of part orientation within the feeder, prior to discharge at the outlet point. Misoriented parts are removed and returned to the feeder for recirculation. The photo-optic device adds extra flexibility to the feeder in that the controller memory can retain and recognize entire families of parts with minimal or no changeover required.

Parts having exited through the outlet point usually travel down an inclined chute

[6]W. H. Schwartz, "Robot PC Board Assembly," *Assembly Engineering,* 32, no. 6 (June 1989), pp. 35–37.

[7]L. K. Schuch, "Robots, Part Feeders, and Handlers," *Assembly Engineering,* 32, no. 9 (September 1989), pp. 60–65.

[8]R. D. Zimmerman, "Presenting Parts Presentation," *Assembly Engineering,* 32, no. 7 (July 1989), pp. 26–27.

[9]Asfahl, *Robots and Manufacturing Automation,* pp. 164–65.

to a holding fixture, which maintains the proper part position and orientation until the part is removed for assembly by a robot. The holding fixture must allow the robot gripper to have enough clearance to grasp the part.

Magazine feeders are used to present preoriented parts to an assembly workstation. Parts are loaded into tubes or some other type of container in proper orientation as they are manufactured. The magazine feeder escapement mechanisms (allowing only one part to be released at a time) feed the parts from the tube or container to the robot. Typically, part magazine feeders hold fewer parts than the vibratory or centrifugal feeders.

Vibratory, centrifugal, and magazine feeders are usually customized for a particular part configuration. In those cases where a variety of different part geometries is required, matrix trays are often used. In order to use this approach, the parts must be stored in known positions and orientations with respect to specified reference points such as the edges of the tray.

A typical programmable feeder system may be composed of part feeders (vibratory and/or centrifugal), a photo-optic pattern recognition device, a vision controller, mechanical orienting mechanisms, and a conveyor. The vibratory and/or centrifugal feeders move scrambled parts in single file past a photo-optic pattern recognition device, which scans the parts and collects image data. A vision controller determines the proper part orientation by comparing the required part orientation, previously taught to the vision controller, with that of the passing orientation data from the pattern recognition device. Properly oriented parts are conveyed to the assembly workstation by a conveyor, and those rejected because of improper orientation are sent back into the feeders for recirculation.

A programmable feeder system can be further enhanced in its part orientation and sorting capabilities by incorporating mechanical orienting mechanisms. These mechanisms can rotate about two axes to allow only one acceptable orientation for parts arriving in any of four possible attitudes or rotate parts 180 degrees so that reverse orientations can be obtained. The orienting mechanisms aid in attaining higher feed rates, and the degradation of parts is minimized by eliminating recirculation. Programmable feeders can sort parts with minimal differences, parts with different lengths or diameters, and rejected broken parts.[10]

There are two types of end effectors used for grasping an electronic component: mechanical grippers (Fig. 7-5) and vacuum cup/tip grippers (Fig. 7-6). Mechanical grippers have two ways of capturing a part. The first approach involves the gripper fingers physically constraining the part between fingers shaped compatibly with the part geometry. The second approach is for the gripper to hold the part by pressure. Both approaches require gripping the part securely and maintaining sufficient gripper force to hold the part in acceleration, deceleration, and any orientation in the programmed path.

Friction can be increased to better hold the part by fabricating the fingers or contact pads from relatively soft material. Use of softer material for the gripper surfaces can also protect the part from being scratched or dented.

[10]W. H. Schwartz and L. K. Schuch, "Robots, Part Handlers and Feeders," *Assembly Engineering,* 31, no. 9 (September 1988), pp. 66–69.

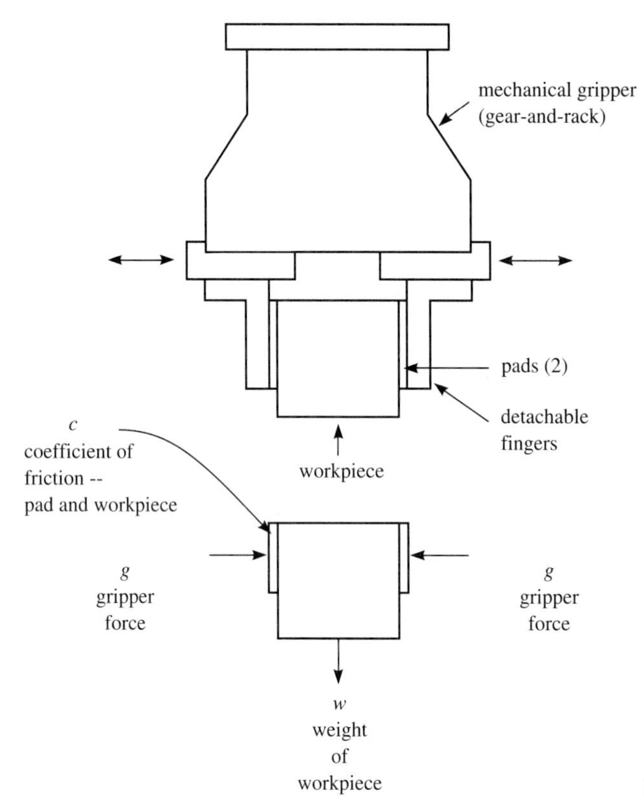

Figure 7-5 Mechanical gripper.

To determine the friction force required to grasp and move an object, the following force equation can be used[11]:

$$g = (w\,f\,s)/(c\,n) \tag{7-1}$$

where

g = Gripper friction force required
w = Weight of component being grasped
f = A factor to account for the combined effect of gravity and acceleration
f = 2.0, if acceleration is applied in a horizontal direction relative to gravity
f = 2.5, if acceleration is applied in the opposite direction of gravity
f = 3.0, if acceleration and gravity are applied in the same direction
s = Safety factor to help compensate for a component not being grasped at its center of mass—a suggested value of 1.5 can be used
c = Coefficient of friction between gripper contact surface and the component surface
n = Number of contacting surfaces

[11]Groover et al., *Industrial Robotics,* p. 120.

Suppose a through-hole electronic component weighing .04 ounce (1.13 g) is grasped and inserted into a PWB by a gripper using friction against two opposing contact surfaces. The coefficient of friction between the component surface and the gripper contacting surfaces is 0.18. The gravitational force on the component is parallel to the contacting surfaces. Acceleration/deceleration is applied in the horizontal direction relative to the gravitational force. With this information, the required gripper force can be determined as follows:

$$g = \frac{.04 \text{ oz} \times 2.0 \times 1.5}{0.18 \times 2} = 0.33 \text{ ounces (9.36 grams)}$$

or approximately nine times the weight of the component.

The vacuum cup is typically a round-shaped elastic material composed of either soft plastic or rubber. If the part to be handled is a soft material, the vacuum cup should

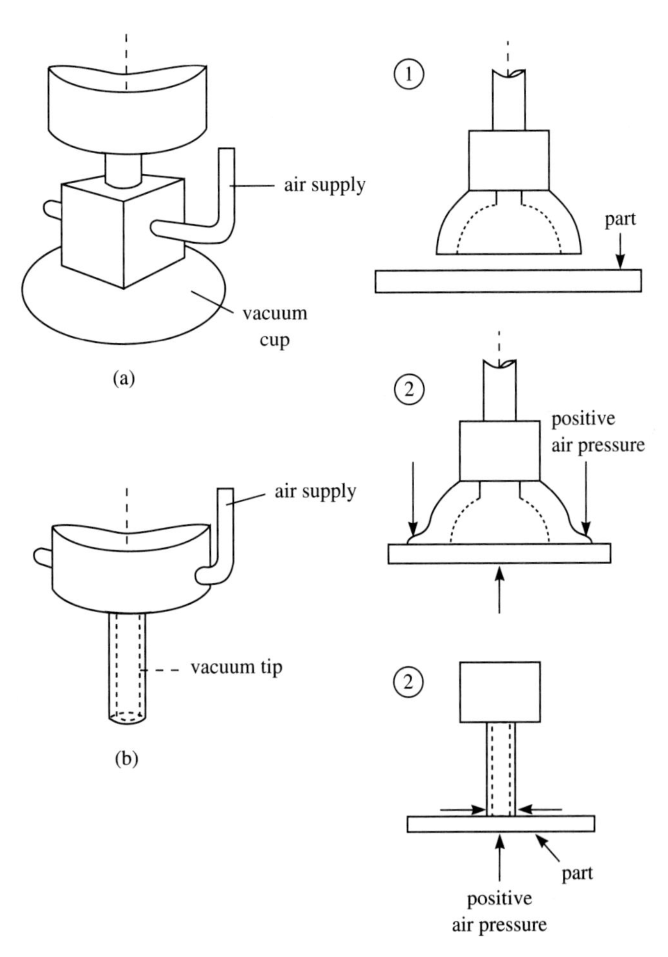

Figure 7-6 Vacuum grippers: (a) vacuum cup; (b) vacuum tip.

be composed of a hard material, such as DuPont Teflon™. The vacuum tip is a round nozzle of a specified diameter to grasp the part and hold it.

Successful application of a vacuum cup/tip requires that the part be smooth, flat, and clean, so that a sufficient vacuum can be created between the part and the vacuum cup/tip. Some characteristics of vacuum cup/tip gripping are that only one surface of a part is used for grasping and a uniform negative air pressure is distributed on the surface of the part.

The lift capacity of a vacuum cup/tip depends on the effective area of the cup/tip and the negative air pressure between the cup/tip and component. The relationship is summarized in the following equation[12]:

$$P = (F\,s)/A \tag{7-2}$$

where

P = Negative pressure (pounds per square inch)
F = Lifting capacity required (pounds)
s = Safety factor to compensate for acceleration, and possible reduction of effective area of vacuum cup/tip
A = Effective area of the vacuum cup/tip used to create the vacuum (square inches)

Suppose a 0.375 inch (0.95 cm) diameter vacuum cup will be used to lift a surface-mount device weighing 0.11 ounce (0.0068 pounds or 3.12 grams) and to place it on a PWB. A safety factor of 1.5 will be used. The negative pressure (compared to a 14.7 lbs/in^2 (101.3 kilopascals) atmospheric pressure) required to lift and place the component will be $P = (F\,s)/A = (0.0068 \times 1.5)/(3.142 \times (0.375/2)^2) = 0.09$ lbs/in^2 (0.62 kilopascals).

There are some approaches in gripper technology which provide a simple solution to the problem of lead and hole alignment. The traditional method of overcoming lead misalignment is a search routine in which a software program instructs the robot to make a number of attempts to insert the component. If a component insertion is detected as having failed, a search subroutine is initiated where the robot makes a specified number of tries to insert the component. If all attempts fail, the component is set aside by the robot, a new component is selected, and the insertion procedure begins again.

Other methods of solving misalignment are compliance systems which are placed between the gripper and the wrist of the robot (Fig.7-7). One of these compliance systems is called a remote center compliance device, which is capable of allowing lateral and angular errors in the component insertion process. Another compliance system is an instrumented remote center compliance device, which uses sensors to measure the forces encountered by a component and allows the robot to compensate for these forces.

SCARA robots can also be used to provide compliance. These manipulators are very stiff in the vertical direction (insertion axis) and relatively compliant horizontally. This feature allows for the assembly of a variety of components which are stacked from the vertical direction.

Gripper design can become quite complex. For example, there is an instrumented

[12]Groover et al., *Industrial Robotics,* p. 128.

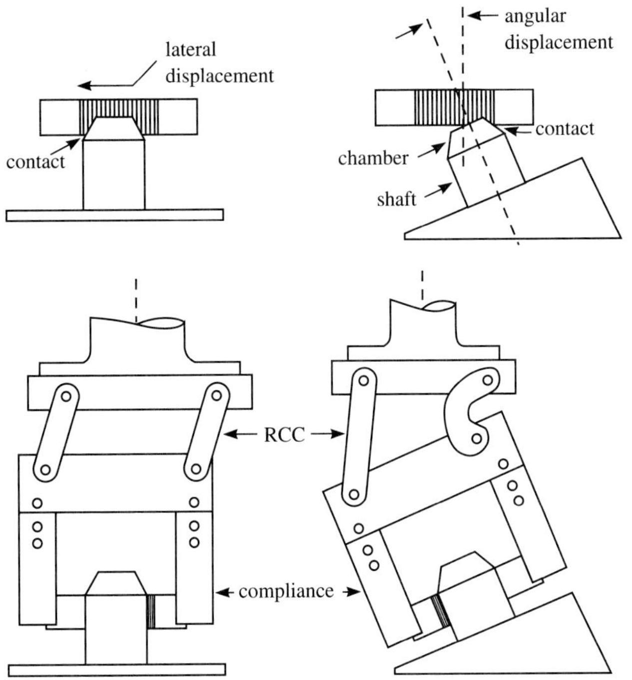

Figure 7-7 Remote center compliance.

gripper which enables a robot to pick up eighteen different types of odd-form components, to locate holes for insertion, and to verify successful insertion. Two finger latches (from a selection of detachable fingers) are used to grasp the odd-form components. Included on the gripper are two force/torque sensing systems which sense successful grasping. If a part is not grasped, a search is initiated for another component from the feeder.

7.3 MACHINE VISION TECHNOLOGY

Machine vision technology is increasingly applied to assembly and inspection tasks traditionally performed by human operators. In the manufacturing of electronic products, miniaturization and increased circuit density make reliable human assembly and inspection virtually impossible.

7.3.1 Automated Machine Vision

Figure 7-8 is a functional block diagram of an automated machine vision system for assembly and inspection. The operation of a machine vision system is composed of four distinct stages, as follows:

- Acquiring the image of an object
- Processing the image data for analysis

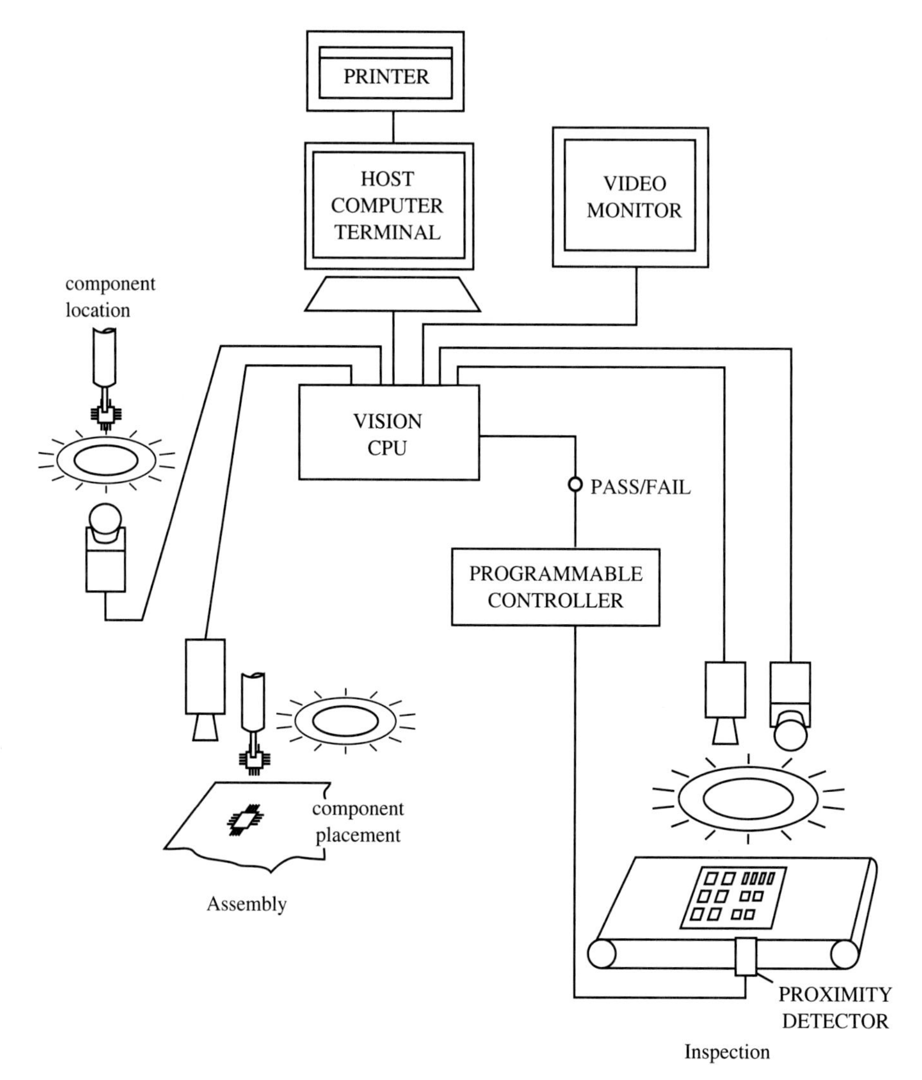

Figure 7-8 Automated machine vision for assembly and inspection.

- Analyzing the image for specific information
- Transmitting resulting information

Image acquisition is the first task of a machine vision system. Acquiring a good quality image is extremely important, since subsequent image processing depends largely on the quality of the image data. Illumination and viewing techniques play a major role in the acquisition of a high-quality image.

There are two techniques of illumination used in machine vision: front lighting and back lighting. Front lighting uses light sources and vision camera located on the same side of the object to be imaged. Back lighting uses a light source directed at the camera and located behind the object. Back lighting is preferred when possible, since it provides the greater image contrast. However, front lighting must be used whenever surface features are to be extracted. Front lighting is commonly used for feature detection, character or pattern recognition, noncontact measurement, and surface flaw detection. The common usage of back lighting is to provide silhouette information, feature detection, and information regarding the internal structure of transparent objects.

There are several other forms of illumination used in industry for various purposes. Structured lighting is a method that provides depth information in a two-dimensional image. The position and orientation of many manufactured parts, including complex curved objects, can be determined. This method is effective in visually noisy picture data typical of many plant environments. Lasers are commonly used as light sources in structured lighting techniques involving line scan or matrix cameras for image acquisition.

The background selection is also important to promote imaging. High-contrast and retro-reflective background can promote object identification and edge detection.

7.3.2 Applications

Machine vision applications can be categorized by task and function in the following manner:

- Assembly is the process of fitting together of two or more discrete parts to form a subassembly. Therefore, the functions of part identification, location, orientation, and part count are important.
- Inspection is the gauging and measuring of the quality characteristics of assemblies. The functions of dimensional accuracy, surface finish, and completeness/correctness of assemblies are important.

Part identification is primarily concerned with recognizing or determining the identity of a part for classification purposes. Identification may involve reading special codes or symbols, such as alphanumeric characters. Identification involves a recognition process in which the part, its position, and/or its orientation is determined. Identification applications arise in assembly, part sorting, palletizing, depalletizing, packaging, and picking parts that are randomly oriented from a bin or from a conveyor.

Recognition involves determining the identity of a part from physical features such as shape, size, or color. For example, recognition of a unique subassembly involves detection of distinguishable features from among many other subassemblies which have otherwise identical features. Such recognition using automated machine vision systems can help to maintain correct inventories of production (counting) or to verify the presence of expected objects.

Another widely used application of machine vision systems is inspection for possible defects. Inspection ensures that specific features or components of a part exist and that undesirable features are absent. This includes checking for gross surface defects such as cracks, pitting, or chipping, discovery of labeling flaws, verification of component presence in assembly, measurement for dimensional accuracy, and checking for the presence of expected features (for example, leads, holes, bolts, nuts) in assembly. In the field of automated inspection, machine vision has become a vital technology in the move from quality insurance (sampling inspection) to quality assurance (100 percent inspection and statistical process control) and then to design improvement as described in Chapter 11.

7.3.3 Machine Vision Applications for PWBs

Machine vision applications are common in the manufacture of PWBs. Machine vision systems are used in the placement of surface-mount devices with lead pitch of 0.025 inch (0.6 mm) or less. There are two steps in the successful placement of surface-mount devices, as follows:

- The location of the device on the vacuum tip which picks and places the device must be determined. Translation and rotational offsets are the difference between the device and the vacuum tip center and rotation. The part orientation in terms of ΔX and ΔY (translational offset) and orientation angle (rotational offset) must be known for accurate placement.
- The location of fiducials (reference marks incorporated into the circuit pattern) must also be determined. The vision processor calculates the required translational and rotational corrections, and the vacuum tip accurately places the device on the PWB. The ability of the vacuum tip to make very small increments in rotation can ensure correct lead-to-pad alignment. Automated machine vision systems typically incorporate the fiducial alignment and precision motion control to compensate for PWB variations and to ensure that the field of view for the solid-state cameras is correct for the specific sequence of component placement.

Critical inspection tasks can be performed by automated machine vision systems either before or after the solder operation. Inspection tasks for PWBs can be classified into three groups: solder paste deposition, component alignment, and solder joint integrity.[13]

Inspection tasks before solder detect

- Component presence and position on the PWB
- Absence or shortages of solder paste on the solder paste screen

Inspection tasks after reflow verify

- Component alignment
- Integrity of solder joints

For component alignment, machine vision systems typically use multiple cameras to view a component. Each camera can have an approximate field of view of 0.5 inch (1.3 cm). Using high-resolution solid-state cameras, machine vision systems can detect

[13] J. P. Kasik, "Who Needs AOI for Assemblies?" *Circuit Assembly,* December 1991, pp. 30–33.

spatial changes of 0.001 inch (0.025 mm) or less. This resolution is sufficient to measure leads and pads of 0.004 inch (0.1 mm) pitch.

Solder joint problems include

- Bridging of solder between solder connections
- Lifted leads due to contamination
- Bent leads
- Shortage/absence of solder in the wetting angle

Machine vision systems can verify solder joint integrity by viewing the wetting angle of each solder joint from various angles using different cameras to obtain three-dimensional data.

Additional information concerning the use of machine vision in surface-mount technology is found in Chapter 9, on surface-mount device placement, including:

- Reasons for implementing machine vision
- Specifications for a machine vision system
- Initial design considerations, such as camera placement and illuminating
- Vision centering

7.4 X-RAY AND LASER INSPECTION TECHNOLOGIES

In addition to machine vision, X-ray and laser inspection technologies are increasing in importance. All **X-ray inspection** systems place an object between the X-ray source and the image recording medium, such as halide film. **Fluoroscopy** is a nondestructive evaluation technique where X-rays are converted to visible light upon striking a phosphor-coated screen. The resulting image is captured by video camera and transmitted to a monitor.

Film images and fluoroscopy lack the clarity of visible light spectrum photography due to scattered radiation that occurs when X-rays impinge upon an object. Image quality is degraded when scattered radiation randomly strikes the image capture medium.

To reduce scatter-induced image degradation, a technique called reverse geometry X-ray was developed for inspection.[14] Typically, X-ray technology uses a point source for X-ray generation and a large detecting medium. The reverse geometry technique provides an opposite approach by using a large source for X-ray generation and a point source detector, as shown in Fig. 7-9. Due to the size and placement of the detector in the reverse geometry technique, scatter radiation is much reduced and increased pictorial clarity is achieved.

Data from the detector are fed directly into a computer without the use of additional equipment. A digitized X-ray image is produced and displayed on a high-resolution monitor.

[14]R. Albert, T. Albert, and J. Fjelstad, "SMT Inspection: Another Choice," *Circuits Assembly,* 2, no. 12 (December 1991), pp.34–37.

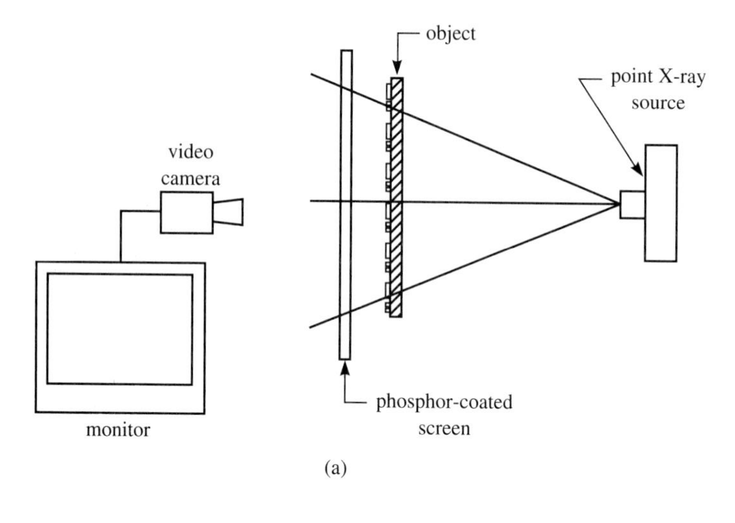

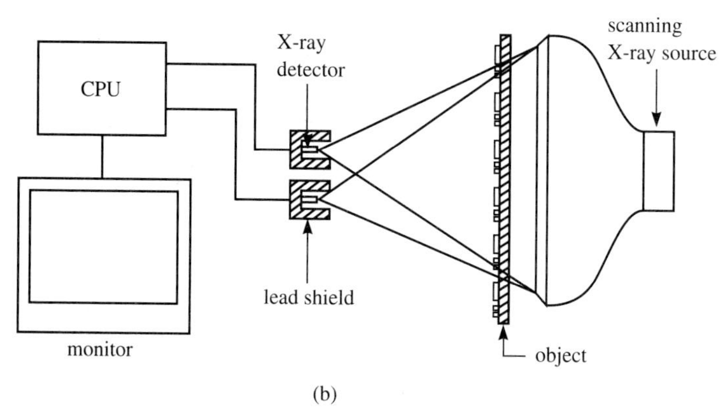

Figure 7-9 X-ray imaging: (a) fluoroscopy; (b) three-dimensional reverse geometry imaging.

Reverse geometry X-ray systems are used in determining solder thickness uniformity on the leads of surface-mount devices. Without mechanical manipulation, an object can be viewed in its entirety, or features of that object smaller than 0.001 inch (0.02 mm) can also be viewed.

Three-dimensional **stereoscopic inspection** is achieved by positioning two detectors at a distance similar to the distance between two human eyes. With special optics and two offset images, the operator can incorporate depth perception. Three-dimensional viewing aids in locating nonuniformities in solder joints and voids in solder paste deposition for double-sided and multilayer surface-mount PWBs.

In microfocus fluoroscopy, the focal-point size of the X-ray beam is small (for example, 10 microns), which reduces shadowing that can occur with high magnification. The distance between the X-ray source and the object is critical in determining how much geometric magnification can be achieved. Geometric magnification is defined as the ratio between the distance from the X-ray source to the image detector and the distance between the object and the X-ray source. The microfocus technique can be used to detect abnormalities in the interconnections of flip-chips.[15]

Scanned-beam laminography is a three-dimensional X-ray technique used to inspect solder joints by generating horizontal slice images. The images can isolate individual layers of solder joints from other soldered connections regardless of their positions. The fluoroscopy approach to image forming is used. This type of X-ray system can perform single-pass inspection of solder joints on double-sided PWBs.[16]

Laser triangulation provides extremely accurate three-dimensional measurement of area and height. This nondestructive technique aids in the evaluation of improper solder paste deposition and component placement (before a PWB enters the reflow operation in which solder joints are formed and rework costs increase significantly).[17]

Improper solder paste deposition must be identified prior to component placement. The criteria for successful solder paste screening include:

- Correct solder paste volume dispensed in the printing
- Proper paste-to-pad registration (which includes the verification that no pad-to-pad bridging exists)

Optimal screen printing requires the feedback of critical process variables immediately after deposition of solder paste. The critical solder paste parameters of height, volume, and registration of solder paste position, relative to the pad on the PWB, require measurement of both area and height.

Laser triangulation involves the use of a low-powered laser similar to those found in optical disc players. A light beam is projected onto the PWB surface, as shown in Fig. 7-10. Light reflected from the surface is focused onto a sensor. If the distance to the PWB surface changes, the reflected light strikes a different point on the sensor. By measuring the change in position on the sensor, the change in surface height can be determined. The position and intensity of the reflected light on the sensor produce both height and area measurement. By using four sensors oriented at 90 degree intervals and viewing from four directions, the preferred scanning direction can be achieved without moving the sensor or the PWB.[18]

A three-dimensional reproduction of solder paste deposition or component topology is created by the rapid pulsing of a laser beam across the inspection area at high resolution. High-speed image processors analyze the three-dimensional data obtained to determine critical measurements. The comparison between data-defined measurements and user-defined tolerances reveals process variations and defects. Due to continuous parallel

[15] O. A. Lijap, "X-ray Fluoroscope Inspection of Flip-Chips," *Circuits Assembly,* 2, no. 8 (August 1991), pp. 66–67.

[16] A. Jones, M. Lamar, and M. Strand, "X-Ray Inspection of SIMMs," *Circuit Assembly,* 2, no. 8 (August 1991), pp. 44–46.

[17] R. J. Nieves, "Fine-Pitch Technology: Meeting the Challenge," *Circuit Assembly,* 2, no. 5 (May 1991), pp. 40–48.

[18] S. Case, "Front-End Process Control," *Circuits Assembly,* 3, no. 1 (January 1992), p. 46.

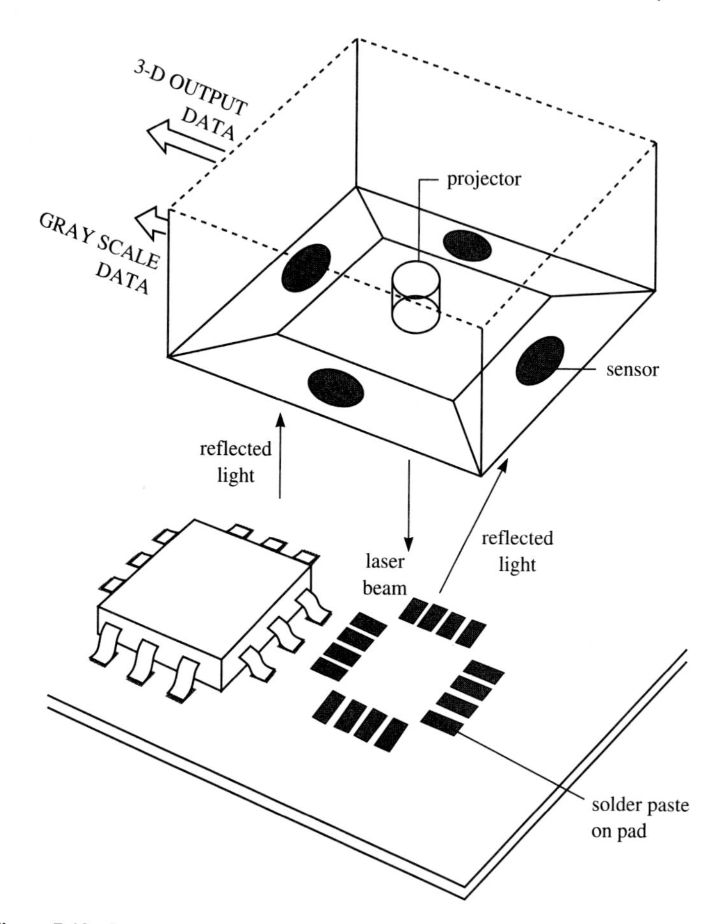

Figure 7-10 Laser triangulation.

processing the data can be monitored in real time and defects can be corrected immediately. The resolution of laser triangulation systems typically exceeds 0.001 inch (0.02 mm) in the *X* and *Y* axes (area) and 0.00025 inch (0.006 mm) in the *Z* axis (height).

Laser-infrared inspection is applied after the reflow soldering operation has been completed, to inspect individual solder joints and verify adequate solder paste deposition. The solder joints are heated by a laser beam, and the thermal rise rate is compared to a reference profile. Solder joints with insufficient solder exhibit rapid temperature rise, while joints with adequate solder exhibit much slower temperature increase.[19]

7.5 SUMMARY

This chapter has provided an introduction to the robotic and sensory technologies available for automaton of assembly and inspection processes. Chapters 8 and 9 describe the

[19]Albert et al., "SMT Inspection: Another Choice," pp. 34–37.

applications of these technologies to through-hole insertion and surface-mount placement, respectively.

KEY TERMS

Accuracy (local and global)
Compliance
End effector
Fluoroscopy
Inspection
Laser triangulation
Machine vision
Manipulator
Mechatronics
Pick-and-place
Repeatability
Precision
Robotics
Scanned-beam laminography
Selective compliance assembly robot arm (SCARA)
Sensors
Stereoscopic inspection
X-ray inspection

EXERCISES

1. What robot configurations are preferred for light-duty assembly tasks?
2. Why would pneumatic drive be limited to simple, smaller robots for pick-and-place operations?
3. What is the difference between accuracy and repeatability?
4. Define compliance, and explain the difference in high and low compliance.
5. What type of sensors would be considered for indicating presence or absence of an object?
6. If we use a 0.5 inch diameter vacuum cup to place a surface-mount device of 0.625 ounces on a PWB, what is the negative pressure (in psi) using a safety factor of 1.5?
7. Which inspection technologies would be the best for determining solder joint integrity in various applications?
8. What design parameters for an end effector must be considered for the insertion, end-wise to a depth of 12 mm into a subassembly, of a cylindrical object, composed of an aluminum alloy, with a diameter of 38 mm, a length of 75 mm, and a weight of 229 grams?
9. Explain what effects electronic component geometry can have on component selection, orientation, and attachment onto a PWB.
10. If machine vision is used for the detection of distinguishable features of a group of hollow cylinders, what features should be detected and gauged for proper recognition?

BIBLIOGRAPHY

1. Asfahl, C. Ray, *Robots and Manufacturing Automation* (2nd ed.). New York: John Wiley & Sons, Inc., 1992.
2. Clark, R., *Printed Circuit Engineering, Optimizing for Manufacturing.* New York: Van Nostrand Reinhold, 1989.
3. Fu, K. S., R. C. Gonzalez, and C.S.G. Lee, *Robotics: Control, Sensing, Vision, and Intelligence.* New York: McGraw-Hill Book Company, 1987.
4. Groover, M. P., M. Weiss, R. N. Nagel, and N. G. Odrey, *Industrial Robotics: Technology, Programming, and Applications.* New York: McGraw-Hill Book Company, 1986.

8 Leaded Component Insertion

8.1 INTRODUCTION

In this chapter we will consider through-hole assembly (**leaded component insertion**). The following topics are discussed:

- Insertion procedure for axial, radial, and dual in-line package components
- Types of packaging used for handling large quantities of components
- Component sequencing and preparation for insertion into a PWB
- Recommended standards for formed lead length and insertion hole size
- Insertion area and optimal insertion pattern for each component type
- Special tooling for component insertion
- Similarities and differences between semiautomatic and automatic insertion machines

8.1.1 Types of Insertion Components

There are three major types of insertion component configurations: axial, radial, and dual in-line package. These three component types were introduced in Chapter 3 and are shown in Fig. 8-1.

8.1.2 Automatic and Semiautomatic Insertion

In both automatic and semiautomatic insertion the components are loaded onto the machine on tapes/reels or in various types of carriers, such as tubes. The components are then picked up from the feeders, the leads are formed for insertion, the component is inserted

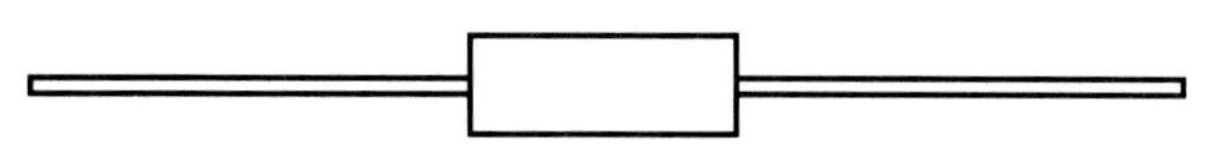

Typical axial component

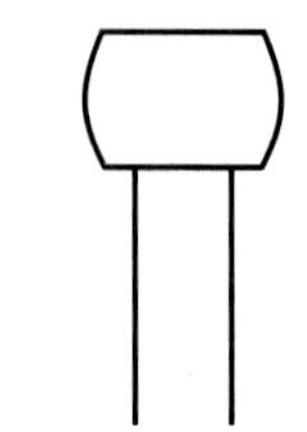

Typical radical component

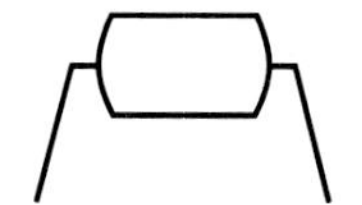

Typical DIP (Dual In-line Package) component

Figure 8-1 Types of insertion components.

in the PWB, and the leads are cut and clinched. It is essential for the PWB to be correctly positioned under the insertion head and for the components to be prepared and properly sequenced before they are fed to the machine.

An operator loads the component carriers onto the machine. In semiautomatic insertion the operator positions the PWB on the insertion machine and initiates the insertion sequence. In automatic insertion, an operator only monitors the machine functions and corrects any problems that arise.

8.1.3 Tooling Holes

The locations of insertion holes are referenced relative to **tooling holes** or to **datum points** or lines. Tooling hole diameter and position are controlled to an order of magnitude closer positioning than are insertion holes. Three tooling holes are selected at three corners of the PWB as far away from each other as possible to minimize the effects of variation. As shown in Fig. 8-2, insertion holes are located on the PWB in reference to **datum lines** constructed through the tooling holes.

Alternatively, arbitrarily chosen insertion holes may serve as reference points through which the datum lines are drawn. However, this method is not very accurate and should only be used in an emergency.

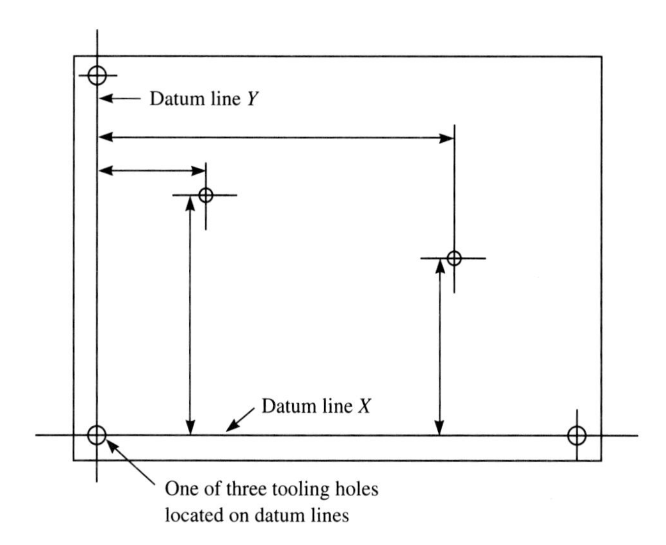

Figure 8-2 Tooling holes and datum lines.

8.1.4 Types of Component Orientation

Random orientation of components, as shown in Fig. 8-3, results in excessive production cost and creates problems in testing (Chapter 12). If possible, **component orientation** should be limited to single axis or double axis, as shown in Fig. 8-4. Three or four axes orientation is acceptable depending on the application. Polarity orientation should also be kept uniform if possible. Additional guidelines for producible design are presented in Chapter 10 and in reference 2 at the end of this chapter.

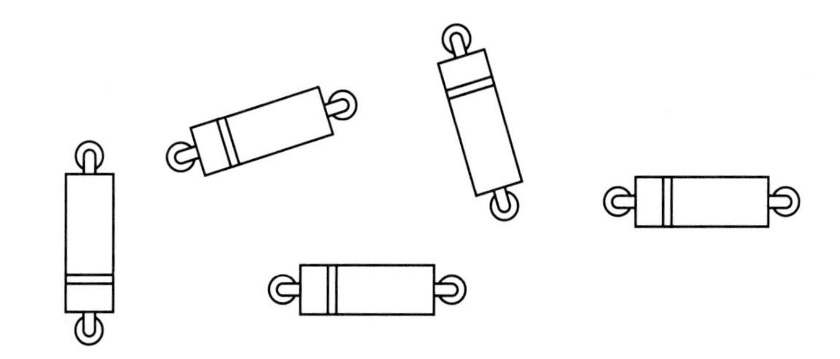

Figure 8-3 Random orientation.

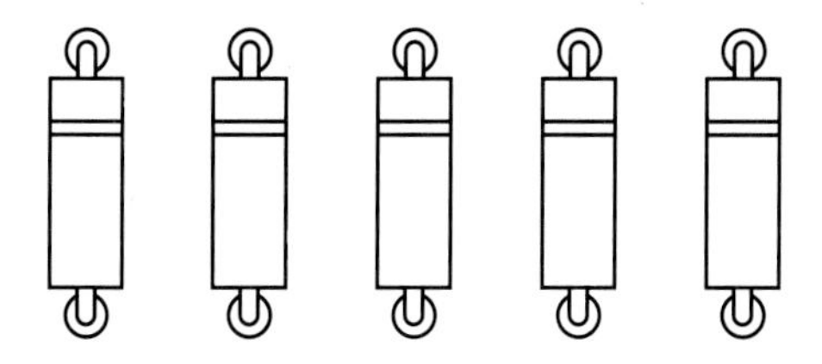

Figure 8-4 Single-axis orientation.

8.2 AXIAL-LEAD COMPONENTS

This section deals with axial-lead components which constitute a major portion of insertion components. All the examples, diagrams, specifications, and design guidelines are based on the Dynapert variable center distance insertion unit as an example, unless otherwise specified. These guidelines are helpful in understanding the concepts of component insertion. The requirements of insertion equipment may vary, but the concepts are consistent.

8.2.1 Variable Center Distance Machine

The **variable center distance** (VCD) insertion machine is widely used in the electronics industry. The VCD machine has the capability to insert components with different lead center distances (Fig. 8-5).

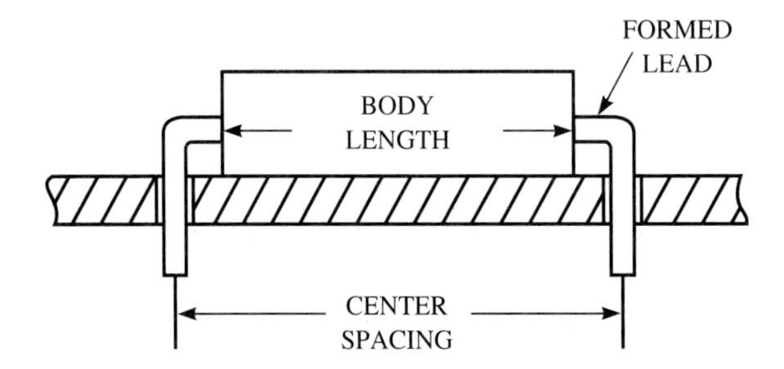

Figure 8-5 Center spacing.

Programmed variable tooling provides automatic adjustments for different center spacings between insertion holes, different lead wire diameters, and different component body diameters. Insertion tooling is able to handle center spacings from 0.3 inch (7.6 mm) to 1.3 inch (33 mm). Tooling for other insertion machines accommodates different ranges of lead center spacing.

A typical VCD insertion unit is shown in Fig. 8-6. The insertion unit has two symmetrical sides. Each side includes a driver, shear bar, outside former, inside former,

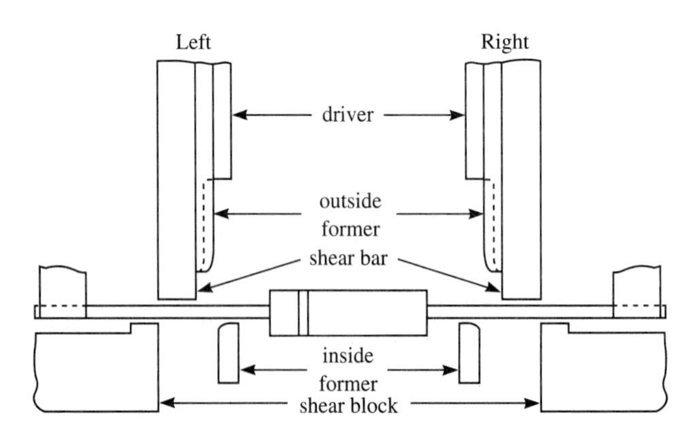

Figure 8-6 VCD insertion unit.

shear block, and indexing wheel. A programmable ball screw mechanism varies the spacing between left and right portions to accommodate the different insertion spacings.

A typical cut-and-clinch mechanism is shown in Fig. 8-7. The unit automatically adjusts to the center spacing. The length of cut can be adjusted from 0.04 inch (1.0 mm) to 0.09 inch (2.3 mm). The angle of clinch is adjustable from 0° to 45° relative to the plane of the board.

8.2.2 Insertion Area

Insertion area (Fig. 8-8) is the region within which the insertion machine can be programmed to insert components. The programmable insertion area of a VCD machine is 18 × 18 inches (0.46 m × 0.46 m). The *X* axis is measured from the center of the insertion unit.

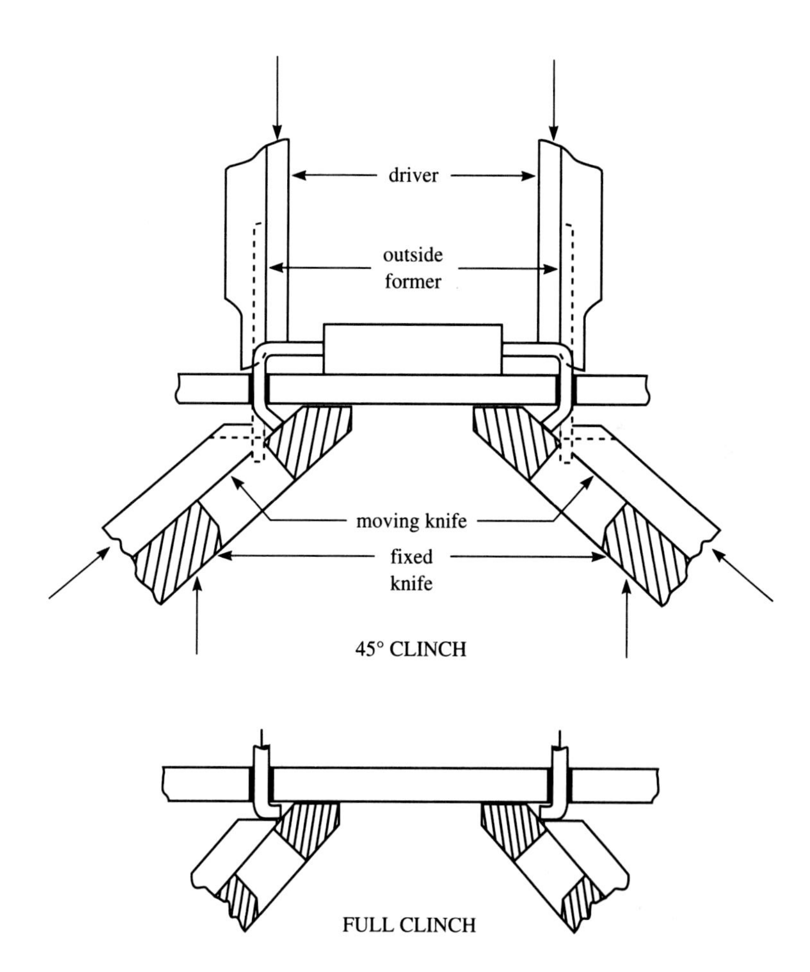

Figure 8-7 Cut-and-clinch mechanism.

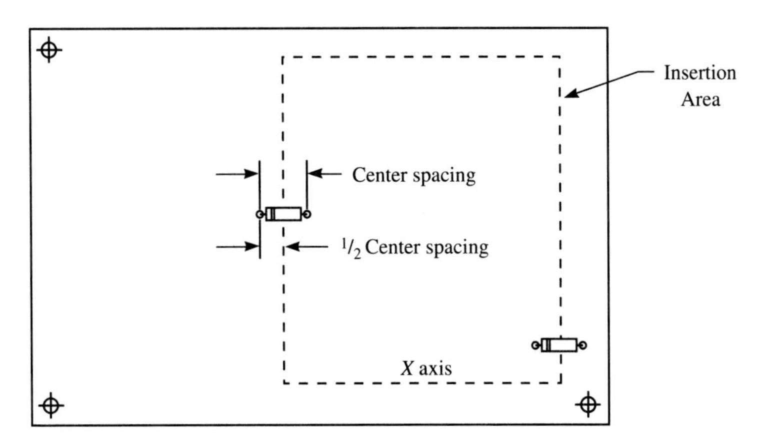

Figure 8-8 Insertion area for axial lead.

8.2.3 Insertion Hole Diameter and Formed Lead Length

Insertion hole diameter must be large enough to accommodate the sum of both insertion machine tolerances and PWB tolerances. A recommended standard is to have the hole diameter 0.015 inch (0.38 mm) more than the diameter of the lead to be inserted, as shown in Fig. 8-9. Larger or smaller holes are possible but have different limitations. Larger holes are limited by the clinch and soldering requirements, while smaller holes result in higher cost and more insertion problems.

Lead forming (Fig. 8-10) is another operation in PWB assembly. The **formed lead length** (*FLL*) should be 0.275 inch (7.0 mm) for all tooling except long cut-length tooling,

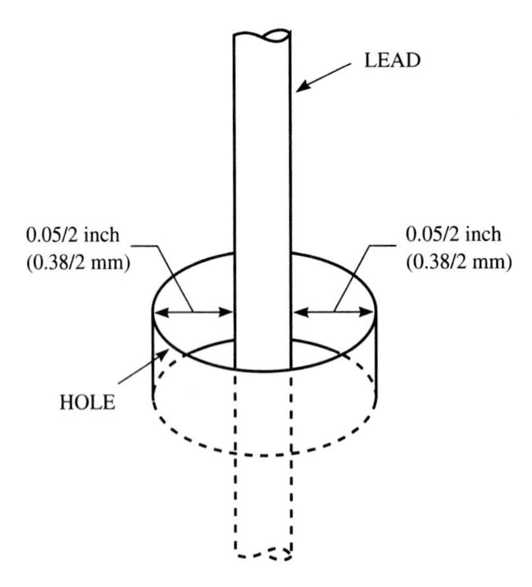

Figure 8-9 Recommended standard for axial lead.

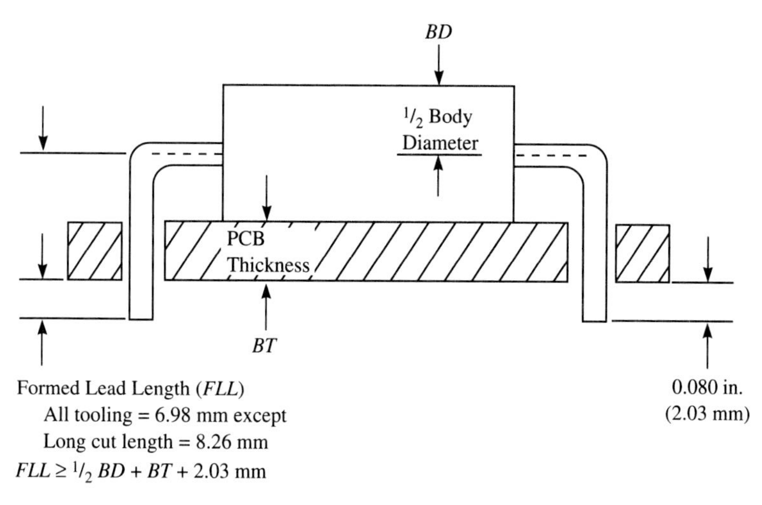

Figure 8-10 Formed lead length.

for which it should be 0.325 inch (8.3 mm). The following formula gives a good estimate of *FLL*:

$$FLL \geq \tfrac{1}{2}\, CBD + BT + 0.080 \qquad (8\text{-}1)$$

where

CBD is the component body diameter
BT is the board thickness

The least allowable length below the board for cut and clinch is 0.080 inch (2.0 mm).

8.2.4 Tooling Footprint

The **tooling footprint** is the projected area of PWB relative to the tooling, during the insertion cycle. To avoid interference of the insertion tooling with the components, insertion tooling is located 0.80 inch (20.3 mm) above the PWB. The obstruction of the outside formers with the components already inserted should also be taken into account during insertion. Since the outside formers guide the component leads down to the holes in the PWB, enough space must be available to avoid the collision of the outside formers with the inserted components. Figure 8-11 is a top view showing the minimum clearance needed between the lead to be inserted and the adjacent components already on the PWB. Table 8-1 gives the dimensions of the clearance for different lead diameters. Figure 8-12 shows front and side views of the outside former.

Consideration must be given to the worst-case locations of the component lead in the insertion hole (Fig. 8-13). The outside formers may hit the inserted components on the PWB if the component lead is not centered in the insertion hole. To avoid the

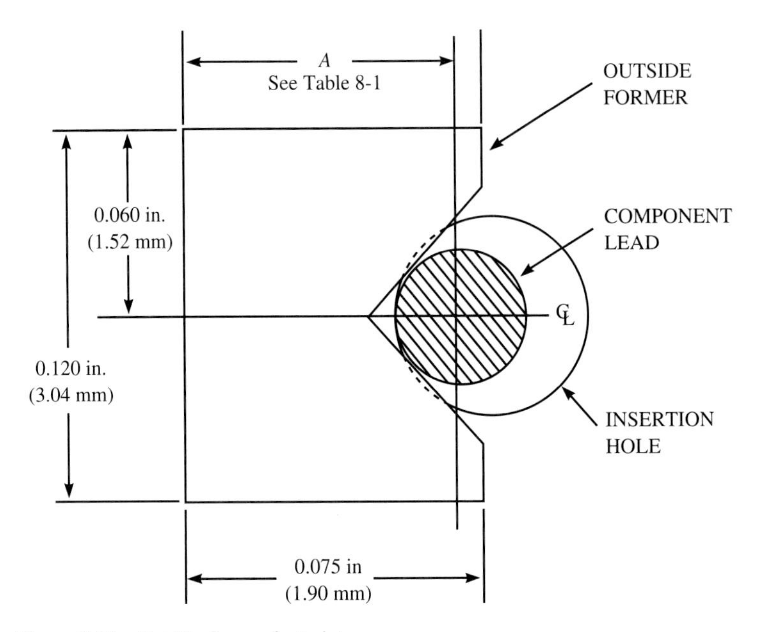

Figure 8-11 Outside former footprint.

(*Source:* Adapted from *Design Guidelines, Vol. One: Axial Lead Components* (Beverly, Mass: Dynapert Corp., 1988), p. 10).

obstruction of the cut-and-clinch unit with the insertion component edges or any devices below the board, the unit is located 0.526 inch (13.4 mm) below the PWB.

The **cut-and-clinch unit** (Fig. 8-14) moves toward the board from below as the insertion tooling moves down to insert the component. As the component is inserted in the PWB, the moving knifes are actuated to cut and clinch the leads. The raised cut-and-clinch unit also stops the board from downward movement during insertion.

Figure 8-15 shows front and side views of the cut-and-clinch unit. Dimensions are

TABLE 8-1 DIMENSION *A*

Lead wire diameter (inch)	Dimension *A*	
	Millimeter	Inch
Standard tooling		
0.019–0.036	1.69–2.01	0.067–0.097
0.020 (diodes)	1.70	0.067
0.025 (¼ W)	1.80	0.071
5mm tooling		
0.016–0.032	1.65–1.92	0.065–0.076

(*Source:* Adapted from *Design Guidelines, Vol. One: Axial Lead Components,* p. 11).

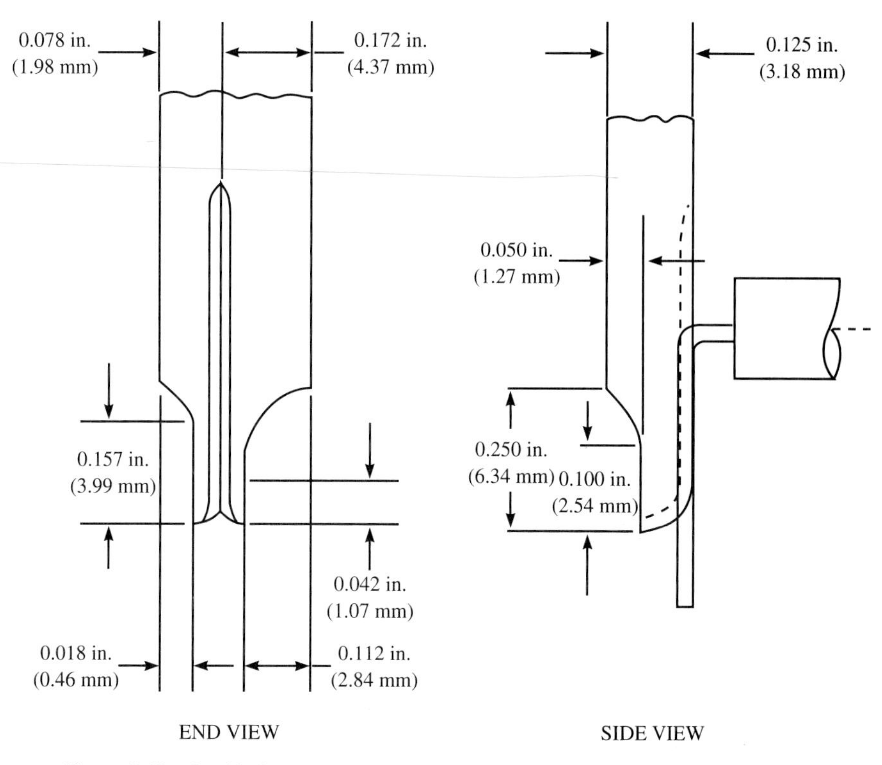

Figure 8-12 Outside former.

(*Source:* Adapted from *Design Guidelines, Vol. One: Axial Lead Components,* p. 11).

also given to calculate the required clearance between the cut-and-clinch unit and the insert edges or any other devices mounted on the bottom of the board.

8.2.5 Center Spacing

The **center spacing** *(CS)* depends on the particular component, insertion tooling, and input tape separation. The minimum center spacing is a function of the component body length, lead diameter, and tooling, and it can be calculated by the following formula:

$$CS(min) = L - WC + 2K \qquad (8\text{-}2)$$

where

CS(min) is minimum center spacing
L is the length of the body
WC is the wire correction
K is the tooling correction

The **wire correction** *(WC)* feature adjusts the center spacing based on position of the component lead against the outside former. Table 8-2 gives different *WCs* for different

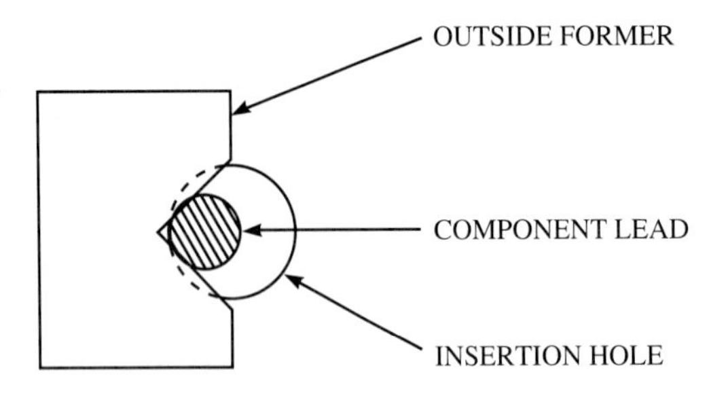

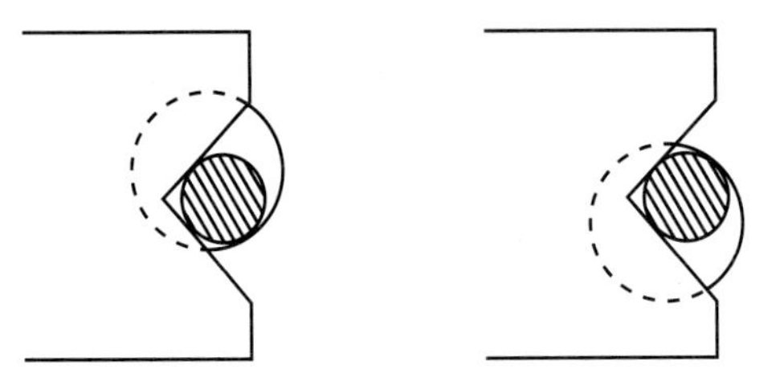

Figure 8-13 Worst-case component lead locations.

(*Source:* Adapted from *Design Guidelines, Vol. One: Axial Lead Components,* p. 12.)

lead wire diameters. **Tooling correction,** *K,* ensures that there is enough bare lead on the component to avoid the contact of the driver with the component body and depends on the tooling used. Table 8-3 gives different *K* values for different tooling.

The maximum center spacing is mainly a function of the VCD machine input tape separation. It can be calculated by the following formula:

$$CS(max) = TS - 2(FLL) - 0.30 \tag{8-3}$$

where

CS(max) is the maximum center spacing
TS is the VCD machine input tape separation
FLL is the formed lead length

The *FLL* is 0.275 inch (7.0 mm) for all tooling except long cut-length tooling, for which it is 0.325 inch (8.3 mm). The constant 0.30 inch (7.6 mm) takes into account the lead length required for indexing wheel and sheer bar on each side, as illustrated in Fig. 8-16.

8.2.6 Component Sequencing for Insertion

Axial-leaded components must be fed to the insertion machine in the proper sequence. The **component sequencing** machines are designed to cut and dispense components from a reeled tape format. The components are then organized in a predetermined sequence on a conveyor chain. The conveyor transports the sequenced components to a taping unit where they are centered, retaped in the required sequence, and wound onto tape reels which are then mounted on the VCD insertion machines. Maximal and minimal component dimensions for the Dynapert sequencing machine are shown in Fig. 8-17.

The **input tape separation** *(ITS)* for the sequencing machine affects the center spacing of a component. *CS* is a function of the *smallest ITS* for the sequencing machine. For all insertion tooling except long cut-length tooling the maximum center spacing is given as

$$CS(max) = smallest\ ITS - 1.2\ \text{inch} \quad (8\text{-}4)$$

For long cut-length tooling the maximum center spacing is given as

$$CS(max) = smallest\ ITS - 1.3\ \text{inch} \quad (8\text{-}5)$$

where

CS(max) is maximum center spacing
ITS is the sequencer input tape separation

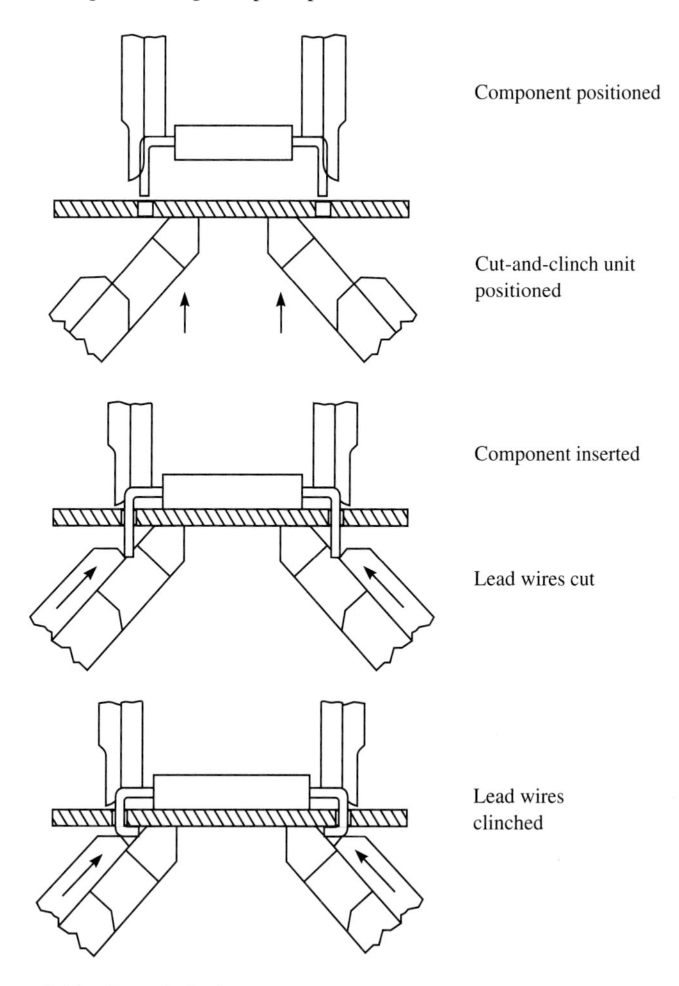

Figure 8-14 Cut-and-clinch.

(*Source:* Adapted from *Design Guidelines, Vol. One: Axial Lead Components,* p. 11.)

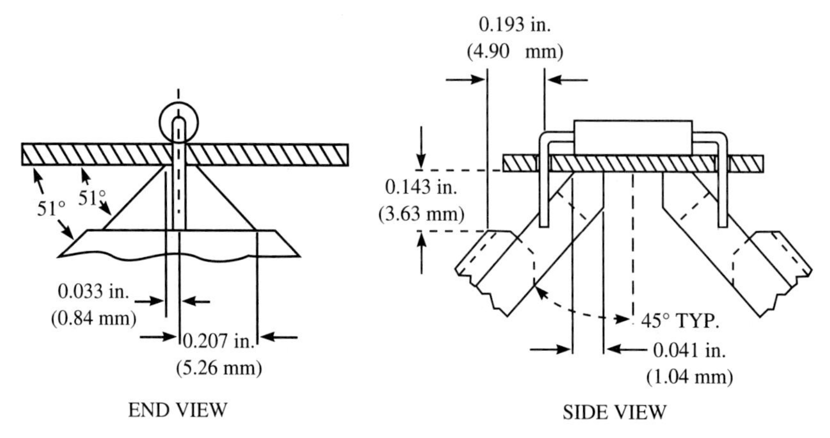

Figure 8-15 Cut-and-clinch clearances.

(*Source:* Adapted from *Design Guidelines, Vol. One: Axial Lead Components,* p. 12.)

TABLE 8-2 WIRE CORRECTION

Lead wire diameter (inch)	Wire correction	
	Millimeter	Inch
Standard Tooling		
0.019–0.036	−0.43, +0.18	−0.017, +0.007
0.020 (diodes)	−0.41	−0.016
0.025 (¼ W)	−0.23	−0.009
0.031	0.00	0.000
5 mm tooling		
0.016–0.032	−0.53, +0.03	−0.021, +0.001

(*Source:* Adapted from *Design Guidelines, Vol. One: Axial Lead Components,* p. 12.)

TABLE 8-3 TOOLING CORRECTION

Tooling	Tooling correction	
	Millimeter	Inch
Standard	1.62	0.064
⅛ W resistor	1.62	0.064
Long cut-length	2.26	0.089

(*Source:* Adapted from *Design Guidelines, Vol. One: Axial Lead Components,* p. 13.)

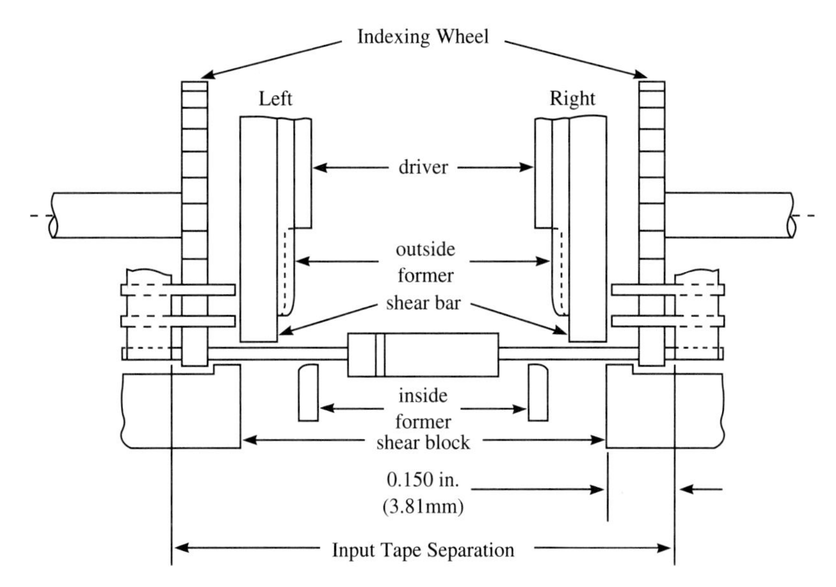

Figure 8-16 Insertion tooling.

(*Source: Adapted from Design Guidelines, Vol. One: Axial Lead Components,* p. 13.)

The constants 1.2 or 1.3 (inches) are the approximate sums of lead losses due to sequencer input cutting, sequencer output taping, VCD machine input cutting, and formed lead length for the particular tooling.

8.3 RADIAL-LEAD COMPONENTS

A variety of different devices (for example, capacitors and transistors) are available in the market in radial-lead configuration. For the purpose of illustration, we will deal with the specifications, tolerances, and guidelines for Universal radial-lead insertion equipment, unless otherwise specified.

8.3.1 Insertion Area

As for axial components, the maximum insertion area is usually 18 $\times$ 18 inches (0.46 m $\times$ 0.46 m).

8.3.2 Insertion Hole Diameter

The insertion hole diameter should be 0.013 inch (0.33 mm) more than the lead diameter. For maximum reliability, the hole diameter can be as high as 0.015 inch (0.38 mm), as shown in Fig. 8-18. Larger and smaller holes have advantages and disadvantages. Maximum hole size is limited by clinch and soldering requirements, while smaller holes require greater precision in machine operation and can cause more bent leads during insertion.

8.3.3 Tooling Footprint

Refer to Fig. 8-19. The radial-lead insertion head consists of an insertion jaw which positions the component on the PWB. Once the component is positioned, the insert pusher unit exerts the pressure on the top of the component envelope to insert it into the PWB. To make sure that the insertion is complete, the insert pusher is allowed to over-travel beyond the insertion jaw tip. The insertion jaw can be rotated among discrete stops at $+90°$, $0°$, and $-90°$, relative to the front of the machine. The insertion jaw retracts away from the component while the component is being inserted. The insertion board layout and insertion sequence must be designed to ensure that the insertion jaw does not hit other previously inserted components. Figure 8-20 illustrates the minimum clearances required between the component being inserted and the components already on the PWB.

The cut-and-clinch unit (Fig. 8-21) accepts the leads coming through the board and

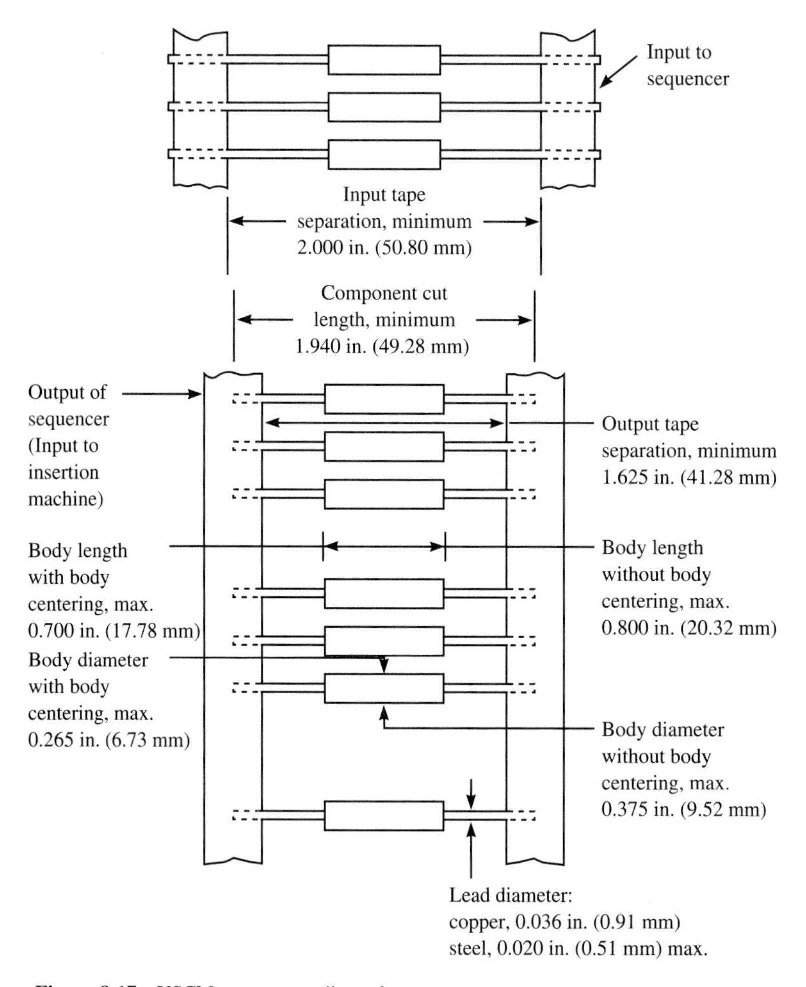

Figure 8-17 USCM component dimensions.

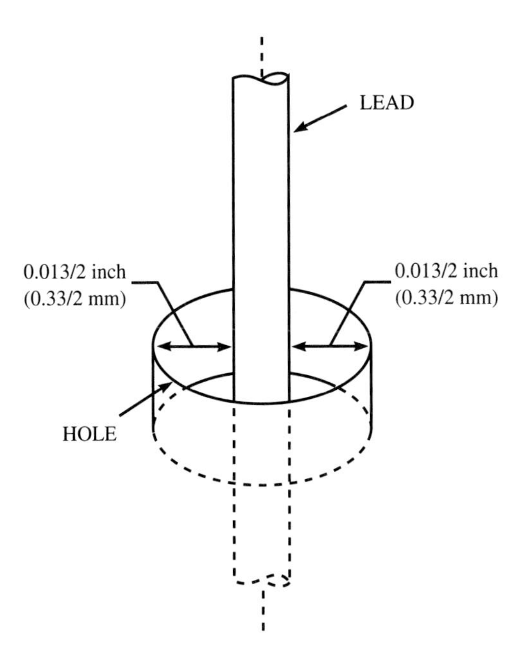

Figure 8-18 Recommended standard for radial lead.

also supports the board from the bottom to avoid any bending of the board and to ensure uniform insertion. Once the leads are through the PWB completely, two cutters trim the leads to their finished lengths and clinch them to the underside of the PWB to secure the component to the board. After the cut-and-clinch process is completed, the unit is lowered to await the next component insertion.

8.3.4 Taping Considerations

Taping considerations are important for reliable insertion of components on the PWB. Figure 8-22 gives specifications for taping of sequenced components by the sequencer.

8.4 DUAL IN-LINE PACKAGE COMPONENTS

Dual In-line package (DIP) components are another important category of insertion components. Since there are a lot of leads in these type of components, they are relatively more challenging to insert. There are different sizes of DIPs with different lead counts which require the use of strict guidelines to ensure reliability of DIP insertion. There are different specifications, tolerances, and guidelines for different machines. To illustrate the concepts for DIP insertion we will refer to specifications and guidelines for a Dynapert DIP component insertion machine, unless otherwise specified.

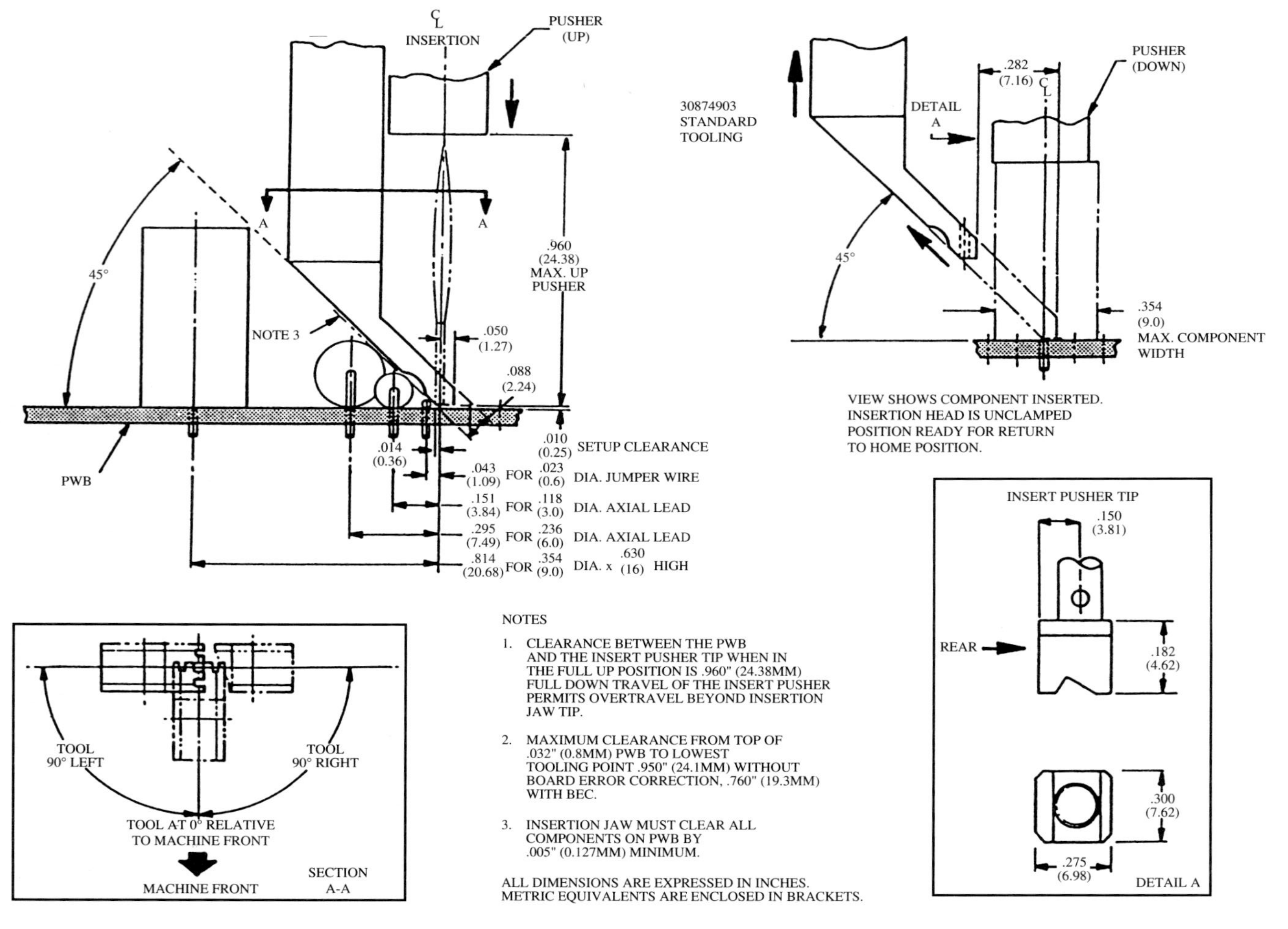

Figure 8-19 Radial-lead insertion tooling footprint.

(*Source: Design Guidelines for Leaded Component Insertion, 15MI/85* (Binghamton, N.Y.: Universal Instruments Corp., 1985), p. 4-4.)

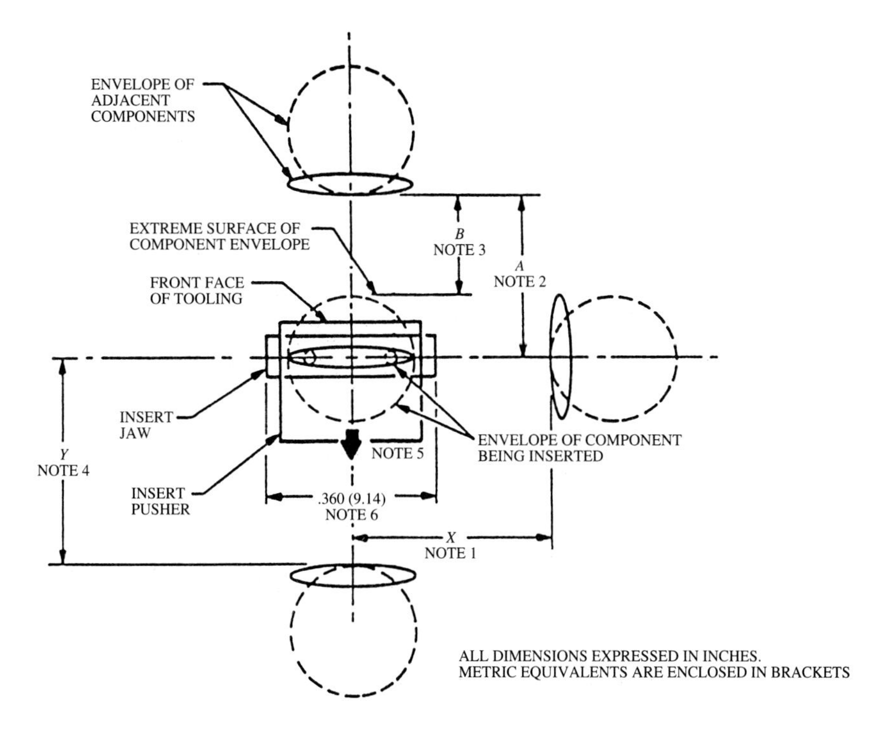

NOTES:

1. DIMENSION *X* IS .197 (5) WHEN WIDTH OF COMPONENT BEING INSERTED IS UP TO .375 (9.52). WHEN WIDTH OF COMPONENT BEING INSERTED IS GREATER THAN .375 (9.52), THE MINIMUM *X* DIMENSION BECOMES ONE HALF THE MAXIMUM WIDTH OF THE COMPONENT PLUS (+) 0.008 (0.2).

2. WHEN THE BODY OF THE COMPONENT BEING INSERTED DOES NOT EXTEND BEYOND THE FRONT FACE OF THE INSERTER PUSHER, THE *A* DIMENSION MAY NOT BE LESS THAN .150 (3.81) EXCEPT UNDER THE FOLLOWING SPECIAL CONDITIONS.

   A. THE COMPONENT BEING INSERTED IS AT LEAST HIGHER THAN THE PART THAT IS ALREADY IN THE PWB, OR

   B. THE COMPONENT ALREADY ON THE PWB HAS A TAPERED OR THIN-TOPPED CROSS SECTION.

   UNDER THESE CONDITIONS THE *A* DIMENSION MUST BE .020 (0.51) GREATER THAN THE HALF THICKNESS OF THE PART BEING INSERTED BUT MAY NOT BE LESS THAN .090 (2.29).

3. WHEN THE BODY OF THE COMPONENT BEING INSERTED EXTENDS BEYOND THE FRONT FACE OF THE INSERT PUSHER. THE *B* DIMENSION MUST BE A MINIMUM OF .020 (0.51) FOR RELIABLE INSERTION.

4. THE *Y* DIMENSION DEPENDS UPON THE GEOMETRY OF BOTH THE INSERTION TOOLING AND THE COMPONENT(S) ALREADY ON THE PWB.

5. INSERTION LIFTS AT 45° IN THE DIRECTION INDICATED BY ARROW TO CLEAR THE COMPONENT BEING INSERTED.

6. JAW IN UNCLAMPED (OPEN) CONDITION.

IMPORTANT NOTE: CLEARANCES INDICATED PROVIDE CONTIGUITY BETWEEN THE ENVELOPE OF THE COMPONENT ALREADY ON THE PWB AND THE ENVELOPE OF THE COMPONENT BEING INSERTED (OR THE FACE OF THE INSERTION TOOLING) AND SHOULD BE CONSIDERED AS THE ABSOLUTE MINIMUM OBTAINABLE UNDER IDEAL CONDITIONS. IN THE INTEREST OF OVERALL INSERTION RELIABILITY. THE MINIMUM CLEARANCES SHOULD BE USED ONLY WHEN ABSOLUTELY NECESSARY AND WITH THE UNDERSTANDING THAT INPUT COMPONENT TOLERANCE VARIATIONS WILL HAVE AN EFFECT ON SYSTEM RELIABILITY.

Figure 8-20 Insertion clearances.

(*Source: Design Guidelines for Leaded Component Insertion,* p. 4-6.)

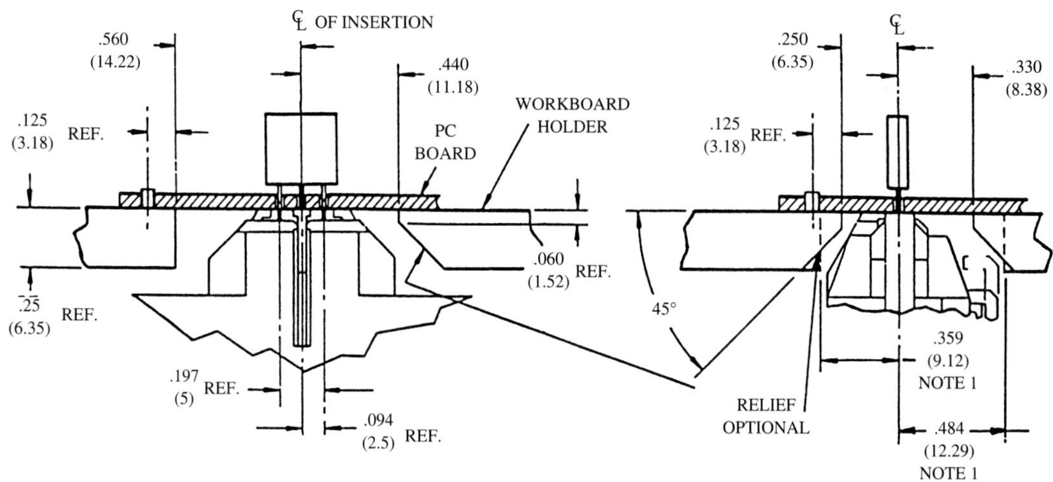

NOTES:
1. CLEARANCE REQUIRED WHEN OPTIONAL RELIEF IS NOT USED.
2. ALL DIMENSIONS ARE EXPRESSED IN INCHES; METRIC EQUIVALENTS ARE ENCLOSED IN BRACKETS.

Figure 8-21 Cut-and-clinch shown without pass-thru.

(*Source: Design Guidelines for Leaded Component Insertion,* p. 4-8.)

8.4.1 DIP Insertion Machines

Conventional DIP insertion machines are designed to handle components with 0.30 inch centerpiece (7.62 mm) and lead separation of 0.10 inch (2.54 mm) with lead count from 6 to 20 (high-performance machines can insert DIPs with lead count as high as 40). Up to 90 different DIP components and sockets can be loaded into channels to be selected and inserted in a programmable sequence.

8.4.2 DIP Component Specifications

Reliable insertion of DIP components on the PWB requires that the component conform with certain dimensions. This is especially important in the case of DIPs because of the high count of the leads. Figure 8-23 gives the dimensional requirements for standard DIP components, and Fig. 8-24 gives the dimensional requirements for side-brazed DIP components. Sockets mounted on the PWB should also conform with certain requirements. The dimensional requirements for sockets are given in Fig. 8-25.

8.4.3 Carrier Stick Specifications

The DIP components must be supplied to the machine in plastic tubes called **DIP sticks,** as shown in Fig. 8-26. Specifications for the DIP stick are shown in Fig. 8-27.

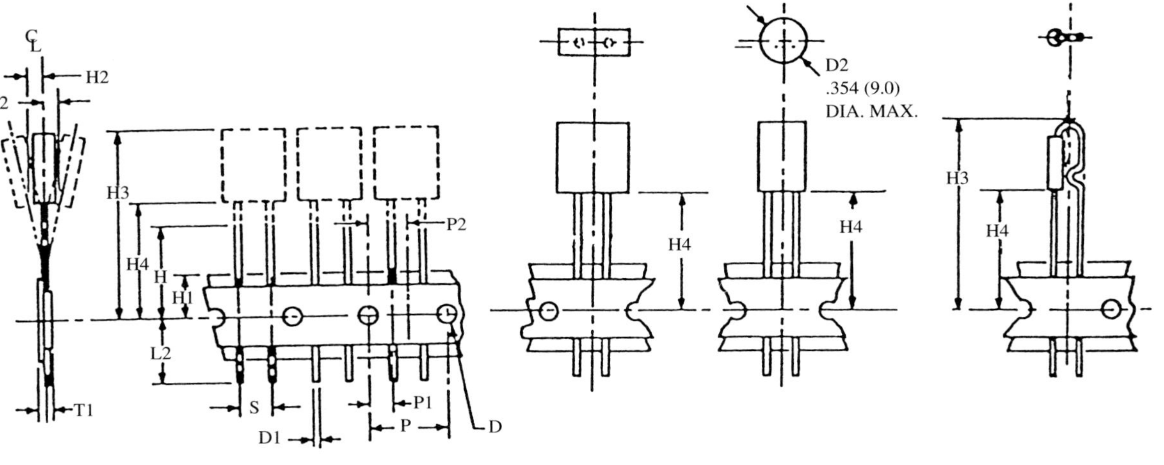

SYMBOL	ITEM	MINIMUM		MAXIMUM	
		INCH	MM	INCH	MM
D	FEED HOLE DIAMETER	.146	3.7	.169	4.3
D1	LEAD DIAMETER	.014	0.36	.028	0.7
D2	COMPONENT BODY DIAMETER			.354	9.0
H	HEIGHT OF SEATING PLANE	.610	15.5	.650	16.5
H1	FEED HOLE LOCATION	.335	8.5	.384	9.75
H2	FRONT-TO-REAR DEFLECTION	.000	0	.031	0.8
H3	COMPONENT HEIGHT			1.26	32
H4	FEED HOLE TO BOTTOM OF COMPONENT	.610	15.5	.886	22.5
L2	LEAD PROTRUSION	.000	0	.44	11.2
P	FEED HOLE PITCH	.488	12.4	.512	13
P1	LEAD LOCATION	.124	3.15	.179	4.55
P2	CENTER OF SEATING PLANE LOCATION	.234	5.95	.266	6.75
S	COMPONENT LEAD SPAN	.165	4.2	.228	5.8
T1	TOTAL TAPED PACKAGE THICKNESS			.056	1.42

Figure 8-22 Input specifications.

(*Source: Design Guidelines for Leaded Component Insertion,* pp. 6-4, 6-5.)

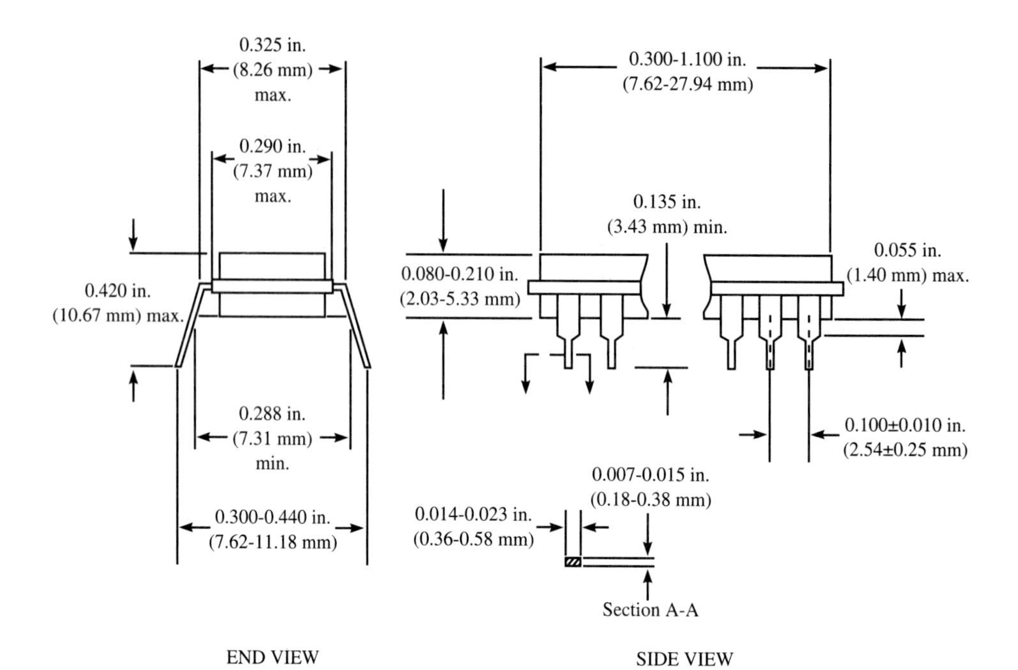

Figure 8-23 Standard DIP component.

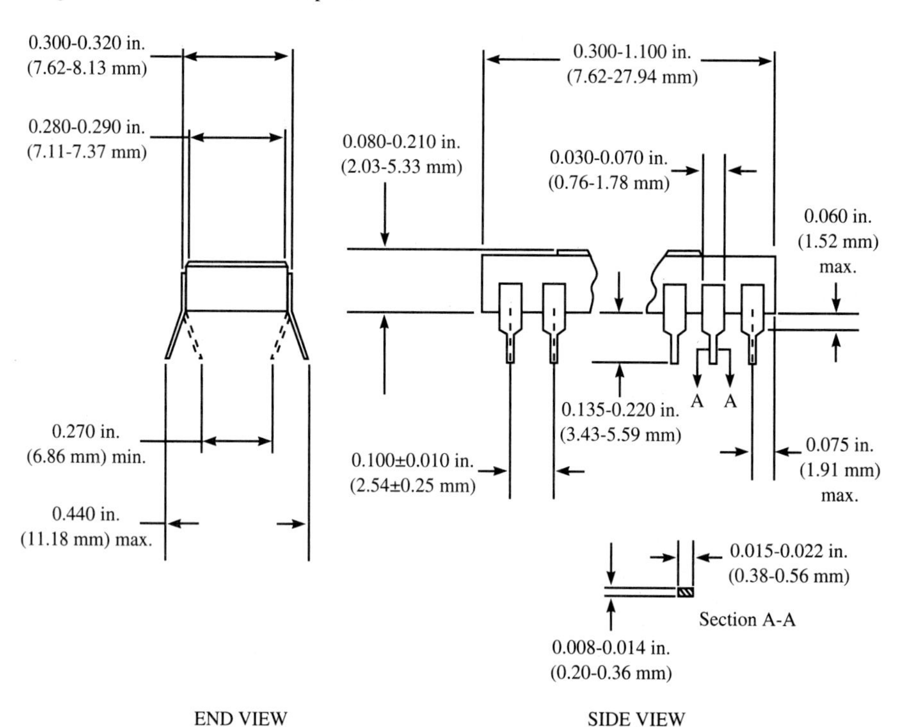

Figure 8-24 Side-brazed DIP component.

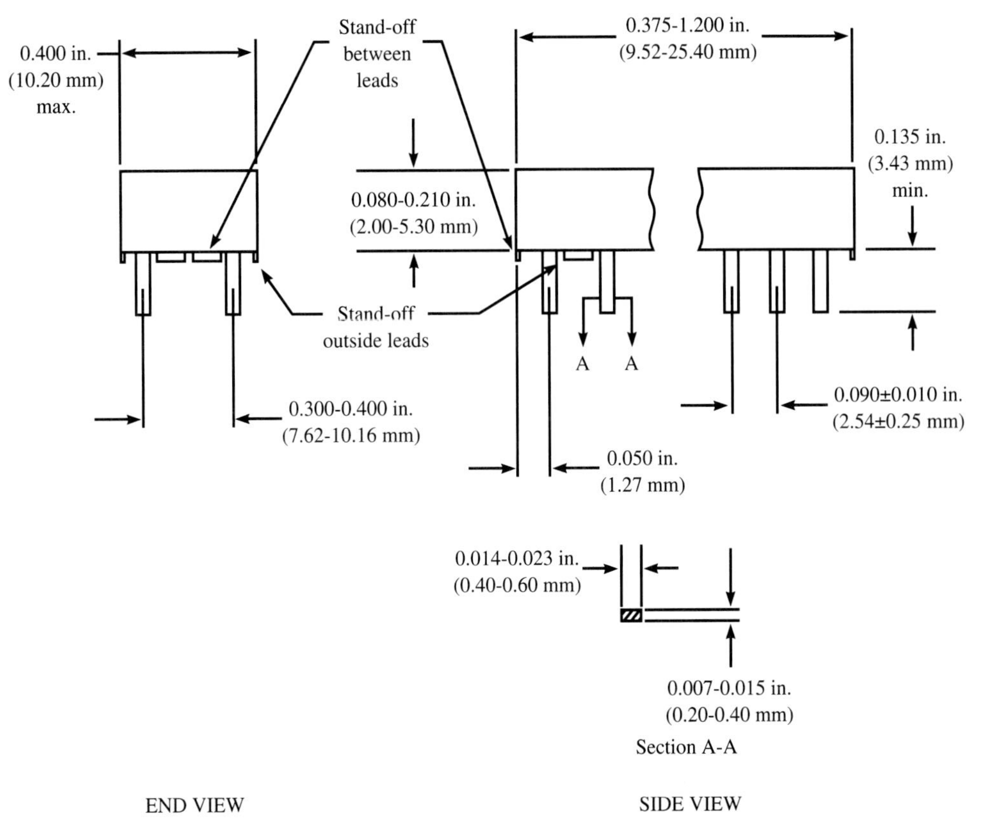

Figure 8-25 Socket.

8.4.4 Insertion Area

As for axial and radial insertion machines, the maximum programmable insertion area for DIP insertion machines is 18 × 18 inches (0.46 m × 0.46 m). Since the dimensions are measured from the center of the fingers, the programmable insertion area allows for extension up to one-half of the component width or length in the *X* axis and *Y* axis. See Fig. 8-28.

8.4.5 Insertion Hole Diameter

The diameter of the insertion holes should be sufficiently larger than the diameter of the insertable lead, to accommodate the tolerances of the insertion machine lead and PWB. The recommended standard is for the insertion hole diameter to be 0.015 inch (0.38 mm) more than the lead diameter. Smaller or larger hole sizes can be used (within limits) for closer and looser tolerances, respectively. Smaller holes can result in less reliable insertion of components, and maximum hole size is limited by clinch and soldering requirements. Insertion holes for socket components must be between 0.037 inch (0.94 mm) and 0.041 inch (1.04 mm).

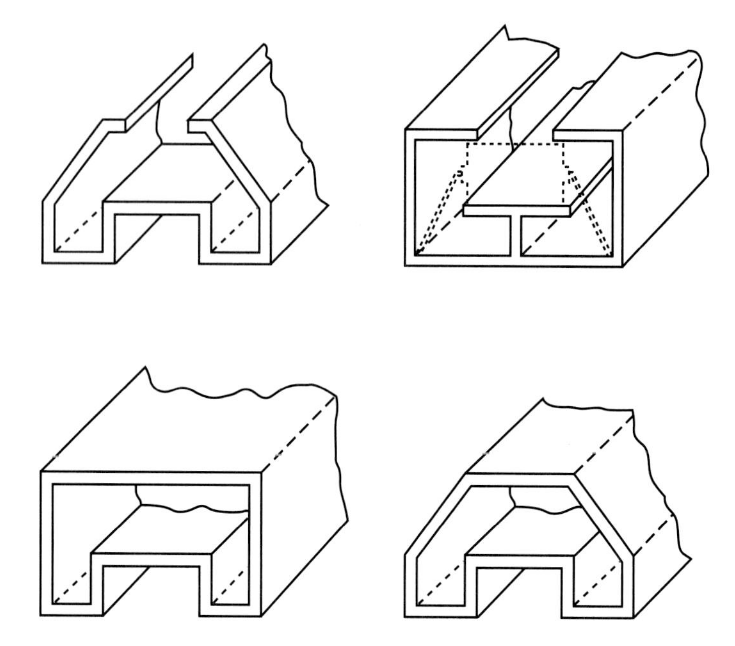

Figure 8-26 DIP sticks.

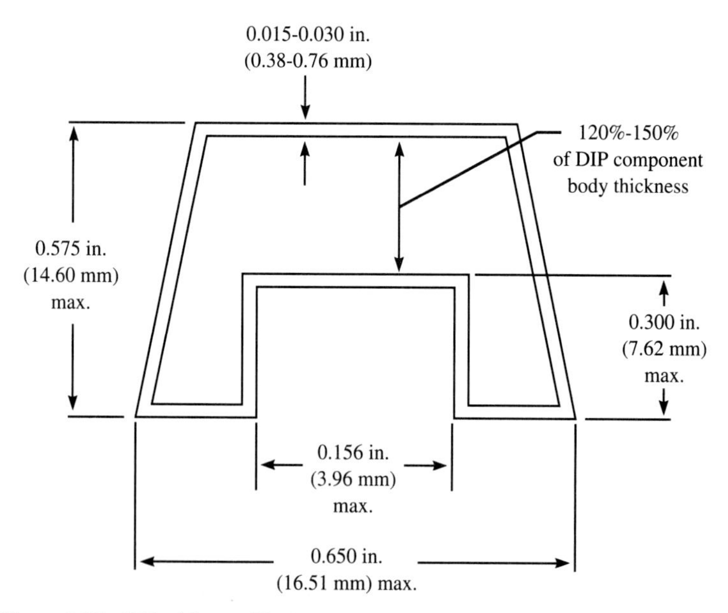

Figure 8-27 DIP sticks specifications.

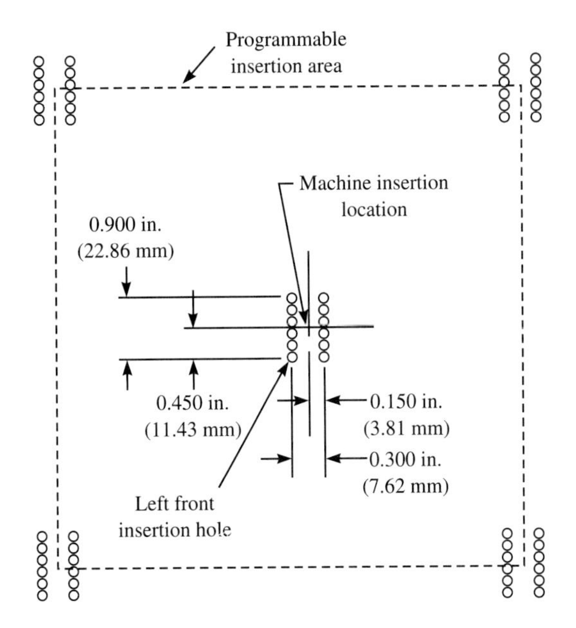

Figure 8-28 Insertion area for DIP.

8.4.6 Tooling Footprint

Clearance is required between the insertion tooling and the preinserted components to avoid interference during the insertion process. For this reason, home position for insertion tooling is approximately 0.60 inch (15.2 mm) above the PWB. Since the insertion fingers need some space to place the component on the PWB and then move back to the home position, considerations must be given to the minimum clearance (see Fig. 8-29) needed between the insertion fingers and the other components on the PWB. Figure 8-30 shows the front and side views of different insertion tooling and gives the accompanying dimensions. These dimensions can be used to determine the clearances required between the insertion tooling and the components above the PWB.

There are two types of cut-and-clinch units available: inward and outward. At the bottom of the insertion stroke, both types allow 0.562 inch (14.3 mm) clearance to avoid any obstruction with the insert edges or any devices below the PWB.

As the insertion tooling starts to move down from home position toward the top of the board, the cut-and-clinch unit moves up toward the bottom of the board from below. As the component gets placed on the board, the moving knives of the cut-and-clinch unit are actuated, cutting and clinching the component leads. The raised cut-and-clinch unit also serves the purpose of supporting the board from the bottom during insertion of the component. The insertion of a component into a socket (instead of directly into the PWB) disables the cut-and-clinch mechanism, but the tooling still supports the PWB from below during insertion.

Figures 8-31 and 8-32 show the inward and outward cut-and-clinch units and the required dimensions, respectively. The dimensions can be used to calculate the clearances between the cut-and-clinch units and the insert edges or any other obstructions on the bottom of the PWB.

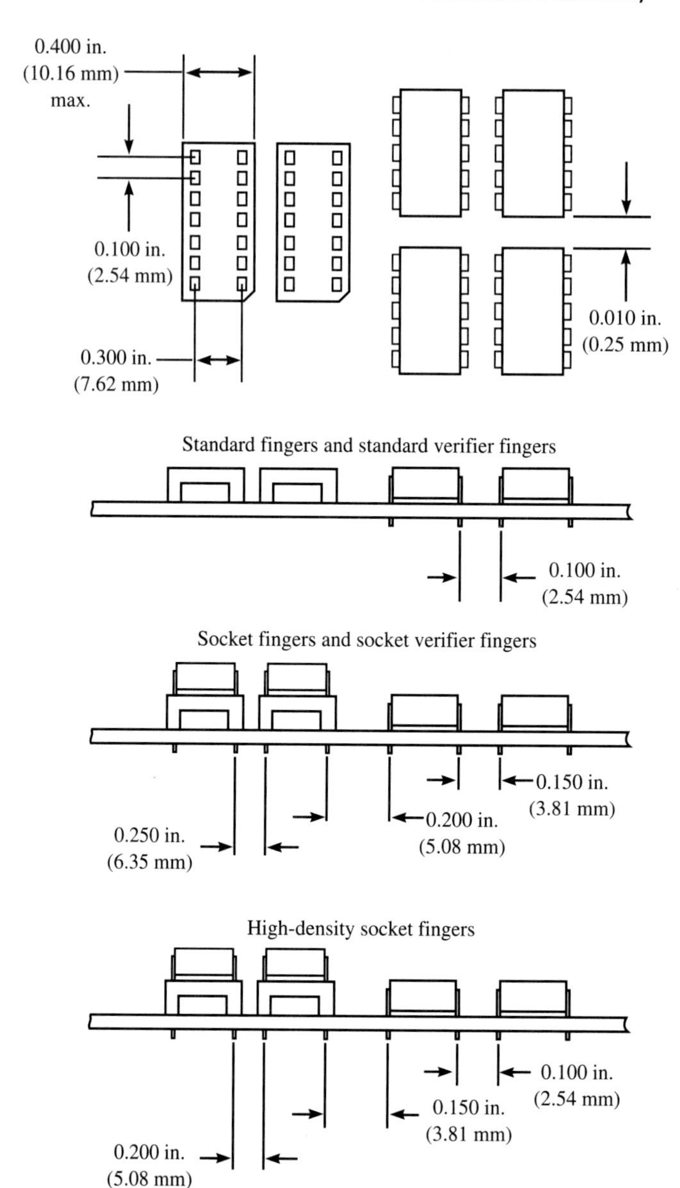

Figure 8-29 DIP component and socket clearances.

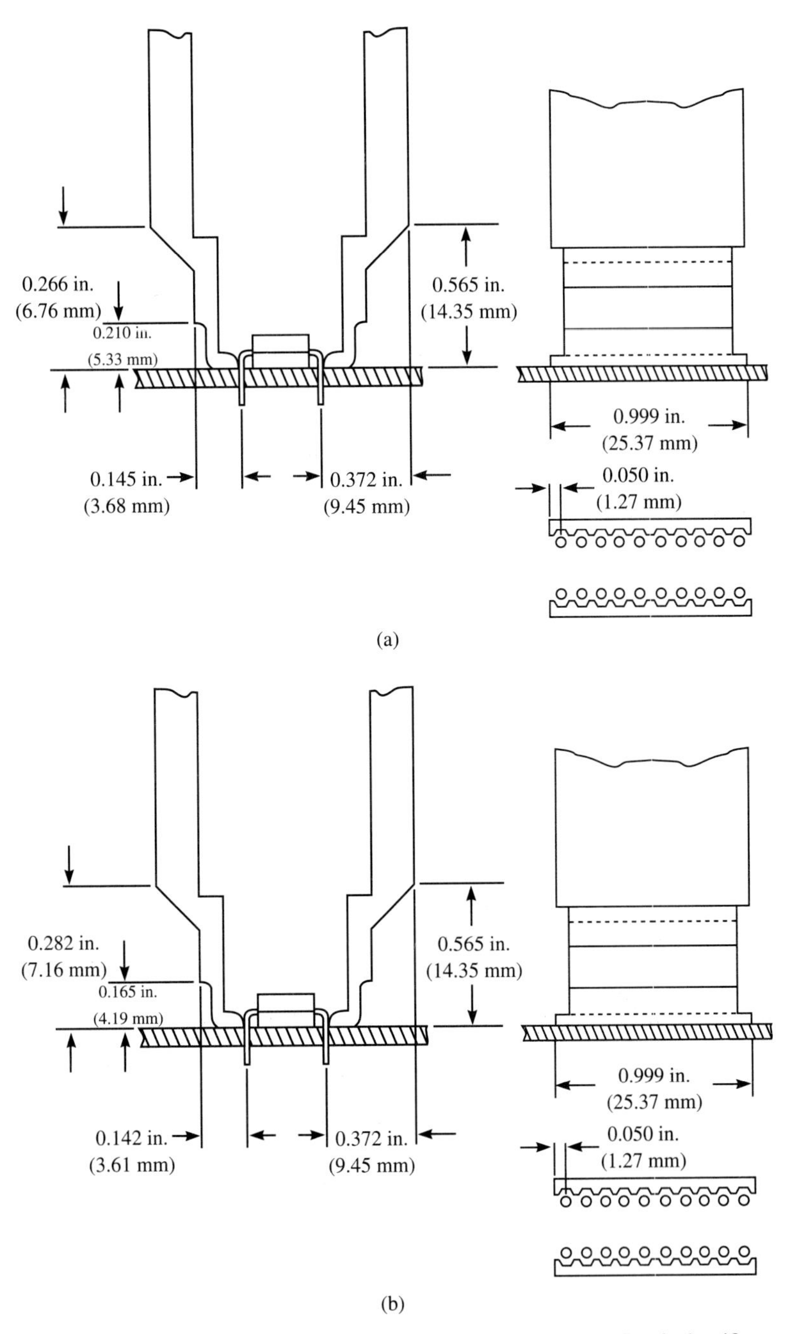

Figure 8-30 Insertion tooling. (a) DIP insertion fingers. (b) Socket. (c) Standard verifier. (d) Socket verifier. (e) High-density socket.

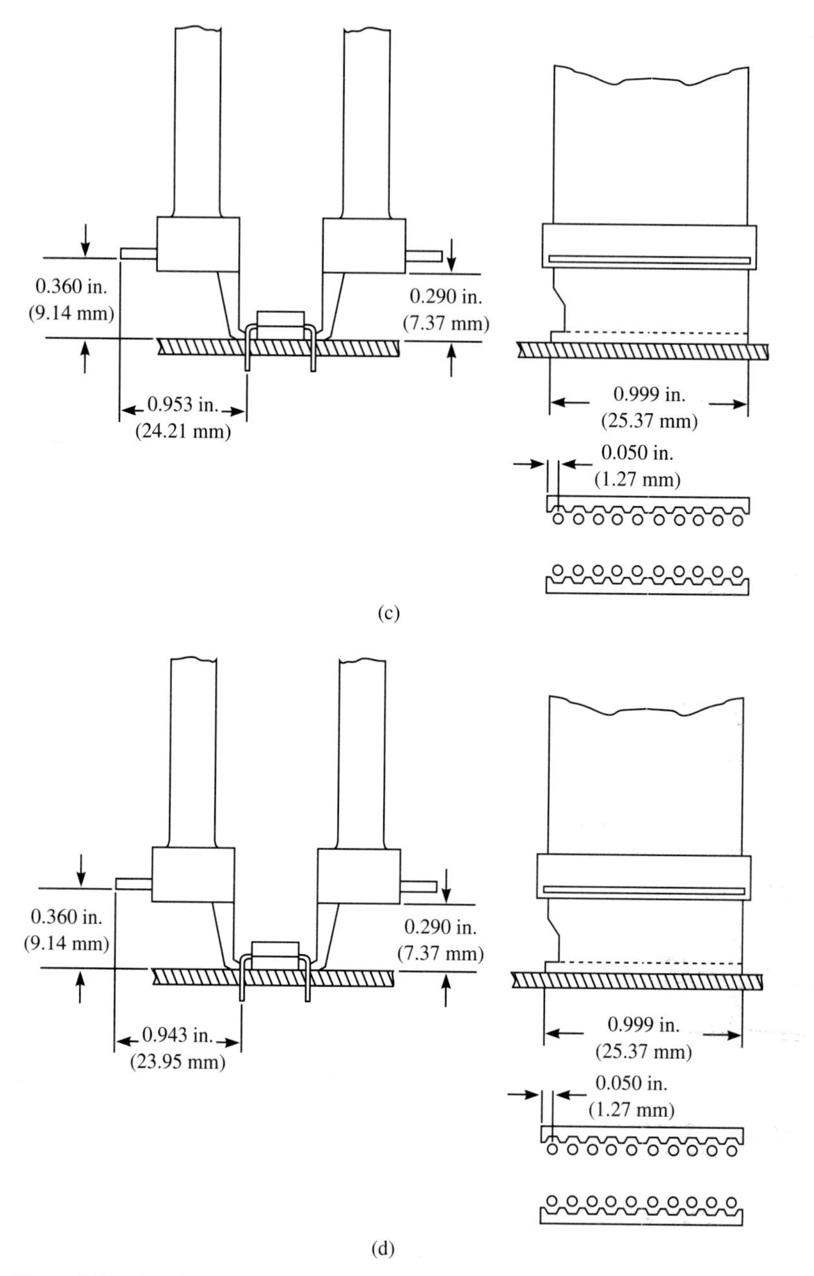

Figure 8-30 (continued)

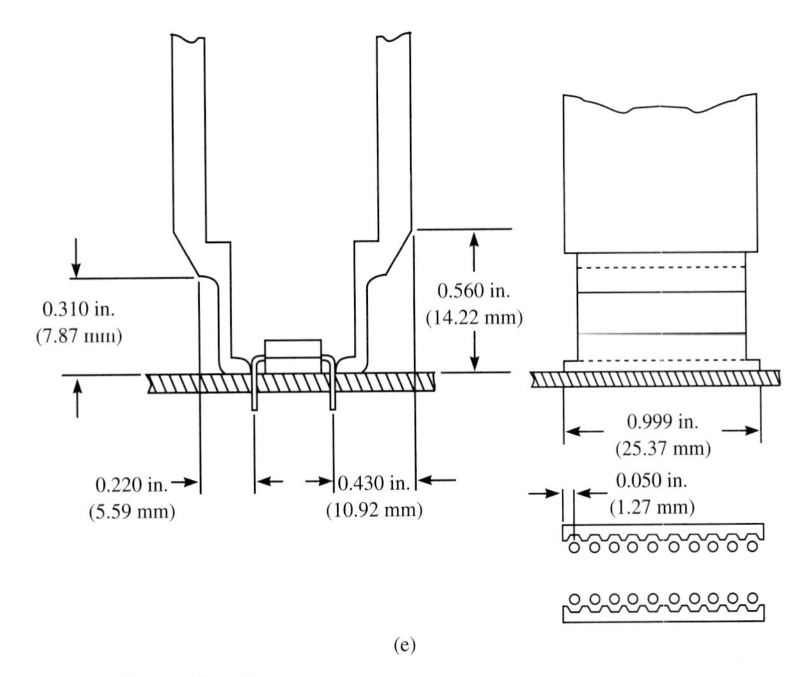

Figure 8-30 (continued)

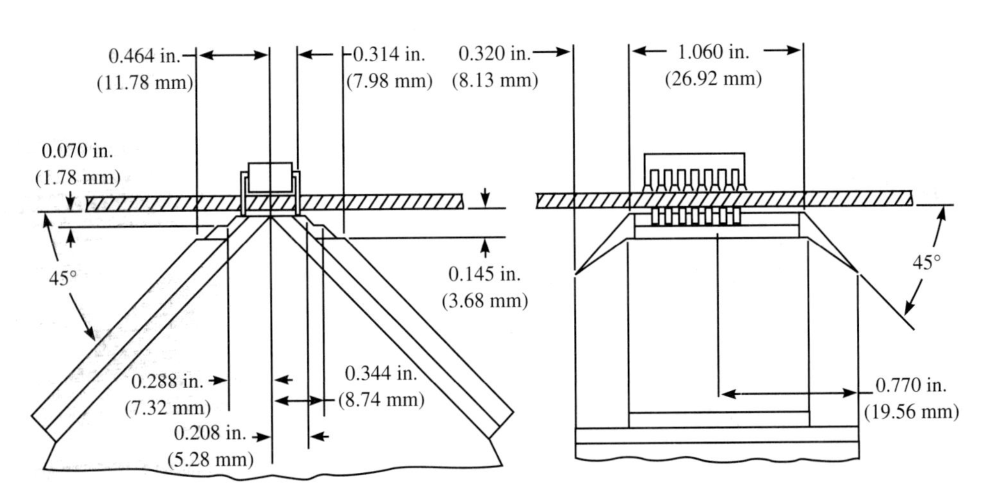

Figure 8-31 Inward clinch unit.

8.5 OPTIMAL INSERTION PATTERN

Optimal insertion patterns make the best use of PWB insertion space with minimal insertion head movement and less chance of obstruction between the insertion head and the inserted components.

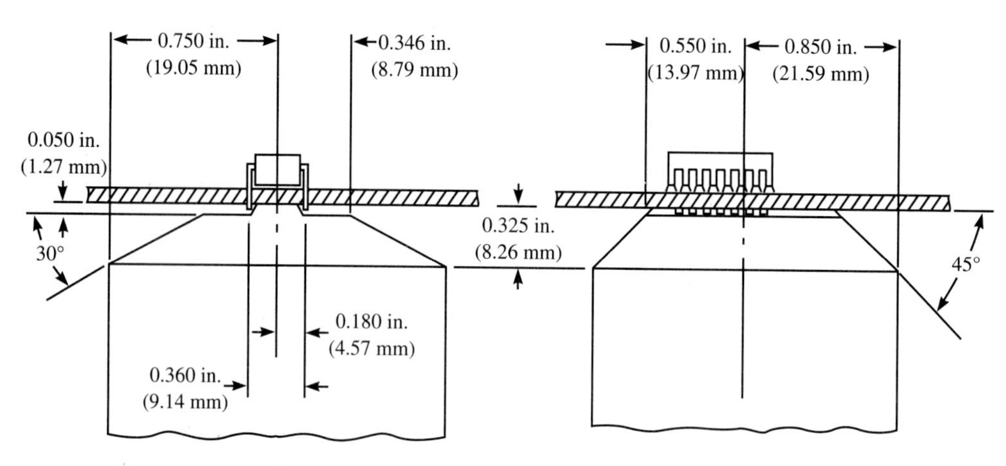

Figure 8-32 Outward clinch unit.

The following guidelines[1] for leaded component insertion illustrate the principles of optional insertion patterns:

- *X* axis movements should be minimized, and movements should be in plus or minus *Y* axis directions (Fig. 8-33).
- For radial-lead insertion equipment, component insertion should proceed from back to front of the PWB.

[1]*Design Guidelines for Leaded Component Insertion,* 15MI/85 (Binghamton, N.Y.: Universal Instruments Corp. 1985), pp. 3-10, 4-8/9, 5-5.

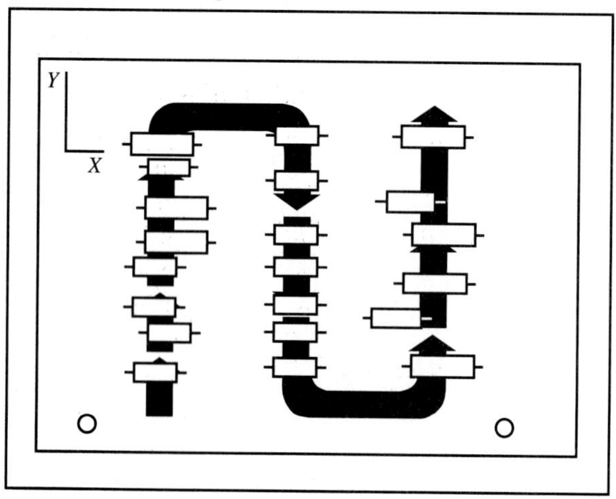

Figure 8-33 Axial lead optimal insertion pattern.

(*Source: Design Guidelines for Leaded Component Insertion,* p. 3-10.)

- For DIP components, movement should be minimal in the *Y* axis and maximum in the plus or minus *X* axis. To avoid interference with the components already on the board, components should be inserted row by row (Fig. 8-34).

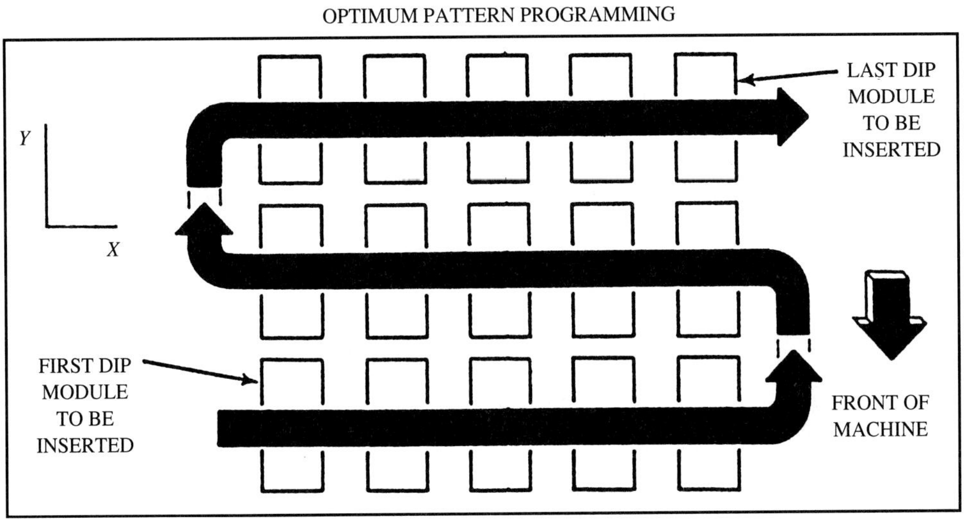

OPTIMUM PATTERN PROGRAMMING

Y
X
LAST DIP MODULE TO BE INSERTED
FIRST DIP MODULE TO BE INSERTED
FRONT OF MACHINE

Figure 8-34 DIP optimal insertion pattern.

(Source: Design Guidelines for Leaded Component Insertion, p. 5-6.)

8.6 SUMMARY

In this chapter, the three major types of through-hole components (axial, radial, and DIP) have been considered as to their handling, packaging, preparation, and sequencing for insertion and the optimal insertion pattern on a PWB.

KEY TERMS

Center spacing
Component orientation
Component sequencing
Cut-and-clinch unit
Datum lines (points)
DIP sticks
Formed lead length (FLL)
Input tape separation (ITS)
Insertion area
Insertion hole diameter
Leaded component insertion
Optimal insertion patterns
Tooling correction
Tooling footprint
Tooling holes
Variable center distance (VCD)
Wire correction (WC)

EXERCISES

1. What are the reference points called to position components on PWBs?
2. What are the minimum and maximum center spacing for a component with axial leads and a body length of 8 mm? Assume a lead diameter of 0.031 inch, with standard tooling and with VCD input tape separation of 3.81 cm.
3. What are the optimal insertion patterns for axial, radial, and DIP components?
4. What is the advantage to be gained by having components oriented in a single or double axis layout?
5. What are the functions required in handling packaging of large quantities of components?
6. What physical characteristics of the DIP component are used to develop insertion tooling?
7. What are the advantages and disadvantages, from an insertion viewpoint, of using DIP sockets instead of inserting the DIPs directly into a PWB?

BIBLIOGRAPHY

1. CLARK, R., *Printed Circuit Engineering, Optimizing for Manufacturability.* New York: Van Nostrand Reinhold, 1989.
2. KEAR, F. W., *Printed Circuit Assembly Manufacturing.* New York: Marcel Dekker, Inc., 1987.
3. NOBEL, P. J. W., *Printed Circuit Board Assembly.* New York: Halsted Press, 1989.
4. STILLWELL, H. R., *Electronic Product Design for Automated Manufacturing.* New York: Marcel Dekker, Inc., 1989.
5. TUMMALA, R. R., AND E. J. RYMASZEWSKI, eds., *Microelectronics Packaging Handbook.* New York: Van Nostrand Reinhold, 1989.

9 Surface-Mount Device Placement

9.1 INTRODUCTION

Surface-mount devices (SMDs) were introduced in Chapter 3. Chapter 6 discussed methods of soldering SMDs to printed wiring boards. In this chapter, we discuss surface-mount technology (SMT), which encompasses the placement and attachment of active and passive electronic components directly onto the prepared surface of a circuit medium (PWB or ceramic substrate).[1]

SMT presents technical challenges different from those of through-hole insertions. A major difference between surface-mount and through-hole technologies is that SMT placement does not provide a secure mechanical attachment of the component to the board prior to soldering. Reflow soldering may be used in place of or in addition to wave soldering. Additionally, the miniaturization, fine lead pitches, and packaging density of SMT necessitate automated processes to achieve required low cost and high quality.

Relatively high cost, concerns about reliability, rework and field repair problems, and a lack of industry standardization in component and process equipment have somewhat hindered adoption of SMT. However, the following advantages ensure that SMT will be a major component of the electronics packaging field for the foreseeable future[2]:

- *Components*
  - — Miniaturization
  - — Light weight

[1]*Guidelines for Surface Mount Technology* (Binghamton, N.Y.: Universal Instruments Corp., 1989), p. 1-1.

[2]Otmar Hintringer and Werner Maiwald, *An Introduction to Surface Mounting* (Munich, Germany: Siemens Aktiengesellschaft, 1986), p. 9.

— Lead forming eliminated
— Packaging permits greater lead density
— Closer tolerances on electrical parameters

- *Printed Wiring Board:*
  — Substantial size reduction
  — Hole drilling eliminated
  — Lower power requirements
- *Assembly processes:*
  — Replaces multiple machines (for example, for axial, radial, and DIP insertion)
  — Assembly speed and accuracy
  — Mixed technology (use of both SMDs and through-hole components)

Table 9-1 summarizes surface-mount device types, available package formats, required relative placement accuracies, and available component handling systems. It should be noted that as the number of leads for a SMD increases, the placement accuracy becomes more critical. Tape-and-reel feeders for component handling can be used for most SMDs, except for those where leads can be easily damaged.

Chapter 3 described the device types shown in Table 9-1. Section 9.2 overviews the three main types of SMT board assemblies. Section 9.3 discusses component pickup, centering, and placement. Chapter 5 discussed PWBs, and Sec. 9.4 provides board requirements for SMT assembly. Section 9.5 covers the topic of placement accuracy for the device types shown in Table 9-1. Section 7.3 introduced theory of machines. Machine vision technology is discussed in Sec. 9.6. Section 9.7 discusses the handling systems noted in Table 9-1.

9.2 SURFACE-MOUNT ASSEMBLIES

Refer to Fig. 9-1. The Type I assembly uses SMDs on one or both sides of the board. Type II assemblies have through-hole components on top of the PWB and SMDs on the bottom. Type III assembly permits through-hole components on the top and SMDs on both sides of the board. Assembly processes, corresponding to the three board designs, are flowcharted in Fig. 9-2. The mixed-technology approach of utilizing surface-mount and through-hole components on the same board (that is, Types II and III), has facilitated the adoption of SMT. The Type III approach involves the most complex assembly process flow.

9.3 PLACEMENT

SMT assembly requires pickup, centering, and placement of components. Centering is accomplished either by mechanical means or with the aid of machine vision. The placement accuracy required determines the type of centering to use.

TABLE 9-1 DEVICE SELECTION CHART

Device Function	Capacitor	Resistor	Diode	Transistor		Integrated Circuit			
Available Package Forms	Polarized Chip	Chip	Cylindrical	SOT	Bare IC or Flip-Chip	SOIC	Plastic Leaded Chip Carrier	Flat-Pack	Leadless Chip Carrier
Relative Placement Accuracy Group (Note 1)	A	A	A	A	C	B	B,C	B,C	B,C
Available Handling Systems									
Tape–8mm and Larger	●	●	●	●		●	●	●	●
Sticks–Rigid and Flexible			●			●			●
Waffle Trays					●			●	
Bulk–Vibratory Feeder	●	●	● (Note 2)						

Notes:
1. Place accuracy required.
   Group A—Moderate.
   Group B—Medium.
   Group C—High.
2. For Resistors only (MELF); Diodes are polarized.

Source: Guidelines for Surface Mount Technology, p. 2-1.

TYPE I

Surface-Mounted
Devices Only

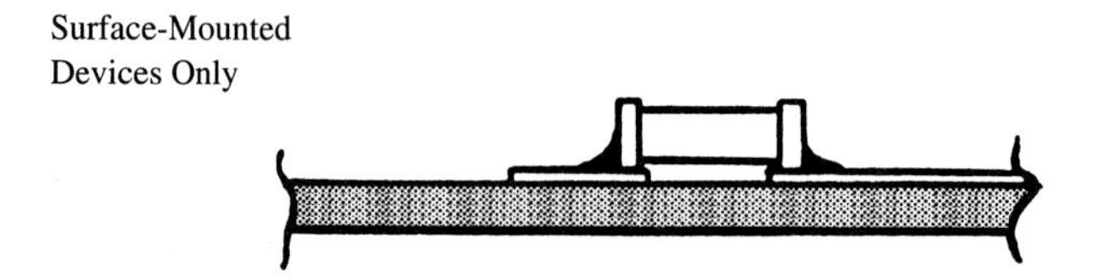

TYPE II

Inserted Components on Top,
Surface-Mounted Devices
on Bottom

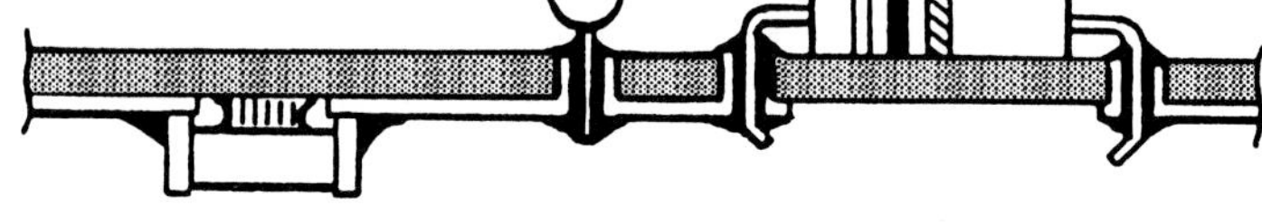

TYPE III

Inserted Components on Top,
Surface-Mounted Devices
on Top and Bottom

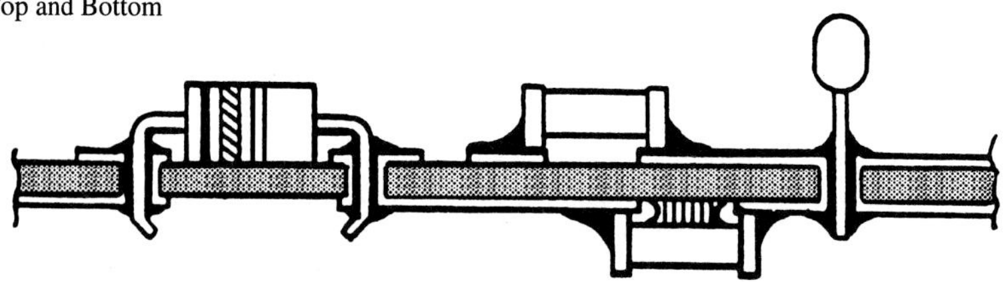

Figure 9-1 Types of SMT Assemblies

(*Source: Guidelines for Surface Mount Technology,* p. 1-6. Reprinted by permission of Universal Instruments Corp.)

9.3.1 Pick-and-Place Spindle

Figure 9-3 is a schematic diagram of the typical **pick-and-place spindle.** The head carries a vacuum spindle and can rotate the component, in increments as large as 90° or as small as 0.01°, depending on the machine. Rotation is important for two reasons. Although design guidelines recommend layout of components on the board in 90° increments (Chap. 10), rotation capability allows the circuit designer additional freedom to design the board layout. Also, for parts with many leads, machine vision can direct the very small rotational corrections required to ensure correct lead-to-pad alignment.

The spindle has variable placement pressure, which permits the SMD to be pressed onto the substrate with sufficient controlled force for required penetration into the solder paste or glue.

9.3.2 Mechanical Centering

The placement head picks up a component from the feeder. The **mechanical centering device** centers the part based on either the body outline or on the lead patterns. The

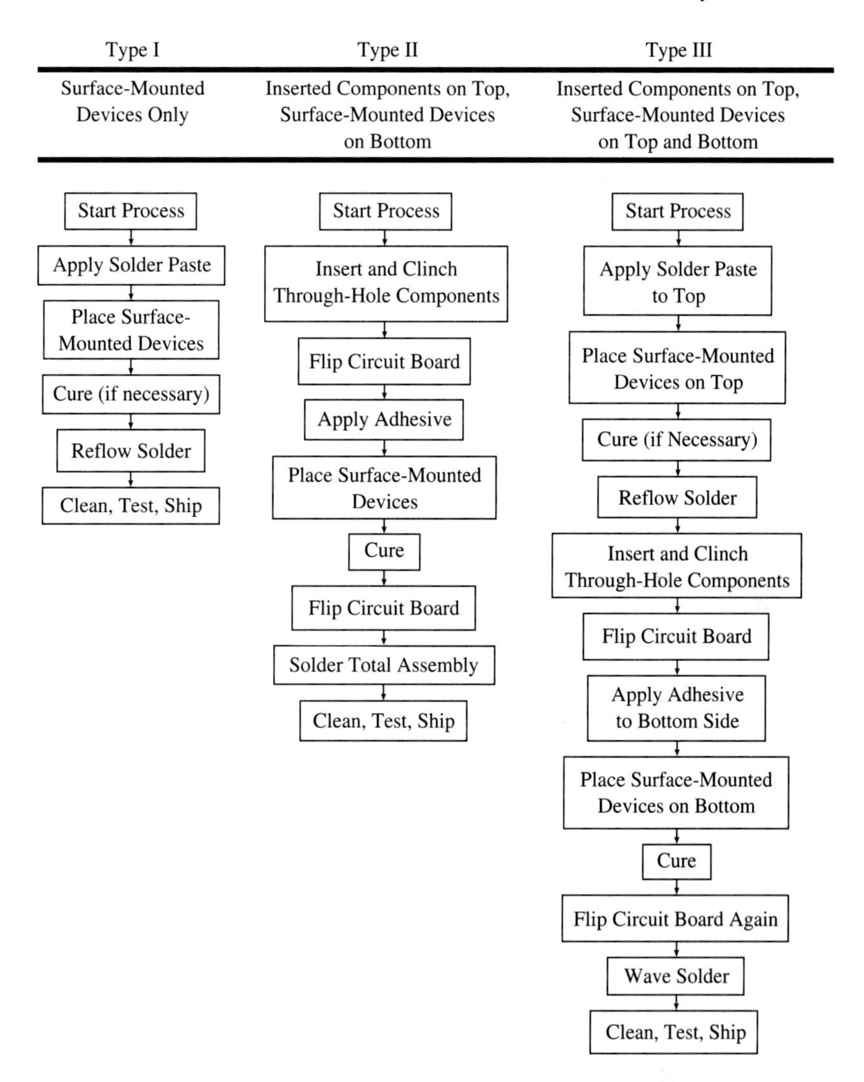

Figure 9-2 SMT assembly processes.

(*Source: Guidelines for Surface Mount Technology,* p. 1-5.)

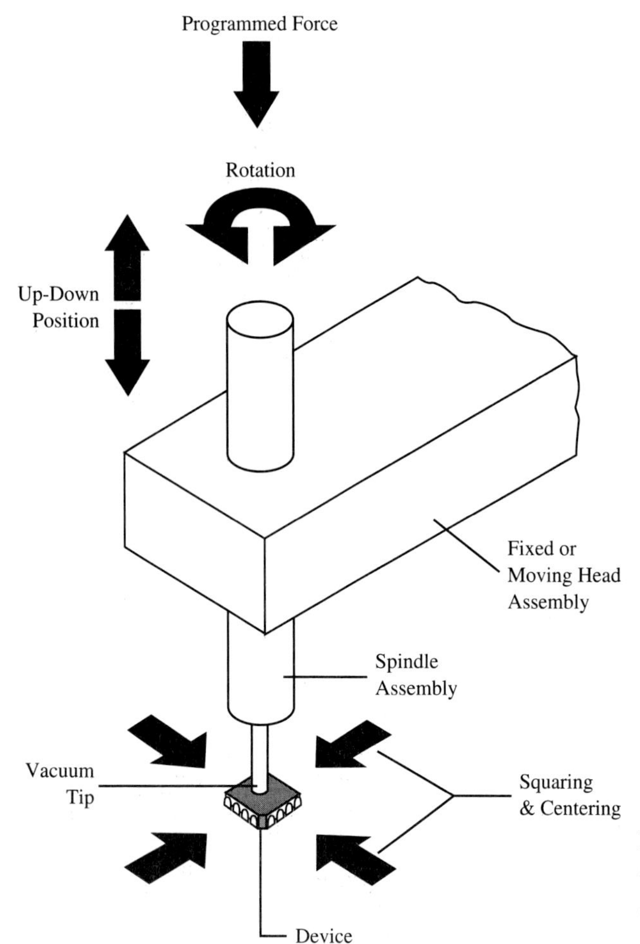

Figure 9-3 Typical pick-and-place spindle functions.

(*Source: Guidelines for Surface Mount Technology,* p. 2-11.)

centering mechanism centers either the leads or the body of the part on the nozzle, or moves the part to a consistent position, known as the placement center. The mechanical centering mechanism uses the part edge surfaces for centering and thus depends on the part edge-to-lead tolerances. The accuracy of actual placement on the board also depends on the coefficients of thermal expansion, the pad-to-pad tolerances, the adhesive or solder paste influence, and the machine placement tolerances (translational and rotational).

9.3.3 Vision Centering

Machine vision provides valuable assistance in centering. In **vision centering,** the vision camera locates board placement sites relative to the fiducials (Fig. 9-4) and identifies the part location. Control algorithms determine the part translations and rotations necessary for precise placement.

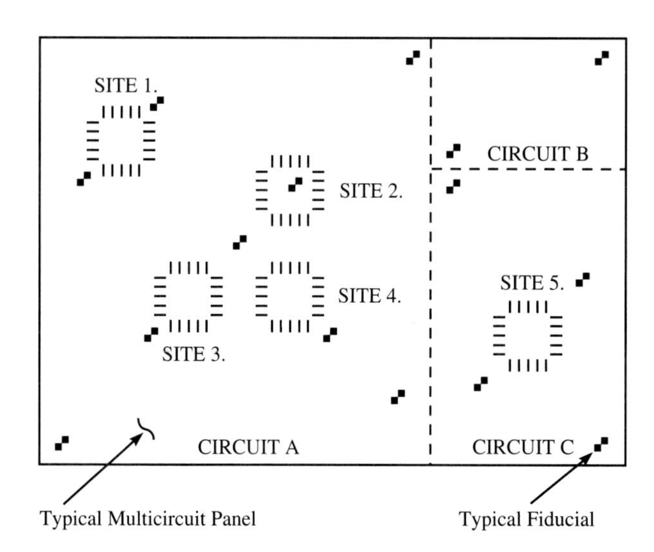

Figure 9-4 Recommended fiducial locations.

(*Source: Guidelines for Surface Mount Technology,* p. 3-3.)

9.4 SUBSTRATE REQUIREMENTS

Figure 9-5 gives the land patterns for different SMDs. Industry and vendor data generally show the lead and body dimensions and suggested pad dimensions and locations.

If the application involves automatic board handling, the boards should conform to specifications for edge parallelism, squareness, and flatness, as shown in Fig. 9-6.

First-quadrant dimensioning is preferred, as most machines use positive X and positive Y positioning. If the medium is a PWB with drilled tooling holes (as discussed in Chapter 8), the two reference lines should include the centers of both reference holes in one axis, and the other axis should intercept the center of one of these holes. This approach is much preferred to having an offset as reference coordinate system because it allows two sets of tolerances between the reference holes and the insertion holes and/or the placement pad. If the reference holes coincide with the coordinate reference system, only one tolerance is allowed to the insertion holes or placement pads.

When locating holes are used and automatic board handling is not, the relationship of the locating holes to the outside board edges is not highly critical. If automatic board handling is used, the relationship of locating holes to board edges becomes critical.

For pure SMT (Type I boards), there may be no need for any holes in the board. In this case edge locating must be used instead of hole locating, and the same locating points must be used for all operations. Usually two locations on the long side and an edge location near the reference side are used. The edges must be parallel and square (Fig. 9-6(a)). The finish of the edge must be sufficiently smooth so as to not affect the locating means.

There are several reasons for specifying the flatness of the medium; one is for handling reasons, particularly for an automatic board handling system. Fig. 9-6(b) shows

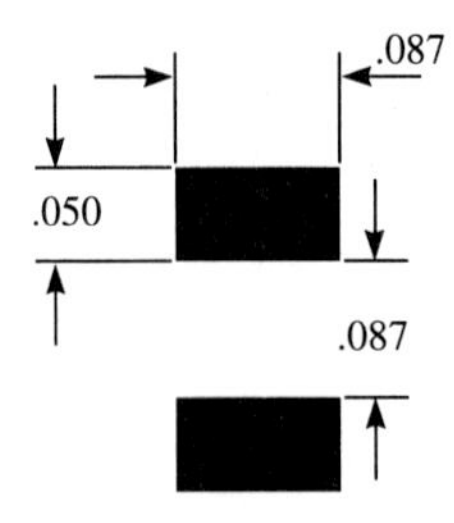

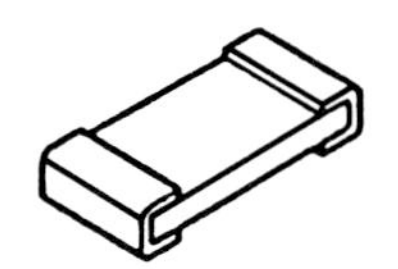

View A–For Passive Device Land Pattern Design

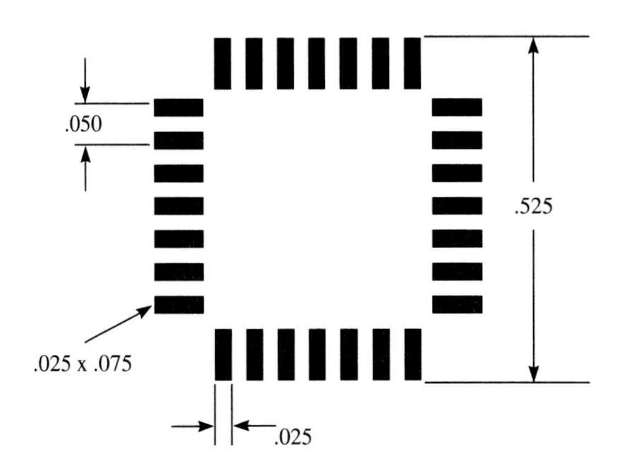

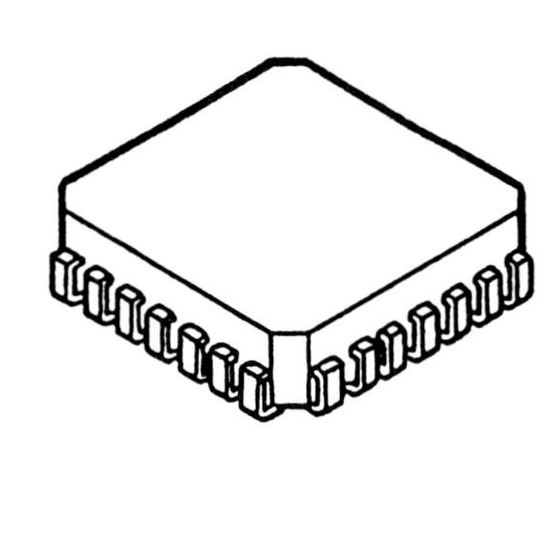

View B–For PLCC and LCC Device Land Pattern Design

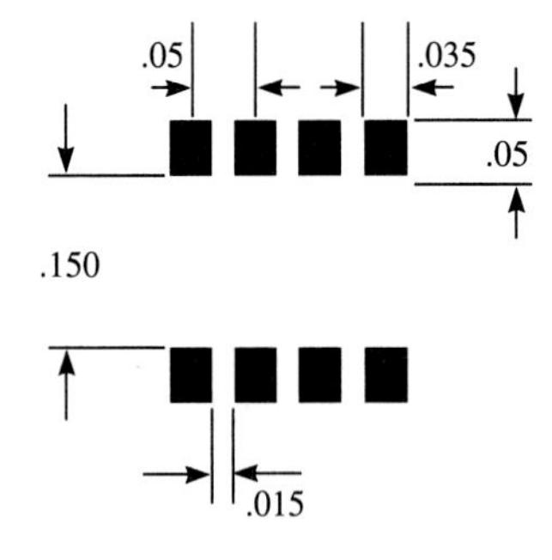

View C–For SOIC Device Land Pattern Design

Figure 9-5 Suggested land pad and pattern data.

(*Source: Guidelines for Surface Mount Technology,* p. 3-1.)

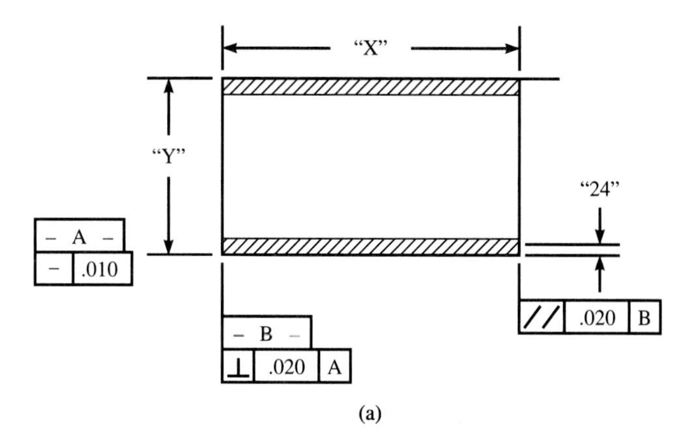

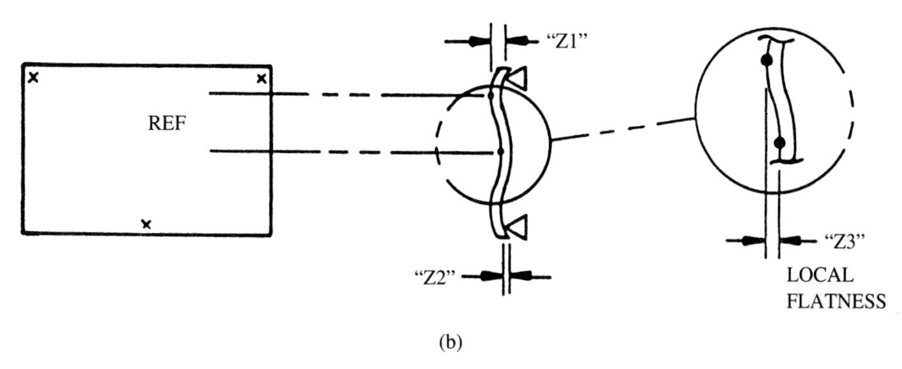

Figure 9-6 SMD design guidelines data: (a) Edge parallelism and squareness; (b) Flatness (measured with board resting on three points of the long edge, one at midpoint of opposite long edge).

(*Source: Guidelines for Surface Mount Technology,* p. 3-5.)

a suggested setup for measuring the overall flatness. For good handling, the peak-to-peak maximum deviation should not exceed 0.007 inches per inch (7 mils/inch or 0.07 mm/cm) of the maximum diagonal, with the board mounted on three points as shown. Also, regardless of the overall size, the maximum deviation should not exceed 0.1 inch (2.54 mm).

An additional measure of flatness is local flatness, defined as the maximum peak-to-peak deviation, obtained with the medium mounted on three points over any 2 inch by 2 inch (51 mm by 51 mm) area. This criterion is a requirement for the placement of SMDs where the coplanarity of the component lead plane and the mounting plane is critical. The actual limit is totally dependent on the components to be placed, but 7 mils per inch (0.07 mm/cm) is a good reference standard. Larger components requiring vision-assisted placement may require precise local flatness.

9.5 PLACEMENT ACCURACY REQUIREMENTS

Placement accuracy is a significant characteristic of any placement system. It affects machine cost, speed, and complexity. Table 9-1 shows the categories of placement accuracy required by component types.

Group A components must be placed within ±0.008 inch (0.203 mm) to ±0.015 inch (0.381 mm), depending on the terminal size, pad size, and pad and machine tolerances. Machines which simply pick up, center, and orient parts at 90° increments are adequate for the placement of group A components.

Group B components (including SOICs , PLCCs, flat packs, and LCCs) with fewer than 44 leads and with center spacing of 0.050 inch (1.27 mm) require moderate placement accuracy.

For group C components, high placement accuracy is required. Components with many contact points and close lead or pad spacing typically require machine vision placement. Vision is used for both lead location and orientation and for location of fiducials on the medium. The larger versions of group C components need to be placed to within ±0.002 inch (0.051 mm), or less, in the *X-Y* plane with respect to the mating pads. As shown in Fig. 9-7, rotational correction is also needed and if not taken into consideration, major placement errors can result.

9.6 MACHINE VISION IN SMT

Section 9.3.3 described the use of machine vision for component centering. Section 9.5 noted that machine vision is essential for placement of group C devices. Machine vision can be used to locate reference points on the substrate and to measure both linear displacements (ΔX, ΔY) and angular displacement (θ) of the SMD. Positional data may be logged and monitored for necessary corrections. Machine vision also provides a means of compensating for wear in the assembly machine components.

The machine vision system should[3]

- Inspect all SMD types
- Check for component presence and absence
- Check for translation and rotation
- View board sizes up to 18 inch by 18 inch (0.46 m by 0.46 m)
- Inspect at a rate of at least two boards per minute
- Be capable of processing CAD data
- Be capable of storing data bases for several boards
- Keep statistical measurement and error data and upload these to a host computer
- Be capable of rapidly changing from one board type to another

[3]Donald C. Mead, "Machine Vision in SMT," *Assembly Engineering*, September 1987, pp. 40–42.

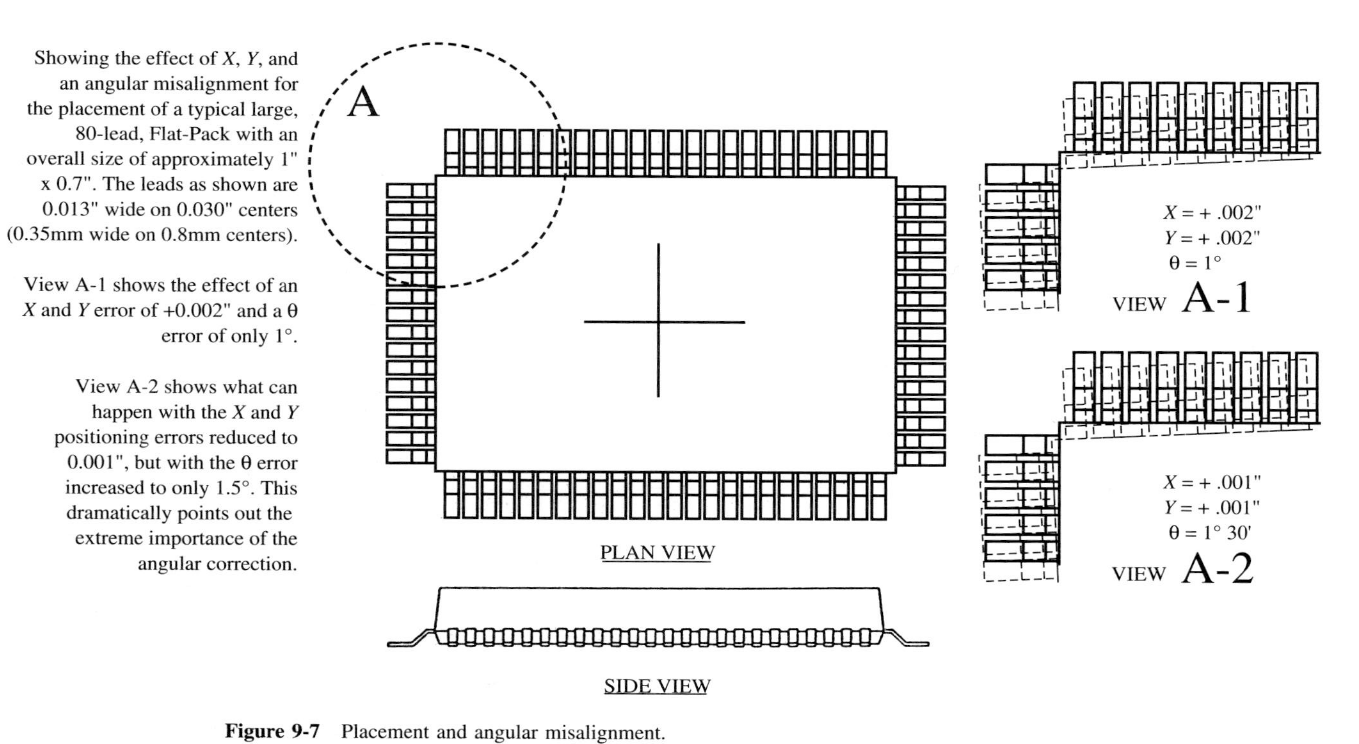

Figure 9-7 Placement and angular misalignment.

(*Source: Guidelines for Surface Mount Technology,* p. 2-10.)

9.7 COMPONENT HANDLING

Component handling is critical to overall productivity of the SMT assembly process. The component handling approach should

- Carry a reasonable quantity of parts
- Provide protection to the parts during shipment and storage
- Provide identification of the components
- Permit easy handling by one person
- Allow simple replacement when exhausted
- Allow access to the part for testing while still in the handling system.

The following subsections describe major SMD handling systems.

9.7.1 Tape-and-Reel Feeders

Tape and reel is a method used for packaging and feeding of chip resistors, capacitors, metal electrode leadless faces (MELFs), small-outline transistors (SOTs), small-outline integrated circuits (SOICs), and small-outline components with J-type leads (SOJs).[4] When large quantities of components are to be used, tape-and-reel feeders are preferred. Paper and blister tapes are available for use with these feeders. Paper tapes have punched holes fitted to the component size, and the components are held in place by two adhesive foils. The price of paper tapes is favorable, but applicability is restricted by the tape and component thickness and by the possible abrasive damage by the tape paper. Blister tapes have compartments which correspond to the component size and are covered by adhesive tape. Blister tapes consist of either plastic or plastic-clad aluminum foil. The tape and/or foil is peeled back at the time of component placement (Fig. 9-8).

Tape configurations have been standardized internationally. The tape is wound on reels with diameters of 180 to 325 mm. The number of components per reel depends on component thickness and reel diameter. Usual tape widths are 8 to 56 mm. The reels protect the components from damage during shipping and handling, save on loading and unloading time on pick-and-place equipment, and prevent both mix-up of components (by compensating for the lack of part markings on many SMDs) and incorrect orientation of components.

Tape is used with high-speed automatic placement machines because of small feeder footprint and the large available on-line inventory which enables continuous delivery of components and reduces the frequency of parts replenishments required during production. The large variety of standard components that are currently available on tape and reel is a good reason for the use of this packaging.

The use of tape-and-reel packaging also presents a few problems. Not all components are available on tape and reel, and not all pick-and-place machines have tape-and-reel

[4]Ray P. Prasad, *Surface Mount Technology: Principles and Practices* (New York: Van Nostrand Reinhold, 1989), p. 403.

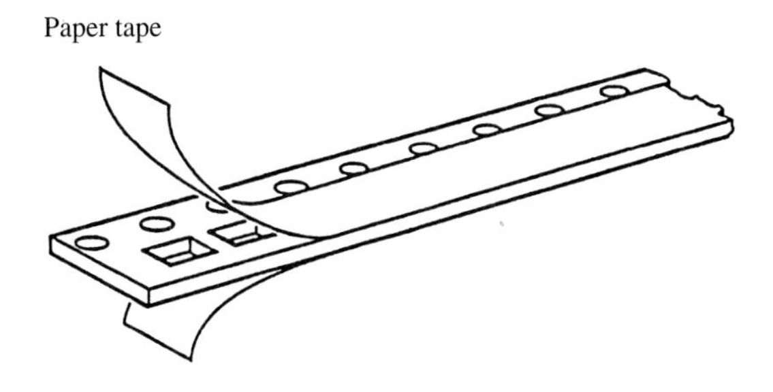

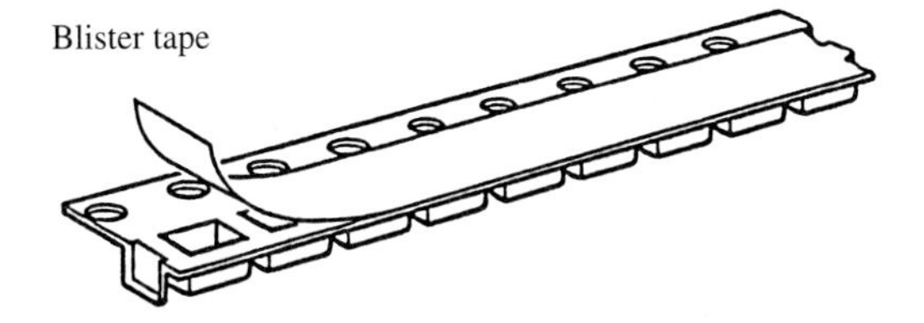

Figure 9-8 Tape packing of SMDs.

(*Source:* Otmar Hintringer and Werner Maiwald, *An Introduction to Surface Mounting* (Munich, Germany: Siemens Aktiengesellchaft, 1986), p. 19.)

handling capability. Also, tape-and-reel feeders require more slot space than do tube feeders, and as a result, the machine occupies more floor space.

EIA-RS-481 is a standard that controls the tape-and-reel construction. The tolerances in this specification are not very tight, which makes reel quality vary substantially among suppliers and lots. Due to poor tape tolerance, the center-to-center distance of sprocket holes may differ, resulting in problems of component pickup (Fig. 9-9).

Tape provides only rough location of the components within the pocket, making it desirable to use registration means such as machine vision in order to precisely place

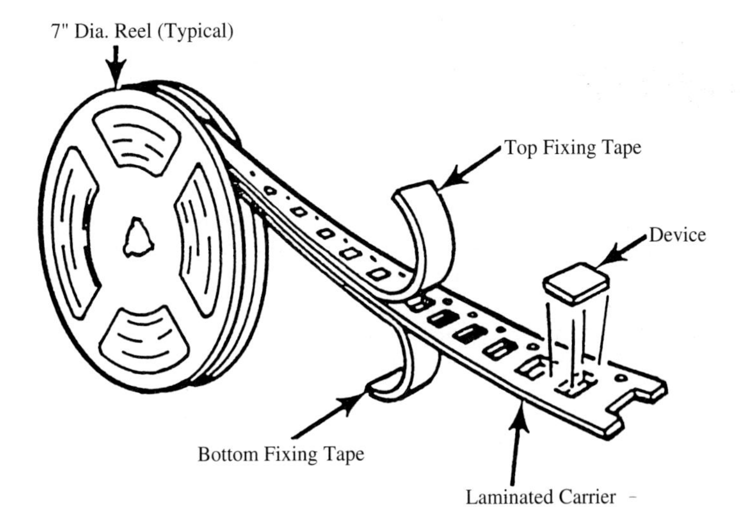

Figure 9-9 Sprocket-driven tapes.

(*Source: Guidelines for Surface Mount Technology,* p. 2-7.)

the component on the PWB.[5] The registration time increases the overall cycle time for component placement.

Components to be placed on EIA-pocketed tape must fit within the size constraints that current EIA standards permit. These standards only address very small and flat components that fit on tape from 8 to 56 mm in width. Many of the larger fine-pitch SMDs cannot be packaged on EIA tape, and therefore an alternate packaging and delivery method should be used for automatic placement.

Another EIA specification allows only ±30 pounds (13.6 kg) in peel strength. Thus, there are problems in cover removal due to the wide variation in pull-force required to peel off the cover tape to expose components for placement.[6]

Tape and reel is best suited for large quantities. The capacity of a tape-and-reel system can exceed the quantity desired for a particular assembly batch. The EIA specification establishes maximum reel size at 13 inches (33 cm) without mentioning any minimum size. In actuality, the standard reel sizes for passive and active components have become 7 inch (17.8 cm) and 13 inch (33 cm) diameter. Since active components are only available in 13 inch (33 cm) diameter reel size, inventory cost may be a significant problem for small users. Thus, the desired package quantity and work-in-process inventory level are important considerations when selecting the feeder equipment.

9.7.2 Tube Feeders

Tube and stick feeders are a popular form of packaging because they require less space and lower inventory levels at the machine, compared to tape and reel. The tube feeders may be divided into three categories: horizontal, stick-slope, and ski-slope.

Horizontal tube feeders have multiple tracks which accommodate components of different widths, and the number of tracks per feeder depends on the width of the component. Typically a row of horizontal tube feeders would be mounted on the pick-and-place machine.

In a stick-slope feeder the plastic tube can be directly loaded into an individual track of the horizontal feeder. Hence the components need not be removed from their plastic shipping tubes. Electromagnetic induced vibration is used to move components to the pickup position in horizontal as well as stick-slope feeders.

Ski-slope feeders use spring action to move components and are otherwise similar to stick-slope feeders. Unlike horizontal and stick-slope feeders, ski-slope feeders do not have machined tracks for different parts but the tracks are integrated into the feeder itself. These feeders require tighter component tolerances and will jam easily. The ski-slope feeders are more expensive and require more space on a machine due to their length. Ski-slope feeders can take more components than either horizontal or stick-slope feeders, hence requiring less frequent loading and unloading. Nevertheless, the ski-slope feeders are not as popular as are horizontal and stick-slope feeders.

When the placement of large odd-form SMDs is required, tube feeders are sometimes used. Tube feeders require minimal space on the machine and can be used in conjunction

[5]W. A. Mahoney and K. G. Murphy, "Feeders for SMD, an Overview of Current and Emerging Technology," *Expo SMT 88 Technical Proceedings*, IPAC, Calif., p. 167.

[6]Prasad, *Surface Mount Technology*, p. 406.

with other types of feeders in the same work cell. Since tube feeders contain few parts, they require frequent replenishment. Also, since tube feeders are long and narrow, they are susceptible to warping, bending, twisting, and other problems. Tubes do offer variety in the range of components that can be accommodated, but each different tube shape is limited in use, being specifically tooled for a particular component. Tubes do not provide exact location of components in the package, and thus components must be indexed out of the tube to the exposing area and registration station before being picked up by the placement head. The registration step increases cycle time and thus reduces productivity.

9.7.3 Bulk Feeders

Bulk feeders are the most simple and low-cost mode of delivery for SMDs. Generally, only passive devices are bulk fed. Bulk components are fed into assembly machines via a vibrating bowl and a track. The track width limits the width of component that can be fed. If required, bulk feeders permit a large component quantity to be available.

Bulk feeding is not a good choice for many SMT devices due to potential damage to terminations and leads, and misplacements and danger of interchangeable components. Another problem is the jumping out of components from the feeder due to improper intensity of vibration.

9.7.4 Waffle Packs

Waffle packs (matrix trays) are flat cases with formed pockets or plastic carriers. Waffle packs consume a large input slot space. Whereas the pickup position is the same for all components in the tube or tape, the head must access a different position for each component pickup from the waffle pack.

As with tube feeders, waffle packs have small capacity, and frequent replenishment is required. One solution to this problem is the use of waffle pack handlers. These handlers deliver all parts from waffle packs to a common pickup position for the head. This added step increases machine complexity, cost, and, potentially, cycle time since two sequential pickups are required: one by the handler and the other by the placement head. Waffle packs are the only handling option for certain delicate components such as bumperless fine-pitch components where part movement within the tubes can easily result in lead damage.

9.7.5 Magazine Feeders

Linear, waffle, stick, and stack **magazine feeders** are available. Refer to Fig. 9-10. Stick and stack magazines are quite frequently used. Surface-mount devices with pins on all four sides (quad packs) require linear or waffle magazines. Magazines have the disadvantage that they can only accommodate a small amount of components.

9.7.6 Selection of Feeder Systems

Feeders are frequently supplied or recommended by the manufacturer of the automatic placement machine. EIA tape feeders are the best choice for high-speed automatic assembly machines. For placement equipment designed for custom components, any of the feeders with the exception of bulk feeders can be used subject to constraints of system cost, component

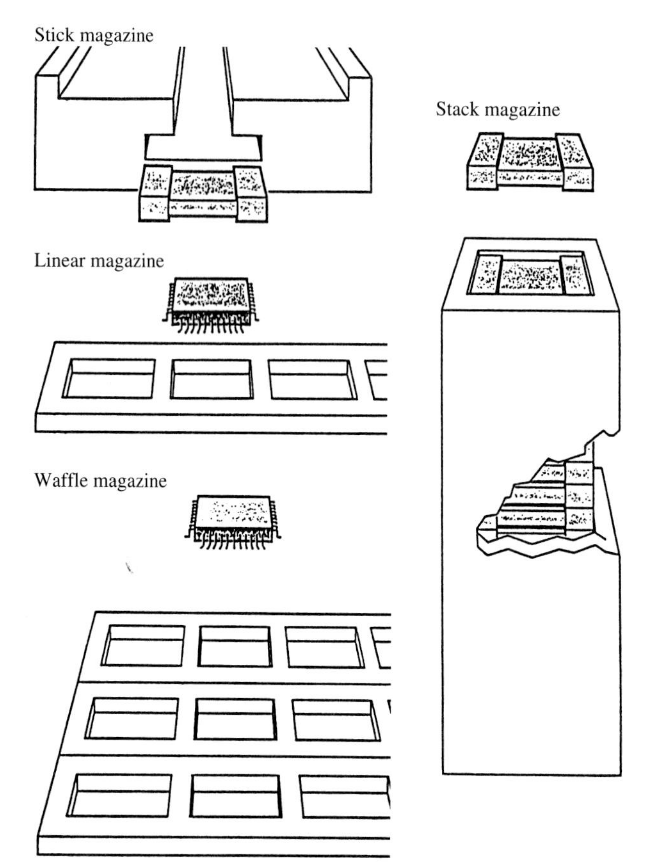

Figure 9-10 Magazines for SMDs.

(*Source:* Hintringer and Maiwald, *An Introduction to Surface Mounting,* p. 21.)

package availability, and throughput. The size and shape of the components being handled affect feeder selection (for example, odd-form components require tube feeders or waffle packs, while flat and smaller sized components may be easily handled by tape and reel). A high-volume, low-mix application favors a large feeder capacity and fewer feeders with a rapid cycle time, whereas a low-volume, high-mix application favors several moderate-capacity feeders for numerous components. Bulk and waffle packs should be avoided due to the longer cycle times required, unless dictated by the parts handling requirements.

9.8 PLACEMENT MACHINES

There are five general categories of placement machines based on placement speeds[7]:

- *Level I:* Machines which can place fewer than 2,000 components per hour (approximately 1.8 seconds per placement). Table-top machines used for pilot runs and educational purposes fall into this category.

[7]*AT&T Surface Mount Technology: Two Day Advanced Design & Manufacturing Workshop* (San Jose, Calif.: International Quality Technologies, Inc., 1988), np.

- *Level II:* Machines capable of handling a wide range of boards and component sizes with relatively good accuracy. These machines are capable of placing 2,000 to 3,000 components per hour (1.8 to 1.2 seconds per placement). Although most Level II machines are manually loaded, some have been adapted to in-line conveyor systems. During operation, one operator per machine is needed for overseeing board assembly as well as assisting board and component feeding.
- *Level III:* These machines represent the fastest growing market segment for placement machine manufacturers. Level III machines handle a wide range of boards and component mixes at high speed and precision. The range of placement rates is 3,000 to 10,000 placements per hour (1.2 to 0.36 seconds per placement). Machines in this category employ a high degree of automation and are usually integrated in-line with other SMT processes by either conveyor or magazine input. Machine vision and part verification are standard.
- *Level IV:* These machines can place components in the range of 10,000 to 20,000 per hour, and minimal human intervention is required.
- *Level V:* These machines have the ability to build a complete board in one cycle.

SMD placement systems vary in many respects, including placement rates, accuracy, repeatability, and ability to handle different components. This section briefly discusses factors to be considered in equipment selection.

9.8.1 Positioning Systems

Surface-mount assembly machines incorporate a variety of **positioning systems,** including gantry, split axis, and turret head chip shooters.

In **gantry machines,** the *X*-axis member carries the *Y*-axis member (placement head). This design requires that the *Y*-axis member be light in weight, so repeatability suffers. Coupled-axis movement also results in oscillation of the gantry system. The faster the placement rate, the less time there is for settling and the more likely it is that some components will be placed outside the acceptable envelope. The gantry design does facilitate automatic board handling, which promotes accuracy.

In **split-axis positioning,** the medium and the component move on two separate axes. Each axis uses an independent, massive, rigid structure that permits rapid travel. Split-axis positioning results in better accuracy and repeatability than provided by the gantry mechanism.

Turret-head **chip shooters** are generally designed to achieve high placement rates for a limited mix of components. The board is positioned in the *X* and *Y* axes while components are shuttled to a high-speed rotary placement head. To maximize throughput, components of the same size and type are used. This type of machine is highly efficient but not feasible for placement of a varied component mix.

9.8.2 Examples of SMD Placement Machines

In order to illustrate placement machine features, we describe the following Universal and Siemens machines.

Universal is the manufacturer of a large number of SMD placement systems. The Omniplace II 4621 machine is a dual-placement-head flexible machine having the capability of placing 25 mil components.[8] The machine accommodates 40 inches (1.0 m) of feeder input area as standard. This input area can be expanded to over 200 inches (5 m) by using the optional feeder input modules which Universal refers to as the Extended Input Modules (EIMs).

The Omniplace II machine can place components with sizes ranging from 0.04 inch (1.01 mm) to 1.25 inches (31.75 mm) square with standard tooling. These components can be fed to the machine by 8 mm to 56 mm tape-and-reel holders and by bulk, stick, and matrix trays.

The Omniplace II vision processor can be used to read the fiducials on the board and make the required placement coordinate corrections, check for any damaged or bent leads, and populate a multiboard panel with a bad-board check option.

Placement speed of up to 5,000 components per hour can be achieved by using both of the placement heads simultaneously. The accuracy and repeatability of the positioning system are ±0.002 inch (0.051 mm) and ±0.0005 inch (0.013 mm), respectively.

The Siemens SP-120 is a chip-shooter type of machine[9] that can handle only a limited size and type of different components compared to the Omniplace II machine. The standard dual SP-120 (which Siemens refers to as the Basic Version of SP-120) can achieve a placement rate of 14,400 components per hour. The placement accuracy and repeatability are ±0.003 inch (0.076 mm) and ±0.0008 inch (0.020 mm), respectively.

The SP-120 uses a 12-position revolver head for fast placement of the SMDs. Before placement, the machine verifies the electrical value of the component, which greatly reduces the chances of placing a bad or wrong component on the PWB.

The vision system is capable of reading the fiducial marks and then making the appropriate corrections to the placement coordinates. Vision also checks for bad boards in a multiboard panel and automatically skips any bad board.

The SP-120 (standard dual) can accommodate 120 inputs of 8 mm tape. The feeder input area can be expanded to a maximum of 360 inputs of 8 mm tape. The components can be fed to the machine by 8 mm, 12 mm, and 16 mm tape and reel holders and by bulk and stick magazines.

9.9 SUMMARY

In this chapter we have discussed the rapidly developing field of SMT assembly. SMT includes both pure designs, with SMDs on both sides of the PWB, and mixed-technology designs combining surface-mount and through-hole devices. The trend in commercial electronics is toward SMT, although through-hole PWBs will continue to be a factor, particularly in the aerospace/defense industry. The circuit density and fine lead pitches

[8] *Omniplace II 4621 SMC Placement System* (Binghamton, N.Y.: Universal Instruments Corp., 1989).

[9] *SMD-Speed Placer SP-120 for high volume placement and a wide component spectrum* (Germany: Siemens Aktiengesellschaft, 1989).

of SMT necessitate automated assembly with integrated computer vision. SMDs are packaged in tape and reel, tube, and waffle pack. SMT placement machines range in price up to several hundred thousand dollars and can place over 20,000 SMDs per hour.

KEY TERMS

Bulk feeders
Chip shooters
Gantry machines
Magazine feeders
Mechanical centering
Pick-and-place spindle
Positioning systems
Split-axis positioning
Tape and reel
Tube and stick feeders
Vision centering
Waffle packs (matrix trays)

EXERCISES

1. What are the advantages of SMDs?
2. What are some considerations when looking at SMDs?
3. Define surface-mount technology.
4. When faced with large production quantities, what kind of feeder would be most suitable for SMDs?
5. Why should a vision system be implemented in the beginning, even though the placement equipment has high accuracy and repeatability?
6. What are two means of centering a part, and what does each system use to center the part?
7. Discuss the placement accuracy required for a SMD as a function of its number of leads.

BIBLIOGRAPHY

1. Clark, R., *Printed Circuit Engineering, Optimizing for Manufacturability.* New York: Van Nostrand Reinhold, 1989.
2. Nobel, P. J. W., *Printed Circuit Board Assembly.* New York: Halsted Press, 1989.
3. Prasad, R. P., *Surface Mount Technology: Principles and Practices.* New York: Van Nostrand Reinhold, 1989.
4. Traister, John E., *Design Guidelines for Surface Mount Technology.* San Diego, Calif.: Academic Press, Inc., 1990.
5. Tummala, R. R., and E. J. Rymaszewski, eds., *Microelectronics Packaging Handbook.* New York: Van Nostrand Reinhold, 1989.

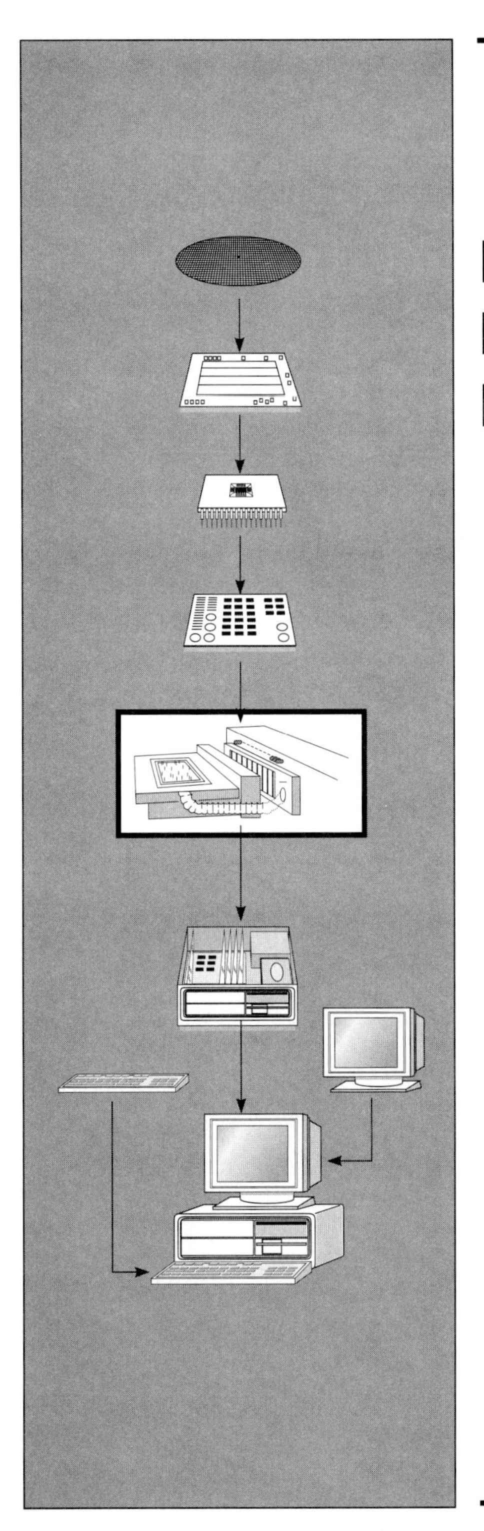

PART FOUR
LIFE-CYCLE ENGINEERING

10 Design for Assembly

10.1 INTRODUCTION

There have been important advances in manufacturing productivity in recent years, including numerical control (NC), flexible manufacturing systems (FMS), computer-integrated manufacturing (CIM), and just-in-time (JIT). However, these approaches have primarily addressed the production function in the manufacturing enterprise. Concurrent engineering (Chap. 2) and quality function deployment (Chap. 11) are concepts that emphasize integration of all business functions to achieve competitive success. Producibility, design for manufacturability, and design for assembly are disciplines that promote integration of design and manufacturing functions.

Priest defined **producibility** as "the engineering discipline directed toward achieving a design compatible with the realities of manufacturing."[1] **Design for manufacturability** (DFM) is synonymous with producibility. Both producibility engineering and DFM imply integration of design and manufacturing to ensure that product designs are producible, and thereby speed the successful introduction of high-quality, low-cost products to the marketplace.

We define DFM as encompassing all phases of the production process which, as noted in Chapter 2, includes both fabrication of components and assembly of components into products. For example, in the design of PWB substrates (Chap. 5), decisions about through-hole tolerances for drilling and plating affect the PWB fabrication process as well as the assembly process of soldering leads through the holes.[2]

[1]John W. Priest, *Engineering Design for Producibility and Reliability* (New York: Marcel Dekker, Inc.), p. 35.

[2]Ron Daniels and Pete Waddell, "Design for Assembly," *Circuits Assembly,* 2, 7 (July 1991), pp. 24–26.

This chapter is devoted to **design for assembly** (DFA). DFA is a subset of DFM and focuses on design principles compatible with assembly processes such as those discussed in Part II of this text.

In Sec. 10.2 we discuss management principles of DFA. Section 10.3 addresses general DFA design rules and DFA for mechanical designs such as chassis. Sections 10.4 through 10.6 deal with electrical designs such as PWBs. Section 10.5 focuses on through-hole technology, while Sec. 10.6 pertains to surface-mount technology. Section 10.7 addresses interconnections. DFA rating methods are discussed in Sec. 10.8.

10.2 MANAGEMENT PRINCIPLES

Design for assembly is an application of the concurrent engineering approach to the design of products and assembly processes. Design and process engineers cooperate with assembly operators, test technicians, and maintenance personnel to achieve designs that are both producible and maintainable.

Recall from Sec. 2.1.3 that as much as 75 percent to 90 percent of manufacturing cost is determined in the design phase. By coordinating product and process design early in the product life cycle, the DFA approach helps to deliver good products to market faster. Studies have shown that companies in the United States continue to process design changes very late in the design phase and even well into production and field use (Fig. 10-1). Conversely, Japanese companies tend to process the majority of design changes well before the start of production.[3]

The concurrent engineering approach, including DFA, depends on teamwork. Unfortunately, the reality for many companies is that some people tend to resist cooperation, sharing of knowledge, and loss of power. Egos, territorialism, and fear are likely initial reactions to DFA. Consequently, upper management must provide strong leadership for DFA and create new, team-oriented, long-term goals, which reward cooperation.

Some design engineers express concern that the extra effort and coordination involved in DFA may adversely impact product development schedules. The initial implementation of DFA may increase development time, but this delay is usually offset by the time saved later in investigating problems and processing engineering changes.

The DFA approach ensures that designers understand the assembly sequence, operations, and equipment capabilities. DFA methods and tools also permit engineers to examine design alternatives, identify opportunities for improvement, and cost-justify those improvements. Proposed designs are rated in terms of cost, standardization, potential for automation, and potential for assembly problems.

DFA is a disciplined approach considering the capabilities of the product and process and seeking a design that is simple and easily assembled. DFA stresses simplicity through reduced parts count and maximum use of standardized parts. Assembly time is substantially reduced through elimination of parts such as fasteners.

A typical DFA analysis might compare an assembly containing several parts joined by fasteners with another design fabricated as an integral unit. For example, Texas Instru-

[3] L. P. Sullivan, "Quality Function Deployment," *Quality Progress,* 19, 6 (June 1986), pp. 39–50.

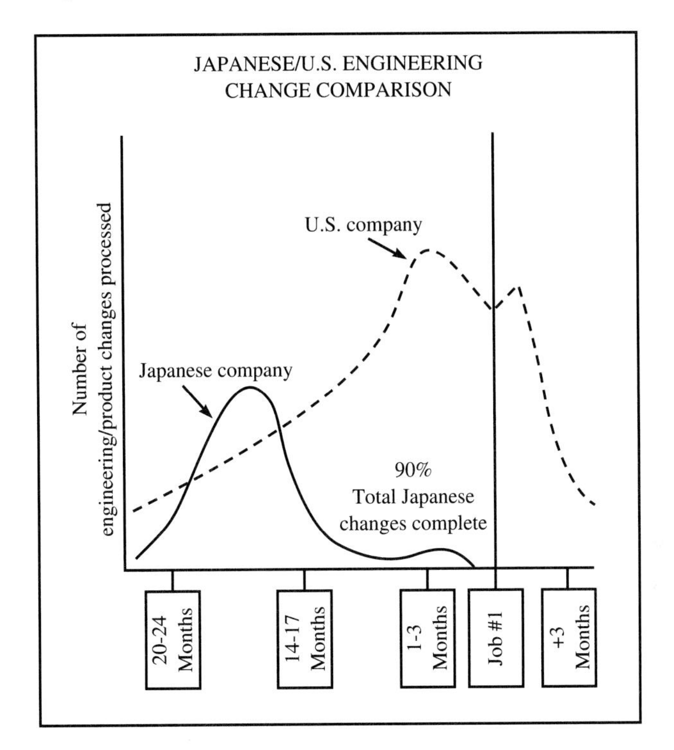

Figure 10-1 Design changes in the product development cycle.

(*Source:* L. P. Sullivan, "Quality Function Deployment," *Quality Progress,* 19, 6 (June 1986), p. 39.)

ments' Defense Systems & Electronics Group (DSEG) performed a mechanical DFA analysis on an optical reticle assembly (Fig. 10-2). DFA revealed that the reticle assembly required too many fasteners and that assemblers were required to awkwardly turn the assembly around often during the assembly process.

The new design used a cam instead of a gearbox, which created smoother motion and eliminated some assembly operations. Almost all of the fasteners were eliminated by using self-securing parts, and the connecter bracket was incorporated into the housing. Cast aluminum was chosen for the two main metal parts, which eliminated machining and thus reduced fabrication time. Table 10-1 summarizes the design improvements achieved through DFA.[4]

Since products are designed with fewer parts, companies can save labor, inventory, and capital costs. In many cases, reductions in floor space, purchasing, documentation, and administration can be at least as significant as the direct savings in material and assembly costs.

DFA can help increase production without expensive automated assembly tech-

[4]Therese R. Welter, "Designing for Manufacture and Assembly," *Industry Week,* September 4, 1989, pp. 79–82.

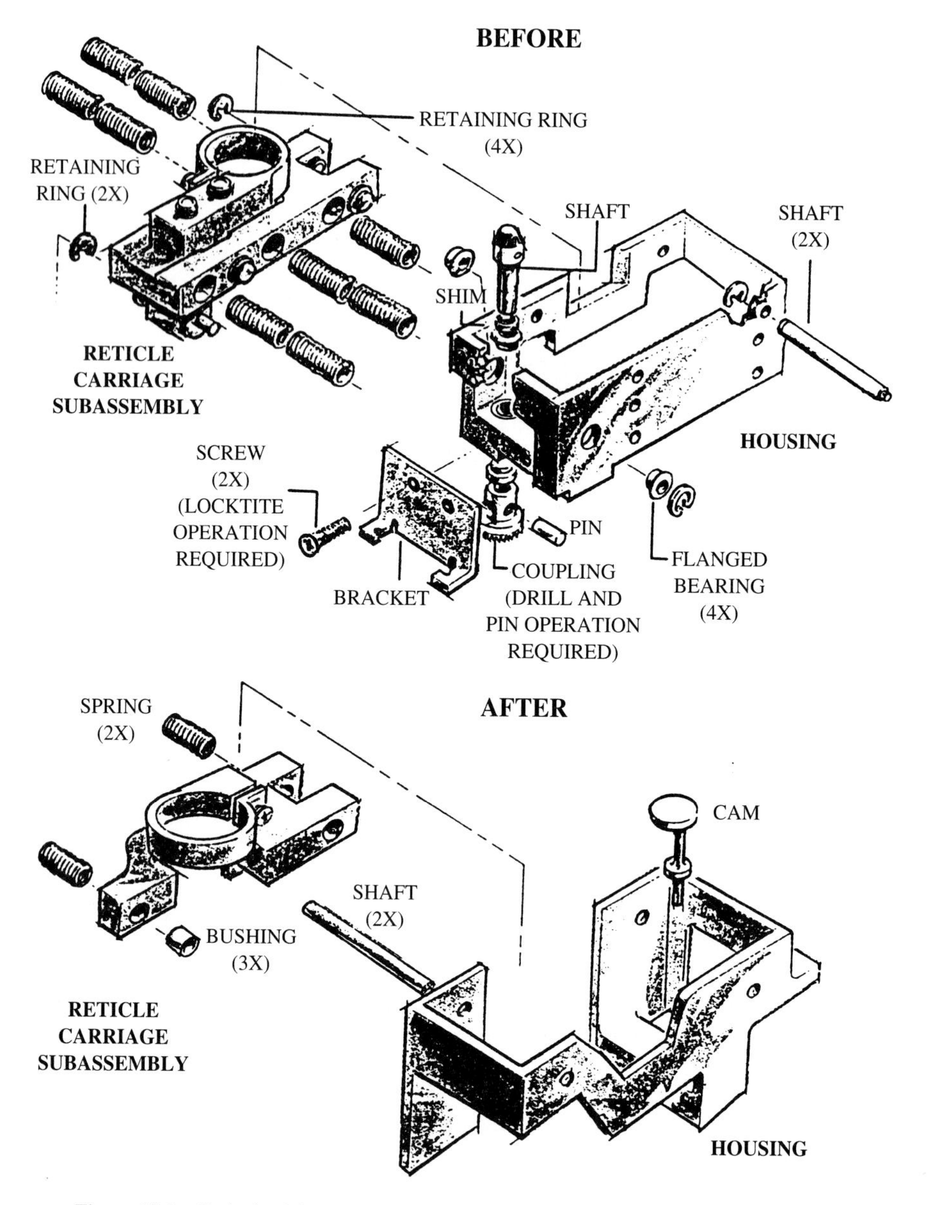

Figure 10-2 Optical reticle assembly.

(*Source:* Therese R. Welter, "Designing for Manufacture and Assembly," *Industry Week,* September 4, 1989, p. 81.)

TABLE 10-1 DFA IMPROVEMENT OF OPTICAL RETICLE ASSEMBLY

	Original design	Redesign ▼
Assembly time (hrs)	2.15	0.33
Number different parts	24	8
Total number parts	47	12
Number operations	58	13
Fabrication time (hrs)	12.63	3.65
Weight (lbs)	0.48	0.26

Source: Therese R. Welter, "Designing for Manufacture," p. 81.

niques. When the design is simplified to best facilitate automation, we often find that the product can be manually assembled for less cost and with greater flexibility. Where automation is appropriate, DFA is essential to ensure that the full benefits of automation are achieved.

10.3 GENERAL AND MECHANICAL DESIGN RULES

Design rules for assembly can be conveniently divided into two categories: product design for ease of assembly and design of parts for feeding and orienting (ease of component handling).

10.3.1 Rules for Ease of Assembly

The following design rules promote ease of assembly and are relevant to both mechanical and electronic assembly processes:

- *Minimize Part Count:* Reduce the total **parts count** by eliminating unnecessary parts, combining parts, and standardizing required parts. The following constraints can limit reduction in parts count:
  — Parts move relative to each other
  — Different materials are required (for example, insulators)
  — Maintenance access requires certain parts (for example, fasteners)
  Figure 10-3 shows several examples of redesign to eliminate parts. The designer should avoid separate fasteners and substitute snap fits where possible (Fig. 10-4). Washers should be captive and fasteners should be self-threading and have conical or oval points.
- *Minimize assembly surfaces:* Simplify the design so that fewer surfaces need be processed and so that all processes on one surface are completed before processing other surfaces. An assembly surface should include error-proof features (for example, keyed connectors, self-alignment, guide pins, captive fasteners).

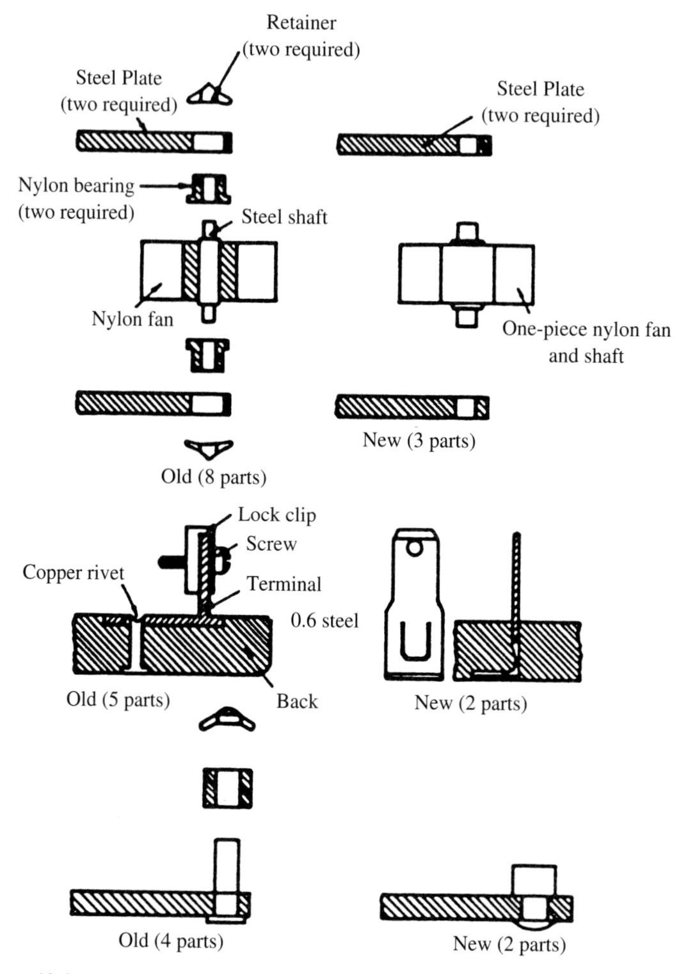

Figure 10-3 Parts reduction.

(*Source:* Geoffrey Boothroyd, Carrado Poli, and Laurence E. Murch, *Automatic Assembly* (New York, N.Y.: Marcel Dekker, 1982), p. 257.)

- *Design for Top-Down Assembly:* Provide assembly access from the top down, and plan the assembly process from the bottom up. Allow for **top-down assembly** in sandwich, or layer, fashion, with each part being placed on top of the previous one, using straight-line motions. Gravity assists in holding the layers in place, including during movement. Parts should be designed to prevent dislocation. In automated assembly, a work carrier, or platform, may be provided with tapered pins that facilitate quick and accurate location of the base part (Fig. 10-5).
- *Maximize Part Compliance:* Provide chamfers, tapers, grooves, guide surfaces, and proper tolerances for mating parts (Fig. 10-6).

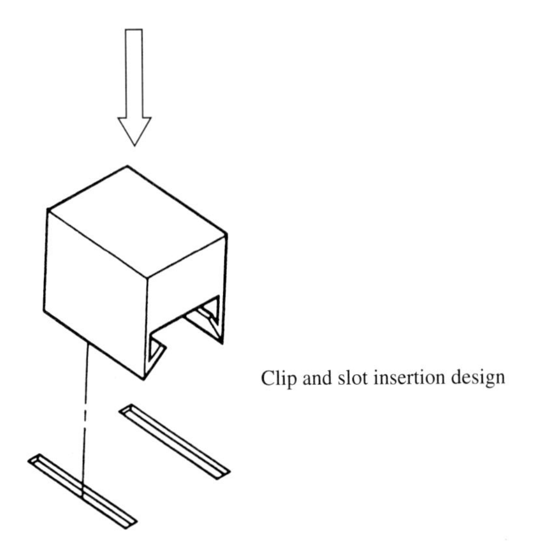

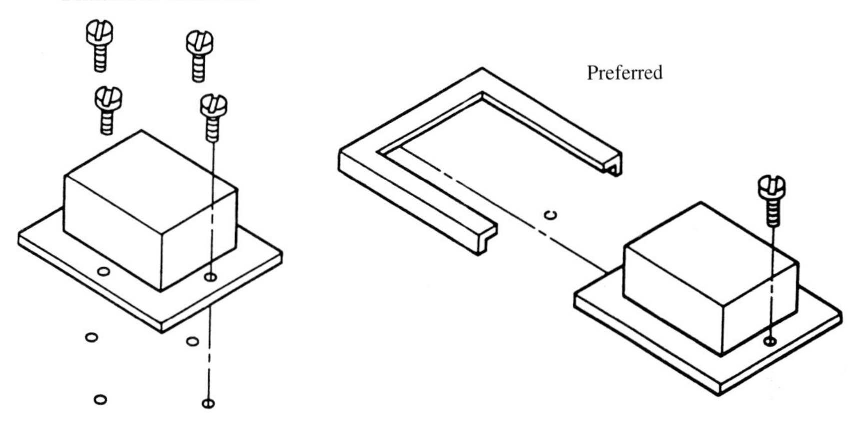

Minimize the number of screw-type fasteners required.

Figure 10-4 Design alternatives for minimizing fasteners.

(*Source:* J. Ronald Bailey, "Product Design for Robotic Assembly," *Proceedings of the 13th Symposium on Industrial Robots and Robots 7,* April 1983, p. 11-52. Reprinted with permission of the Society of Manufacturing Engineers.)

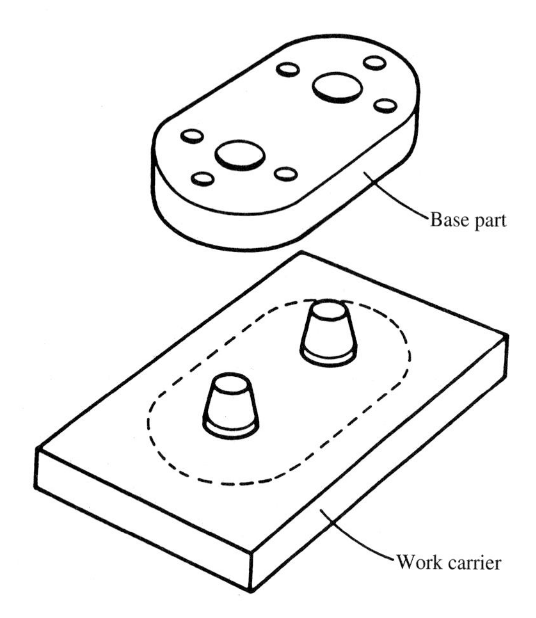

Figure 10-5 Work platform with tapered alignment pegs.

(*Source:* Boothroyd et al., *Automatic Assembly,* p. 261.)

- *Improve Assembly Access:* Accessibility involves easy physical access, unobstructed view, and adequate clearance for standard tooling.
- *Design for Modularity:* Use standard modules for common functional requirements and standard interfaces for easy interchangeability of modules. **Modularity** permits more options, faster updates of designs, and easier testing and maintenance. Incorporate parts into subassemblies.

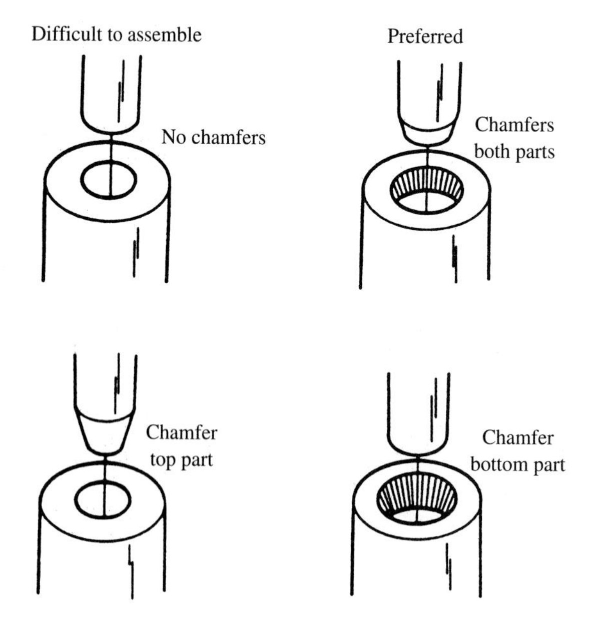

Figure 10-6 Chamfering facilitates part insertion.

(*Source:* Bailey, "Product Design for Robotic Assembly," p. 11-51.)

- *Design for Standard Tooling:* Design all components of the assembly to be handled with the same tooling (for example, robot end-effector); otherwise, ensure that tooling can be changed quickly and easily.

10.3.2 Rules for Ease of Part Handling

Parts should be designed for ease of handling, feeding, and orientation in assembly. The following design principles promote ease of part handling:

- *Handling and Feeding:* Avoid designing parts that are difficult to handle and feed. Design parts to be rigid, rather than flexible. A small change in part form can prevent handling problems without affecting fit or function. The part design must provide adequate surfaces for mechanical gripping and prevent shingling, tangling, nesting, or interlocking (see Fig. 10-7).
- *Symmetry:* Symmetrical parts are easier to orient and handle, and their use avoids the need for extra orienting devices with corresponding loss in feeder efficiency.
- *Recognition:* If symmetry is not possible, design in obvious asymmetry or alignment features to facilitate orientation (Fig. 10-8). Parts with distinguishing features (for

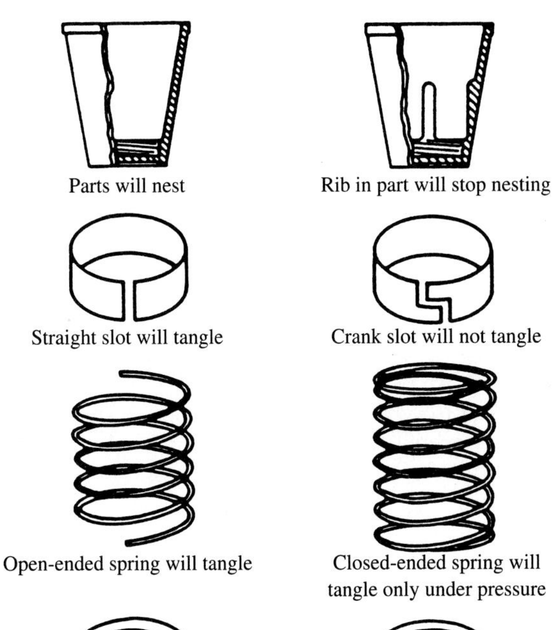

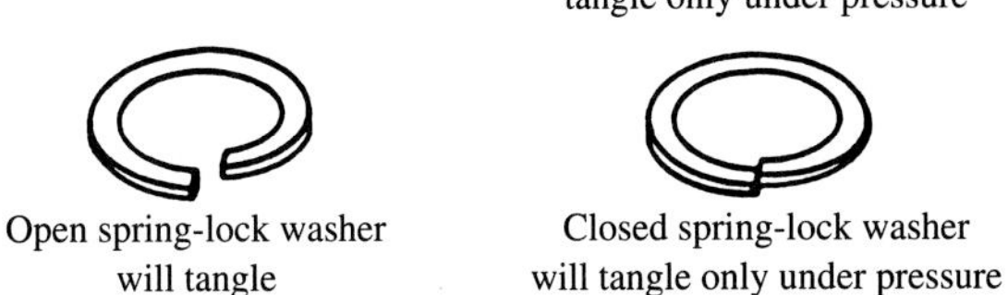

Figure 10-7 Redesign of parts to prevent nesting and tangling.

(*Source:* Boothroyd et al., *Automatic Assembly,* p. 264.)

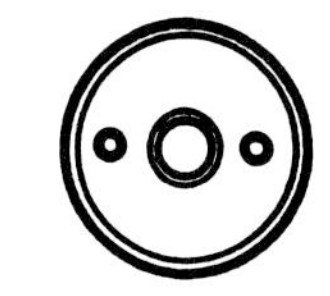

Difficult to orient with respect to small holes

Flats on the sides make it much easier to orient with respect to the small holes

No feature sufficiently significant for orientation

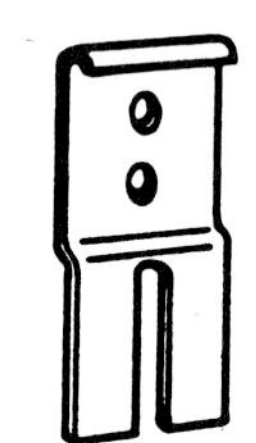

When correctly oriented will hang from rail

Triangular shape of part makes automatic hole orientation difficult

Nonfunctional shoulder permits proper orientation to be established in a vibratory feeder and maintained in transport rails

Figure 10-8 Asymmetry assists in identification and orientation.

(*Source:* Boothroyd et al., *Automatic Assembly,* p. 265.)

example, integral self-locks, tabs, indentations, or projections on mating surfaces) are easier to identify and orient.

10.3.3 Case Study

The IBM Proprinter is a classic case study in mechanical DFA, illustrating the design principles discussed in this section. IBM formed a task force of five engineers and managers to develop the product and process concepts for manufacturing in a new plant. The following objectives were established:

- Modular design
- Minimum part count
- Self-aligning parts

Sixty engineers were organized into 20 subsystem design teams. Numerous design reviews were held over a 6-month period. Among the many DFA accomplishments were the elimination of all fasteners, springs, cables, and harnesses. Manual assembly of the Proprinter in three minutes has been demonstrated.[5]

Professors Geoffrey Boothroyd and Peter Dewhurst studied the Proprinter in comparison to its predecessor (Fig. 10-9), a dot-matrix printer manufactured in Japan and marketed by IBM.[6] They found that the predecessor design included 103 subassembly parts and 49 final assembly components (parts and subassemblies). Many of the functional parts were observed to hang off of other components during the assembly and to not be secured, resulting in many assembly problems. Using their DFA methodology, Boothroyd and Dewhurst estimated that the final assembly process required 57 separate operations taking over 500 seconds to complete. The assembly time for the printer subassembly was estimated to be over 1300 seconds per unit.

Boothroyd and Dewhurst also performed a DFA analysis on the Proprinter, which has a total of 32 components and no subassembly operations (Fig. 10-10). The total assembly time was estimated to be 170 seconds. The dramatic improvement in printer design is apparent in the comparison of Figs. 10-9 and 10-10.

10.4 PWB DESIGN RULES

For the design of PWB assemblies, we should consider how the PWB will be manufactured, assembled, tested, reworked, and maintained. The designer must consider such factors as the grouping and orientation of components on the PWB, the minimum spacings among components for tooling clearance (Chaps. 8 and 9), and the solderability principles discussed in Chapter 6. The designer should also adhere to a standard panel size to facilitate conveyability on standard-width tracks (see Fig. 14-14) among machines in a product-focused layout.

MIL-STD-2000 (reference 2 at the end of this chapter) provides extensive guidelines for PWB processes and producibility. Manufacturers in the electronics industry typically also maintain proprietary, in-house DFM/DFA tools and methodologies.

There are similarities and basic differences between the DFA requirements for through-hole and surface-mount applications. Section 10.5 discusses DFA for through-hole PWBs, and Sec. 10.6 covers SMT.

10.5 DFA FOR THROUGH-HOLE APPLICATIONS

Many of the DFA principles for PWBs serve to facilitate automated insertion (Chap. 8) and achieve better flow solderability (Chap. 6).

[5]Sammy G. Shina, *Concurrent Engineering and Design for Manufacture of Electronics Products* (New York: Van Nostrand Reinhold, 1991), pp. 62–65.

[6]Peter Dewhurst and Geoffrey Boothroyd, "Design for Assembly in Action," *Assembly Engineering,* January 1987, pp. 64–68.

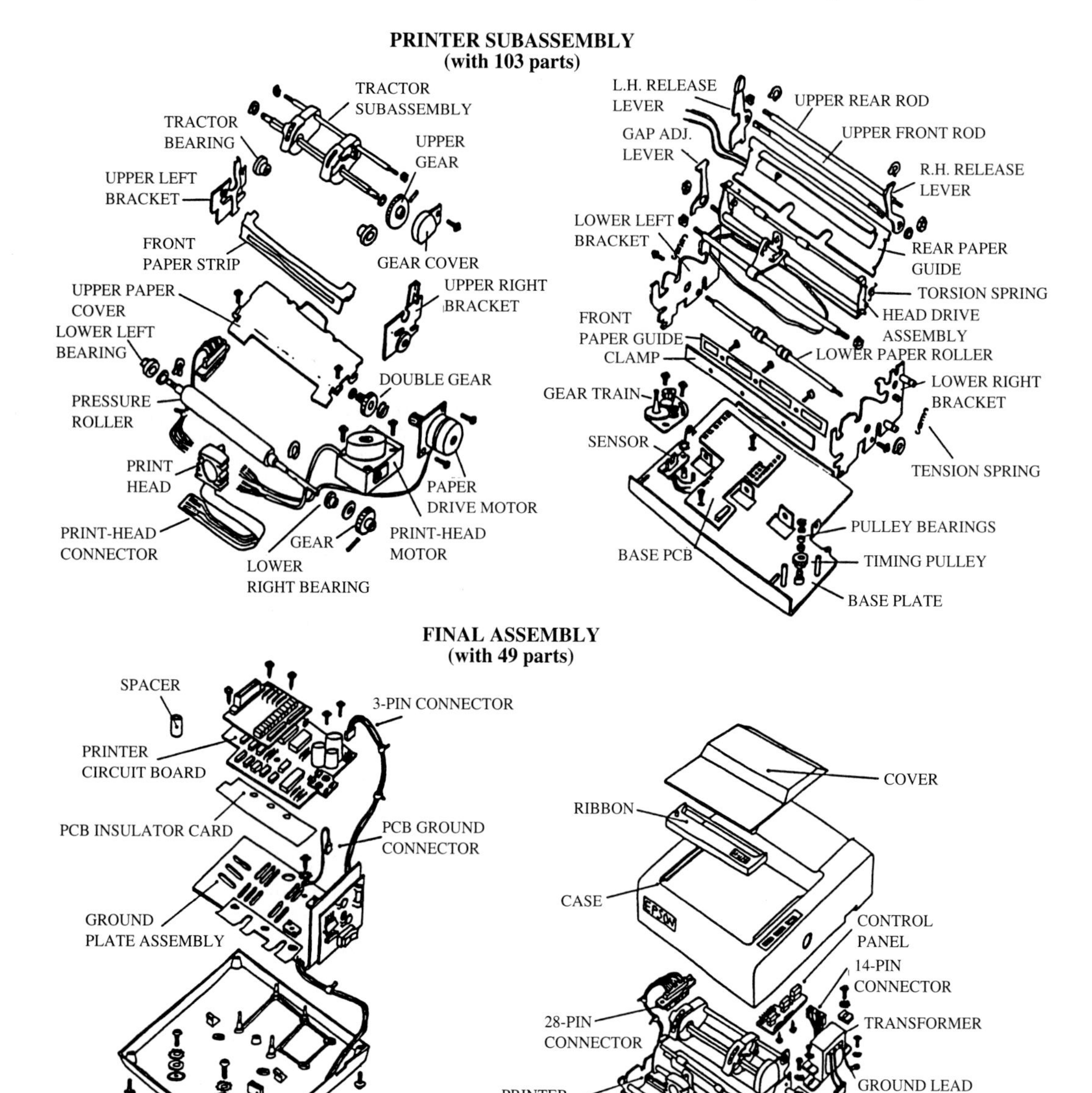

Figure 10-9 Predecessor to the IBM Proprinter.

(*Source:* Excerpted from *Assembly Engineering,* January 1987, pp. 65–66. By permission of the Publisher © 1987. Hitchcock Publishing Co. All rights reserved.)

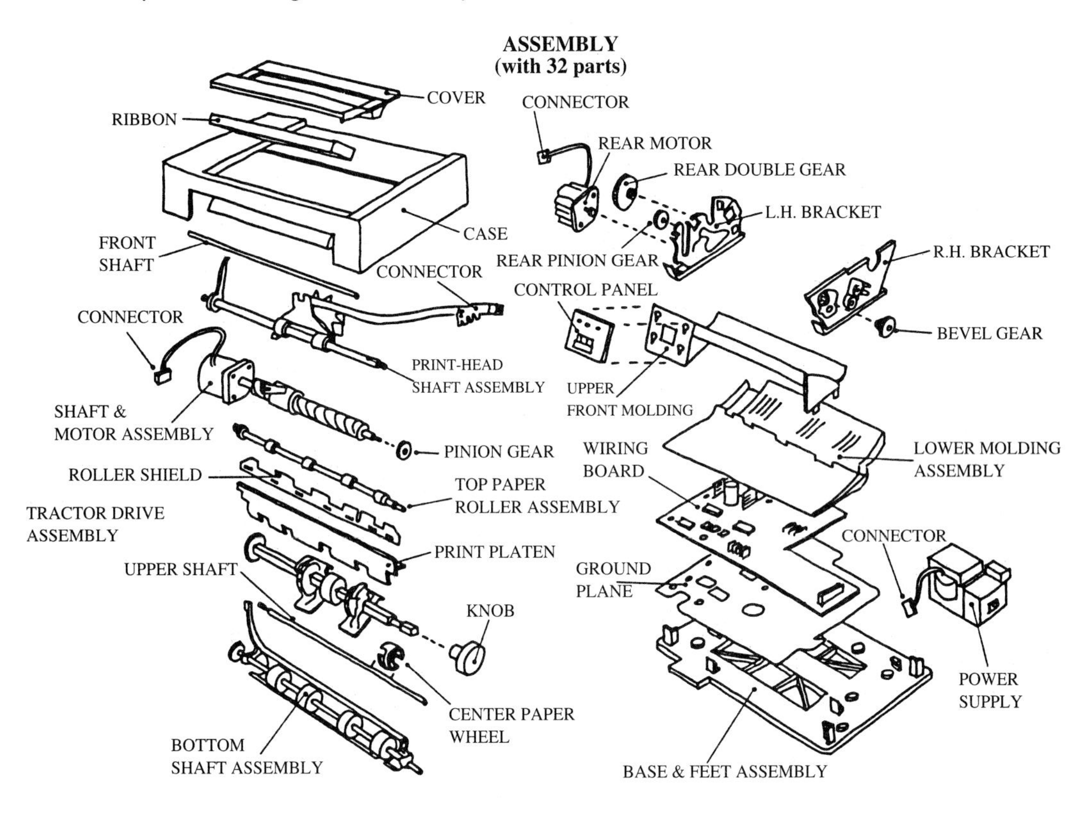

Figure 10-10 IBM Proprinter.

(*Source: Assembly Engineering*, p. 67.)

10.5.1 Design Rules

In general, the board should be designed for automated insertion of components in the following sequence:

- DIPs
- Axials
- Radials

Components should be located on the top side of the board on, for example, 0.100 inch (2.54 mm) grid increments and mounted as illustrated in Fig. 10-11. **Stand-up axials** cause assembly and reliability problems as well as increasing the board maximum-height envelope. Axials in a stand-up orientation tend to short against adjacent components and float or tilt during wave soldering. If the component density on a board necessitates use of stand-up axials, then take the following precautions:

- Turn bare-lead sides away from the metal bodies of adjacent components
- Allow adequate spacing among components

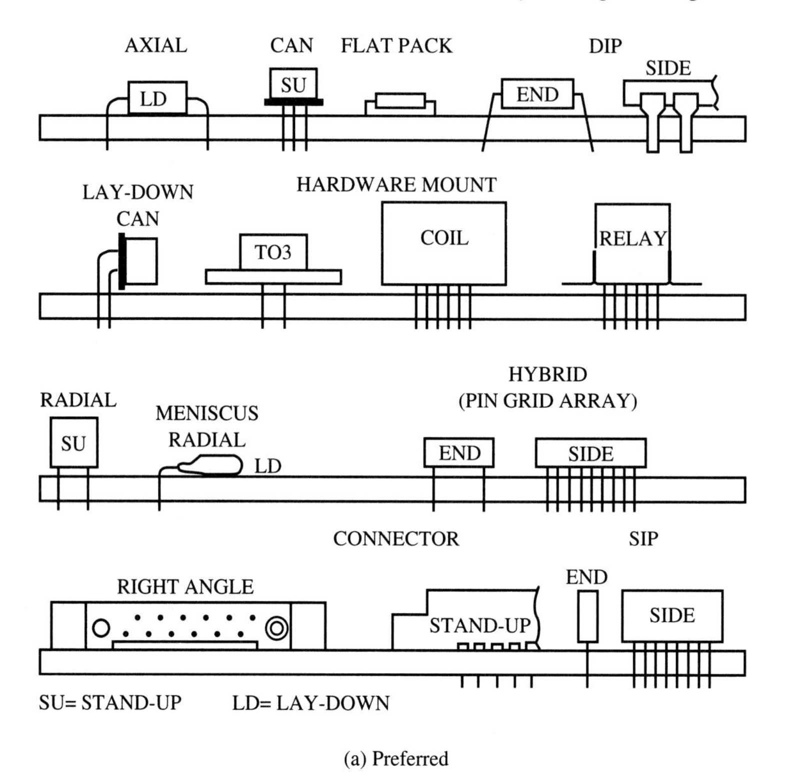

(a) Preferred

(b) Reduced Producibility

Figure 10-11 Preferred mounting of through-hole components.

(*Source:* Reprinted by Permission of Texas Instruments.)

- Sleeve the bare lead

The following additional design rules should be observed:

- Group similar components together. However, this DFA guideline should not override electrical requirements for grouping of components to achieve proper performance.

- Select proper hole sizes and tolerances.
- Utilize global references (tooling holes or fiducials) for dimensioning.
- Arrange components (for example, DIPs) in well-ordered rows, parallel to the board-edge connector. If axial components cannot be located on common mounting centers, arrange axials in the row by length from shortest to longest mounting centers.
- Mount components with polarity in the same direction.
- Avoid placements that obscure the termination or prevent the removal of another part.
- Ensure that no part overhangs the board edges.
- Avoid mounting components beside connectors to facilitate manual test probing.
- Do not locate components near masked areas, heat sinks, support areas, or tooling holes.
- Maximize the auto-insertability of the design.

Figure 10-12 illustrates the principles of good board design for through-hole technology.[7] The features noted in Fig. 10-12 arc discussed further as follows:

1. *Orientation of Components:* The connector is mounted on the short side of the PWB. Components of a given type (for example, DIPs) are mounted in well-ordered rows. DIPs are parallel to the connector, allowing the connector and DIPs to be perpendicular to the solder wave and thus minimizing bridging. Most axial and radial components are perpendicular to the connector, allowing them to be parallel to the flow solder wave and thereby minimizing flotation.
2. *Staggering of Stand-up Axials:* Staggered arrangement of stand-up axials reduces the risk of shorts.
3. *Clearance Between Stand-up Axials and Heatsinks:* Use a minimum of 0.150 inch (3.81 mm) clearance between stand-up axials. Turn the bare lead side away from all metal bodies or leads.
4. *Location of Stand-up Axials:* Provide a minimum clearance of 0.150 inch (3.81 mm) between stand-up axials and can-type devices.
5. *Placement of Lay-down Axials/Radials:* Placing lay-down axials and radials between can devices reduces potential of shorts and facilitates insertion.
6. *Clearances for Conformal Coating Mask:* Maintain a 0.100 inch (2.54 mm) clearance around connectors, all hardware-mounted components, jack tips, and variable components for conformal coat masking.
7. *Standardized Polarity of Components:* Polarity should be in the same direction for like components. Can devices should all have tabs oriented in the same direction. DIP polarity should be standardized for all boards in a unit.
8. *Mounting Configuration and Center-to-Center Distance:* Like components should have the same mounting configuration and center-to-center dimensions across a board.
9. *Clearances for Auto-Insertion and Tooling Holes:* As discussed in Chapter 8, **auto-**

[7]*Design for Producibility Guide for PWB Design, Vol. I* (Dallas, Tex.: Texas Instruments Defense Systems and Electronics Group, March 1988), p. 122.

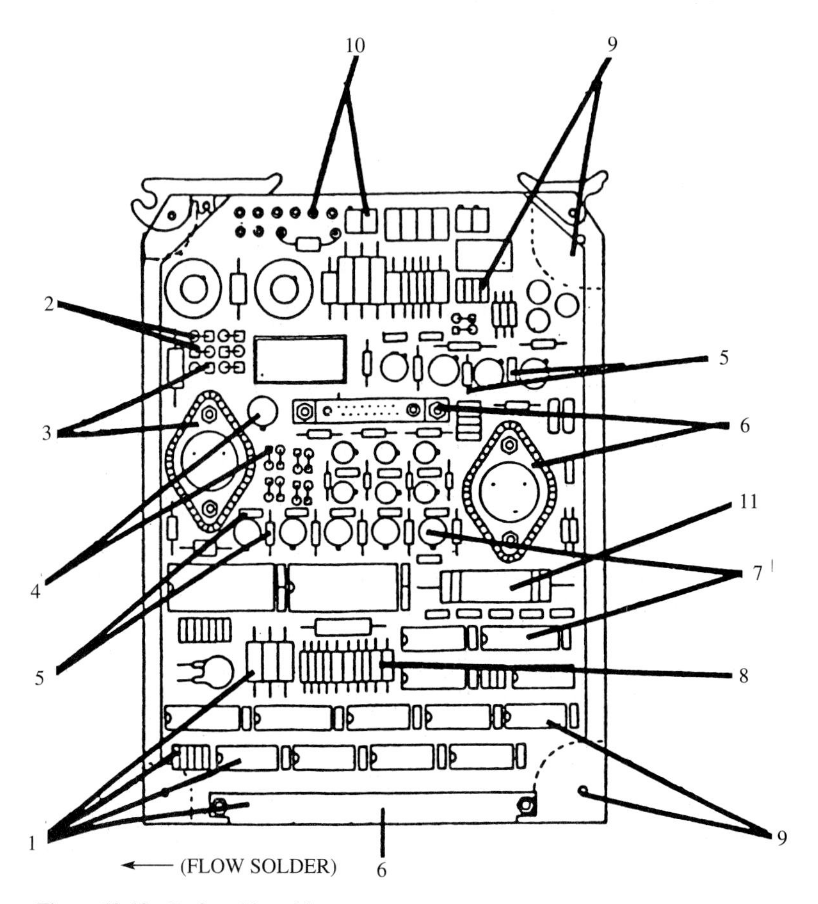

Figure 10-12 Preferred board layout.

(*Source:* Reprinted by Permission of Texas Instruments.)

insertion requires that adequate clearances be provided for components relative to the tooling holes, other board features, and cut-and-clench tooling.

10. *Test Points and Variable-Value Resistors:* Locate test points and variable-value resistors at the edge of the board for ease of testing and adjusting of potentiometers, as well as for conformal coat masking clearance.
11. *Tie Wraps:* Use tie wraps on components weighing 0.025 ounce (0.7 g) or more per lead, in order to provide additional mechanical support.

10.5.2 Case Study

Texas Instruments DSEG used the DFA methodology in the transition from prototype design to production design of a high-volume built-in-test (BIT) board. The DFA evaluation criteria were as follows:

- *Density:* Percent of board real estate required for components, including tooling clearances (as shown by the dashed outlines in Fig. 10-13).
- *Height:* Number of parts with height problems relative to the board height envelope.
- *Automation Potential:* Percent of components that can be auto-inserted.
- *Quality/Labor Problems:* Percent of components that have experienced quality and/ or labor (manual assembly) problems in previous designs.

The prototype and production board layouts are shown in Fig. 10-14. The following design changes were made:[8]

- A design review revealed that some components were redundant and were thus eliminated from the design.
- All stand-up axials were converted to lay-down configuration.
- Polarity was standardized for all axial components.
- Component orientations were changed to improve flow solderability and spaced to reduce solder bridging.
- All variable resistors were located on the PWB perimeter.
- Tab orientation on all can-type devices was standardized.

Table 10-2 summarizes the extent of design improvements. Parts count was decreased by 18 percent, and the use of standard parts was increased. Because the BIT board was a high-volume product, automation of the insertion process and elimination of quality/ labor problems were high priorities. The designers selected auto-insertable components and laid out the components with proper spacing and orientation as required by the automatic insertion machines. The 210 percent improvement in auto-insertability led to a 29 percent reduction in assembly cycle time. Height problems were solved and potential quality problems were reduced by eliminating stand-up axials and halving the number of can-type components.

10.6 DFA FOR SURFACE-MOUNT APPLICATIONS

Many of the DFA principles for through-hole technology (Sec. 10.5) also apply to SMT, particularly for mixed-technology boards. Compared to the design rules for through-hole technology, the design rules for SMT are complex and relatively less standardized. Chapter 9 discussed requirements for automated placement, and Chapter 6 discussed solderability requirements for both flow and reflow solder applications. This section does not attempt to give a comprehensive coverage of surface-mount DFA, but rather an overview of placement machine considerations, pad and SMD tolerances, and soldering. Prasad (reference 3 at the end of this chapter) gives a thorough presentation of SMT design.

[8]Priest, *Engineering Design,* pp. 37–41.

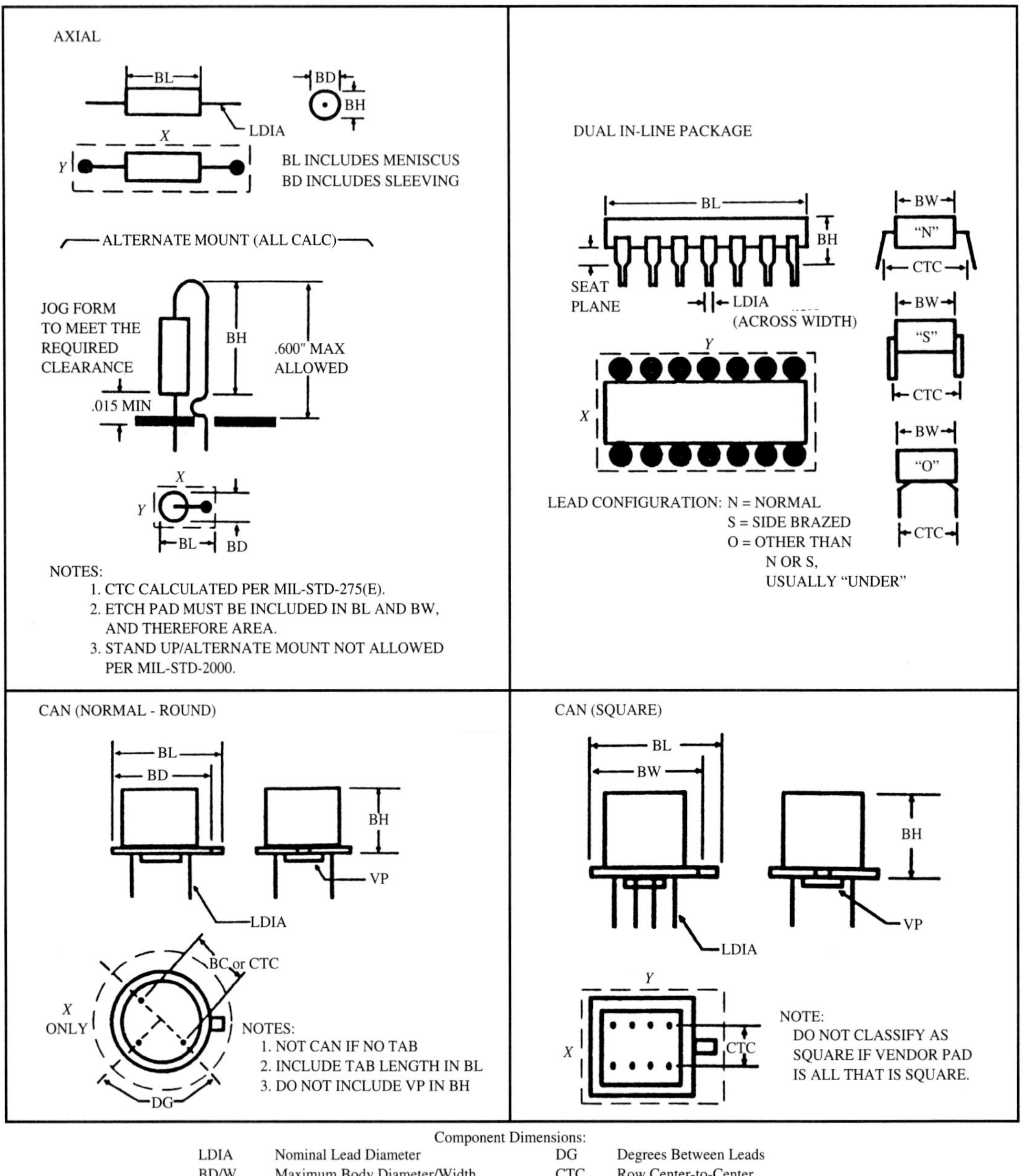

Component Dimensions:

LDIA	Nominal Lead Diameter	DG	Degrees Between Leads
BD/W	Maximum Body Diameter/Width	CTC	Row Center-to-Center
BL	Maximum Body Length		
BH	Maximum Body Height		
VP	Have Vendor Pad/Feet?		

Note: The dashed line indicates the producibility X/Y space outline.

Figure 10-13 Through-hole component space envelopes.

(*Source:* Reprinted by Permission of Texas Instruments.)

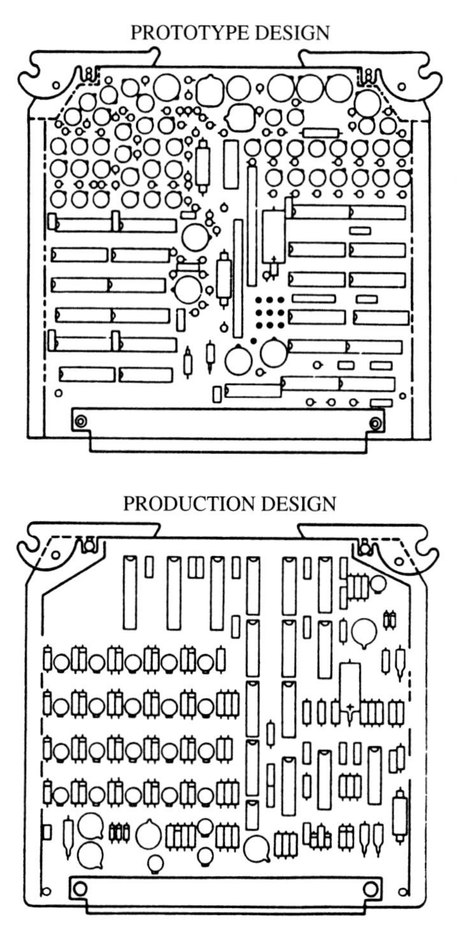

Figure 10-14 Built-in-test (BIT) board.

(*Source:* John W. Priest, *Engineering Design for Producibility and Reliability* (New York: Marcel Dekker, Inc.), p. 40.)

10.6.1 Placement Machine Considerations

The placement equipment used can substantially influence the layout of components on the PWB, as follows:

- *Layout:* Placement machines with multiple heads require a fixed grid for precise component locations and may also require specified distribution and orientation of components on the PWB to achieve maximum placement rates.
- *Orientation:* Options for component orientation on the PWB may be restricted (for example, only certain angular positions are possible).
- *Components:* Some placement machines process only certain components (for example, only cylindrical or cubic components).

TABLE 10-2 DFA IMPROVEMENT OF BUILT-IN-TEST (BIT) BOARD

Parts	Before	After	Percent improvement
Number of components	186	153	18%
Density %	97	90	8%
Number of height problem parts	21	0	100%
Auto insert %	25	77	210%
Quality/labor %	78	43	45%
Total code letters	259	124	52%
Part qty with problem	145	66	55%
Qty cans	57	29	50%
Qty stand-up axials	78	0	100%
Assembly time analysis:			
Amount of time to assemble the PWB design	34.95	10.30	29.4%

Source: Reprinted by Permission of Texas Instruments.

- *Board Size:* Travel range of the insertion or placement head limits the real estate for placement.
- *Location Referencing:* Machines vary in method of reference location (tooling hole or fiducial) and, accordingly, achievable precision. As shown in Fig. 10-15, several types of fiducials are used in the industry. One standard for location of fiducials is the **left-hand rule:** holding the board in the left hand, targets should be in the top left, bottom left (CAD datum), and bottom right corners.
- *Board Clamping:* Workpiece real estate must be allowed for the holding requirements of the particular machine.
- *Tooling Clearance:* Adhere to minimum spacings among adjacent components to allow clearance for placement heads which use jaws to place the components.

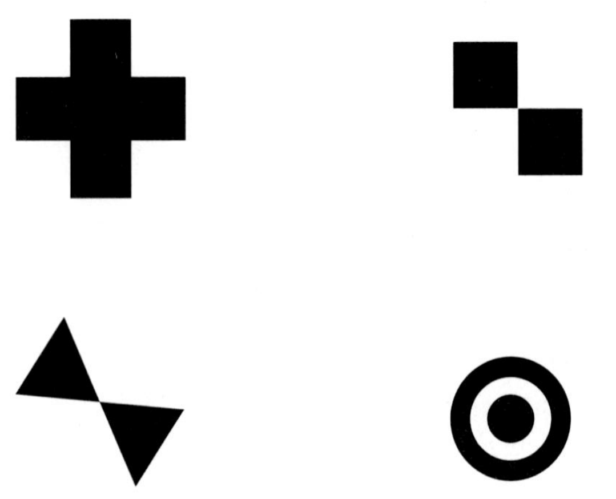

Figure 10-15 Fiducial patterns.

10.6.2 Pad and SMD Tolerances

Adequate solder attachment of SMDs depends on the SMD land patterns and the pads. Insufficient solder fillet zone can result in component tombstoning. The lengths of the component body and the solder terminal(s) determine the required pad area. For example, in Fig. 10-16(a) the pads must be enlarged outward, because the metallization on the underside of the SMD is inadequate, and the solder joint can only be made using the side areas.

Conversely, it is not necessary that the pads in Fig. 10.16(b) cover the entire

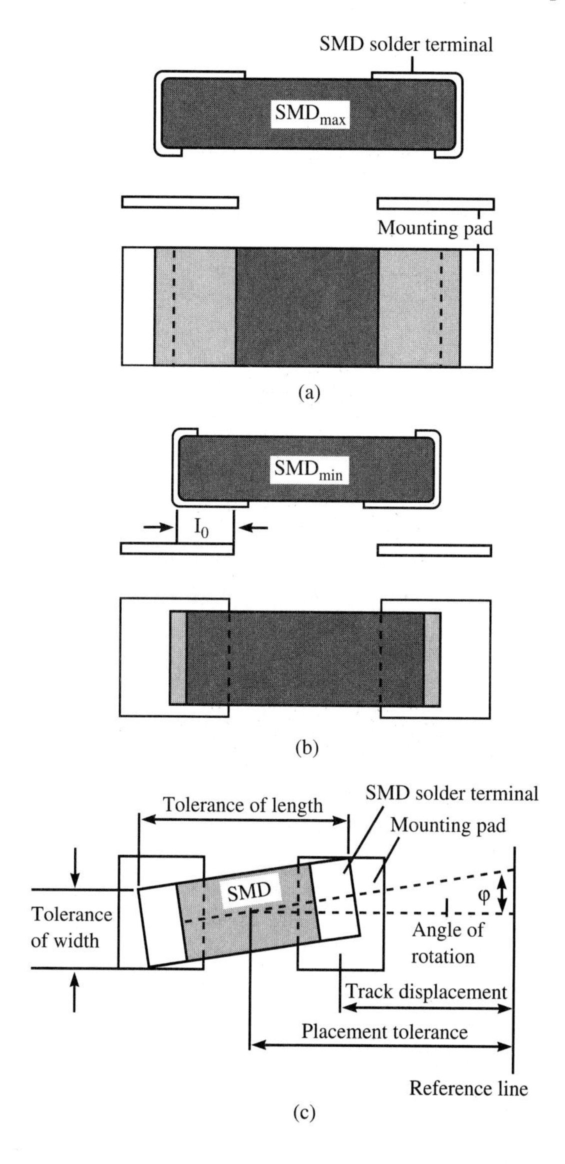

Figure 10-16 Placement of SMDs on pads: (a) insufficient contact; (b) sufficient contact; (c) imperfect registration.

(*Source:* Werner Maiwald, *PCB Layout Recommendations* (Munich, Germany: Siemens Aktiengesellschaft, September 1986), pp. 4–5.)

metallized area of the SMD; because of the SMD contact configuration, a small pad provides adequate surface for the solder joint. Where component contacts permit, it is preferred to minimize the pad extensions outside the part footprint and maximize the pad extension under the part to make the part self-centering and avoid swimming during the soldering operation. Taller solder columns also promote reliability.

In spite of component standardization, it is not always possible to use the same pad sizes for SMDs from different suppliers, because of component tolerances. The board should be designed for a specific make of SMDs or for well-controlled substitutes.

Under ideal conditions (Fig. 10-17), a good solder connection between the solder pad and cubic or bent-lead contact can be obtained. On leaded components, longer leads are more compliant and provide better reliability. For cylindrical SMDs the projected area of the SMD connection should be large enough. In practice, however, various factors (for example, tolerances and potential for imprecise registration—see Fig. 10-16 (c)) necessitate the enlargement of the pads.

10.6.3 Soldering

Soldering techniques have a marked effect on the layout. For mixed-technology boards a wave-soldering process is used (Sec. 6.4.1), and the same solderability issues discussed in Sec. 10.5 apply. Reflow-soldering techniques (Sec. 6.4.2) are used on both pure SMT and mixed-technology boards.

SMDs should be pretinned with a Pb/Sn coating to inhibit oxidation and improve solderability. Gull-wing leads are preferred over J-leads because J-lead forming requires more dies and control of the pretinning process is more difficult. Some SMDs will require a specified number of solder pedestals underneath.

The soldering method also influences the PWB layout. Sufficient spacing must be provided between adjacent printed circuits and components (Fig. 10-18) to allow solder

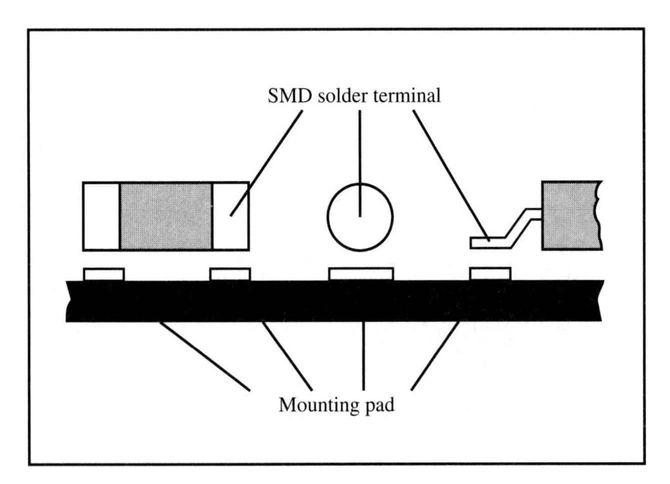

Figure 10-17 SMD-to-pad contact surfaces.

(*Source:* Maiwald, *PCB Layout Recommendations,* p. 4.)

CHIP (WITH AND WITHOUT LEADS)

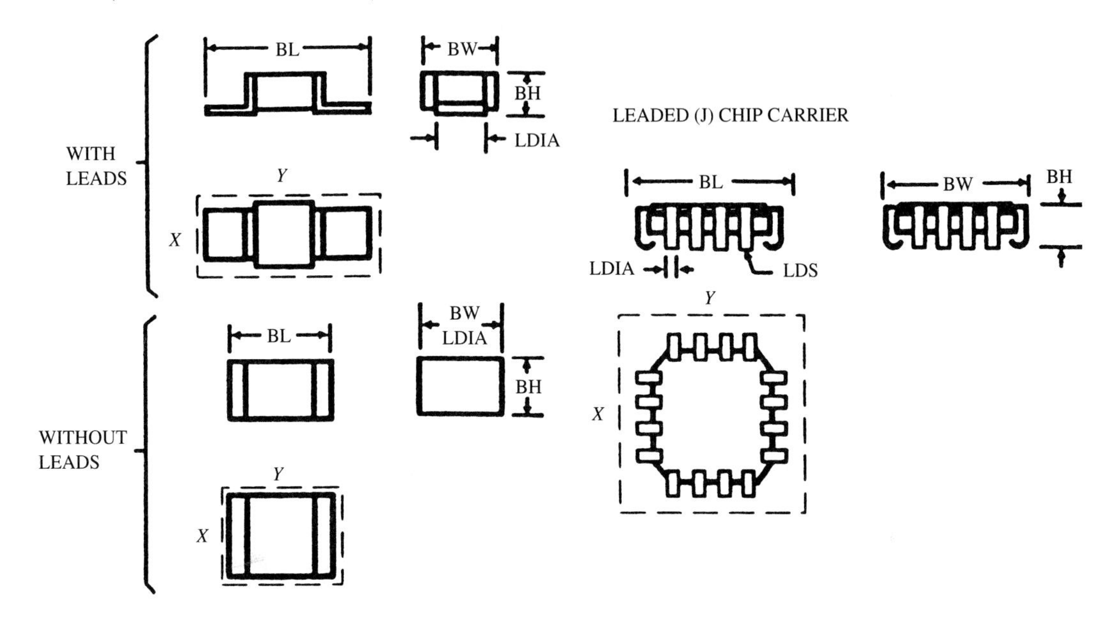

COMPONENT DIMENSIONS:

LDIA	Nominal Lead Diameter
DD/W	Maximum Body Diameter/Width
BL	Maximum Body Length
BH	Maximum Body Height

Note: The dashed line indicates the producibility X/Y space outline.

Figure 10-18 Surface-mount component space envelopes.

(*Source:* Reprinted by Permission of Texas Instruments.)

flow in wave soldering and avoid the shadowing effect of adjacent components. Shadowing effects also occur in near-IR reflow soldering because some regions of the board are not exposed directly to the radiative heat source.[9] Certain minimum distances must also be kept to avoid solder bridges.

10.7 DFA FOR INTERCONNECTION

Interconnections are often considered as an afterthought in the design. Consequently, the DFA process for interconnection designs may be less technically sophisticated and procedurally formalized in many companies. Some companies, such as Texas Instruments, have developed proprietary expert systems for internal use in design of cables, harnesses,

[9]Ray P. Prasad, *Surface Mount Techonology: Principles and Practices* (New York: Van Nostrand Reinhold, 1989), p. 448.

flex circuitry, and motherboards. These systems imbed guidelines for routing and for use of standard connectors. Flags are provided for potential producibility, quality, or reliability problems.

10.7.1 Flex Circuitry

There is a trend toward use of flex circuitry for modularity and reliability. Flex circuitry substantially reduces component count and simplifies assembly. The flex circuitry in Fig. 10-19 provides card-edge connectors for several PWBs and routes interconnection to circular connectors on the chassis.

10.7.2 RF Interconnects: A Case Study

Interconnections for radio-frequency (RF) transmission require sophisticated production processes. Stripline is a planar-type transmission line fabricatable by a photolithographic process and used in microwave integrated circuitry. Texas Instruments DSEG conducted a DFA study to replace a weld and seal process of fabricating stripline, involving the following 12 steps and requiring over 30 hours of touch labor per unit:

- Clean and prepare all surfaces
- Gold ribbon weld all RF connections
- Inspect
- Weigh and mix silicon
- Package silicone and clean up
- Add catalyst and apply to cavities
- Touch up and clean up
- Mix ground foil adhesive
- Load and apply adhesive
- Position ground foil
- Inspect
- Complete all paperwork

Forty-six percent of the touch labor was devoted to welding and sealing.

TI replaced the stripline with coplanar flex, involving a five-step process and about 18 hours of touch labor per unit as follows:

- Clean and prepare all surfaces
- Reflow solder

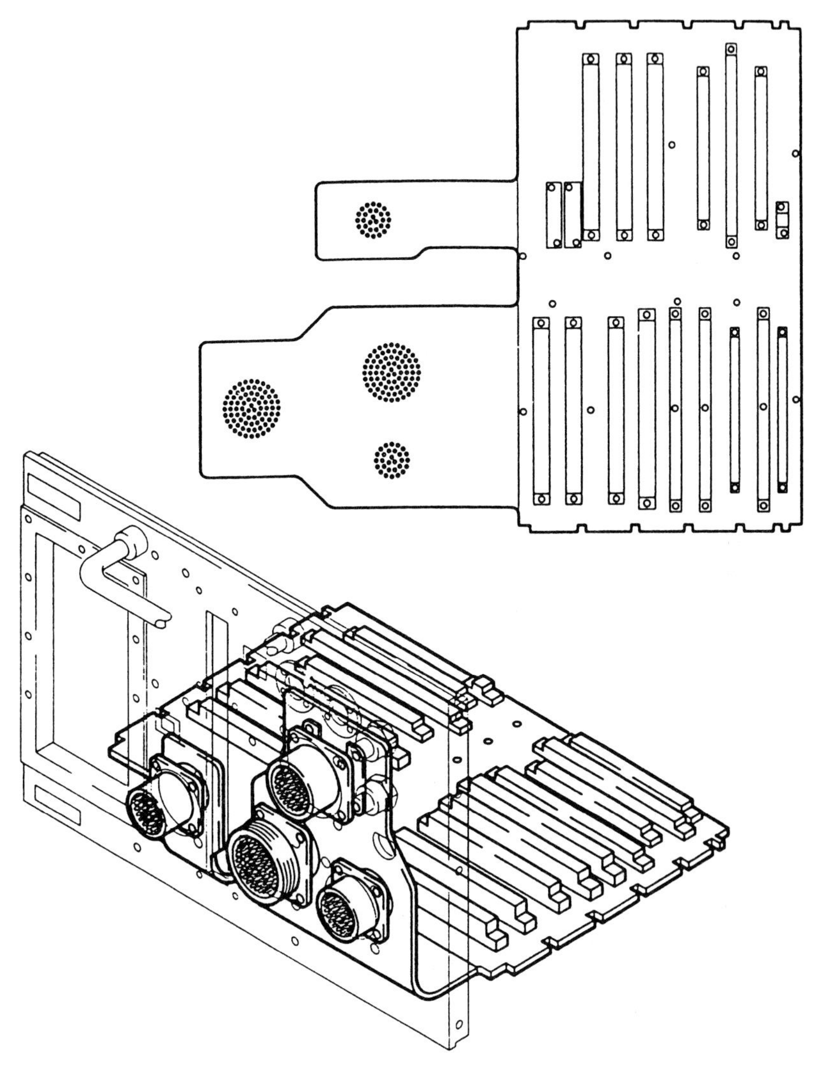

Figure 10-19 Flex circuitry.

(*Source:* Reprinted by Permission of Texas Instruments.)

- Clean and degrease assembly
- Inspect
- Complete all paperwork

Installing the coplanar flex required only 10 percent of the touch labor.

10.8 DFA RATING

The DFM and DFA methodologies involve tools and techniques for **rating** the producibility of designs and evaluating the manufacturing costs of alternative designs. Most approaches, including checklists, are rule-based. It is essential to perform these analyses early in the design process, when the DFA information can impact the product design. Furthermore, this information should be presented in such a way that the designers are guided toward design alternatives that reduce the manufacturing and assembly costs.

10.8.1 Rating Factors

Section 10.5.2 presented a case study in which a PWB design was rated in terms of the component density, height interference, quality/labor problems, and potential for automated assembly. These criteria are typical of DFA rating schemes used in industry. The utilization of standard parts is another criterion often rated.

10.8.2 Cost Estimates

DFA cost models generally estimate the costs of material and assembly labor, based on the following inputs:

- Parts list
- Component data base
- Assembly size and shape
- Technology and functionality
- Other characteristics

The tool should be automated, with a good user interface, to encourage use by the design engineer, and should

- Identify potential problems
- Break down costs by product partition and part type
- Estimate the cost penalties of design decisions
- Permit rapid what-if studies.

10.8.3 PWB Assembly Cost Models

A PWB **assembly cost model** might use the following information:

- PWB type (for example, number of layers, hole size, surface-mount and/or through-hole mount components to be used)

- Functionality of the board (analog versus digital)
- Component quantities, costs, quality levels, and problem history
- Mix of components (cost trade-offs between standard and nonstandard parts, special-purpose versus stock parts)
- Assembly costs (component insertion/placement done automatically or manually, assembly of mechanical items, paste application, soldering, cleaning, and rework)
- Preparation costs (setup, kitting, and sequencing)
- Testing and inspection costs
- Cleaning costs (including treating, recycling, or disposal of chemicals)

10.9 SUMMARY

This chapter has emphasized the importance of design for assembly and has given an overview of the DFA tools and techniques for mechanical, through-hole, surface-mount, and interconnection designs. Although the DFA methodology can lengthen the time-phase for early design, the benefits accrue when problems and resulting design changes are avoided later. DFA tools range from design-rule checklists to automated expert systems with sophisticated cost-modeling capability. The trend is to embed DFA tools in the designer's workstation, making the DFA process easier, faster, and more effective.

KEY TERMS

Assembly cost model
Auto-insertion
Design for assembly (DFA)
Design for manufacturability (DFM)
Design rules
Left-hand rule
Modularity
Parts count
Producibility
Rating
Stand-up axials
Top-down assembly

EXERCISES

1. What is a likely reaction to DFM/DFA from people in the company? What can be done to deal with this reaction?
2. When large savings are claimed due to automation, what other consideration might be responsible for the cost savings?
3. What factors should be considered for PWB layout in through-hole technology? In SMT?
4. How does top-down assembly make it easier to assemble a product?
5. Choose a product used in everyday life and perform a qualitative DFA study to make the product easier to assemble.
6. How does parts standardization promote DFA?

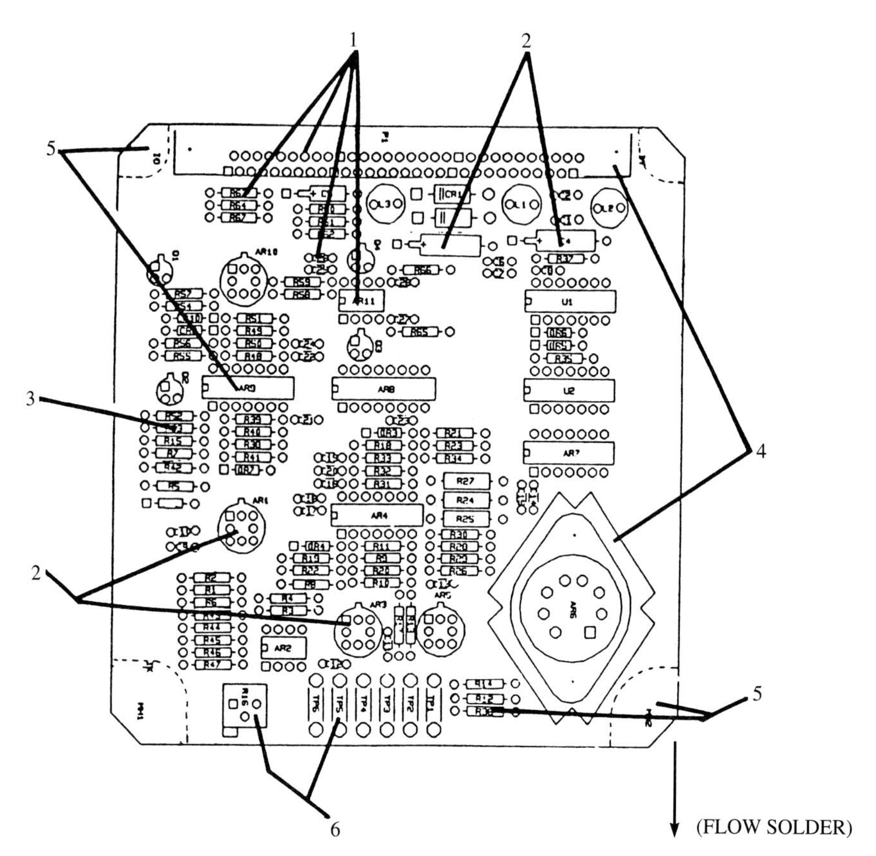

Figure 10-20 Analog board.

(*Source:* Reprinted by Permission of Texas Instruments.)

7. Several DFA features are noted on the analog board in Fig. 10-20. Discuss each feature (1 through 6).
8. Comment on the PWB DFA features of the digital board in Fig. 10-21. What direction should the board be run through wave soldering?
9. Obtain a PWB from a discarded piece of electronic equipment and apply the DFA rules to recommend improvements.
10. Obtain a mechanical assembly and perform a DFA analysis on it.

BIBLIOGRAPHY

1. ASFAHL, C. RAY, *Robots and Manufacturing Automation.* New York: John Wiley & Sons, 1985.
2. BOOTHROYD, GEOFFREY, CARRADO POLI, AND LAURENCE E. MURCH, *Assembly Engineering.* New York: Marcel Dekker, Inc., 1982.
3. PRASAD, RAY P., *Surface Mount Technology: Principles and Practices.* New York: Van Nostrand Reinhold, 1989.

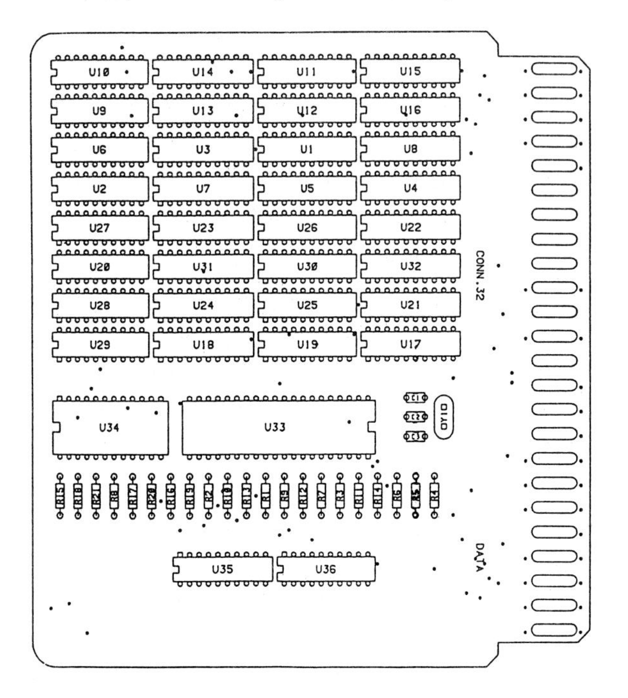

Figure 10-21 Digital board.

4. PRIEST, JOHN W., *Engineering Design for Producibility and Reliability.* New York: Marcel Dekker, Inc., 1988.

5. SHINA, SAMMY G., *Concurrent Engineering and Design for Manufacture of Electronics Products.* New York: Van Nostrand Reinhold, 1991.

6. STILLWELL, H. RICHARD, *Electronic Product Design for Automated Manufacturing.* New York: Marcel Dekker, Inc., 1989.

7. United States Department of Defense, *MIL-STD-2000A: Standard Requirements for Soldered Electrical and Electronic Assemblies.* Washington, D.C.: U.S. Government Printing Office, 1991.

11 Quality and Reliability

11.1 INTRODUCTION

The world electronics industry is characterized by fierce competition on the basis of cost, performance, quality, and market availability. This chapter focuses on the management concepts (Sec. 11.2) and technologies (Sec. 11.3–11.5) proven to promote quality and reliability.

Quality has been defined as fitness for use.[1] In this definition, quality is a subjective characteristic which changes along with customer expectations. Increasingly, the tendency in industry is to define quality from the viewpoint of the customer. Quality characteristics are the peculiar (distinguishing) features of a product or the grade of a product, including

- Appearance
- Dimension
- Performance
- Length of useful life
- Dependability
- Reliability
- Durability
- Maintainability

In order to objectively measure product quality in manufacturing, we have traditionally defined quality in terms of **conformance,** meaning the degree to which the product

[1]Joseph M. Juran, *Quality Control Handbook* (New York: McGraw-Hill Book Company, 1988), p. 2-2.

conforms to a design or specification. This concept of quality led to the use of inspection as the primary approach to quality improvement. Products were manufactured and inspected to verify conformance, and then nonconforming units were reworked or scrapped.

As shown in Fig. 11-1, our approach to quality improvement has evolved since the 1920s, when manufacturers simply inspected for conformance. Beginning with World War II, statistical process control has been used to continuously monitor the manufacturing process and thereby build in quality, rather than inspecting out defects. In the 1980s, we began to emphasize improvement of product and process design (Sec. 11.3). The 1990s have seen quality improvement become ingrained in the overall management of companies and industries, with a trend toward formal registration of quality programs under international standards organizations.

Section 11.4 introduces concepts, principles, and tools of quality control. **Quality control** encompasses the entire collection of activities through which we achieve quality. **Statistical quality control** (SQC) applies mathematics and statistics to measure and control quality:

> With the help of numbers, or data, we study the characteristics of our process in order to make it behave the way we want it to behave.[2]

Reliability (Sec. 11.5) may be thought of as quality demonstrated over time, and it is defined as the probability that a component, device, equipment, or system will be performing its intended functions for a specified period of time under a given set of conditions. Alternatively, reliability may be defined as the probability that capacity (strength) exceeds load (stress).

In the context of electronics manufacturing, we are interested in reliability from two perspectives: reliability of products and reliability of the (manufacturing) process.

[2]AT&T Technologies, *AT&T Statistical Quality Control Handbook,* 11th ed. (Charlotte, N.C.: Delmar Printing Company, 1985), p. 3.

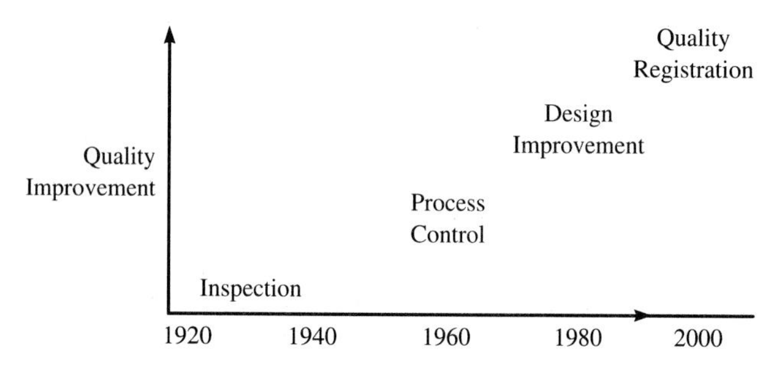

Figure 11-1 Chronology of quality improvement.

(*Source:* ©1988 American Society for Quality Control, Reprinted by permission.)

11.2 QUALITY MANAGEMENT

The quality assurance process spans the three phases of the product life cycle shown in Fig. 11-2. Historically, the various functions of quality assurance have been assigned to staff specialists and organizations involved in limited time phases. The modern trend is toward a team approach. An effective quality program is much broader than production quality control. Quality requires top-level management commitment and the participation of all organizations, from marketing, through design and manufacturing, and finally in field support. This approach is referred to as **total quality management** (TQM) [5] or **companywide quality control** (CWQC) [5, 8]. Juran's [9] philosophy of quality emphasizes that most problems relate to the management systems and most improvements are the responsibility of management, rather than the workers.

Deming is noted for articulating quality management through his fourteen points, which further place the responsibility for quality management:[3]

- Create constancy of purpose for improvement of product and service.
- Adopt a new philosophy.
- Cease dependence on inspection to achieve quality.
- End the practice of awarding business on the basis of price alone. Instead, minimize cost by working with a single supplier.
- Improve constantly and forever every process for planning, products, and service.
- Institute training on the job.
- Adapt and institute leadership.
- Drive out fear.
- Break down barriers between staff areas.
- Eliminate slogans, exhortations, and targets for the workforce.
- Eliminate numerical quotas for the workforce and numerical goals for management.
- Remove barriers that rob people of pride of workmanship. Eliminate the annual rating or merit system.
- Institute a vigorous program of education and self-improvement for everyone.
- Put everybody in the company to work to accomplish the transformation.

[3] W. Edwards Deming, *Out of Crises,* 2nd ed. (Cambridge, Mass.: Massachusetts Institute of Technology, 1986), pp. 23–24.

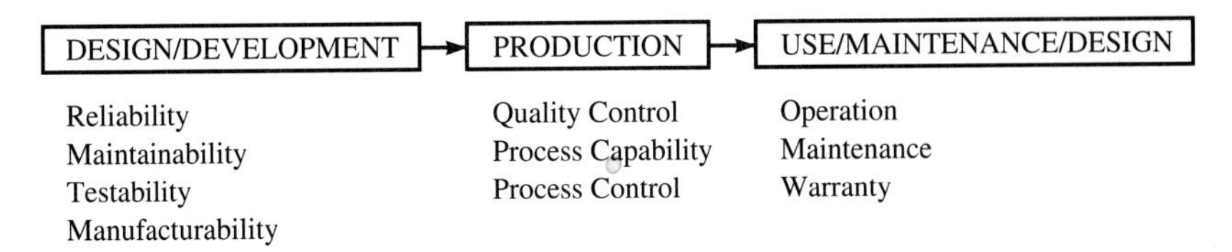

Figure 11-2 Quality assurance in the product life cycle.

11.2.1 Total Quality Management

Continual improvement in quality and reliability requires disciplined closed-loop failure analysis and corrective action. An adaption of the Shewhart cycle,[4] called the Deming Cycle (Fig. 11-3) is a simple conceptual model for the process of continuous improvement.[5] There is a never-ending process of planning (design and prediction), doing (manufacturing), checking (test and data analysis), and acting (redesign).

Figure 11-3 Quality improvement cycle.

To more effectively accomplish TQM/CWQC, companies in the electronics industry are emphasizing **life-cycle engineering** (LCE). LCE is an engineering approach that considers system performance, organizational procedures, and total costs of a product over its full life cycle. Historically, product design and process design have been functions segregated both organizationally and temporally in the product life cycle. As noted in Chapter 2, the modern trend is to fully integrate product and process design though concurrent engineering.

The objectives of concurrent engineering also extend to the usage phase for a product. Increases in quality and reliability lead to lower operating and maintenance costs and indirectly to increased satisfaction for the customer and greater market share for the supplier. Figure 11-4 shows that **life-cycle cost** (LCC) is a function of the design reliability. We want to achieve the right level of quality and reliability to minimize the LCC, through the application of proven design rules, breakthrough technologies, and test-analyze-fix.

[4]Walter A. Shewhart, *Statistical Method from the Viewpoint of Quality Control* (Graduate School, U.S. Department of Agriculture, Washington, 1939), p. 45.

[5]Deming, *Out of Crises, p. 88.*

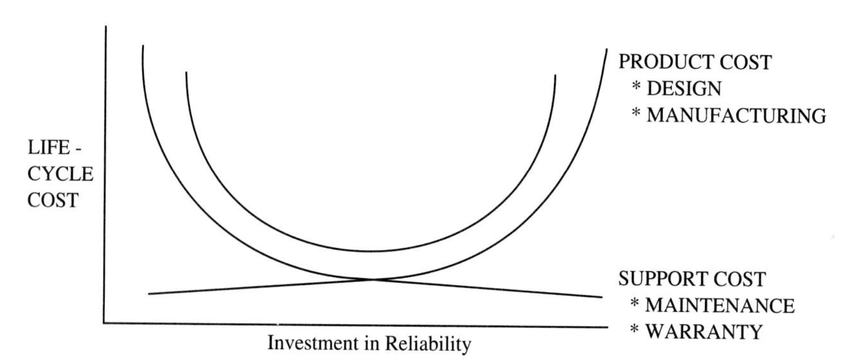

Figure 11-4 Life-cycle cost.

The experience of a Japanese manufacturer of computer controllers for diesel engines demonstrated the benefits of LCE. Investment in environmental (thermal and vibration) testing resulted in a 75 percent reduction in warranty expense.

11.2.2 Quality Function Deployment

Industry experience has indicated that the majority of LCC is determined in design and development (Fig. 11-5). Comparative studies of the engineering process in Japanese and American companies have indicated that Japanese companies tend to finalize the design much earlier in the product life cycle than do American counterparts.

Recall from Chapter 10, Fig. 10-1, that the typical U.S. company tends to process the bulk of engineering changes very late in the design phase, and the change process extends into the production and field use phases. Conversely, Japanese companies process most design changes at least a year before production, and 90 percent of design changes are processed one to three months prior to production.[6] In order to accomplish comparable results, U.S. companies adopted the **Quality Function Deployment** (QFD) methodology first employed by the Mitsubishi shipyards. QFD seeks to define customer requirements and systematically disseminate these throughout the design organizations and processes in a disciplined, well-documented manner. The QFD methodology makes use of several different tools to map customer requirements into product parameters that are controlled by the producer, including design, production, and service.

QFD is part of the emerging trend to focus quality improvement activities in design. QFD has been defined as

> a systematic means of ensuring that customer or marketplace demands (requirements, needs,

[6] L. P. Sullivan, "Quality Function Deployment," *Quality Progress,* 19, no. 6 (1986), pp. 39–50.

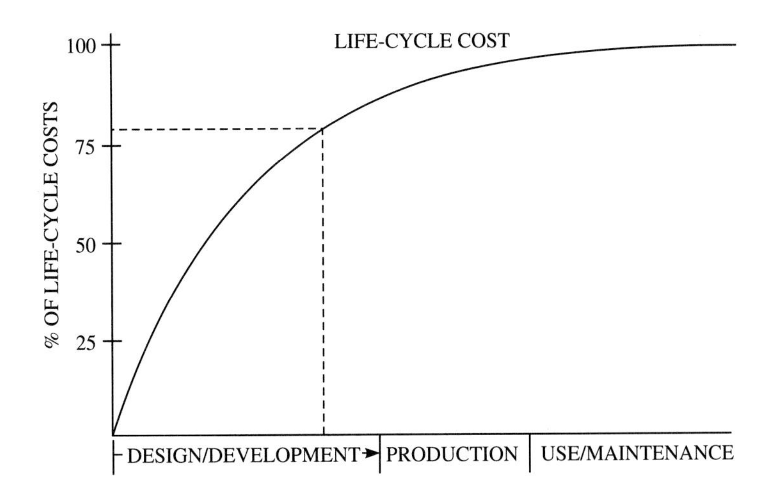

Figure 11-5 LCC in the product life cycle.

wants) are accurately translated into relevant technical requirements and actions throughout each stage of product development.[7]

Management should emphasize early investment in quality (marketing people properly defining customer requirements and engineers selecting the right technology to achieve customer requirements, then working with manufacturing to ensure manufacturability). The design function has the greatest leverage on product quality. When an assembly operator makes an error in manufacturing, one unit of the product is defective. However, when an engineer makes a design error, such as failing to properly interpret the customer's needs and failing to make the design producible, 100 percent of the products are defective![8]

11.2.3 Vendor Quality

There are major trends in the electronics industry which have made management of vendor (supplier) quality an important topic. As discussed in Chapter 2, material costs have become an increasingly larger percentage of total direct cost (material and labor). There is also a trend toward JIT manufacturing (refer to Chaps. 2 and 15). One of the basic principles of JIT is the focused factory, a manufacturing plant that specializes in a few processes and products and is therefore able to achieve very high product quality and excellent delivery performance. Rather than setting up a department to perform a new process, a company seeks a supplier that can do the job better. Then the supplier and customer work together for continuing improvement.

Vendor quality is a broad topic which cuts across organizational lines. Traditionally, vendor relations were the responsibility of a purchasing or procurement organization. Quality evaluation was done by the quality department, primarily through receiving inspection. However, experience has proven that inspection is relatively expensive, because it is labor-intensive. Also, inspection is largely ineffective because it occurs so far removed from the original source of defects (in both geographical and temporal terms) that information feedback and corrective action are hindered.

The modern approach is to place responsibility for quality at the source, where group problem-solving and statistical process control ensure that defects are not passed along to external customers. This process requires much more involvement of engineers. Both supplier and customer engineers work together and with manufacturing personnel to set specification requirements and solve problems. The customer's engineers may make visits to the supplier to inspect and certify the processes or to communicate problems and participate in problem-solving. Conversely, supplier engineers visit the customer to better understand how the product is used.

The principles of vendor quality have been summarized as follows:

- Quality is conformance to valid requirements.
- The supplier is responsible for product quality.

[7]R. M. Fortuna, "Beyond Quality: Taking SPC Upstream," *Quality Progress,* 21, no. 6 (1988), pp. 23–28.

[8]John Mayo, "AT&T: Management Questions for Leadership in Quality," *Quality Progress,* 19, no. 4 (1986), pp. 34–39.

- Quality is obtained by emphasizing prevention.
- The supplier must strive for ever-improving quality.
- The cost of quality should be minimized.[9]

Design and manufacturing engineers serve as members of interdisciplinary teams in vendor quality management. They are involved in organizing and establishing policy for vendor quality programs. An example is the Purchased Product Quality Plan (PPQP) Core Team at AT&T in Little Rock, Arkansas, which includes engineers from the following disciplines:

- Quality engineering
- Manufacturing engineering
- Reliability engineering
- Applications engineering

Engineers also serve on commodity teams which specialize in problem-solving related to product classes, such as purchased microelectronic devices or mass storage devices.

Vendor surveys are events when a team from the customer organization visits a supplier's facilities to evaluate supplier capability for quality manufacturing. New suppliers are surveyed for potential to provide the desired product quality, while old suppliers are evaluated for capability to sustain quality levels. A survey of a problem vendor is an investigation to determine causes and needed corrective actions. As the problem vendor improves, follow-up surveys may be needed to evaluate progress and guide future improvement. Vendor surveys may require broad engineering involvement. Some vendor surveys focus on systems and procedures (operating practices, and the like). Such systems evaluations are usually performed by quality professionals, and design or manufacturing engineers are not involved. The vendor survey typically involves an opening conference, a period of fact-finding, a closing conference, preparation of a written final report, and follow-up.

Effective cooperation between customer and supplier often depends upon specifications and standards. Specification is the process of establishing valid requirements and is one of the most important engineering functions in the vendor quality program. The supplier's design, manufacturing, and quality efforts will be based on the specification requirements. Engineering participation is essential to tailor specifications and purchase contracts for specific applications and vendor circumstances. There are several different types of specifications serving different purposes.

The **product specification** defines what is required for a product to perform as expected by the customer. The product specification is the basis for a purchase agreement and may include any of the following items:

- Product name
- Product description

[9]J. L. Pence, "Customer Satisfaction: The Focus Is on Value," *ASQC Quality Congress Transactions*, American Society for Quality Control, 1989, np.

- Performance requirements
- Required characteristics
- Drawings, standards, or references to drawings or standards
- Verification requirements
- Tests
- Inspections
- Supporting data
- Sampling plans
- Accept/reject criteria
- Safety
- Noise and environmental restrictions
- Government regulatory requirements
- Packaging and shipping
- Handling and storage
- Labeling, warning, identification, and traceability
- Shelf life
- Reliability
- Maintainability
- Supportability
- Standardization[10]

A **performance specification** is included as part of or in lieu of the product specification and describes what an item is required to do and under what conditions. A performance specification gives the supplier substantial freedom in design of the item, provided the performance is achieved.

Engineering drawings are usually included in the product specification, and they define the product to be purchased and/or equipment with which the product must interface. An engineering drawing is a diagrammatic and narrative description of an item, including dimensions and other descriptive information. The dimensions include tolerances and are given in sufficient detail to avoid possible misinterpretation. Use of a common drafting approach (such as specified by the American National Standards Institute) will eliminate potential interpretation errors.

When the customer does not want to allow the flexibility of a performance specification, a **material specification** is used. Material specifications dictate in exacting detail the product characteristics. For raw materials, even the composition and chemical and physical properties are defined.

A **process specification** defines parameters of the manufacturing process which must be controlled to attain a desired quality level.

When product quality is to be evaluated by an independent (third-party) agency,

[10]James L. Bossert, ed. *Procurement Quality Control* (Milwaukee, Wisc.: ASQC Quality Press, 1988), np.

such as an analytical laboratory, the **analytical specification** defines what values or attributes to measure and to what level of accuracy.

Standards tend to be more generic than specifications and apply across a company or industry. **Workmanship standards** define characteristics of acceptable workmanship such as the following:

- Solder joints
- Burrs
- Breakaway
- Concentricity
- Flatness
- Squareness
- Surface roughness
- Taper
- Blow holes
- Coloration
- Crazing
- Appearance[11]

Commercial standards are developed and adopted by industry/trade groups and by national or international standards institutes. The Electronics Industry Association (EIA), American National Standards Institute (ANSI), American Society for Testing and Materials (ASTM), Institute for Packaging and Production of Printed Circuits (IPC) and the International Standards Organization (ISO) are examples.

11.2.4 Quality Registration

In order to formalize quality management practices, the International Standards Organization (ISO) has developed the ISO 9000 Quality Management and Quality Assurance Standard. ISO 9000 has been adopted by the European Community (EC) to promote quality, excellence, and cooperation in the flow of goods and services among EC countries. Electronics manufacturers in the United States and elsewhere are rapidly adopting ISO 9000 practices for the following reasons:

- Internal improvement
- Market positioning
- Supplier control
- Customer or regulatory requirements[12]

[11] R. H. Johnson and R. T. Weber, *Buying Quality* (Milwaukee, Wisc.: ASQC Quality Press, 1988), np.

[12] Elizabeth Potts, "A Quality Registrar's View," *Circuits Assembly,* 3, no. 12 (1992), p. 46.

ISO 9000 consists of a series of standards which give guidance on processes, management, and implementation. Companies usually retain an accredited registrar who facilitates the process of earning and maintaining ISO 9000 registration through application, document review, preassessment, assessment, registration, and surveillance.[13]

11.3 DESIGN

Design excellence is essential to achievement of high quality. Among the techniques of quality design are

- Robust design
- Design rules
- Environmental stress analysis and testing
- Design evaluation

11.3.1 Robust Design

Genichi Taguchi has proposed that it is not sufficient to manufacture products that are within specification. Taguchi recommends use of statistically designed experiments, in the product development phase, to help designers find the parameter settings that will result in a product whose important characteristics are consistently close to the ideal target despite the presence of manufacturing variations or the effects of age [13].

In **robust design,** we want to design a product which is insensitive to variation in

- Manufacturing process
- Operator and maintenance procedures
- Usage environment

11.3.2 Design Rules

Design rules capture our accumulated knowledge and experience. Every engineering discipline has numerous design rules that are specific to technologies and applications. A general set of design rules should include the following:

- Include quality and reliability in specifications.
- Select high-quality components.
- Seek simplification.
- Maximize standardization.
- Judiciously apply safety factors.
- Test over a wide range of parameter values under a wide range of conditions.

[13]Potts, "A Quality Registrar's View," pp. 47–55.

- Minimize preventive maintenance requirements.
- Design for maintenance.
- Provide overload protection.
- Avoid overdesign.
- Avoid short-life materials.
- Perform stress analysis.
- Cooperate with manufacturing, maintenance, user.
- Use mature manufacturing processes.
- Use designs of proven reliability where possible.[14]

11.3.3 Environmental Stress Analysis and Testing

Vibration and temperature cycling can lead to fatigue failures of components and solder joints. Also, failures of many electrical components and materials, such as wiring insulation, are accelerated by elevated temperature providing activation energies for degrading chemical reactions (Fig. 11-6). Humidity contributes to failures of electrical contacts and connectors by the corrosion mechanism. Integrated circuits are particularly vulnerable to particulate contamination during the production process (see Sec. 14.6), and dust can also cause failures of electrical connections.

Studies by the United States Air Force Flight Dynamics Laboratory have identified temperature and vibration as the major environmental factors leading to airborne electronics (avionics) failures. Figure 11-7 shows the sources of stress and the locations of failures

[14]Sidney Anderson, "Design for Product Reliability," *Agricultural Engineering,* August 1973, pp. 13–14.

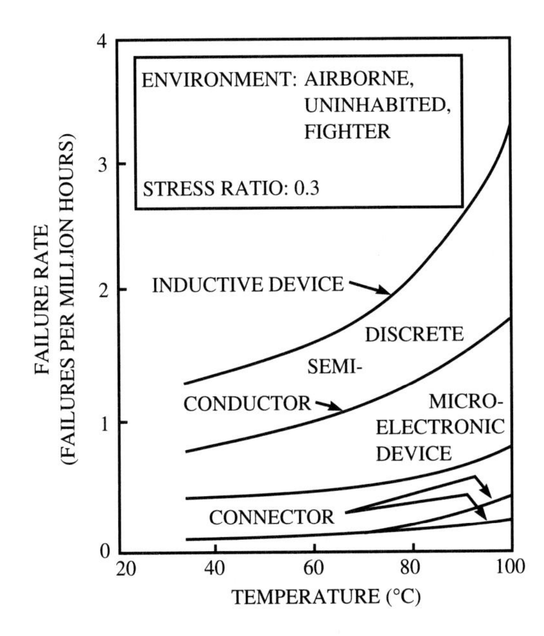

Figure 11-6 Failure rates are accelerated by high temperatures.

(*Source:* Douglas Holzhauer, "Thermal Analysis in Design and Reliability," unpublished presentation (Rome Air Development Center, Griffiths Air Force Base, N.Y., 1987), np.)

for avionics. Chapter 13 discusses how these environmental stresses are used in the factory to reveal quality defects.

In order to improve the thermal designs for electronic equipment, mechanical engineers perform thermal analyses to identify potential hot spots that would accelerate component failures. Figure 11-8 illustrates finite-element thermal modeling of a very-high-speed integrated circuit (VHSIC).

Electronic equipment is exposed to different forms of shock and vibration, over wide ranges of frequency and acceleration. Typical failure modes include:

- Broken leads
- Loose screws
- Broken artwork traces

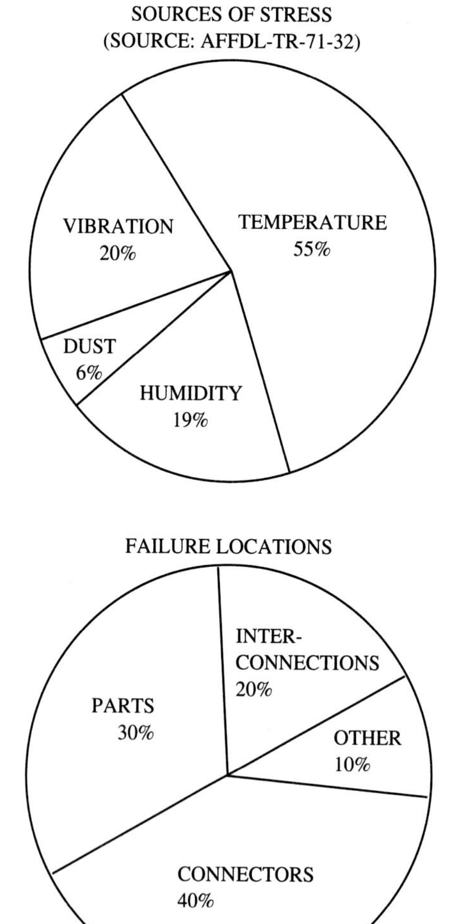

Figure 11-7 Sources and locations of environmentally induced failures.

(*Source:* AFFDL-TR-71-32 (Wright Patterson Air Force Base, Ohio: Air Force Flight Dynamics Laboratory, 1973), np.)

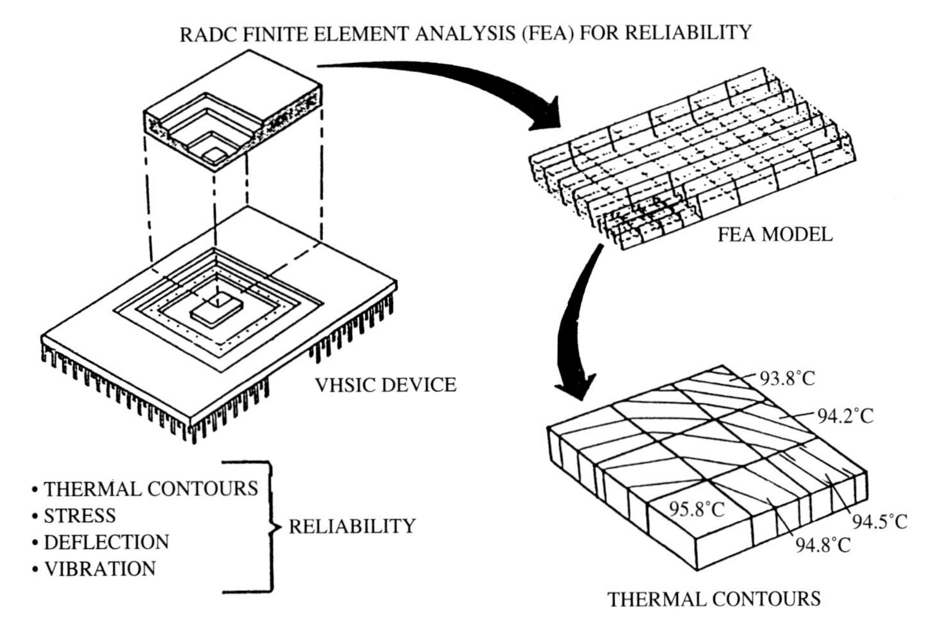

Figure 11-8 Finite element modeling of thermal stresses on IC.

(*Source:* Holzhauer, "Thermal Analysis in Design and Reliability," np.)

- Cracked solder joints
- Broken wires

The typical failure mechanisms for vibration are

- Progressive deterioration of elements whose frequencies fall within the excitation envelope
- Fatigue resulting from repeated stress reversals

Shock loading failures can occur when stress exceeds the ultimate strength of the product, during handling and shipping.

Finite-element modeling is also useful to analytically assess mechanical stresses before hardware designs are committed. Figure 11-9 illustrates finite element modeling of mechanical stresses in a surface-mount solder attachment. The upper diagram shows the finite element grid, and the lower diagram shows a computer-graphics model of the shear stress contours in a simulation.

11.3.4 Design Evaluation

The design engineer uses several tools and techniques to enhance designs, including design review, margins and derating, and experimentation.

Prior to the advent of concurrent engineering, the primary approach for integrating

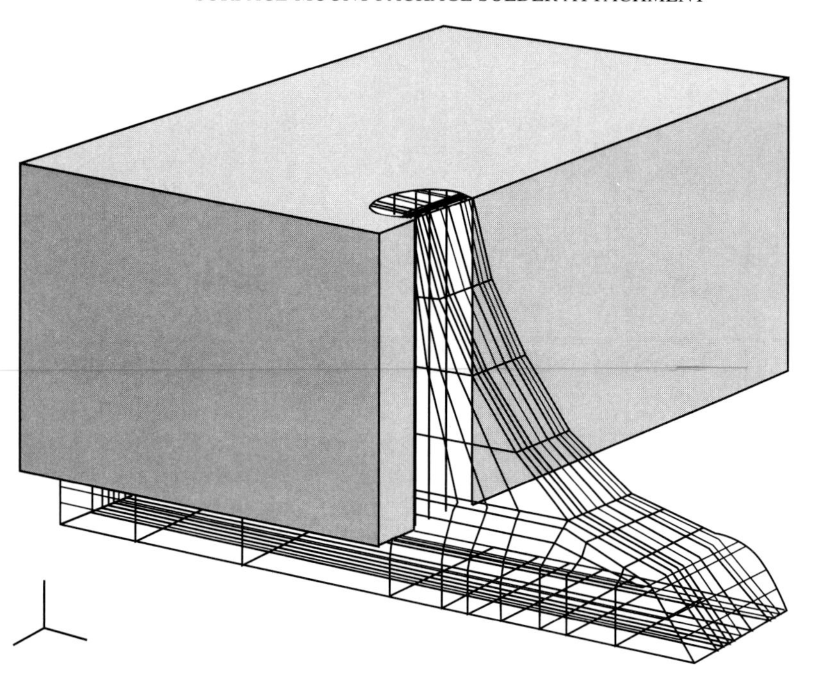

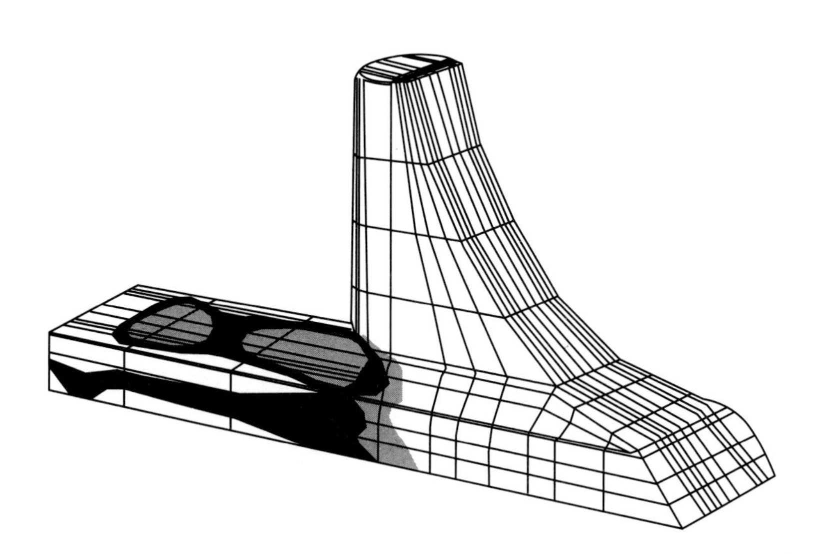

Figure 11-9 Finite element modeling of shear stresses in SMT solder attachment. (*Source:* Holzhauer, "Thermal Analysis in Design and Reliability," np.)

engineering disciplines has been the design review. A design review involves the use of disciplined procedures and systematic methods to evaluate intermediate designs, identify problems, and define corrective actions. Design review is an iterative process involving both the consultation of technical specialists and multidisciplinary team meetings. The configuration control board or engineering change review panel is a type of design review meeting where engineering changes are controlled to ensure the integrity of the design and accuracy of the design documentation, such as drawings and specifications.

Engineers have developed several useful tools to analyze the reliability and safety of designs. The resulting documentation is carefully evaluated in the design review process. The two main categories of tools are inductive and deductive tools.

Failure modes, effects, and criticality analysis (FMECA) is an inductive approach to exhaustive enumeration of the failure mechanisms and their effects, with particular emphasis on safety consequences. The FMECA is summarized in a spreadsheet format, with columns for the following information:

- Item
- Failure mode
- Cause of failure
- Possible effects
- Probability of occurrence
- Criticality

Sneak circuit analysis (SNA) is an inductive method wherein the engineers study a design to detect unintended events, such as reverse current or fluid flow paths with safety implications.

In the **fault tree analysis** (FTA) method of deductive analysis, the engineer identifies an undesirable event such as equipment failure and/or safety hazard, then traces the possible chain of events leading to that undesirable top-level event.

Engineers routinely use **safety factors** (design margins and derating). However, the typical approach is deterministic, assuming fixed loads and capacity as follows:

$$\nu = c/l \tag{11-1}$$

where ν = Safety factor
c = Capacity (strength)
l = Load (stress)

This approach to margins, where deterministic point values are assumed for the load and capacity, is simplistic and tends to result in overdesign. A more robust approach to margins and derating is probabilistic design, which considers the stochastic nature (variability) of stresses and strengths (loads and capacities). As an example, consider the case of **normally distributed** load and capacity, as shown in Fig. 11-10.

A load L is specified as being some number n_l of standard deviations σ_l above the mean μ_l for load distribution $f_l(l)$:

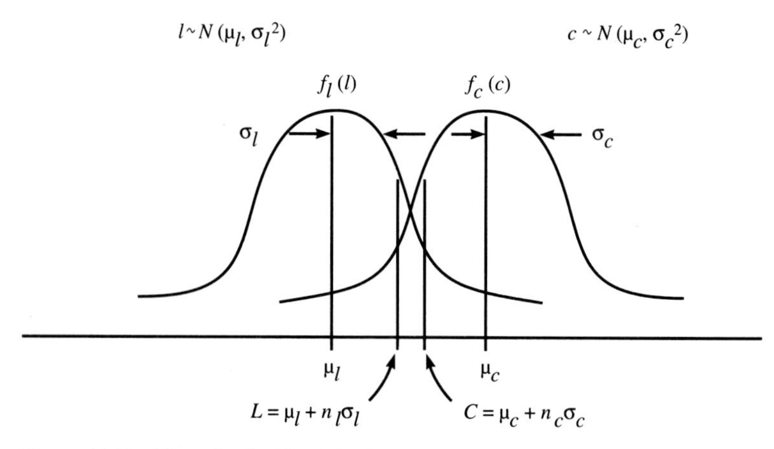

Figure 11-10 Normally distributed load and capacity.

$$L = \mu_l + n_l\sigma_l \tag{11-2}$$

A capacity C is specified as being some number n_c of standard deviations σ_c below the mean μ_c for a capacity distribution $f_c(C)$:

$$C = \mu_c - n_c\sigma \tag{11-3}$$

The policy variables n_c and n_l are determined based upon economics and management's willingness to accept risk. Define the safety factor as

$$\boldsymbol{\nu} = \frac{C}{L} = \frac{\boldsymbol{\mu}_c - n_c\boldsymbol{\sigma}_c}{\boldsymbol{\mu}_l + n_l\boldsymbol{\sigma}_l} \tag{11-4}$$

We want to find the mean capacity (μ_c) of a component such that a prescribed safety factor can be achieved for a given $f_c(C)$. Let z be a **random variable** denoting the difference between load and capacity:

$$z \sim N(\mu_c - \mu_l, \sigma_c^2 + \sigma_l^2) \tag{11-5}$$

Reliability is defined as the probability that capacity exceeds load for all possible loads:

$$r = P\{c > l\} = P\{z > 0\} \tag{11-6}$$

$$= \frac{(\boldsymbol{\mu}_c - \boldsymbol{\mu}_l) - 0}{\sqrt{(\boldsymbol{\sigma}_c^2 + \boldsymbol{\sigma}_l^2)}}$$

Example:

Find the mean strength of a part for which we desire

$\upsilon = 1.2$, given

$n_l = 2 \qquad \sigma_l = 10 \qquad \mu_l = 60$

$n_c = 2 \qquad \sigma_c = 15$

$$\mu_c = n_c\sigma_c + \upsilon(\mu_l + n_l\sigma_l)$$
$$= 2(15) + 1.2\,[60 + 2(10)]$$
$$= 126$$

Calculate the reliability for the resulting design

$$R = P\{c < l\} = \Phi\left[\frac{(126 - 60) - 0}{\sqrt{15^2 + 10^2}}\right] \tag{11-7}$$
$$= \Phi(3.66) = 0.99987$$

Derating is the design principle of selecting parts which have a high probability $r = P\{l < c\}$. Design rules often involve guidelines for parts derating.

Experimentation is another important approach to design improvement. Traditionally, a trial-and-error approach has been used similar to the phases of the Deming Cycle (Fig. 11-3). We develop a design, build a prototype, test the prototype under various design and over-stressed conditions, analyze the failures, and take corrective action (that is, redesign). This process is called **test-analyze-and-fix** (TAAF). The statistically designed experiment is a more formal and systematic approach to testing. Critical design factors and the levels (values) of those factors are determined. Then a sample of units is tested at carefully selected combinations of factor levels. This approach is discussed in Sec. 11.4.6.

11.4 STATISTICAL QUALITY CONTROL

Statistical quality control (SQC) or **statistical process control** (SPC) is the application of the scientific method to control and improve the manufacturing process.

11.4.1 The Seven Problem-Solving Tools

The seven problem-solving tools are simple to use, yet powerful enough to solve many quality problems. The following paragraphs discuss the following tools:

- Pareto chart
- Histogram
- Check sheet
- Data stratification
- Cause-and-effect diagram
- Scatter plot
- Control chart

The **Pareto chart** is a simple method of descriptive statistical analysis, wherein data is categorized by cause and vertical bars are constructed with the height of each bar proportional to the number of observations for the cause. Figure 11-11 is a Pareto chart

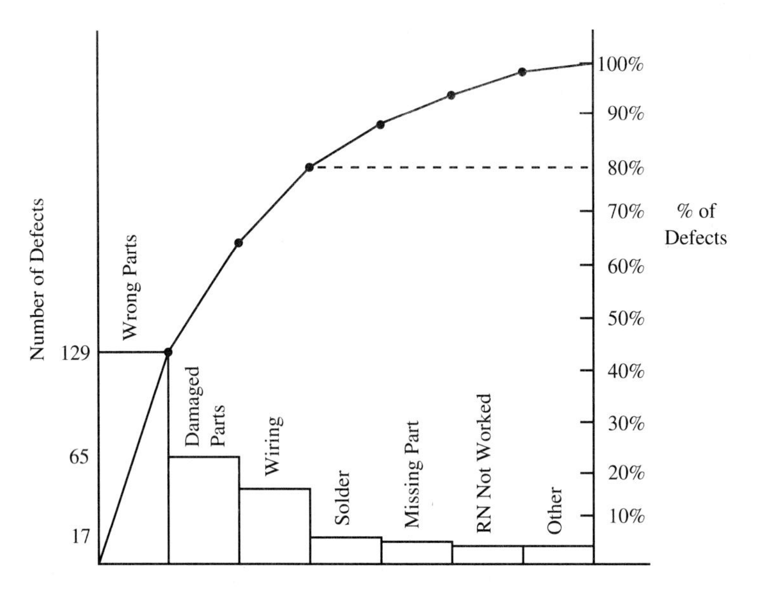

Figure 11-11 Pareto analysis of defects in PWBs.

(*Source:* W. A. Huff, "Implementing Yield Improvment Programs," in *Proceedings QIE* (Milwaukee, Wis.: American Society for Quality Control, 1982), pp. 48–52.)

for the distribution of workmanship defects in a study of PWBs. A bar chart shows the categories of defects ordered from most important to least important based on number of defects. The Pareto chart carries out one of the basic principles of quality: make problems visible. The histogram indicates what the priorities should be in investigating and correcting the quality problems. In this case study, insertion of wrong parts was the most serious problem, and the one which should be worked on first.

Figure 11-11 also includes a line graph of the cumulative percent defects. Two of the seven problem categories (or 28 percent of the problems) account for almost 70 percent of the defects. This phenomenon, where approximately 20–30 percent of the problems account for 70–80 percent of the defects, is typical. The Pareto rule (also called the 80-20 rule) was first observed by the Italian economist Pareto, who studied the distribution of wealth in the population and concluded that about 20 percent of the people held approximately 80 percent of the wealth. The Pareto rule has proven applicable to a wide range of engineering, science, and social contexts. Figure 11-12 contains a Pareto chart for defects. The leading category of defects was cracks.

A **histogram** is a type of bar chart with the vertical bars ordered horizontally by value of a random variable. The histogram gives a visual indication of the probability distribution for a variable. The heights of the rectangles (bars) are proportional to the corresponding probabilities. Figure 11-13 continues the quality analysis of defects and shows the distribution of crack size. The crack size varies from 1.0 to 4.0, and the mode (most commonly occurring crack size) is in the range of 2.0 to 2.5. The histogram gives indications of central tendency, dispersion, and symmetry.

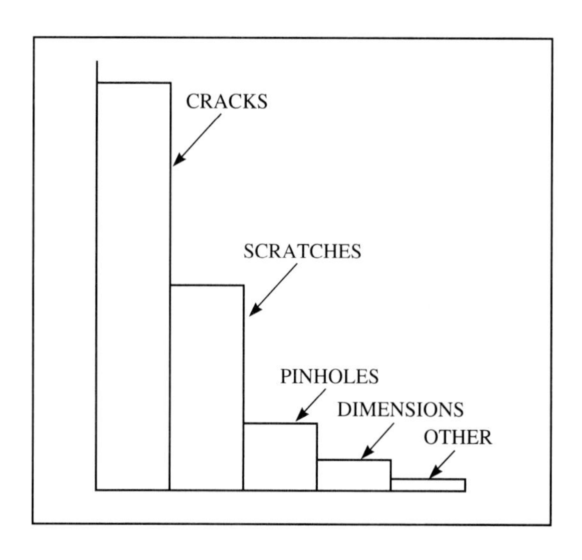

Figure 11-12 Pareto analysis of defects.

(*Source:* ©1987 American Society for Quality Control, Reprinted by permission.)

A **check sheet** (Fig. 11-14) is a data collection tool from which a histogram or Pareto chart may be constructed. A check sheet has areas assigned to categories. Employees working in the production process tally events occurring by category during the process. In an automated process, the frequency data is logged automatically.

Data stratification is an extension of the histogram tool. The distribution is stratified based upon some criteria, in order to gain more specific insight about source(s) of problems. In the example of crack defects, after the distribution of crack sizes was assessed, data was analyzed for samples off of two different production machines. The data was stratified by machine; that is, a histogram was constructed for the cracked parts from each machine. Figure 11-15 shows the results of data stratification, revealing that machine type A produces parts with more frequent cracks. In addition, the cracks tend to be larger and the distribution of crack size tends to be more widely dispersed than does the distribution for type B. This information helps us to focus the investigation on the higher-priority problem: machine type A.

Once we have isolated the source of undesirable variation, by means of the Pareto chart, histogram, and data stratification, the next step is to enumerate the possible causes

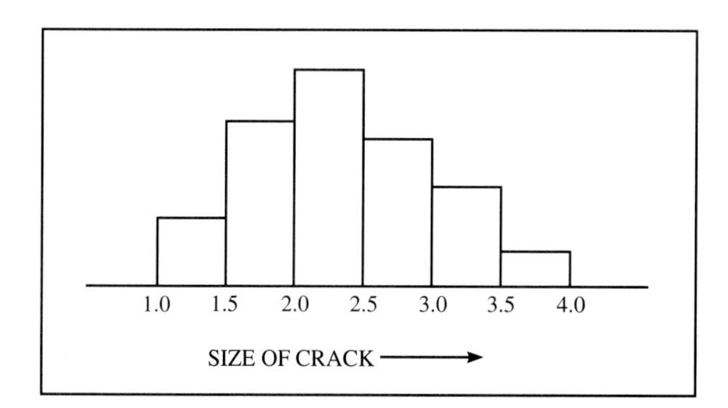

Figure 11-13 Histogram of crack size distribution.

(*Source:* ©1987 American Society for Quality Control, Reprinted by permission.)

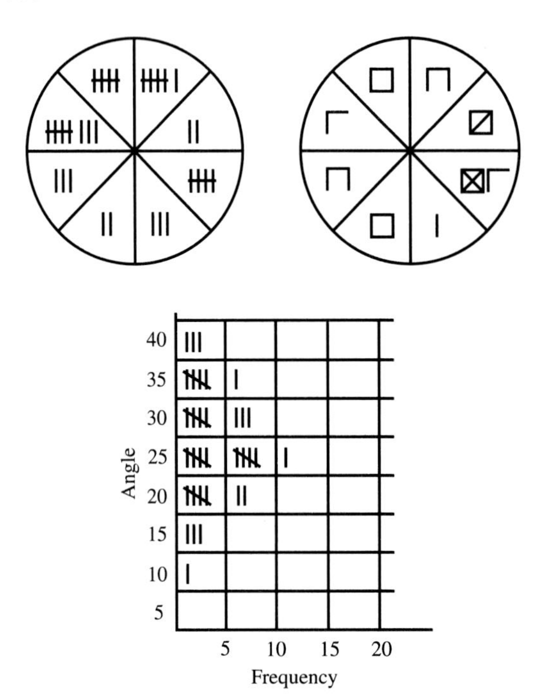

Figure 11-14 Check sheets.

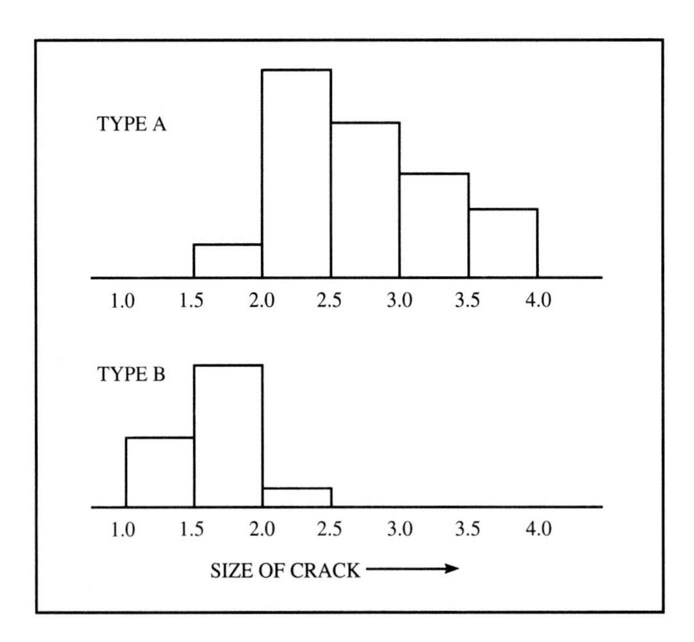

Figure 11-15 Data stratification by machine for crack size.

(*Source:* ©1987 American Society for Quality Control, Reprinted by permission.)

of the problem (effect). There are various formats for the **cause-and-effect diagram,** including the tabular format in Fig. 11-16 for the workmanship defects on PWBs.

An alternate format for cause-and-effect analysis is the fishbone, or Ishikawa, diagram. In an investigation of quality defects in windings of miniature 2-pin DIP, wirewound capacitors, the Pareto analysis showed nonflat ends as the predominant category of defects. The engineers proceeded to develop a fishbone chart of causes for nonflat ends (Fig. 11-17).

A **scatter plot** is a presentation of data on an *X*-*Y* coordinate system. The purpose of the scatter plot is to obtain some visual indication of the relationship between two variables. The scatter plot does not necessarily establish any cause-and-effect relationship. Scientific principles, engineering judgment, and experience are the basis for hypothesized relationships. If there is an indication of some relationship between the variables, the statistical techniques of regression and correlation are used to better understand the relationship and build a parametric model. Figure 11-18 illustrates some typical scatter plots. Diagram (a) shows a scatter plot indicating a strongly positive linear relationship (with

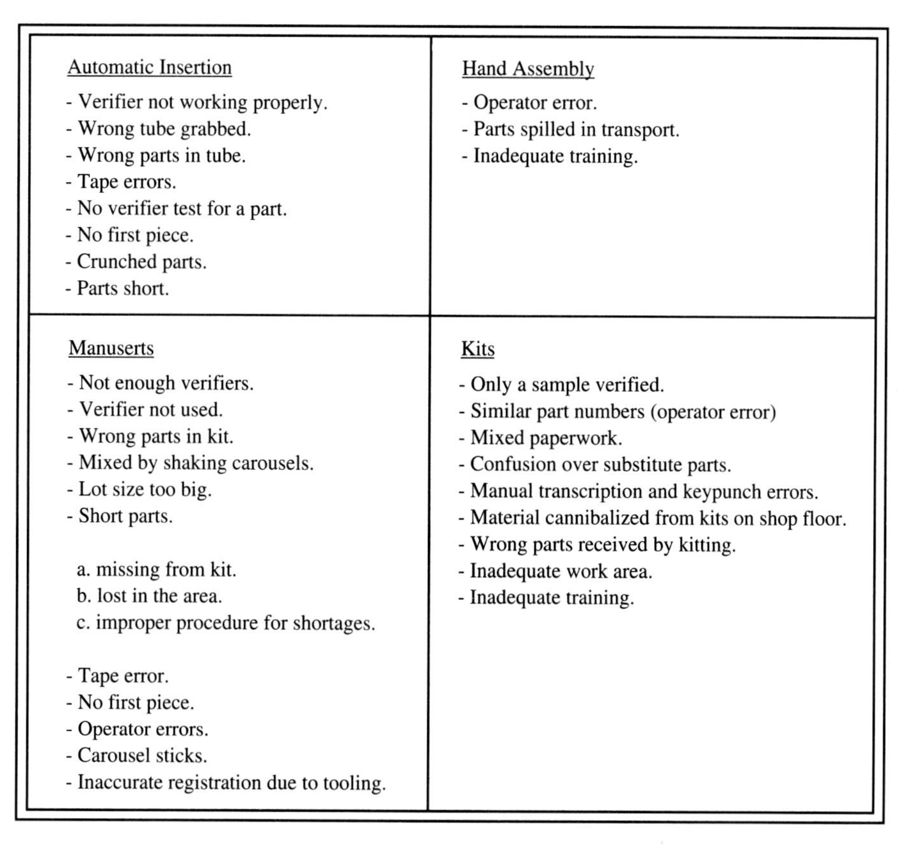

Automatic Insertion	Hand Assembly
- Verifier not working properly. - Wrong tube grabbed. - Wrong parts in tube. - Tape errors. - No verifier test for a part. - No first piece. - Crunched parts. - Parts short.	- Operator error. - Parts spilled in transport. - Inadequate training.
Manuserts	**Kits**
- Not enough verifiers. - Verifier not used. - Wrong parts in kit. - Mixed by shaking carousels. - Lot size too big. - Short parts. a. missing from kit. b. lost in the area. c. improper procedure for shortages. - Tape error. - No first piece. - Operator errors. - Carousel sticks. - Inaccurate registration due to tooling.	- Only a sample verified. - Similar part numbers (operator error) - Mixed paperwork. - Confusion over substitute parts. - Manual transcription and keypunch errors. - Material cannibalized from kits on shop floor. - Wrong parts received by kitting. - Inadequate work area. - Inadequate training.

Figure 11-16 Cause-and-effect diagram for PWB defects.

(*Source:* W. A. Huff, "Implementing Yield Improvement Programs," in *Proceedings QIE* (Milwaukee, Wis.: American Society for Quality Control, 1982), 48–52.)

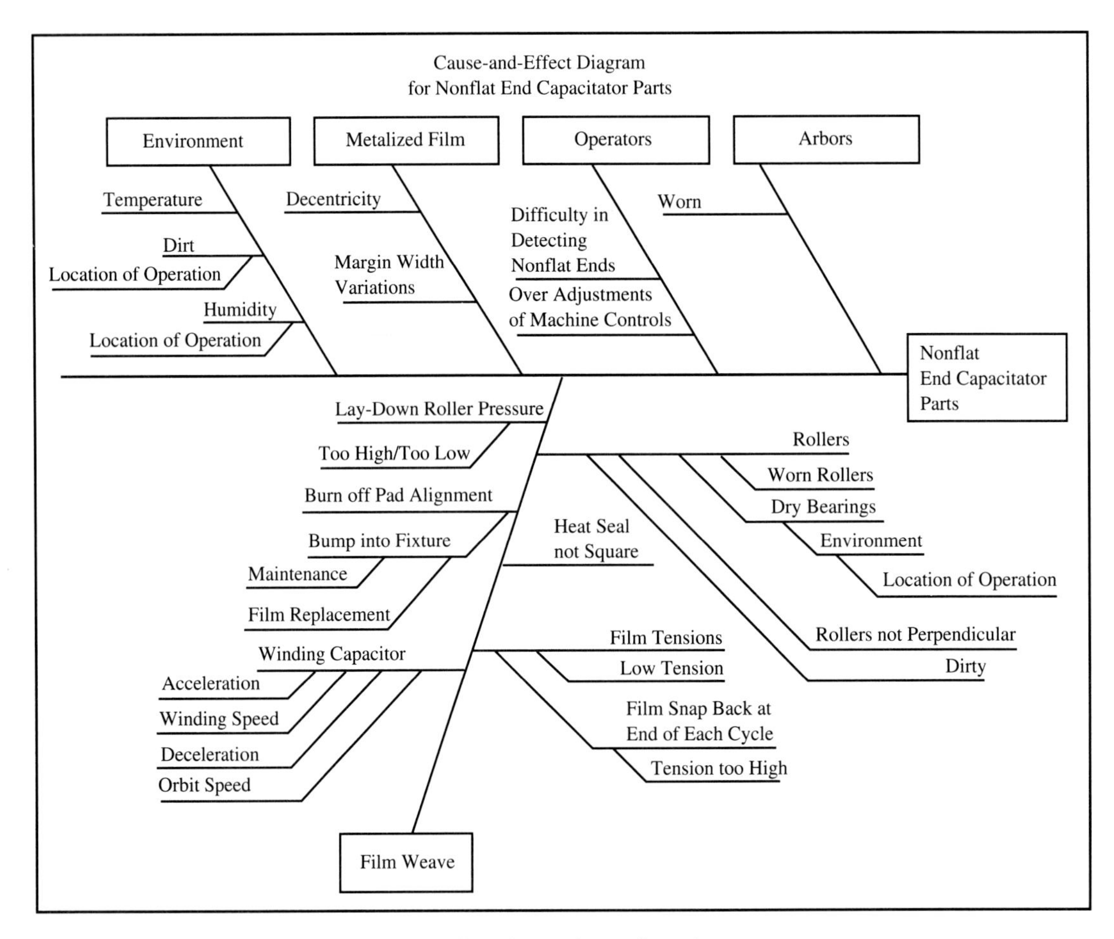

Figure 11-17 Cause-and-effect diagram for nonflat ends.

(*Source:* M. F. Echevarria and S. J. Kanellos, "A Team Approach to Process Optimization," in *Proceedings of the Juran Conference* (Chicago, Ill.: 1988), np.)

correlation coefficient of $r = 1$). Similarly, diagram (d) illustrates a strongly negative linear relationship (with correlation coefficient of $r = -1$). Scatter plots (b) and (e) are typical of data with some linear tendency, but also random scatter. Scatter plot (c) indicates no relationship between X and Y, and scatter plot (f) suggests a possible quadratic form. Scatter plots (c) and (f) have correlation coefficients of $r = 0$, meaning that there is no indication of a linear relationship.

Statistical process **control charts** are used to monitor data from a process. There are two types of process data. Variables data are obtained by measuring values of a continuous variable, such as length or resistance. The X-bar chart (Fig. 11-19) monitors variables data. The R chart (Fig. 11-19) measures variability of the data. Attributes data record the presence or absence of an attribute (for example, defective or nondefective). The p-chart (Fig. 11-20) monitors attributes data.

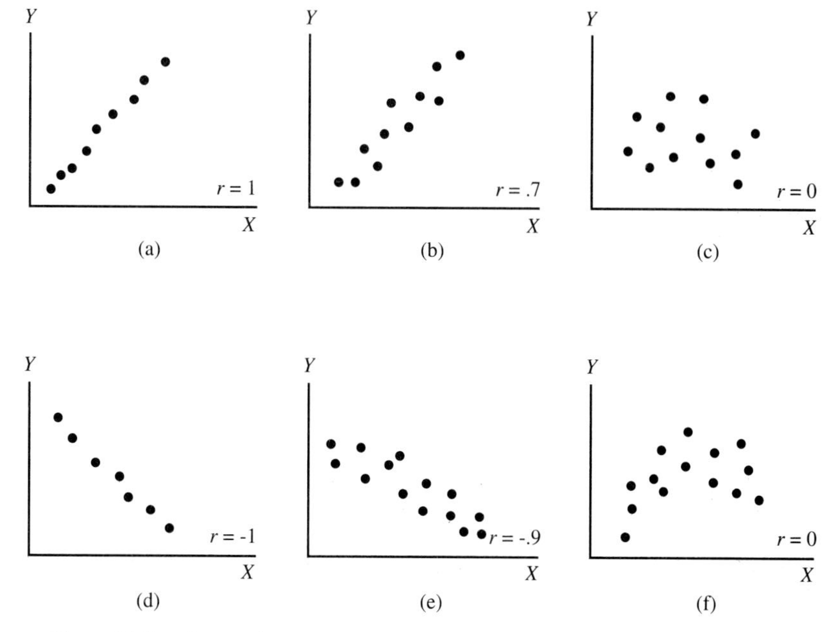

Figure 11-18 Scatter plots.

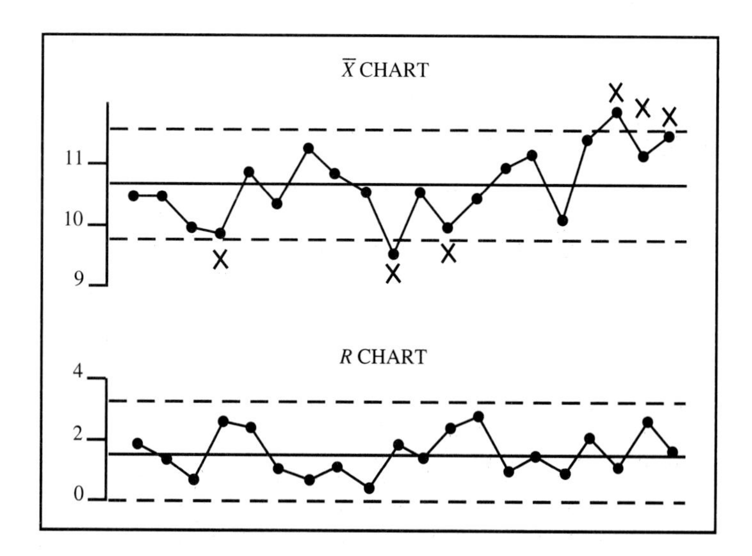

Figure 11-19 *X*-bar and *R* charts for amplifier data.

(*Source:* Reproduced with permission of AT&T ©1986.)

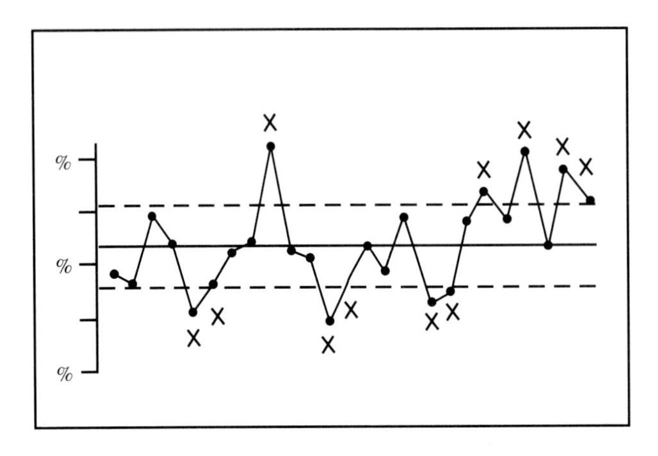

Figure 11-20 *P*-chart for attributes data.

(*Source:* Reproduced with permission of AT&T ©1986. All rights reserved.)

The control chart is designed to signal process control problems, and it consists of a center line (at the nominal value of a control variable), bounded by control limits beyond which variability is considered unacceptable. Actual data from the process are plotted on the chart to make visible the variability and trends indicating problems. Since control charts are somewhat more complicated than the other six problem-solving tools, the following two subsections are devoted to further discussion and applications of variables (Sec. 11.4.2) and attributes (Sec. 11.4.3) control charts.

11.4.2 Variables Control Charts

The **variables control chart** monitors status of a continuous variable, such as a measurement. Let $x_1, x_2, \ldots, x_n$ be successive measurements on a process. If that process is in a state of statistical control, then the x_i's should be a random sample of size n on some distribution. In other words, if the process is in statistical control, then the measurements are observations on **independent identically distributed** (IID) random variables. Let θ be a statistic computed from samples. In successive samples, the values of $(\theta_1, \theta_2, \ldots, \theta_n)$ will be independent random samples having identical distributions. The common distribution will have mean μ_θ and variance σ_θ^2

Consider the case where the x_i's are IID normal. If a sample of size n is drawn from the process, and the variables x_i ($i = 1$ to n) are measured, then a statistic formed by linear combination of the x_i's is also distributed normally.

If $x_1, \ldots, x_n$ are independent random variables with means μ_i and variances σ_i^2 $(i = 1, \ldots, n)$, then a linear combination of the x_i

$$L = \sum_{i=1}^{n} c_i x_i \tag{11-8}$$

has

$$\mu_L = E(L) = \sum_{i=1}^{n} c_i \mu_i \tag{11-9}$$

and

$$\sigma_L^2 = E(L - \mu_L)^2 = \sum_{i=1}^{n} c_i^2 \sigma_i^2 \tag{11-10}$$

From this sample of size n, we form a statistic that is a linear combination of the x_i (i = 1, . . . , n). The sample mean

$$\bar{x} = \sum_{i=1}^{n} \frac{1}{n} x_i \tag{11-11}$$

is such a linear combination, where $c_i = 1/n$, for all i. Then assuming the process is in control, the mean of the variable $\bar{x}$ is

$$\mu_{\bar{x}} = \sum_{i=1}^{n} c_i \mu = \sum_{i=1}^{n} \frac{1}{n} \mu = \frac{n}{n} \mu = \mu \tag{11-12}$$

The variance of $\bar{x}$ is

$$\sigma_{\bar{x}}^2 = \sum_{i=1}^{n} c_i^2 \sigma_i^2 = \frac{1}{n^2} \sum_{i=1}^{n} \sigma^2 = \frac{\sigma^2}{n} \tag{11-13}$$

Then the probability density function is

$$f(x) = \frac{1}{\sigma_{\bar{x}} \sqrt{2\pi}} \exp\left[-\frac{1}{2\sigma_{\bar{x}}}(x - \mu_{\bar{x}})^2 \right], \quad -\infty < x < \infty \tag{11-14}$$

When x (i = 1, . . . , n) is a random sample of size n from a nonnormal distribution, the Central Limit theorem states that, if n is large, then $\bar{x}$ is a random variable distributed approximately normally with error of approximation tending to zero as $n \to \infty$.

Typically we are concerned not only that the sample averages are centered on the target (nominal specification) value but also that the variation is minimal. Variation is measured by the variance or standard deviation. However, in control charting, variability is usually measured by the range of values (difference between smallest and largest). Such charts are called R charts.

Control charts are designed to follow the principal of management by exception, meaning that action is taken only to correct exceptional problems. Consequently, control limits are generally set at values three standard deviations above and below the mean value, to provide reasonable assurance that when an observation falls outside the limits, the process is truly out of control. We do not want to take actions in response to random noise inherent to the process, but we do want to take action when the process changes by either a shift of the mean or an increase of the variation (standard deviation).

A type-I error occurs when we take action to investigate the process when in fact the process is in control. The symbol α is often used to denote the probability of a type-I error. Conversely, a type-II error (with probability β) occurs when the process is out of control, but we fail to detect the change. Control charts should be designed to limit the probabilities (α and β) of both type-I and type-II errors, respectively.

Empirical control charts are used when process parameters are not known with certainty, such as when the manufacturing process is being started up but is not yet in statistical control. These charts are based on the assumption that we take a sample of n objects closely spaced in time. This procedure is repeated for a total of k samples. Let $\bar{x}_1, \bar{x}_2, \ldots, \bar{x}_k$ be the sample means. Then

$$\bar{\bar{x}} = \frac{1}{k}\sum_{i=1}^{k} \bar{x}_i \tag{11-15}$$

is the mean of the sample means. Also let $R_1, R_2, \ldots, R_k$ be the sample ranges (for example, $x_{1,\max} - x_{1,\min}$ for sample 1). Then

$$\overline{R} = \frac{1}{k}\sum_{i=1}^{k} R_i \tag{11-16}$$

is the mean of the sample ranges.

Constants A_2 have been tabulated for which A_2R is an unbiased estimator of $3\sigma/\sqrt{n}$. Then the control limits are estimated by

$$UCL_{\bar{x}} = \bar{\bar{x}} + A_2\overline{R} \tag{11-17}$$

and

$$LCL_{\bar{x}} = \bar{\bar{x}} - A_2\overline{R} \tag{11-18}$$

Constants have also been tabulated for construction of empirical R charts. The control limits are

$$LCL_R = D_3\overline{R} \tag{11-19}$$

and

$$UCL_R = D_4\overline{R} \tag{11-20}$$

The factors A_2, D_3, and D_4 vary with sample size and are given in Table 11-1.

The R chart is constructed by the following procedure:

- Select the sample size (n) to be collected at intervals.
- Collect a series of at least 10 (preferably > 30) samples (of size n).
- Calculate the range R_i for each sample and the $\overline{R}$ for all samples.
- Calculate the control limits by employing Eqs. (11-19) and (11-20).
- Plot the successive values of R and connect them with a line graph.

The X-bar chart is constructed by the following procedure:

- Calculate $\bar{x}$ for each sample and $\bar{\bar{x}}$ for all samples.
- Plot a horizontal straight line at the value of $\bar{\bar{x}}$ (the centerline).
- Calculate the control limits by employing Eqs. (11-17) and (11-18).
- Plot the control limits as dashed lines about the centerline.
- Plot the successive values of $\bar{x}$ from the samples.

TABLE 11-1 FACTORS FOR *X*-BAR AND *R* CHARTS

Number of Observations in Sample	$\overline{X}$ Chart. Factors for control limits	*R* Chart. Factors for control limits	
(n)	(A_2)	(D_3)	(D_4)
2	1.88	0	3.27
3	1.02	0	2.57
4	.73	0	2.28
5	.58	0	2.11
6	.48	0	2.00
7	.42	.08	1.92
8	.37	.14	1.86
9	.34	.18	1.82
10	.31	.22	1.78

Source: Reproduced with the permission of AT&T © 1986. All rights reserved. Reprinted with permission.

The following example illustrates construction of *X*-bar and *R* charts.

Example:

Table 11-2 contains data collected on a process variable (amplifier gain in dB) over a 9-day period, March 31 through April 8. Three samples of five amplifiers were tested on each of six days, and two samples of the five amplifiers were tested on the final day. The following calculations lead to construction of the control charts.[15]

The *R* chart lines are

$$\overline{R} = \frac{1}{k}\sum_{i=1}^{k} R_i = \frac{31.8}{20} = 1.59$$

$$UCL_R = D_4\overline{R} = 2.11(1.59) = 3.35$$

$$LCL_R = D_3\overline{R} = 0\,(1.59) = 0$$

The *X*-bar chart lines are

$$\bar{\bar{x}} = \sum_{i=1}^{k} \bar{x}_i = \frac{213.20}{20} = 1.59$$

$$UCL_{\bar{x}} = \bar{\bar{x}} + A_2\overline{R} = 10.66 + 0.58(1.59)$$

$$= 10.66 + 0.92$$

$$= 11.58$$

$$LCR_{\bar{x}} = \bar{\bar{x}} - \mathrm{A}_2\overline{R} = 10.66 - 0.92$$

$$= 9.74$$

[15]*Statistical Quality Control Handling,* AT&T Technologies (Charlotte, N.C.: Delmar Printing Company, 1988), p. 14.

TABLE 11-2 AMPLIFIER DATA

	GAIN IN DB.									
	3/31 1	2	3	4/1 4	5	6	4/4 7	8	9	4/5 10
	11.1	9.6	9.7	10.1	12.4	10.1	11.0	11.2	10.6	8.3
	9.4	10.8	10.0	8.4	10.0	10.2	11.5	10.0	10.4	10.2
	11.2	10.1	10.0	10.2	10.7	10.2	11.8	10.9	10.5	9.8
	10.4	10.8	9.8	9.4	10.1	11.2	11.0	11.2	10.5	9.5
	10.1	11.0	10.4	11.0	11.3	10.1	11.3	11.0	10.9	9.8
	52.2	52.3	49.9	49.1	54.5	51.8	56.6	54.3	52.9	47.6
$\bar{X}$	10.44	10.46	9.98	9.82	10.90	10.36	11.32	10.85	10.58	9.52
R	1.8	1.4	.7	2.6	2.4	1.1	.8	1.2	.5	1.9
	11	12	4/6 13	14	15	4/7 16	17	18	4/8 19	20
	10.6	10.8	10.7	11.3	11.4	10.1	10.7	11.9	10.8	12.4
	9.9	10.2	10.7	11.4	11.2	10.1	12.8	11.9	12.1	11.1
	10.7	10.5	10.8	10.4	11.4	9.7	11.2	11.6	11.8	10.8
	10.2	8.4	8.6	10.6	10.1	9.8	11.2	12.4	9.4	11.0
	11.4	9.9	11.4	11.1	11.6	10.5	11.3	11.4	11.6	11.9
	52.8	49.8	52.2	54.8	55.7	50.2	57.2	59.2	55.7	57.2
$\bar{X}$	10.56	9.96	10.44	10.96	11.14	10.04	11.44	11.84	11.14	11.44
R	1.5	2.4	2.8	1.0	1.5	.8	2.1	1.0	2.7	1.6

Source: Reproduced with the permission of AT&T © 1986. All rights reserved.

Figure 11-19 contains the X-bar and R charts with observations plotted and anomalies marked by "X".

11.4.3 Attributes Control Charts

An **attributes control chart** monitors the status of a discrete variable (attribute), such as the number defective in a lot of parts. The binomial distribution arises for quality attributes, which are discrete random variables with two possible states (for example, 0, 1 or good, bad). Let x be a discrete random variable:

$$x = \begin{Bmatrix} 0, \, good \\ 1, \, bad \end{Bmatrix}$$

Let

p = Probability a part is bad
$1 - p$ = Probability a part is good

In one trial $n = 1$ and

$$p(x) = p^x(1 - p)^{1-x} \tag{11-21}$$

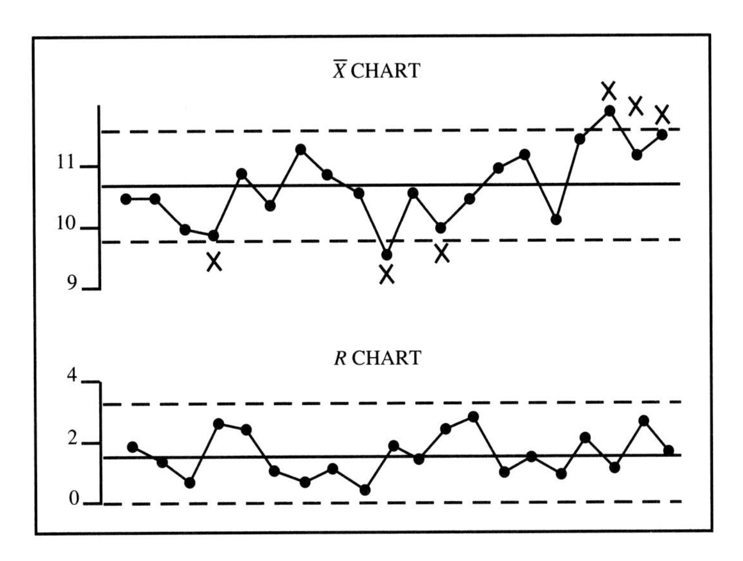

Figure 11-19 *X*-bar and *R* charts for amplifier data.

(*Source:* Reproduced with permission of AT&T ©1986. All rights reserved.)

If $x = 1$

$$p(1) = P^1(1 - p)^0 = p$$

If $x = 0$

$$p(0) = p^0(1 - p)^1 = (1 - p)$$

Generalizing, there are

$$\begin{bmatrix} n \\ x \end{bmatrix} = \frac{x!}{n!\,(n - x)!}$$

ways in which to get x bad out of a sample of size n. The binomial probability mass function, giving the probability of x-out-of-n, is

$$p(x) = b(x) = \begin{bmatrix} n \\ x \end{bmatrix} p^x (1 - p)^{n-x}, \qquad x = 0, 1, 2, \ldots \tag{11-22}$$

where p is the probability of a bad item. For the binomial distribution, the mean and variance are:

$$E(x) = \mu = np \tag{11-23}$$

$$E(x - \mu)^2 = \sigma^2 = np(1 - p) \tag{11-24}$$

The procedure for constructing a p-chart is as follows:

- Collect at least 10 (preferably > 30) samples of size n.
- Calculate the percent defective p in each sample.
- Calculate $\bar{p}$, the average of the sample percent defective.
- Calculate the control limits using the following equations:

$$UCL_p = \bar{p} + 3\sqrt{\frac{\bar{p}(1-\bar{p})}{n}} \tag{11-25}$$

$$LCL_p = \bar{p} - 3\sqrt{\frac{\bar{p}(1-\bar{p})}{n}} \tag{11-26}$$

- Draw the centerline at the value of $\bar{p}$, and draw the control limits about it using dashed lines.
- Plot the successive values of p.

Example:

The following example illustrates construction of a p-chart.[16] Table 11-3 contains data on percent defective for a process observed over the period August 11 through September 15.

The sample percent defective is:

$$p = \frac{\Sigma\ defects}{\Sigma\ items} = \frac{2103}{36060}$$

$$= 0.0583$$

Since the sample size varies from day to day, it is necessary to calculate an average n as follows:

$$n = \frac{\Sigma\ Numbers\ in\ samples}{\Sigma\ Days} = \frac{36060}{25}$$

$$= 1442$$

The control limits are

$$UCL_p = \bar{p} + 3\sqrt{\frac{\bar{p}(1-\bar{p})}{n}}$$

$$= 0.0583 + 3\sqrt{0.0583\frac{(1-0.0583)}{1442}}$$

$$= 0.0583 + 0.0186$$

$$= 0.0769$$

$$LCL_p = \bar{p} - 3\sqrt{\frac{\bar{p}(1-\bar{p})}{n}}$$

$$= 0.0583 - 0.0186$$

$$= 0.0397$$

Figure 11-20 contains the p-chart with anomalies marked "X."

[16] *Statistical Quality Control Handling,* p. 18.

TABLE 11-3 PERCENT DEFECTIVE DATA

Date	Number in sample	Number defective in sample	% defective in sample
8/11	1524	70	4.59
8/12	1275	53	4.16
8/15	1821	132	7.25
8/16	1496	91	6.08
8/17	1213	32	2.64
8/18	1371	55	4.01
8/19	1248	69	5.53
8/20	1123	67	5.97
8/22	1517	159	10.48
8/23	1488	94	6.32
8/24	2052	105	5.12
8/25	1696	37	2.18
8/26	1427	58	4.06
8/29	1277	75	5.87
8/30	1613	73	4.53
9/2	1987	145	7.30
9/6	1360	41	3.01
9/7	1439	50	3.47
9/8	1723	118	6.85
9/9	2035	169	8.30
9/10	1314	88	6.70
9/12	215	24	11.16
9/13	1384	77	5.56
9/14	1995	185	9.27
9/15	467	36	7.71
Total	36060	2103	

Source: Reproduced with the permission of AT&T

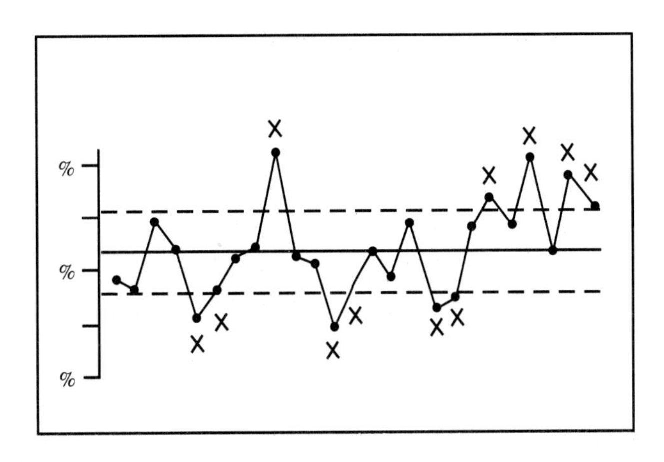

Figure 11-20 *P*-chart for attributes data.

(*Source:* Reproduced with permission of AT&T ©1986. All rights reserved.)

11.4.4 Interpretation of Control Charts

Much of problem-solving occurs when an informed observer witnesses a significant event. The scientific method uses experimentation and data collection to increase the probability of the observer witnessing the significant event. Control charts are useful tools for this purpose. When trained, experienced analysts examine a control chart, they can identify **unnatural patterns** indicating process changes.

In the **natural pattern,** plotted points fluctuate at random and obey the laws of chance. Most of the points should be near the centerline and evenly distributed about the centerline. In addition, a few of the points should approach the control limits but none should exceed the control limits.

When the points fluctuate too widely and/or fail to balance around the centerline, an unnatural pattern is indicated. If one or more of the characteristics of a natural pattern is absent, then the pattern is unnatural.

When we observe an unnatural pattern, we then investigate to determine an assignable cause of variation. **Assignable causes** can be traced to some disturbance and tend to result in control chart behavior which is

- Unstable
- Nonhomogeneous
- Erratic
- Shifted
- Unpredictable
- Inconsistent

Conversely, the causes of natural variability cannot be assigned and are thus referred to as nonassignable causes.

11.4.5 Process Capability Studies

A major use of control charts is to determine the capability of the process (the natural variability due to **common, or nonassignable, causes**). A process capability study is a systematic investigation of a process by means of statistical control charts in order to:

- Discover whether it is behaving naturally or unnaturally
- Determine assignable causes
- Eliminate assignable causes[17]

If the process capability is inadequate to meet specification requirements, then management action (such as investment in better equipment) is required to improve the

[17] *AT&T Statistical Quality Control Handbook,* 11th ed., p. 34.

process capability. It is important to recognize that a process, and particularly the workers in the manufacturing process, cannot be expected to achieve product specifications that exceed the process capability.

11.4.6 Off-Line Quality Control

The use of statistical control charts is often referred to as on-line quality control, since data is collected in real-time during the production process. This terminology is particularly appropriate as computer software and network communications facilitate data logging of process variables and control chart plotting in automated systems. Conversely, statistical studies done in pilot plants, in laboratories, or in the shop during nonproductive time are called off-line studies.

Statistically designed experiments are an increasingly important tool for off-line quality control. Statisticians have been designing efficient experiments for many years. These techniques have found wide application in agriculture and in health sciences, but they have not been adequately utilized by engineers. Recently, interest in experimental design has been stimulated among engineers by the work of Genichi Taguchi [13]. These studies employ design of experiments (DOE), a topic usually taught in a secondary or advanced course on probability and statistics and therefore beyond the scope of this text. Rather than presenting the mathematical background for DOE, we will here present an example case study illustrating the potential power of the method. The Taguchi method also uses a cost-of-quality concept called the Loss Function. Taguchi's studies have shown that the total cost to society of imperfect quality often approximates a quadratic function of the deviation from target value of the quality parameter.

Phadke and colleagues at AT&T Bell Laboratories published a case study on the application of Taguchi methods to achieve:

- Process design optimization for routing of printed wiring boards
- Optimal design of differential operation amplifier circuits used in telecommunications[18]

The focus of this text is manufacturing processes, so our discussion is limited to the case study on PWB routing, as described in Chapter 5.

Phadke discusses the Taguchi method for optimizing process design, referred to as robust design. Robust design involves identifying and controlling the sources of process variation. The sources of variation are

- External (factors outside the process, such as environment)
- Imperfections (essentially random variation due to nonassignable causes)
- Deterioration (degradation in performance characteristics with wear and age)

[18]M. S. Phadke, "Design Optimization Case Studies," *AT&T Technical Journal,* 65 (March/April), pp. 51–67.

Phadke illustrated the design optimization problem, as shown in Fig. 11-21. The terminology in Fig. 11-21 is defined as follows:

- *Signal Factors:* Set by the user/operator of a system to attain an intended output with a desired level of performance.
- *Control Factors:* Product or process factors controllable by the design engineer.
- *Scaling/Leveling Factors:* A special type of control factors that are easily adjusted to achieve a desired functional relationship between signal factors and desired output.
- *Noise Factors:* Uncontrolled factors due to technical or economic limitations.

Stock panels of 18 inches × 24 inches (46 × 61 cm) must be cut down to designed sizes of PWBs, such as 8 × 4 inches (20.3 × 10.2 cm). The desired size boards are either stamped out (blanked) or cut out using a high-speed router bit. Routing is preferred because of superior dimensional control, smooth edges, and process advantages. However, routing is considerably slower than stamping, resulting in lower productivity. When router bits begin to dull, excessive dust is produced, caking to the board edges and increasing surface roughness. Caking creates the need to either perform an additional cleaning operation or change bits more frequently; both of these options decrease productivity.

A routing machine has four synchronized spindles, with one router bit per spindle. The spindles are synchronized in three degrees of freedom:

- Rotational speed
- Horizontal feed in the *X-Y* plane
- Vertical feed (in-feed) in the *Z* axis

The stock panels are stacked from two to four high to increase production rate. The process sequence is to lower the bit into the panel rotating at 30,000 to 40,000 rpm, cutting the board out in the *X-Y* plane, and withdrawing the bit to a point in the *Z* axis above the surface of the panel.

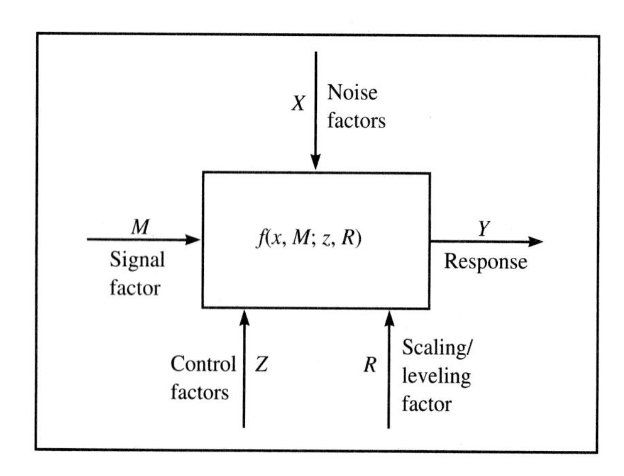

Figure 11-21 Experimental design optimization problem.

(*Source:* Copyright ©1986 AT&T. All rights reserved. Reprinted with permission.)

The study objective was to increase the life of the bits. The researchers studied the process to determine the control factors potentially affecting the lifetime to excessive bit wear. Thus this study deals with off-line quality control for the routing process and deals with reliability for the bits. Table 11-4 summarizes the control factors and levels for the routing process.

One type of classic approach is the complete factorial experiment, wherein multiple trials must be performed for every possible combination of factors and levels. Table 11-4 shows that for this case study there were two factors with four levels each and seven factors with two levels each, so the theoretical number of trials was $4^2 \times 2^7 = 2048$. To perform such an experiment in a production environment would be infeasible in terms of time and expense.

An alternative is to perform a fractional factorial experiment, which is designed to provide information about all of the factors but is not informative about all of the possible interactions among factors. Carefully selected factor combinations called orthogonal arrays permit this approach. One such experimental design is shown in Table 11-5, for this case study.

The fractional factorial experiment required only 32 experiments rather than 2,048. Since there were four spindles on the router machine, the experiment could be completed in eight runs. The researchers used one bit per experiment and inspected bits after every 100 inches of routing. The failure criteria included two failure modes: breakage and excessive dust accumulation. The experiment was terminated after 1700 inches of cut.

The observed life of each bit is shown in the far right-hand column of Table 11-5. Figure 11-22 contains plots of mean life versus various control factors and clearly illustrates which control factors have an effect on the bit life and which levels yield superior life.

TABLE 11-4 CONTROL FACTORS AND LEVELS FOR ROUTING PROCESS

	Level			
Factor	1	2	3	4
A. Suction (in. of Hg)	1	2*		
B. *X*-*Y* feed (in./min)	60*	80		
C. In-feed (in./min)	10*	50		
D. Type of bit	1	2	3	4*
E. Spindle position†	1	2	3	4
F. Suction foot	SR	BB*		
G. Stacking height (in.)	3/16	1/4*		
H. Depth of slot (mils)	60*	100		
I. Speed (rpm)	30,000	40,000*		

† Spindle position is not a control factor. In the interest of productivity, all four spindle positions must be used.

* Denotes starting condition for the factors.

Source:

TABLE 11-5 EXPERIMENTAL DESIGN FOR THE ROUTING PROCESS

Experiment No.	Suction A	X-Y feed B	In-feed C	Bit D	Spindle E	Suction foot F	Stack height G	Depth H	Speed I	Observed life*
1	1	1	1	1	1	1	1	1	1	3.5
2	1	1	1	2	2	2	2	1	1	0.5
3	1	1	1	3	4	1	2	2	1	0.5
4	1	1	1	4	3	2	1	2	1	17.5
5	1	2	2	3	1	2	2	1	1	0.5
6	1	2	2	4	2	1	1	1	1	2.5
7	1	2	2	1	4	2	1	2	1	0.5
8	1	2	2	2	3	1	2	2	1	0.5
9	2	1	2	4	1	1	2	2	1	17.5
10	2	1	2	3	2	2	1	2	1	2.5
11	2	1	2	2	4	1	1	1	1	0.5
12	2	1	2	1	3	2	2	1	1	3.5
13	2	2	1	2	1	2	1	2	1	0.5
14	2	2	1	1	2	1	2	2	1	2.5
15	2	2	1	4	4	2	2	1	1	0.5
16	2	2	1	3	3	1	1	1	1	3.5
17	1	1	1	1	1	1	1	1	2	17.5
18	1	1	1	2	2	2	2	1	2	0.5
19	1	1	1	3	4	1	2	2	2	0.5
20	1	1	1	4	3	2	1	2	2	17.5
21	1	2	2	3	1	2	2	1	2	0.5
22	1	2	2	4	2	1	1	1	2	17.5
23	1	2	2	1	4	2	1	2	2	14.5
24	1	2	2	2	3	1	2	2	2	0.5
25	2	1	2	4	1	1	2	2	2	17.5
26	2	1	2	3	2	2	1	2	2	3.5
27	2	1	2	2	4	1	1	1	2	17.5
28	2	1	2	1	3	2	2	1	2	3.5
29	2	2	1	2	1	2	1	2	2	0.5
30	2	2	1	1	2	1	2	2	2	3.5
31	2	2	1	4	4	2	2	1	2	0.5
32	2	2	1	3	3	1	1	1	2	17.5

*Life was measured in hundreds of inches of movement in *X-Y* plane. Tests were terminated at 1700 inches.

Source:

The following conclusions can be made from analysis of Fig. 11-22:

- 1-inch suction is as good as 2-inch suction.
- Slower *X-Y* feed gives longer life.
- The effect of in-feed (*Z* axis) is small.

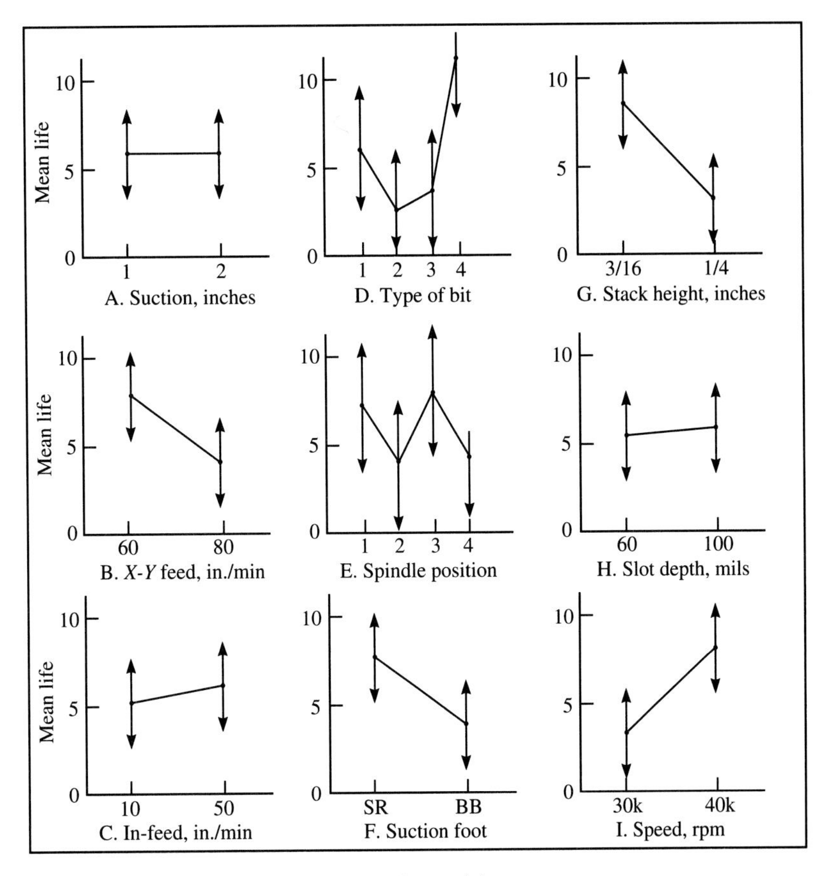

Figure 11-22 Mean life as a function of control factors.

- Bit type 4 has the longest life.
- Spindle position has little effect.
- A solid ring suction foot is superior to a bristle brush.
- Life improves with reduction in stacking height.
- Depth of slot in backup material has little effect.
- Life increases with increase in turning speed.

Figure 11-23 provides insight to the second order (2-factor) interactions. In each graph, one control factor is measured along the horizontal axis. Mean life is plotted on the vertical axis versus the factor at each level of a second control factor. When the lines are parallel, there is no evidence of interaction. The only two factors showing strong interaction are turning speed (rpm) and stacking height.

Figure 11-23 also shows the original settings and optimal settings for the control

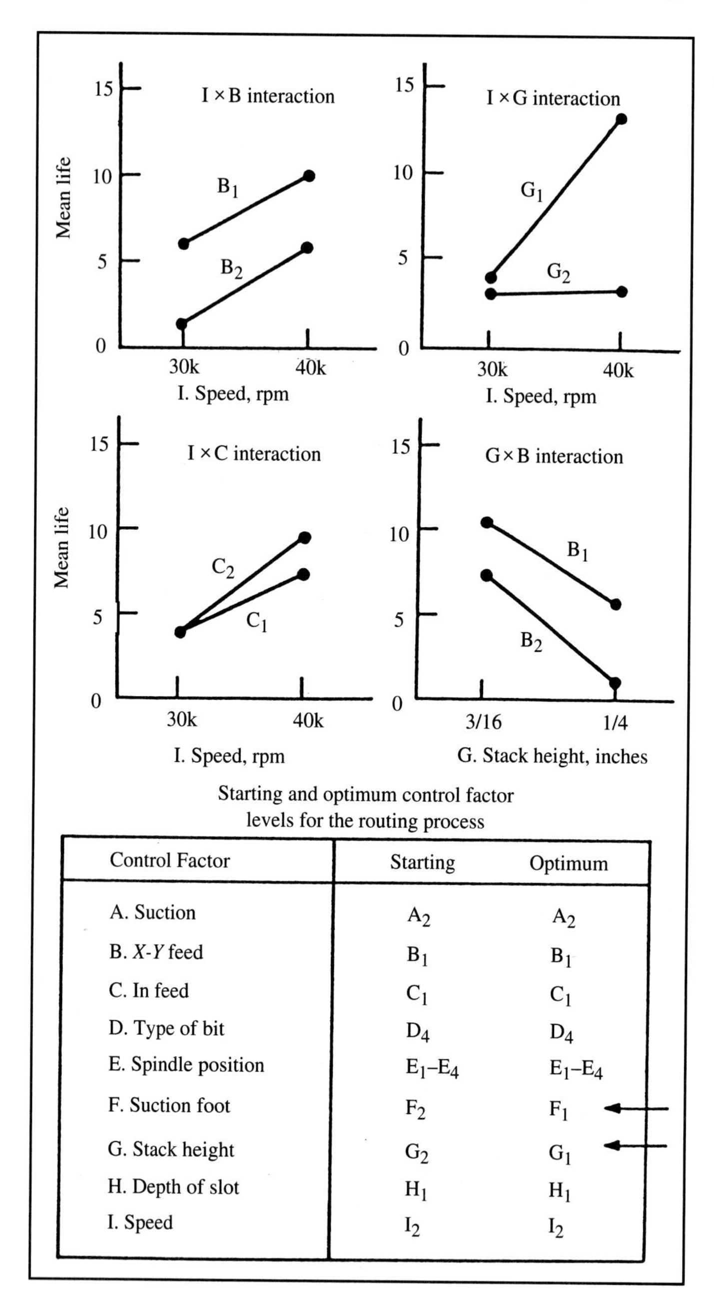

Starting and optimum control factor levels for the routing process

Control Factor	Starting	Optimum
A. Suction	A_2	A_2
B. *X-Y* feed	B_1	B_1
C. In feed	C_1	C_1
D. Type of bit	D_4	D_4
E. Spindle position	E_1–E_4	E_1–E_4
F. Suction foot	F_2	F_1 ←
G. Stack height	G_2	G_1 ←
H. Depth of slot	H_1	H_1
I. Speed	I_2	I_2

Figure 11-23 Two-factor interactions and optimal factor settings.

(*Source:*)

factors, based on the experimental results. The suction foot was changed from the bristle-brush type to the solid-ring type, and the stacking height was reduced from four panels to three, that is, from 1/4 inch (0.63 cm) to 3/16 inch (0.48 cm).

The reduction in stacking height has a negative effect on productivity. As seen in Fig. 11-23, bit life is modestly dependent on *X-Y* feed rate. The researchers decided to increase the feed rate to compensate for the lost productivity due to decreased stacking height, with some penalty in reduced bit life. Here *X-Y* feed rate serves as a scaling/leveling factor. To be an ideal leveling factor, mean life should be insensitive to *X-Y* feed rate (a horizontal straight line in Fig. 11-22).

11.5 RELIABILITY

In the Deming cycle (Fig. 11-3) we can think of the planning function as including design, which involves modeling and predictions. The checking phase involves collection and statistical analysis of data. The functions of modeling, prediction, and statistical data analysis require that we develop stochastic models to quantify reliability. There are two basic cases to be considered: nonrepairable items and repairable items. The following sections discuss analysis techniques for nonrepairable and repairable items, respectively.

11.5.1 Nonrepairable Items

In order to model reliability of a nonrepairable item (or part), we select an appropriate distribution of the time-to-failure random variable, τ. The following functions of a random variable are useful in reliability analysis:

$f(t)$ is the **probability density function** (PDF) and gives the probability of failure in the interval $(t, t+dt)$

$F(t)$ is the **cumulative distribution function** (CDF), and gives the $P\{$failing before some time $t\}$:

$$F(t) = P(\tau < t) = \int_0^t f(\tau)\, d\tau, \qquad t \geq 0 \tag{11-27}$$

$R(t)$ is the **complementary cumulative distribution function** (CCDF), or **reliability function** (R), and gives P{surviving beyond the time t}:

$$R(t) = 1 - F(t), \qquad t \geq 0 \tag{11-28}$$

$h(t)$ is the **hazard function** (HF) and gives the instantaneous rate of failure in the interval $(t, t+dt)$, given (conditioned upon) survival to time t, and

$$h(t) = \frac{f(t)}{R(t)}, \qquad t \geq 0 \tag{11-29}$$

The expected (mean) value of the time-to-failure random variable τ is

$$E(\tau) = \int_0^\infty \tau f(\tau)\, d\tau = \int_0^\infty R(\tau)\, d\tau \tag{11-30}$$

To empirically estimate these functions, we order the observed failure times from smallest to largest. Let i be the number of the ordered event. Also, let

N be the sample size

t_i be the time to failure for the ith event

$\hat{}$ denote an estimate

The empirical estimates of the functions are

$$\hat{F}(t_i) = \frac{i}{N} \tag{11-31}$$

$$\hat{R}(t_i) = 1 - \frac{i}{N} = \frac{N - i}{N} \tag{11-32}$$

$$\hat{f}(t_i) = \frac{1}{(t_{i+1} - t_i)N} \tag{11-33}$$

$$\hat{h}(t_i) = \frac{\hat{f}(t)}{\hat{R}(t)} = \frac{1}{(t_{i+1} - t_i)(N - i)} \tag{11-34}$$

To understand the pattern of failures for nonrepairable items, it is helpful to consider the analogy between failures of nonrepairable parts and deaths in the human population. Figure 11-24 is a plot of the force of mortality (hazard function) for a male population. This type of data is summarized in life tables, which actuaries use to establish life insurance plans. In Fig. 11-24 we can see three distinct patterns, or phases, of mortality:

- Infant (a)
- Childhood through younger adulthood (b)
- Older adult (c)

Failure statistics for nonrepairable items are very similar to the patterns for human mortality. If we plot the hazard function for a large sample of identical parts, a bathtub curve is formed, with three phases corresponding to the following three failure patterns:

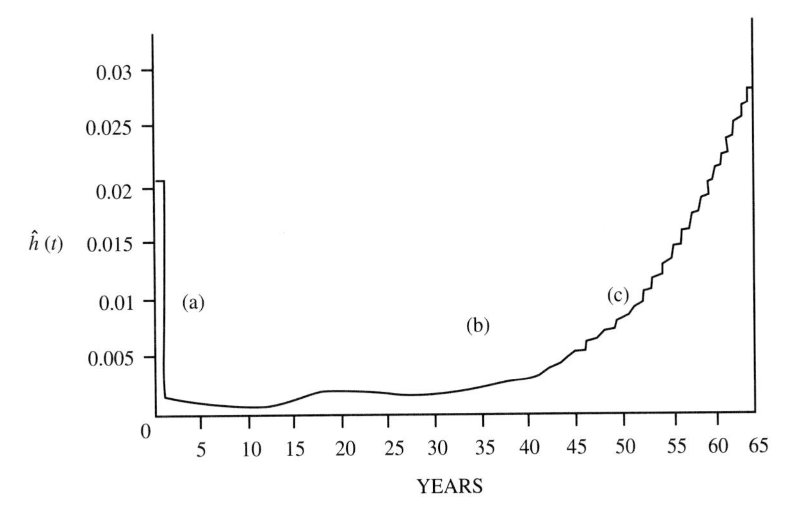

Figure 11-24 Bathtub curve.

- Burn-in (a)
- Random failure (b)
- Wearout (c)

The early phase (a) corresponding to infant mortality serves to eliminate from the population those units with defects in materials and workmanship. Electronics manufacturers frequently perform burn-in tests or environmental stress screens (Chap. 13) on newly manufactured units to purge the defectives. The phase corresponding to childhood through young adulthood (b) can be called the random-failure phase, wherein units that meet specifications fail as random stresses exceed the design strength. The wearout phase (c) for equipment corresponds to aging in the human population. The cumulative effects of wear and material degradation, due to mechanisms such as corrosion, contamination, or fatigue, result in failures as the strength of the units degrades below the design stress level. The increasing hazard function means that the longer a unit survives, the greater its probability of failure in the next interval of time.

The random-failure phase of the bathtub curve is recognized as representative of the HF for many electrical and electronic items. The constant HF arises when the time-to-failure random variable has the exponential probability distribution. The times-to-failure are the independent and identically distributed exponential with PDF

$$f(t) = \frac{1}{\theta} e^{-t/\theta} \tag{11-35}$$

where θ is a scale parameter.

The CDF and CCDF are as follows:

$$F(t) = \int_0^t f(\tau)\, d\tau \tag{11-36}$$

$$= \int_0^t \frac{1}{\theta} e^{-\tau/\theta} = 1 - e^{-t/\theta} \tag{11-37}$$

$$R(t) = 1 - F(t) = e^{-t/\theta} \tag{11-38}$$

The expected (mean) value is

$$E(t) = \int_0^\infty t f(t)\, dt$$

$$= \int_0^\infty R(t)\, dt = \int_0^\infty e^{-t/\theta} dt$$

$$= -\theta\, e^{-t/\theta} \Big|_0^\infty = -\theta(0 - 1) = \theta \tag{11-39}$$

and is called the mean time to failure, or MTTF. The HF is sometimes represented by the symbol λ, where

$$\lambda = \frac{1}{\theta} \tag{11-40}$$

Figure 11-25 contains graphs of the functions for the exponential distribution.

The estimate of θ is

$$\hat{\theta} = \frac{T}{n} \tag{11-41}$$

where

T is the total operating time
n is the number of failures

Example:

A sample of 256K RAMs was placed on life test. At the termination of the test there were 122,280 accumulated operating hours and 5 failures. The estimate of MTTF is

$$\hat{\theta} = \frac{T}{n} = \frac{122{,}280}{5} = 24{,}456 \text{ hours}$$

What is the probability that a 256K RAM chip will fail in the first 5000 hours?

$$\hat{F}(t) = 1 - e^{-t/\hat{\theta}} = 1 - e^{-5000/24{,}456}$$

$$= 0.18$$

What is the probability that a chip will survive beyond its MTTF?

$$\hat{R}(t) = e^{-t/\hat{\theta}} = e^{-24{,}456/24{,}456}$$

$$= \frac{1}{e} = 0.368$$

There is only a 37 percent chance of surviving beyond the MTTF.

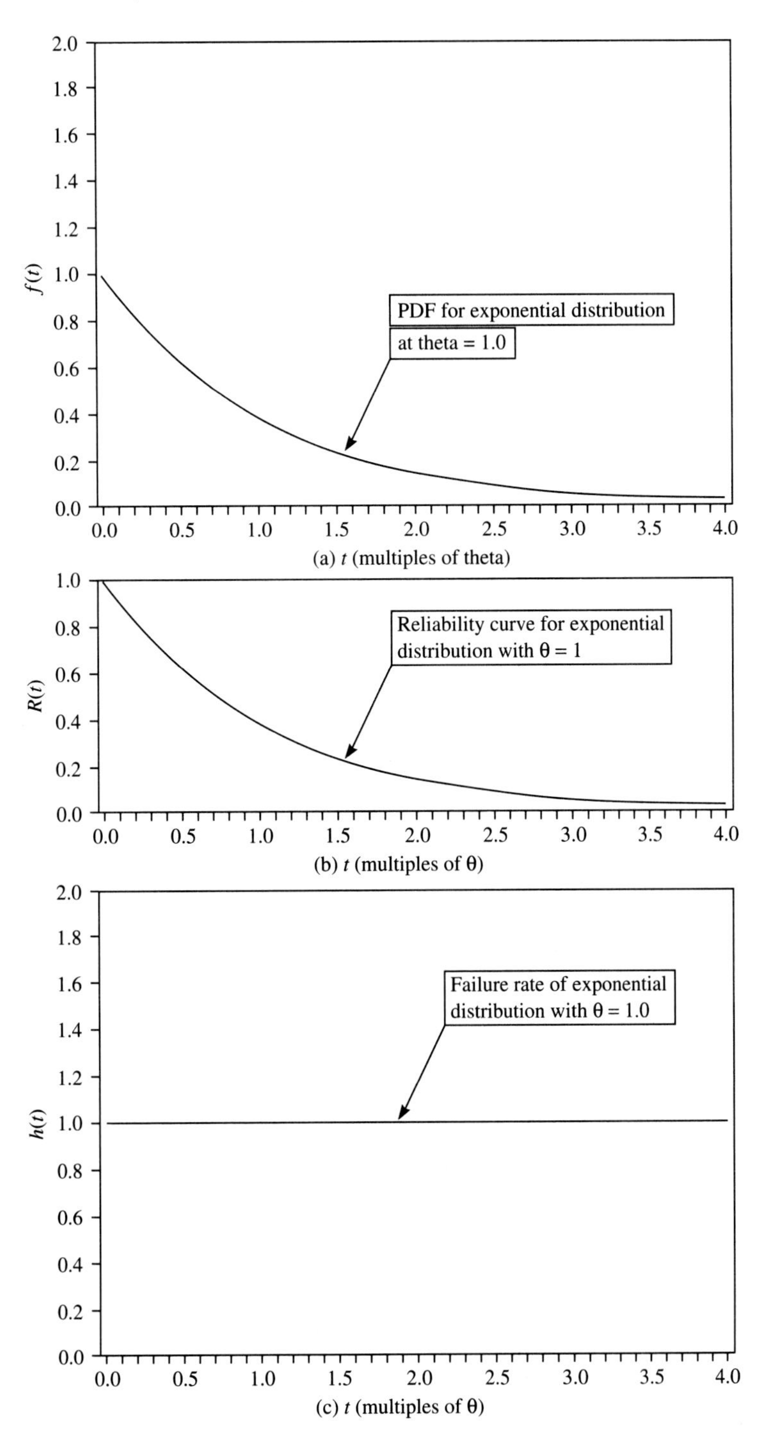

Figure 11-25 Functions of the exponential distribution.

This surprising result ($P\{t > \tau\} = 0.37$) exists because the exponential distribution is not a symmetrical distribution.

The exponential distribution is a special case of the two-parameter **Weibull distribution.** The Weibull distribution is named after a Swedish physicist and mechanical engineer who studied the distribution of time to failure for mechanical components such as ball bearings, which tend to wear out and therefore have an increasing HF. The Weibull is also very relevant to electrical and electronic reliability. The PDF of the Weibull distribution is as follows:

$$f(t) = \frac{m}{\theta}\left(\frac{t}{\theta}\right)^{m-1} \exp\left[-\left(\frac{t}{\theta}\right)^{m}\right] \tag{11-42}$$

where

θ = scale parameter
m = shape parameter

The CDF, CCDF, and HF are as follows:

$$F(t) = 1 - \exp\left[-\left(\frac{t}{\theta}\right)^{m}\right] \tag{11-43}$$

$$R(t) = \exp\left[-\left(\frac{t}{\theta}\right)^{m}\right] \tag{11-44}$$

$$h(t) = \frac{m}{\theta}\left(\frac{t}{\theta}\right)^{m-1} \tag{11-45}$$

The Weibull distribution is very flexible. When $m = 1$, the special case of the exponential distribution arises. When the shape parameter $m > 1$, the HF is increasing and corresponds to segment (c) in Fig. 11-24. Conversely, when $m < 1$, a decreasing HF results, corresponding to the wear-in (burn-in) segment (a) in Fig. 11-24. Figure 11-26 contains graphs of the functions for the Weibull distribution with $\theta = 1$.

11.5.2 Repairable Items

A repairable item (system) is one which, after failure to perform at least one of its required functions, can be restored to performing all of its required functions by any method other than replacement of the entire system.[19]

For repairable systems, we are interested in

- Times between successive failures
- Number of failures in an interval of time
- **Rate of occurrence of failures** (ROCOF) as a function of time

[19]Harold Ascher and Harry Feingold, *Repairable Systems Reliability* (New York: Marcel Dekker, 1984), p. 1.

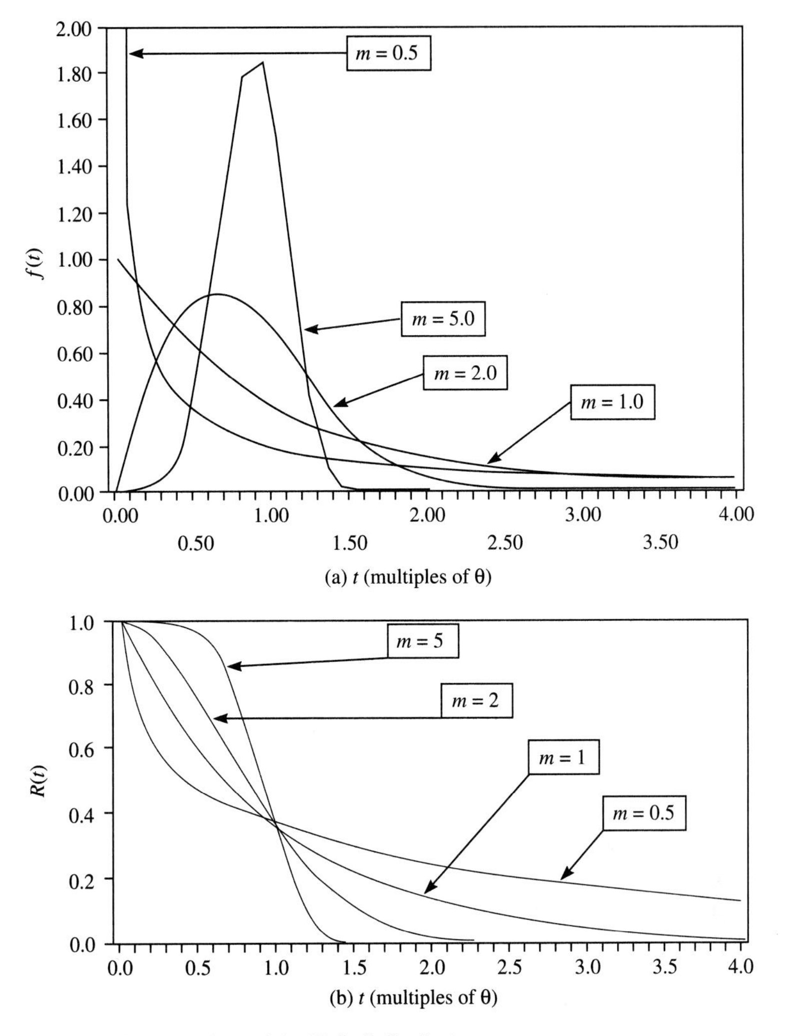

Figure 11-26 Functions of the Weibull distribution.

Just as there was a bathtub curve for the hazard function of the time to failure (nonrepairable items), there is also a bathtub curve for the rate of occurrence of failures for repairable systems. The repairable-item bathtub curve is similar to Fig. 11-24, except that the vertical axis measures ROCOF, rather than hazard function, and the three phases have the following interpretations:

- *Decreasing ROCOF:* Early in the life of the new system (a) there are design and production bugs which must be found and fixed through usage.
- *Constant ROCOF:* After the problems have been corrected, the system undergoes an extended period of useful life (b) where failures occur randomly. Failures usually occur because stresses exceed strength.

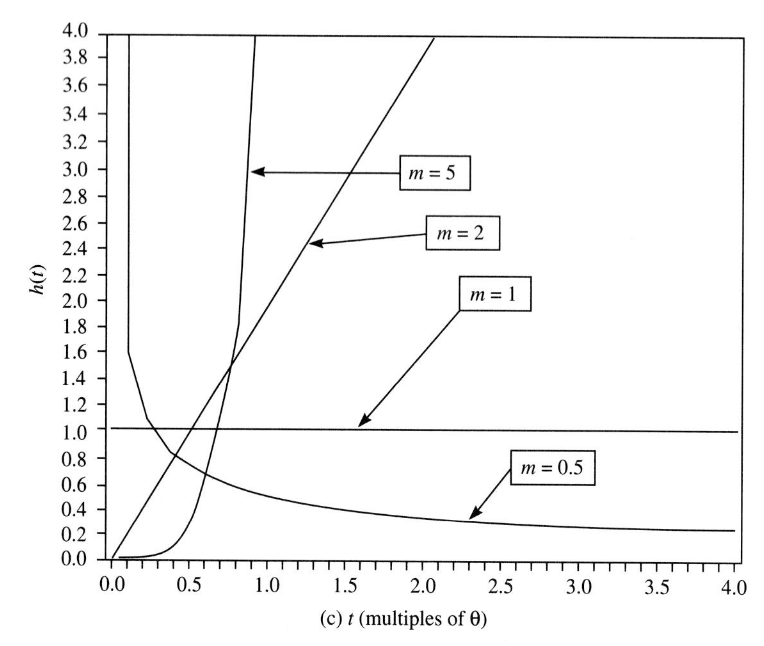

Figure 11-26 (continued)

- *Increasing ROCOF:* Later in the system life (c), the ROCOF begins to increase rapidly as accumulated stresses, deterioration, and wearout lead to component failures.

Figure 11-27 illustrates the failure patterns in the three phases. In the first phase (a) failure becomes less frequent. During phase (b), failures occur randomly at a roughly constant rate. In phase (c) the rate of occurrence of failures increases until disposal.

The pattern of constant ROCOF is called a homogeneous Poisson process (HPP), because the number of failure events in any time interval is a discrete random variable, distributed Poisson. For an HPP, the time between failures is a continuous random variable distributed exponentially. It is important to emphasize that the HPP is a special case of failure pattern for a repairable item and should not be assumed to apply universally. When the ROCOF is time-dependent, as in phases (a) and (c) of Fig. 11-27, the HPP does not apply; instead, a time-varying model such as the nonhomogeneous Poisson process (NHPP) may apply.

11.5.3 Reliability Prediction for Useful Life

It is often assumed that repairable electronic systems fail by the homogeneous Poisson process. This assumption is partially based on theoretical grounds under the reasoning that burn-in or screening (manufacturing test programs) have removed early-life failures due to quality problems and moved the population into the useful-life phase with random failure pattern (b) of Fig 11-27. However, the assumption of an HPP is sometimes used

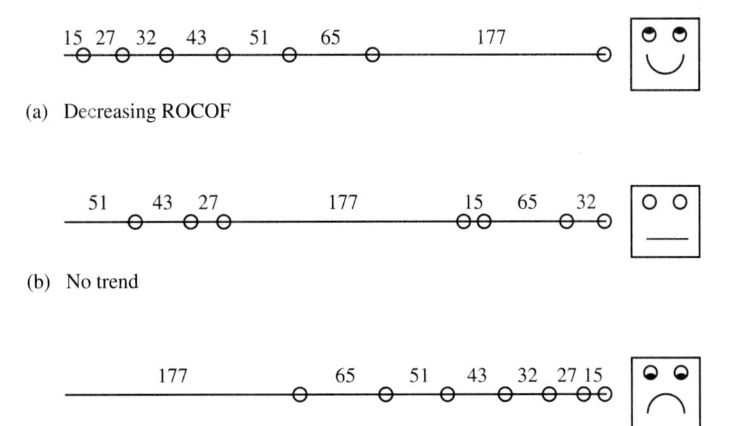

Figure 11-27 Patterns of failure for repairable items (systems).

(*Source:* Harold Ascher and Harry Feingold, *Repairable Systems Reliability* (New York: Marcel Dekker, Inc., 1984), p. 18.)

simply because field failure data is not reported in a way permitting detection of trends such as exhibited in phases (a) and (c) of Fig. 11-27.

Recall that when the stochastic failure pattern in an interval is homogeneous Poisson, the times between failure intervals are IID random variables that are **exponentially distributed.** It can be shown that if a system consists of n components having exponentially distributed times to failure, x_i and constant hazard functions λ_i $(i = 1, \ldots, n)$, then the system has an exponential distribution of time to failure with

$$\boldsymbol{\lambda}_n = \sum_{i=1}^{n} \boldsymbol{\lambda}_i \tag{11-46}$$

This relationship is used in industry-standard methodologies for reliability prediction, such as MIL-HDBK-217. The purpose of MIL-HDBK-217 is to

> provide a common base for reliability prediction during acquisition programs . . . establish a common basis for comparing and evaluating reliability predictions of related or competing designs.[20]

MIL-HDBK-217 provides two methods: parts count prediction and parts stress analysis.

The **parts count prediction method** is applicable during bid proposals and early design phase and uses the following approximation:

$$\boldsymbol{\lambda}_{EQP} = \sum_{i=1}^{n} N_i(\boldsymbol{\lambda}_G \boldsymbol{\pi}_L)_i \tag{11-47}$$

[20]MIL-HDBK-217E, *Reliability Prediction for Electronic Equipment* (Griffiths Air Force Base, New York: Rome Air Development Center, 1986), p. 1-1.

where

λ_{EQP} = Total equipment failure rate
λ_G = Generic failure rate for ith generic part
π_Q = Quality factor
N_i = Quantity of ith generic part
n = Number of generic part types
π_L = Learning factor (optional)
= 10, under the following conditions:
- New device in initial production
- Major design change
- Extended interruption in production
- New and unproven technologies

= 1 otherwise

The π-factors in the MIL-HDBK-217 methodology are coefficients that account for differences in failure rates among strata of the population. For example, π_Q accounts for different quality levels as shown in Table 11-6.

The **parts stress analysis method** is applicable near the end of the design phase and requires a detailed parts list and knowledge of the circuit applications and stresses.

Example:

Silicon NPN general-purpose transistor (Jan Grade) in linear service at 0.4 of its rated maximum power of 1 watt in fixed ground installation (G_F) at 30°C ambient, rated for 1 watt at 25°C with T_{max} = 125°C, and operated at 60% of maximum voltage.

Model:

$$\lambda_p = \lambda_b(\pi_E \times \pi_A \times \pi_Q \times \pi_R \times \pi_{S_2} \times \pi_C) \tag{11-48}$$

TABLE 11-6 QUALITY FACTORS FOR DISCRETE SEMICONDUCTORS

Category	π_Q
Jantxv	0.1
Jantx	0.2
Jan	1.0
Nonhermetic	5.0
Plastic	10.0

Source: MIL-HDBK-217E, *Reliability Prediction for Electronic Equipment* (Griffiths Air Force Base, N.Y.: Rome Air Development Center, 1986), p. 5.2-30.

where

λ_p = Base failure rate (Table 11-7)
= 0.0012×10^{-6}
π_E = Environmental factor (Table 11-8)
= 5.8
π_A = Application factor (circuit function, Table 11-9)
= 1.5
π_Q = Quality factor (Table 11-10)
= 1.2
π_R = Rating factor (maximum power or current, Table 11-11)
= 1.0
π_{S_2} = Voltage stress factor (adjusts model for second electrical stress, application voltage, Table 11-12)
= 0.88

$$VOLTAGE\ STRESS,\ S_2 = \frac{Applied\ (V_{CE})}{Rated(V_{CEO})} \times 100$$

π_C = Complexity factor (multiple devices in single package, Table 11-13)
= 1.0

Multiplying the factors in Eq. (11-49), we obtain the following part failure:

$$\lambda_p = (0.0012 \times 10^{-6})(5.8 \times 1.5 \times 1.2 \times 1.0 \times 0.88 \times 1.0)$$
$$= 0.011 \text{ failures}/10^6 \text{ hours}$$

TABLE 11-7 BASE FAILURE RATES (IN FAILURES PER MILLION HOURS) FOR MIL-S-19500 TRANSISTORS, GROUP I, SILICON, NPN

Temp (°C)	.1	.2	.3	.4	.5	.6	.7	.8	.9	1.0
0	.00049	.00060	.00071	.00084	.00099	.0012	.0014	.0017	.0021	.0027
10	.00056	.00067	.00079	.00093	.0011	.0013	.0016	.0019	.0025	.0034
20	.00063	.00075	.00089	.0010	.0012	.0015	.0018	.0023	.0030	.0043
30	.00071	.00084	.00099	.0012	.0014	.0017	.0021	.0027	.0038	
40	.00049	.00093	.0011	.0013	.0016	.0019	.0025	.0034	.0049	
50	.00089	.0010	.0012	.0015	.0018	.0023	.0030	.0043		
60	.00099	.0012	.0014	.0017	.0021	.0027	.0038			
70	.0011	.0013	.0016	.0019	.0025	.0034	.0049			
80	.0012	.0015	.0018	.0023	.0030	.0043				
90	.0014	.0017	.0021	.0027	.0038					
100	.0016	.0019	.0025	.0034	.0049					
110	.0018	.0023	.0030	.0043						
120	.0021	.0027	.0038							
130	.0025	.0034	.0049							
140	.0030	.0043								
150	.0038									

Source: MIL-HDBK-217E, *Reliability Prediction for Electronic Equipment* (Griffiths Air Force Base, N.Y.: Rome Air Development Center, 1986), p. 5.1.3.1-4.

TABLE 11-8 ENVIRONMENTAL FACTORS

Environment	π_E	Environment	π_E
G_B	1	A_{IB}	35
G_{MS}	1.6	A_{IA}	20
G_F	5.8	A_{IF}	40
G_M	18	A_{UC}	15
M_P	12	A_{UT}	25
N_{SB}	9.8	A_{UB}	60
N_S	9.8	A_{UA}	35
N_U	21	A_{UF}	65
N_H	19	S_F	0.4
N_{UU}	20	M_{FF}	12
A_{RW}	27	M_{FA}	17
A_{IC}	9.5	U_{SL}	36
A_{rr}	15	M_L	41
		C_L	690

Source: MIL-HDBK-217E, *Reliability Prediction for Electronic Equipment* (Griffiths Air Force Base, N.Y.: Rome Air Development Center, 1986), p. 5.1.3.1-1.

TABLE 11-9 APPLICATION FACTORS

Application	π_A
Linear	1.5
Switch	0.7
Si, low noise r.f., <1W	15.0

Source: MIL-HDBK-217E, *Reliability Prediction for Electronic Equipment* (Griffiths Air Force Base, N.Y.: Rome Air Development Center, 1986), p. 5.1.3.1-2.

TABLE 11-10 QUALITY FACTORS

Quality Level	π_Q
JANTXV	0.12
JANTX	0.24
JAN	1.2
Lower*	6.0
Plastic**	12.0

*Hermetic packaged devices.

**Devices sealed or encapsulated with organic materials.

Source: MIL-HDBK-217E, *Reliability Prediction for Electronic Equipment* (Griffiths Air Force Base, N.Y.: Rome Air Development Center, 1986), p. 5.1.3.1-2.

TABLE 11-11 POWER RATING FACTORS

Power Rating (watts)	π_R
≤ 1	1.0
> 1 to 5	1.5
> 5 to 20	2.0
> 20 to 50	2.5
> 50 to 200	5.0

Source: MIL-HDBK-217E, *Reliability Prediction for Electronic Equipment* (Griffiths Air Force Base, N.Y.: Rome Air Development Center, 1986), p. 5.1.3.1-2.

TABLE 11-12
VOLTAGE STRESS FACTORS

S_2 (percent)	π_{S_2}
100	3.0
90	2.2
80	1.62
70	1.2
60	0.88
50	0.65
40	0.48
30	0.35
20	0.30
10	0.30
0	0.30

*$\pi_{S_2} = 0.14(10)(.0133)S_2$ for $S_2 \geq 25$

*$\pi_{S_2} = 0.3$ for $S_2 < 25$

Source: MIL-HDBK-217E, *Reliability Prediction for Electronic Equipment* (Griffiths Air Force Base, N.Y.: Rome Air Development Center, 1986), p. 5.1.3.1-3.

TABLE 11-13
COMPLEXITY FACTORS

Complexity	π_C
Single Transistor	1.0
Dual (Unmatched)	1.5
Dual Matched	1.2
Darlington	0.8
Dual Emitter	1.1
Multiple Emitter	1.2
Complementary Pair	0.7

Source: MIL-HDBK-217E, *Reliability Prediction for Electronic Equipment* (Griffiths Air Force Base, N.Y.: Rome Air Development Center, 1986), p. 5.1.3.1-3.

11.6 SUMMARY

Customer demands and competitive pressures continue to motivate continuous improvement in product quality. This chapter has discussed the management concepts and technical tools available to the electronics manufacturer seeking excellence in product quality. Increasingly, the quality focus is on customer needs, and management techniques such as total quality control and quality function deployment facilitate determination and attainment of customer expectations. The tools of quality improvement include descriptive statistics, cause-and-effect analyses, control charts, and design of experiments.

Reliability is quality demonstrated over time. The design function is decisive in determining potential quality and reliability, while manufacturing and logistics support functions maintain the intended quality level. The concurrent engineering approach provides the cooperation early in the product life cycle to ensure a design that meets customer requirements and is producible. Modeling and statistical analysis techniques aid in the prediction and evaluation of product reliability.

KEY TERMS

Analytical specification
Assignable cause
Attributes control chart
Cause-and-effect diagram
Check sheet
Commercial standards

Common cause
Companywide quality control (CWQC)
Complementary cumulative distribution function (CCDF)
Conformance
Control chart
Cumulative distribution function (CDF)
Data stratification
Derating
Engineering drawing
Exponentially distributed
Failure modes, effects, and criticality analysis (FECA)
Fault tree analysis (FTA)
Hazard function (HF)
Histogram
Independent identically distributed (IID)
Life-cycle cost (LCC)
Life-cycle engineering (LCE)
Material specification
Natural pattern
Nonassignable causes
Normally distributed
Pareto chart (principle)
Parts count prediction method
Parts stress analysis method
Performance specification
Probability density function (PDF)
Process specification
Product specification
Quality
Quality control
Quality function deployment (QFD)
Random variable
Rate of occurrence of failures (ROCOF)
Reliability
Reliability function (R)
Robust design
Safety factor
Scatter plot
Sneak circuit analysis (SCA)
Statistical quality control (SQC)
Statistical process control (SPC)
Test-analyze-and-fix (TAAF)
Total quality management (TQM)
Unnatural pattern
Variables control chart
Weibull distribution
Workmanship standard

EXERCISES

1. Define quality.
2. When are most of the life-cycle costs determined and incurred during an equipment's life cycle?
3. What is QFD, and why is it important?
4. What are the two most common failure sources in electronics, and where are the most common locations of failure?
5. A life test was administered to a sample of 1 kohm resistors. At the end of the test, the resistors had accumulated 240,000 hours of operation and had 8 failures. What is the probability of a resistor failing during the first 8,000 hours of operation?
6. What are the two reliability prediction methods for electronic systems, and what are their differences?
7. Among the seven tools for statistical quality control is control charts. What are the two kinds of control charts, and how are they different?

8. Define specifications in the context of vendor quality, and discuss the different types of specifications. What type of specification would you write to permit the supplier maximum design flexibility?

9. A manufacturer estimates that there is 95 percent probability that its diodes will fail between 61,320 and 78,840 hours of operation. If the PDF is

$$f(t) = \frac{1}{\sqrt{2\pi}\sigma} \exp\left[\frac{(t - \mu)^2}{2\sigma^2}\right]$$

find the MTTF and σ.

10. The data shown in the following table are resistance values for a 1 ohm resistor. Samples of five are collected each hour off an assembly line.

a. Construct a histogram of individual values. Does it resemble any distribution?

b. Construct a histogram of sample means. Does it resemble any distribution?

Data	X_1	X_2	X_3	X_4	X_5	$\overline{X}$
8.00	1.000	1.025	1.050	0.975	0.975	1.005
9:00	0.975	1.025	1.000	1.000	0.975	0.995
10:00	0.975	0.975	1.000	1.075	1.000	1.005
11:00	0.975	0.975	1.025	0.975	0.975	0.985
12:00	1.000	0.975	1.025	0.975	1.000	0.995
1:00	0.975	1.000	1.050	1.000	1.025	1.010
2:00	1.050	1.000	0.975	0.975	1.050	1.010
3:00	1.000	1.050	1.025	0.975	1.025	1.015
4:00	1.000	1.025	1.000	1.025	0.975	1.005

11. General Aviation manufactures the F-123 Tactical Stealth Fighter. The electronic division of General Aviation keeps records of the part number and the source of failure for each failure event. Conduct a Pareto analysis on this data, and accompany it with appropriate graphs. What are your recommendations to management?

Part no.	Source of failure
A413	Temperature
B378	Vibration
D676	Humidity
E432	Dust
A413	Temperature
D676	Humidity
G722	Temperature
F227	Temperature
A413	Temperature
B378	Vibration
A413	Temperature
D676	Humidity
G722	Temperature
B378	Temperature
B378	Vibration

12. Two different designs for a digital circuit are being studied in order to compare the amount of noise. Which design has the least noise?

Sample	Circuit 1	Circuit 2
1	27	26
2	23	23
3	23	25
4	24	23
5	28	28
6	23	22
7	22	24
8	29	29
9	25	24
10	24	25

$$d_i = X_1 - Y_2$$

$$\bar{d} = \frac{\Sigma d_i}{n}$$

13. A resistor has a nominal value of 1 kohm. A sample of size five is taken from a production process. The ten sample groups are listed in the following tables. Each sample is taken at successive intervals of 30 minutes.
   a. Construct X-bar and R charts, and determine whether the process is in control.
   b. If the specifications are 1 kohm ± 10 ohm, what can be said of the process capability?

Sample	$X1$	$X2$	$X3$	$X4$	$X5$
1	999	995	1015	1005	998
2	1010	997	1013	1000	1003
3	991	995	999	998	994
4	1005	1005	997	1003	999
5	1000	1005	995	1002	996
6	997	1004	997	1000	1001
7	996	996	1003	1000	999
8	1002	999	1000	1000	1002
9	1005	1003	1000	1005	998
10	993	996	995	1000	994

14. The following are samples of size three taken from a production process. The measurements are the thickness of a printed circuit board in inches. Construct X-bar and R charts, and determine whether the process is in control.

Sample	$X1$	$X2$	$X3$
1	0.0628	0.0627	0.0622
2	0.0629	0.0631	0.0636
3	0.0620	0.0632	0.0635
4	0.0635	0.0637	0.0621
5	0.0630	0.0633	0.0631
6	0.0623	0.0629	0.0630
7	0.0631	0.0632	0.0631
8	0.0620	0.0629	0.0635

9	0.0624	0.0630	0.0630
10	0.0613	0.0629	0.0625
11	0.0635	0.0630	0.0634
12	0.0627	0.0630	0.0629
13	0.0631	0.0627	0.0632
14	0.0628	0.0633	0.0622
15	0.0636	0.0633	0.0634

15. The following data lists the numbers defective from samples of size 200 for a PWB production process. Construct a *p*-chart, and determine whether the process is in control.

Day	Nonconforming units	Day	Nonconforming units
1	7	11	5
2	8	12	4
3	3	13	7
4	4	14	3
5	1	15	5
6	1	16	0
7	5	17	1
8	2	18	3
9	5	19	2
10	4	20	0

16. The data set in the following table is from an electronic assembly line. It lists the number defective from samples of size 150. Construct an appropriate chart, and determine whether the process is in control.

Day	Nonconforming units
1	6
2	2
3	7
4	5
5	5
6	0
7	1
8	8
9	6
10	3
11	5
12	6
13	9
14	2
15	0

17. The data set in the following table gives the number of defective diodes from parts supplies to be used on a PWB. Sample size varies. Construct the appropriate control chart. Does the process appear out of control?

Day	Sample size	Nonconforming units
1	164	5
2	147	6
3	136	3
4	155	7
5	163	4
6	148	6
7	128	2
8	172	9
9	145	0
10	138	1

18. An integrated circuit has a mean temperature capacity of 43°C with a variance of 9°C. The integrated circuit will operate in an environment with mean temperature of 37°C with temperature variance of 27°C. (a) What is the reliability of this integrated circuit for this environment? (b) What would the mean capacity of the integrated circuit need to be to obtain a reliability equal to 0.9987, assuming variance does not change?

19. The PDF of the lifetime of a capacitor is given by:

$$f(t) = \lambda e^{-\lambda t}$$

Use $\lambda(t) = 0.005$/hr to answer the following questions.
a. What is the probability of failure in the first 10 hours of operation?
b. What is the probability of failure in the time interval of 150 to 160 hours of operation, given that it has successfully operated until $t = 150$ hours?
c. What property was illustrated by parts a and b?

20. The PDF of the lifetime of a capacitor is given by: $F(t) = kt^2$, where $0 \leq t \leq 15$ and t is in years.
a. Determine the value of k such that $f(t)$ is a valid PDF.
b. What is the expected value of the MTTF?
c. What is $R(t)$?
d. What is the hazard function?
e. What is the probability of failure in the interval $3 \leq t \leq 5$, given survival until time $t = 3$?

21. A component has the PDF given in exercise 19. What is the relability of the component at $t = 1/\lambda$?

22. A circuit board has a failure rate characterized by a Weibull distribution with $\theta = 400$ years and $m = 1/2$. The circuit board is required to have a design life reliability of 0.85.
a. What is the design life if there is no burn-in period?
b. What is the design life with a burn-in period of one month?
Hint:

$$R(t/t_0) = \frac{R(t + T_0)}{R(t_0)}$$

where T_0 is the burn-in period in years.

23. The ZPD 2000 computer's CPU is made up of the components listed below with respective failure rate for each component. Use the parts count method to answer the following questions.
a. What is the predicted failure rate of the ZPD 2000's CPU?
b. What is the MTTF?

c. What is the reliability for the first 50 hours of operation of the CPU?

Part	Quantity	Failure rate/ 10^6 hours
Memory board	3	0.0018
I/O board	1	0.0035
Arithmetic board	1	0.0022
Switch	1	0.0004
ROM chips	3	0.0007

24. Consider a silicon NPN general-purpose transistor (JANTX grade) in linear service at 0.2 of its maximum rated power of 2 watts in a ground mobile installation (*Gm*) at 40°C ambient, rated for 2 watts at 35°C with T_{max} = 135°C, and operated at 80% of maximum voltage. Predict system failure rate using parts stress analysis.

25. Obtain a copy of MIL-HDBK-217. Determine, by the parts stress analysis method, the reliability prediction for an 8086, 16-bit microprocessor in a 64-pin ceramic DIP with a glass seal. The device is installed on a multipurpose, all-terrain vehicle and is operating at a case temperature of 80°C. The device has been screened to fully comply with the requirements of paragraph 1.2.1 of MIL-STD-883. It dissipates 0.5 watts and has a case-to-junction thermal resistance of 50°C per watt. The microprocessor has been fabricated using CMOS technology and operates on a maximum supply voltage rating of 5 volts.

BIBLIOGRAPHY

1. ASHER, HAROLD, AND HARRY FIENGOLD, *Repairable Systems Reliability.* New York: Marcel Dekker, Inc., 1984.
2. CROSBY, PHILLIP B., *Quality Is Free.* New York: McGraw- Hill Book Company, 1979.
3. DEMING, W. EDWARDS, *Out of the Crises* (2nd ed.). Cambridge, Mass.: Massachusetts Institute of Technology, 1986.
4. EDOSOMWAN, JOHNSON A., AND ARVIND BALLAKUR, *Productivity and Quality Improvement in Electronics Assembly.* New York: McGraw-Hill Book Company, 1989.
5. FEIGENBAUM, A. V., *Total Quality Control* (3rd ed.). New York: McGraw-Hill Book Company, 1983.
6. FABRYCKY, WOLTER J., AND BENJAMIN S. BLANCHARD, *Life Cycle Cost and Economic Analysis.* Englewood Cliffs, N.J.: Prentice Hall, Inc., 1991.
7. IRESON, W. GRANT, AND W. G. COMBS, eds., *Handbook of Reliability and Management.* New York: McGraw-Hill Book Company, 1988.
8. ISHIKAWA, KAORU, *What is Total Quality Control?* trans., David J. Lu, Englewood Cliffs, N.J.: Prentice Hall, Inc., 1985.
9. JURAN, JOSEPH M., ed. *Quality Control Handbook.* New York: McGraw-Hill Book Company, 1988.
10. LEWIS, E. E., *Introduction to Reliability Engineering,* New York: John Wiley & Sons, 1987.

11. Pyzdek, Thomas, *What Every Engineer Should Know About Quality Control.* New York, N.Y.: Marcel Dekker, Inc., 1989.

12. *Statistical Quality Control Handbook* (AT&T Technologies). Charlotte, N.C.: Delmar Printing Company, 1985.

13. Taguchi, Genichi, Elsayed A. Elsayed, and Thomas C. Hsiang, *Quality Engineering in Production Systems.* New York: McGraw-Hill Book Company, 1989.

12 Testability

12.1 INTRODUCTION

This chapter overviews the testability of electronic products and components. Topics addressed include testing objectives (Sec. 12.2), the cost of testing (Sec. 12.3), testing techniques (Sec. 12.4), and design for testability (Sec. 12.5). It is the goal of this chapter to overview terminology and fundamentals of each of these topics. More detailed information on individual topics is available in the literature sources listed in the bibliography at the end of this chapter.

12.2 TESTING OBJECTIVES

Electronic testing has three objectives:

- Detection of improper product operation
- Isolation of defects such that rework costs are minimized
- Identification of out-of-control manufacturing processes

Detection of improper product operation consists of finding the nondesign failures in electronic products. This is equivalent to locating manufacturing-induced defects or failures and/or defective components. The detection process is accomplished by applying stimuli to the component or product and observing/measuring subsequent electronic responses.

Defect or fault isolation requires first the identification of a test failure and, second, tracing that failure to some probable cause. The test failure is usually observed through an improper response to stimuli applied to the unit being tested. The rework process consists of replacing the failed component and then resubmitting the repaired unit for retest.

Process fault isolation traces test failures to manufacturing processes that are out of control, as opposed to failed components in the **unit under test** (UUT). An example would be a number of open circuit connections on PWBs that have been improperly soldered.

The goals and objectives of testing are somewhat dependent on the engineer's viewpoint. For example, a designer might feel that the objective of testing is to verify that the design works. In contrast, a test engineer might feel that the primary objective to be realized through testing is to verify that all products manufactured will function, regardless of the design configuration that has been implemented. Through the concurrent engineering approach, all members of the team are concerned for the ultimate quality and reliability of the product.

Failure mode analysis is often performed on failed components in order to determine the nature of these quality problems. A commonly used technique to identify long-term reliability difficulties is to eliminate infant mortality component failures, through environmental stress screening (Chap. 13) or burn-in. In these procedures, electronic assemblies are subjected to extremes of temperature, vibration, humidity, and/or input power variation.

Electronic testing is likely to become an increasingly important component of the manufacturing process. As the complexity of electronic products continues to increase, so will the cost to test these products. According to Torino,[1] testing cost in 1970 approximated 10 percent of product costs. This cost had risen to between 30 percent and 40 percent in 1980 and is likely to continue increasing.

12.2.1 Test Terminology

This section overviews a variety of terms typically associated with electronic testing. The unit, assembly, or part being tested is usually called the unit under test (UUT). The corresponding test equipment generally provides **input stimulation.** Stimuli are typically electronic inputs used to energize the UUT. Test equipment is also used to measure UUT responses after input stimuli are applied. Test equipment capable of performing a sequence of electronic tests without human intervention is called **automatic test equipment** (ATE). Figure 12-1 illustrates commercially available test fixturing used to do in-circuit testing on PWBs.

Test equipment measures a variety of UUT responses to stimuli. The first of these is continuity, which infers the presence of an electrical connection between two or more points. The presence of continuity usually indicates zero resistance between these points.

A second type of measurement detects shorts. Short-circuit tests are the opposite of continuity tests; they ensure that an electrical connection does not exist between two or more points in a circuit. Successful completion of shorts tests infers that a significant

[1]Jon Torino, *Testability Practices Today: Seminar/Workshop,* (Campbell, Calif.: Engineering Solutions, Inc., 1982), p. 42.

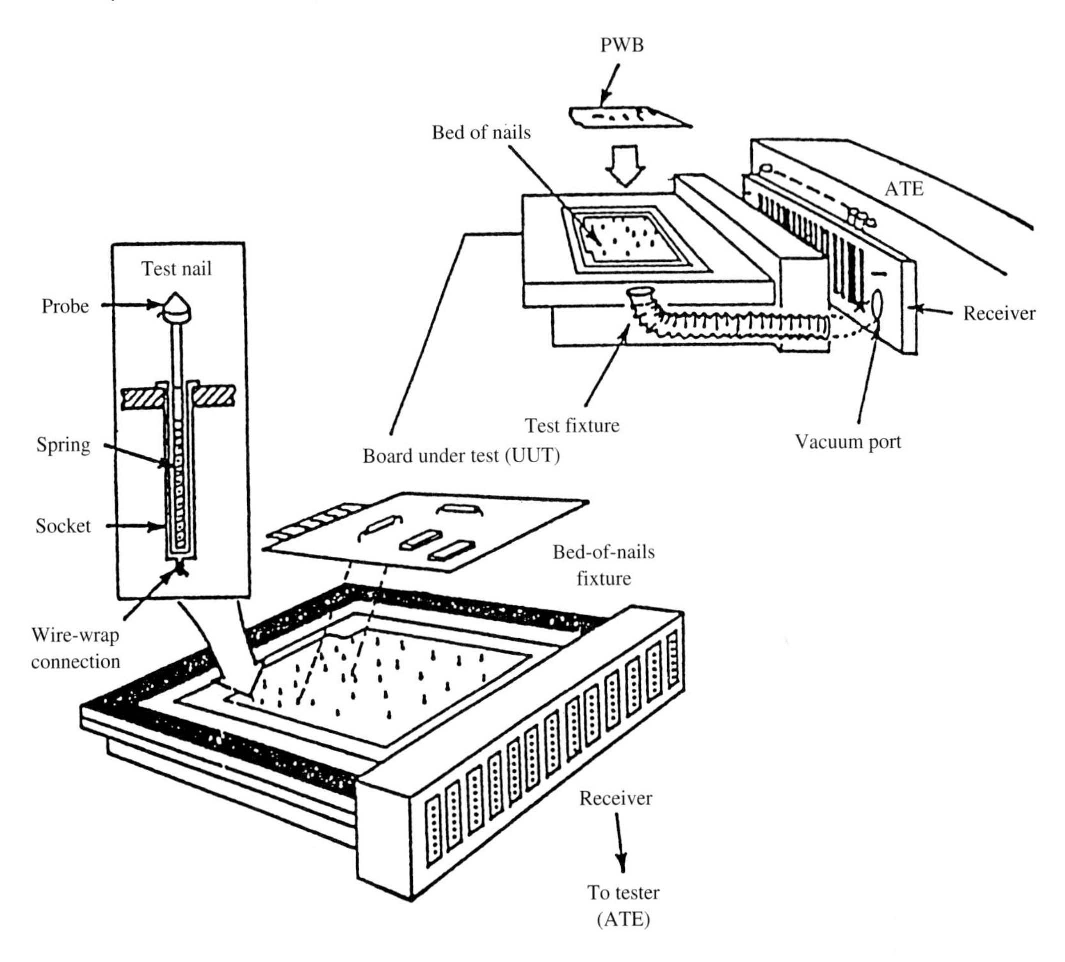

Figure 12-1 In-circuit testing of a printed wiring board.

(*Source:* Frank J. Langley, "Testing in Assembly," in *Printed Circuits Handbook,* Clyde Coombs, ed. (New York: McGraw-Hill Book Company, 1988), p. 21.20. Adapted with the written permission of the publishers.)

amount of electrical resistance is present between the points being measured. Shorts testing is generally performed using in-circuit test devices.

High potential (HIPOT) tests are a separate, but related type of testing that is often completed to ensure Underwriters Laboratory (UL) compliance to AC and high-power inputs/outputs. HIPOT tests use significantly higher voltages than shorts tests. As such, HIPOT tests are usually not performed on assembled PWBs, as the high voltage used can damage many of the components on the boards.

The next category of tests are those that measure resistance and/or impedance. These tests generally ensure that the observed measurement falls within a limited tolerance interval specified in ohms.

Voltage and current are two related test parameters that are checked to be within a specified minimum/maximum range for a given input. Specific values of capacitance and inductance are derived from timed measurements resulting from the application of specified values of either voltage and/or current.

There are three main steps to perform when measuring test parameters. The first step uses input stimulators or **driver circuits.**[2] These devices apply combinations of high-impedance, low-voltage, and high-voltage inputs to the UUT. An enable input allows the data input to drive the output to either a low or high voltage (logic 0 or logic 1). The enable input also can be used to force the output to a high-impedance state that is neither logic 0 nor 1.

The second step is called **backdriving.** The output of the driver circuit is physically connected to a UUT circuit so that the driver signal is forced into the UUT. By using the driver circuit to control the input voltage to the UUT circuit, the test engineer can control the UUT operations. This control is necessary for signal comparison.[3]

The third step is called **signal comparison.** This process entails the use of a receiver circuit to compare a signal obtained from the UUT to a previously defined reference value.[4] This procedure gives the test engineer both control of the UUT and a means of observing its actions, facilitating the testing of circuit operation against expected actions and conditions.

12.2.2 Testing Levels

A variety of product testing levels exist, as described in the following subsections.

12.2.3 Component-Level Testing

Component testing is the lowest testing level. Procedures for component-level testing are straightforward. If the UUT is an integrated circuit, probes may be used in conjunction with input drivers to apply different logic inputs to the UUT. Corresponding probes in conjunction with receivers will measure expected logic outputs from the UUT. All logic combinations must be checked. Logic outputs that do not conform to a predetermined standard may indicate that the unit has failed the test. Where failures occur at the component level, repair is not usually possible.[5]

Component testing of VLSI circuits can require significant investment in automatic test equipment. Equipment cost can be in the millions of dollars. The failure rates of these types of components are low, typically fewer than 100 defects per million parts. Because of the low defect rates and high test costs, manufacturers of VLSI devices may not perform component-level test for applications other than military and space.

[2]Kenneth Parker, *Integrating Design and Test, Using CAE Tools for ATE Programming,* IEEE Publication No. 788 (Piscataway, N.J.: Computer Society Press, Institute of Electrical and Electronic Engineers, 1987), p. 60.

[3]Parker, "Integrating Design and Test," pp. 60–65.

[4]Parker, "Integrating Design and Test," pp. 60–65.

[5] F. Kear, "Testing of Printed Circuit Assemblies," *Printed Circuit Assembly Manufacturing* (New York: Marcel Dekker, Inc., 1987), np.

Component-level testing of less complicated components involves the use of a multimeter or special-purpose commercial equipment. The procedure is often manual and ensures that the component parameters (inductance, resistance, impedance, and so on) fall within predetermined tolerance ranges.

Advantages of component-level testing include low test cost and the fact that fault isolation is also straightforward: the UUT either passes the test or it does not. A disadvantage of component testing is the difficulty of accurately simulating the system interaction in which the component will ultimately operate. Component-level testing fails to consider defects resulting from component insertion and soldering, since these operations have not yet occurred.

12.2.4 Subassembly-Level Testing

The next level of testing addresses the subassembly or printed wiring board level. Subassembly PWB testing is usually of two types: in-circuit and functional. In-circuit testing usually requires commercially available (generally automatic) test equipment. The test equipment is used in conjunction with a **bed-of-nails** test fixture (Fig. 12-1). Functional testing may use commercial test equipment, test equipment that is custom designed, or a **hot mockup** where the PWB is simply connected to a known good system.

PWB functional testing is accomplished by supplying all inputs that the PWB is likely to experience in its operating environment. Corresponding outputs are measured. Measurements are taken at the input/output (I/O) terminations of the PWB and/or at various test points located on the surface of the PWB itself [7, 9, 18].

Advantages of subassembly testing include the identification of failures resulting from component insertion and soldering. It is also possible to identify some failures resulting from faulty system interaction. Disadvantages include increased difficulty of fault isolation. Fault isolation and repair incur both troubleshooting and rework costs for each PWB on which one or more test failures are noted. Subassembly testing does not adequately test all system interactions of the PWB with its operating environment [21, 22].

12.2.5 Systems-Level Testing

The highest level of testing occurs at the systems level. Systems-level testing is equivalent to the evaluation of a group of PWBs and the related interface wiring. At the systems level, the UUT is an entire electronic operating assembly, and the assembly is tested as a whole.

A commonly used procedure for systems-level testing is to start with a system that is initially known to be good. New PWBs are inserted into the system, one at a time. As each PWB is inserted, the system is tested, thereby evaluating the PWB in its electronic operating environment.

The primary disadvantage of systems-level testing is the difficulty of fault isolation. Another major disadvantage is cost. Test times and troubleshooting and repair costs are

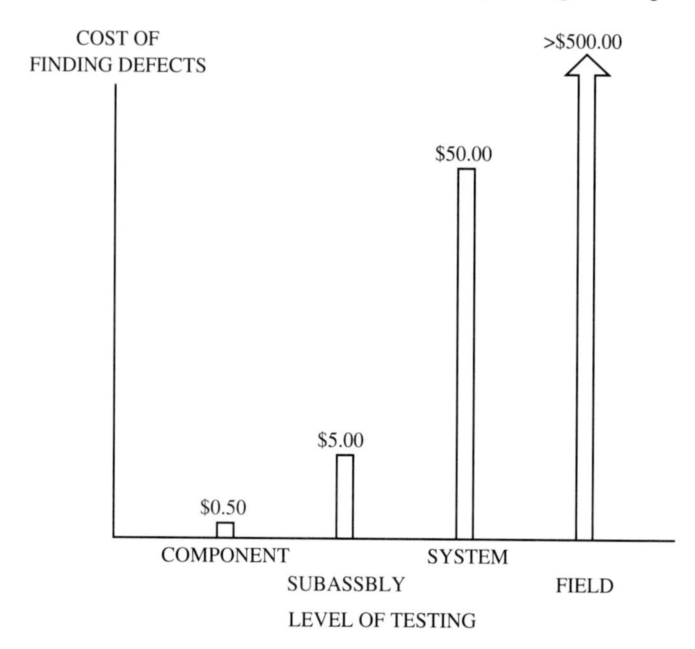

Figure 12-2 Defect location costs.

(*Source:* Adapted from George Hroundas, "Bare Board Testing," in *Printed Circuits Handbook,* Clyde Coombs, ed. (New York: McGraw-Hill Book Company, 1988), p. 17.3.)

highest at the systems level. The primary advantage of systems-level testing is that all operating interactions are evaluated during the testing process.[6]

The costs of finding defects increase dramatically as the test level increases. Representative costs per defect found for component, subassembly, systems, and field tests are illustrated in Fig. 12-2. As might be expected, test complexity increases dramatically with test level.

12.2.6 Impact of New Technology

The use of new technology has had a dramatic effect on miniaturization of electronic products. As this miniaturization has occurred, new difficulties associated with testing these smaller products have arisen. An example is the evolution of multilayer PWBs. With older single-layer boards, testing interface procedures were straightforward. All component leads were accessible on either one board side or the other.

Multilayer boards may have a number of circuit paths embedded between layers. Without special design features, many circuit paths are inaccessible to test equipment. This problem has been addressed through the inclusion of test points on outer surfaces,

[6]Louis Ungar, *The Complete ATE Book: Course Notes* (Reseda, Calif.: Test Engineering Solutions, Inc., 1982), p. 4.3.

which make these inner circuit paths between layers accessible to test equipment [6, 7, 9, 17]. This procedure is illustrated in Fig. 12-3.

Component miniaturization has also increased the difficulty of test equipment interfacing. Smaller contact and lead sizes have required correspondingly smaller test probes. Component miniaturization has reached such dramatic levels that test probes small enough to access contact points are becoming prohibitively expensive [20, 22].

12.2.7 Test Vectors

Test vectors refer to sequences of logic combinations applied to the inputs of circuits to be tested. Consider the three-chip circuit shown in Fig. 12-4. Suppose only two input pins for this circuit exist, as shown in this illustration. Further suppose that the circuit has only one output shown on the right-hand side of the circuit diagram.

If only two input pins A and B exist, the four logic combinations of inputs that may be applied to this circuit include 00, 01, 10, and 11. These four logic combinations comprise four separate test vectors for this circuit.

The test procedure entails the preconditioning of all three chips to a known electrical state. This is usually accomplished by applying a known sequence of test vectors to the circuit's input pins. The next step is to test the circuit by applying vectors of interest. Circuit outputs corresponding to each test vector are measured, evaluated, and compared with known parameters [6, 18].

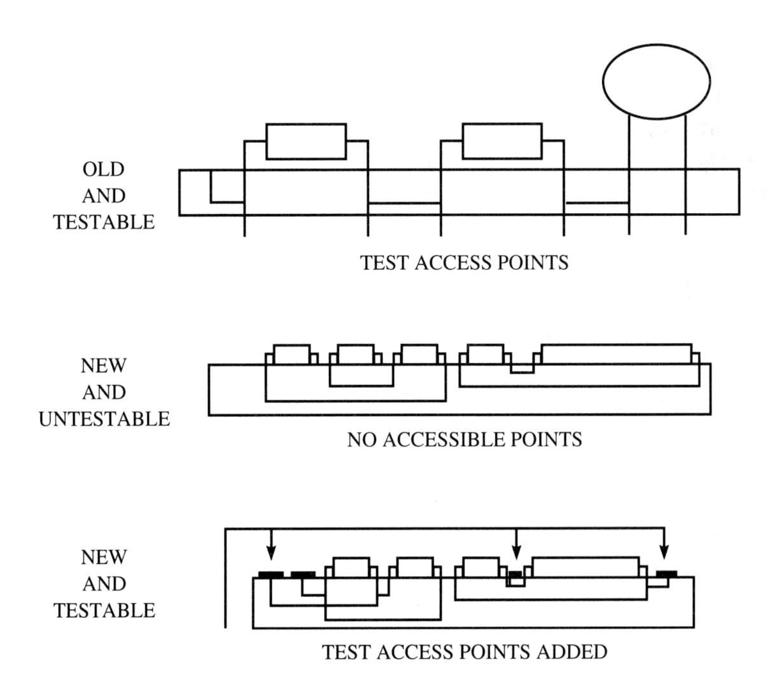

Figure 12-3 Interface requirements for single-layer and multilayer PWBs.

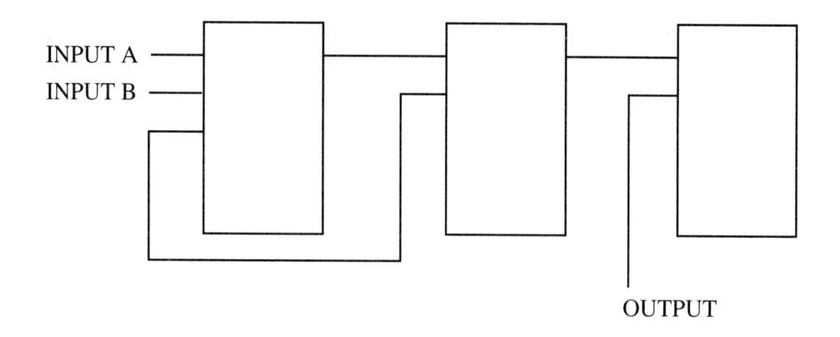

A	B	OUTPUT
0	0	?
0	1	?
1	0	?
1	1	?

Figure 12-4 Use of test vectors on a three-chip circuit.

12.2.8 Automatic Testing of PWBs

Most defects leading to test failures occur at the subassembly or the PWB level. These defects usually result from handling, improper component insertion, or soldering defects. Therefore, manufacturers tend to emphasize testing at the PWB level. Tests conducted at the PWB level are moderate in terms of cost but are likely to yield detection of a maximum number of defects when compared with testing conducted at the component or systems level.[7]

Test sequences for both in-circuit and functional testing of PWBs are too complex and too lengthy to be performed manually. Both types of testing require the use of automatic test equipment. The use of ATE requires a moderate to large production quantity in order to be cost justified.[8] Automatic testing is generally more reliable than are conventional manual test procedures, since it eliminates the chance of operator errors. ATE has the additional advantage of increased repetitiveness with which the tests can be performed. Automatic testing may also handle more complex, more numerous, and more timing-critical tests than may be performed manually [20, 22].

The basic elements of an automated test system are illustrated in Fig. 12-5. The ATE itself usually consists of a computer with connections to vector and test pattern data bases and to the driver/receiver circuits. The second primary element is an interface between the ATE and the UUT, and it may be a test bed, an umbilical cable, or a variety of interface connectors. The third element of the automatic test system is the UUT itself [15, 17].

[7]George Hroundas, "Bare Board Testing," in *Printed Circuits Handbook* (New York: McGraw-Hill Book Company, 1988), p. 17.3.

[8]Ungar, *The Complete ATE Book,* p. 61.

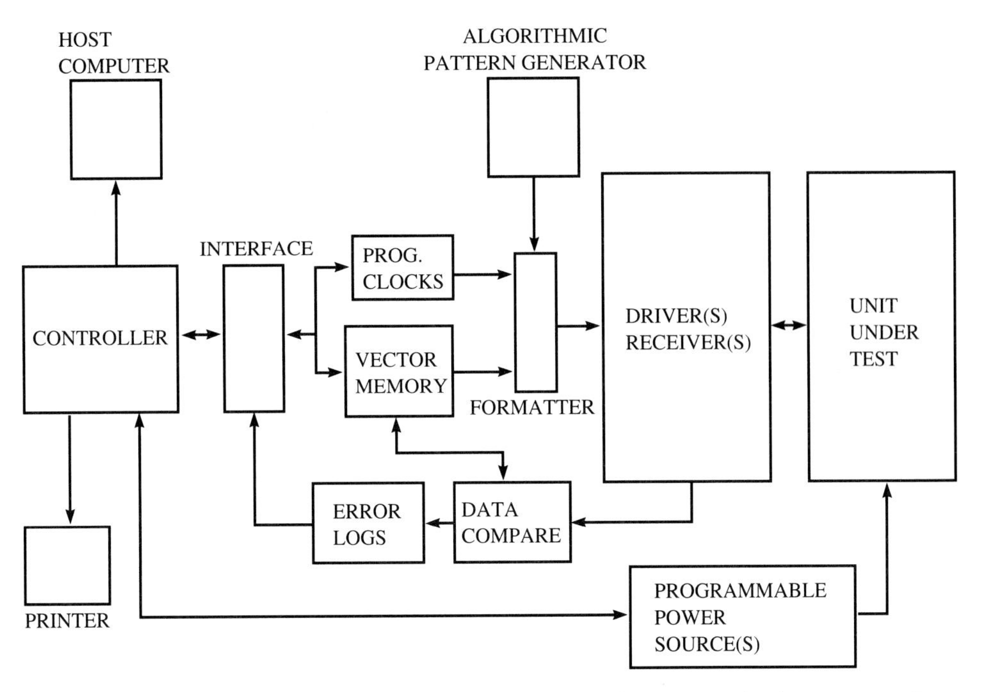

Figure 12-5 Elements of an automatic test system. (Adapted from company brochure, courtesy of Micro Control Company, Minneapolis, Minn.)

12.2.9 Test Programming

Automatic test systems require programming of test routines. The test software controls the inputting of signals to the UUT, circuit backdriving, high impedance measurements, measurement and calculation of UUT responses, and comparison of outputs with specification tolerance ranges.[9]

Test languages that are general in scope include BASIC, Pascal, and C. There are several test-specific languages including ATLAS and CAPS. Test-specific languages contain specialized functions to facilitate the programming of specific electronic measurements [6, 20].

12.3 COST OF TESTING

A variety of costs are associated with the testing process. Many of these costs are nonrecurring and include those to determine which tests should be performed and the related equipment required to perform these tests. Additional nonrecurring costs include those for equipment procurement and installation, the training of test personnel, and the generation of appropriate documentation. The programming of ATE is a significant

[9]Ungar, *The Complete ATE Book,* p. 29.

nonrecurring cost associated with testing. Engineering design changes to the UUT may require modifications of previously developed test programs. In some cases test procedure documentation may also have to be revised accordingly.

Wages of test technicians and salaries of test engineers, test programmers, and failure analysts are all recurring costs. Test equipment maintenance, depreciation, and periodic retraining of engineering and technical personnel all represent additional ongoing costs associated with the testing process.

Recurring testing costs are directly influenced by both the length and number of test cycles that are performed. As these parameters increase, so do the corresponding labor costs and the volume of test equipment that is required.

12.3.1 Unit Testing Costs

Any testing process has an associated test yield. **Test yield** is the probability of a UUT passing an electronic test on any trial. The cost of obtaining a good UUT may then be expressed as follows[10]:

$$C_1 = \frac{\textit{Cost of Manufacturing}}{\textit{Test Yield}} \qquad (12\text{-}1)$$

The cost to find a specific electronic failure and repair it is expressed as follows:

$$C_2 = \frac{\textit{Cost of Test}}{\textit{Failure Rate}} \qquad (12\text{-}2)$$

The cost to find and repair an electronic failure is thus

$$CFR = \frac{\textit{Cost of Test}}{\textit{Failure Rate}} + \textit{Fault Isolation Cost} + \textit{Repair Cost} \qquad (12\text{-}3)$$

12.3.2 Investment Considerations Related to Testing

In many cases, front-end investment in the area of improved design may yield substantial cost savings in testing that may be realized later in the manufacturing process. If savings associated with automating the testing process can be accurately estimated, the procurement of ATE can be justified through payback analyses. This subject is separately addressed in Chapter 16.

Often the rate of payback is a factor in the decision to procure one type of ATE over another. This decision process is illustrated in Fig. 12-6. Suppose a user is considering the procurement of two types of ATE (equipment group B and equipment group C). Equipment group C costs more, but the rate of payback is quicker, due to the fact that this equipment performs the tests more rapidly. Equipment group B has a smaller purchase cost, but the payback time is longer because of the slower payback rate. A decision to

[10]Gregory Illes, "ATE Cost Affectivity: How Much Performance Can You Afford?" *Proceedings of the International Test Conference,* Institute of Electrical and Electronic Engineers, 1982, p. 1005.

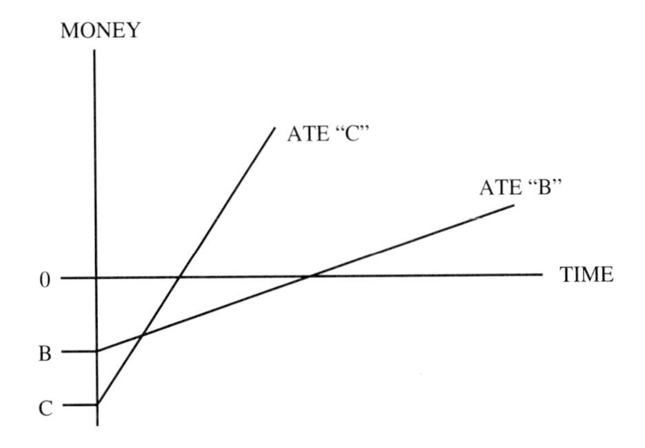

Figure 12-6 ATE payback rates.

procure one equipment group over the other would be based both on rate of return on the dollars invested and the amount of time for which the equipment is estimated to be technologically current [19, 22]

Payback rates on ATE investment may lessen if the user does not exercise care in the procurement decision. It is easy to buy ATE whose technological capabilities in many cases exceed those required by the testing application. These increased technological capabilities manifest themselves in a larger purchase price, thereby reducing the rate at which the equipment investment may be recovered from savings realized in the testing process. According to Torino,[11] a rule of thumb is to buy only those technological capabilities that minimize test programming and execution costs.

Ungar[12] states that another way to identify candidate operations for automatic testing is to compare observed defects or failures of the product being manufactured with all similar products from competitors that are currently on the market. Suppose that defects of types X, Y, and Z in the product being manufactured are significantly greater than those corresponding defects for all other products on the market. Conversely, defects of types A, B, and C of the product being manufactured are the same or less than all other competing products. Prime candidates for automatic testing applications would be the identification of test procedures (and related equipment) to reduce defects of types X, Y, and Z. Reduction of these defects would make the product being manufactured more competitive with other products on the market.

Competitive test equipment costs are illustrated in Fig. 12-7. As seen from this illustration, the least expensive types of test equipment include continuity testers. The most expensive test equipment is a combination of both in-circuit and functional testers. The payback in expending additional funds on more expensive types of test equipment lies in the increased percentage of defects that may be successfully detected (**test coverage**), as illustrated in Fig. 12-8. A combination of in-circuit and functional test equipment is capable of a fault detection rate of nearly 100 percent [6, 9, 15, 19].

[11]Torino, "Testability Practices Today," pp. 43–45.

[12]Ungar, *The Complete ATE Book,* p. 40.

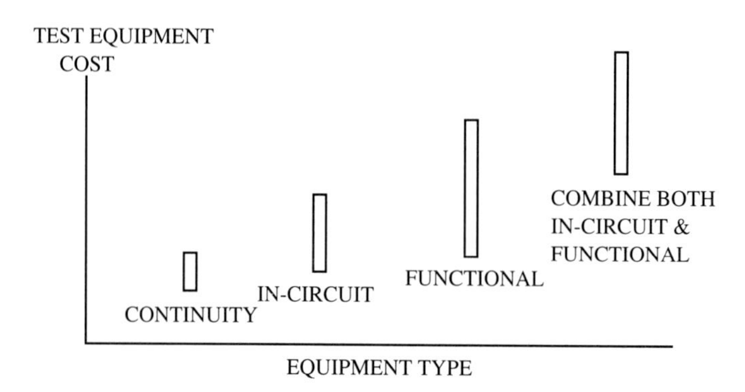

Figure 12-7 Comparative equipment costs.

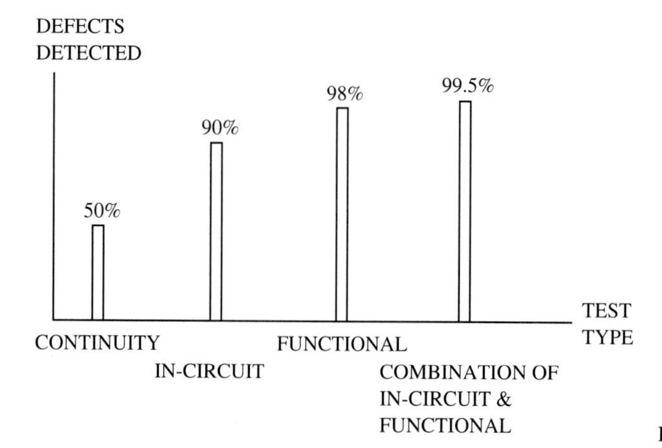

Figure 12-8 Defect detection capability.

12.3.3 Test Yield Determinations

The analysis of test yields has been addressed by Malstrom, McNeely, and Shell [11, 12]. Much of the material in the following sections has been excerpted from these references, with the written permission of the publishers. Low test yields can dramatically increase the number of test cycles required to be performed. A simple test network is illustrated in Fig. 12-9. This illustration indicates a single test station with an off-line troubleshooting/

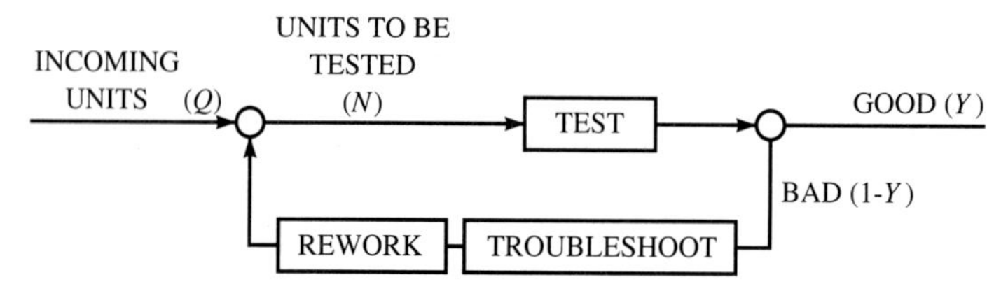

Figure 12-9 A simple test network.

rework feedback loop. Suppose it is desired to process (test) a specific quantity of UUTs through this network. The following notation applies:

Let: Q = Number of UUTs to be processed through the network
N = Number of UUTs arriving at the test station from all outside sources
Y = Test yield = Probability of a test pass on any test trial

It is desired to find the total number of tests that will have to be performed in order to completely process all Q units through the test network. As the first group of Q units is processed, Y percent of these units will pass on the first test trial. $(1 - Y)$ percent of these units will fail. $Q(1 - Y)$ units will then undergo troubleshooting and repair and be resubmitted for test on a second trial. As these $Q(1 - Y)$ units are tested a second time, $(1 - Y)$ percent will again fail. After the second round of tests, $Q(1 - Y)^2$ will be returned a second time for troubleshooting and rework. The total number of test cycles, N, is given by Eq. (12-4):

$$\begin{aligned} N &= Q + (1 - Y)Q + (1 - Y)^2Q + \cdots + (1 - Y)^NQ \\ &= Q[1 + (1 - Y) + (1 - Y)^2 + \cdots + (1 - Y)^N] \\ &= Q\left[\frac{1}{1 - (1 - Y)}\right] \\ &= \frac{Q}{Y} \end{aligned} \tag{12-4}$$

12.3.4 Two-Station Test Networks

A two-station test network is illustrated in Fig. 12-10. In this test process, two test stations are connected in series. Any units failing the second sequence of tests must be returned to the beginning of the test network for additional processing beginning with test station 1. The argument for the preceding simple test network may be extended to determine the number of test cycles required for test stations 1 and 2. These parameters are described by Eqs. (12-5) and (12-6), respectively.

$$N_1 = \frac{Q}{Y_1Y_2} \tag{12-5}$$

$$N_2 = \frac{Q}{Y_2} \tag{12-6}$$

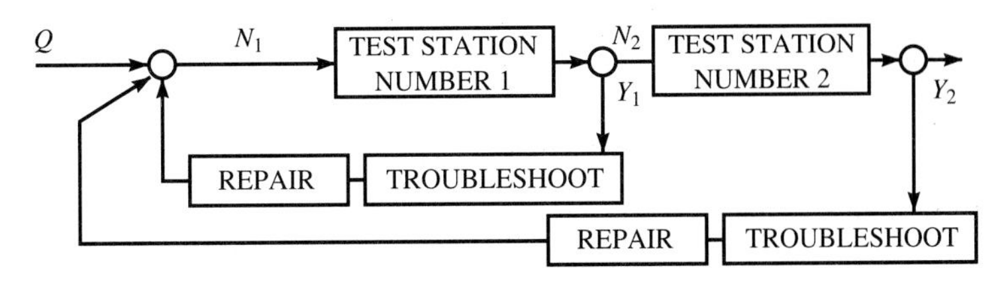

Figure 12-10 A two-station test network with collective feedback.

12.3.5 Networks with Collective Feedback

For a network consisting of N test stations of the format shown in Fig. 12-10, the number of cycles associated with each test station is defined by Eqs. (12-7) through (12-9).

$$N_1 = \frac{Q}{Y_1Y_2 \ldots Y_n} \tag{12-7}$$

$$N_2 = \frac{Q}{Y_2Y_3 \ldots Y_n} \tag{12-8}$$

$$N_n = \frac{Q}{Y_n} \tag{12-9}$$

Once the numbers of test cycles for all test stations in a network are known, it is possible to estimate the total time required to process a group of UUTs through the test network. The total test time is

$$T = (N_1T_1) + (N_2T_2) + (N_3T_3) + \cdots + (N_nT_n) \tag{12-10}$$

where,

T_1 = Length of test 1
T_2 = Length of test 2

⋮

T_n = Length of test n.

12.3.6 Cascade Test Networks

In some test applications, a multiple number of test stations may be used without the requirement that failed units be returned to the first test station when failures occur. This requirement is indicative of a cascade test network and is illustrated in Fig. 12-11. Equation (12-4) may be used to determine the number of test cycles corresponding to any test station. The total test time required to process a group of Q units through the network is again given by Eq. (12-10).

A way to express actual test yields experienced in practice is to divide the cumulative number of UUTs passed through a test network by the cumulative total of test submissions. Test yields are a function of a variety of factors, including the following:

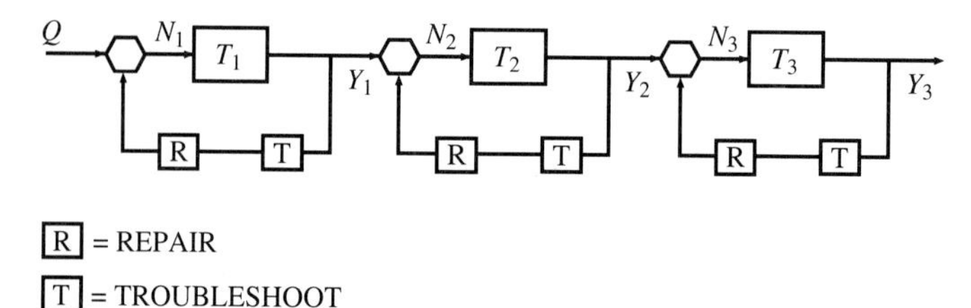

Figure 12-11 Cascade test network.

- Maturity of the design process
- Maturity of the rework process
- Maturity of the manufacturing process
- Test coverage
- Test stress
- Test duration
- Experience and skill of test and fault diagnostic personnel

12.3.7 Estimating Test Yields

Estimating test yields for UUTs that have not been previously manufactured is not always a straightforward process. Elementary reliability theory may be used to estimate the yield that can be expected from a testing process prior to the time that the tests actually begin. Recall from Chapter 11 that for items with an exponential distribution of time to failure, the probability of failure-free operation for t hours is given by Eq. (11-38), repeated here:

$$R(t) = e^{-t/\theta} = e^{-\lambda t} \tag{12-11}$$

where:

$R(t) = P\,\{\text{Component has not failed from time 0 to } t\}$
t = Cumulative unit operating time
λ = Instantaneous failure rate $= 1/\theta$

In Eq. (12-11), values of λ for different component types are tabulated in reliability handbooks such as MIL-HDBK-217, discussed in Chapter 11. The tabulated values reflect steady-state operating conditions and do not consider infant mortality, burn-in, or failures induced by errors that take place during the manufacturing process.

Consider an assembly with n different components. Assuming independence of failures for each component, an aggregate failure rate (λ) for the assembly of n different components may be determined as follows (from Equation 11-46):

$$\lambda_n = \sum_{i=1}^{n} \lambda_i \tag{12-12}$$

The probability that the UUT consisting of these n different components is operational at time t is given by

$$R_n(t) = e^{-\lambda_n t} \tag{12-13}$$

In practice, it is unusual to have a UUT consisting of only one of each of the different n components. Suppose there are j components of type k in an electronic assembly. The probability that all j components of type k are operational at time t is determined as follows:

$$R_k(t) = e^{-j\lambda_k t} \tag{12-14}$$

Equation (12-14) also assumes independence of failures for each of the components within a component type.

Assuming n different component families each with different numbers of components within the family, the probability that a UUT is operational at time t is expressed as follows:

$$R(t) = e^{-j_1\lambda_1 t} e^{-j_2\lambda_2 t} e^{-j_3\lambda_3 t} \ldots e^{-j_n\lambda_n t} \quad (12\text{-}15)$$

where j_1 = Number of components of type 1
j_2 = Number of components of type 2
$\vdots$
j_n = Number of components of type n

Equation (12-15) is likely to estimate test yields that are higher than those actually experienced during the testing process. This is because the λ values in this expression represent steady-state failure occurrences. These steady-state values fail to consider infant mortality failures, those from burn-in, and those resulting from errors that occur during the manufacturing and design processes. The Weibull distribution of time-to-failure (Chap. 11) can be used to model a time-dependent hazard function $\lambda(t)$ such as occurs in burn-in. An alternative method is to escalate the handbook failure rates by a constant factor, L, to account for abnormal failures due to quality problems. Incorporating the escalation factor in Eq. (12-15) yields an approximation of test yield for a specific UUT, as follows:

$$R_{UUT}(t) = e^{-L(j_1\lambda_1 + j_2\lambda_2 + \cdots + j_n\lambda_n)t} \quad (12\text{-}16)$$

12.4 TESTING TECHNIQUES

This section overviews the following test techniques:

- Component
- In-circuit
- Functional
- Systems

12.4.1 Component Testing

Two types of devices are generally tested at the component level. The first group is discrete components, including individual transistors, diodes, resistors, capacitors, and the like. The second group consists of digital devices, including microprocessors, memory chips, and other LSI/VLSI devices. Both groups of components are tested for on/off state, input/output, and high/low failures. The majority of component testers cannot detect timing, sequence, or other interface-related failures. Component testing is generally done by the chip or component manufacturer prior to the time these components are shipped to customers. An example of commercially available component test equipment is shown in Fig. 12-12.

Figure 12-12 Component testers.

(*Source:* Reprinted courtesy SZ Test Systems, Amerang Germany.)

12.4.2 In-Circuit Testing

In-circuit testing checks for continuity or short circuits that may occur due to solder bridges or open circuits (for example, bent or broken leads). In-circuit testing also checks for the wrong component or improper component orientation. Finally, in-circuit testing also screens for components that are damaged by electrostatic discharge.

In-circuit testing applies to populated PWBs using the **bed-of-nails interface.** The bed-of-nails test fixture permits access of numerous test points on the circuit board. The bed-of-nails test fixture consists of several spring-loaded contacts which serve as probes interfacing with board components and contact points, as illustrated in Fig. 12-1 and Fig. 12-13.

Logic circuits on the PWB may be checked using the concept of backdriving. Device inputs are backdriven to an on/off or 1/0 state. Resulting outputs may subsequently be measured by the test equipment. An example is illustrated in Fig. 12-14, where an OR gate is tested by backdriving inputs A and B. Output C is then measured.

A related testing technique is called **guarding.** There are two types of guarding: analog and digital. Analog guarding is used to measure passive devices such as resistors.

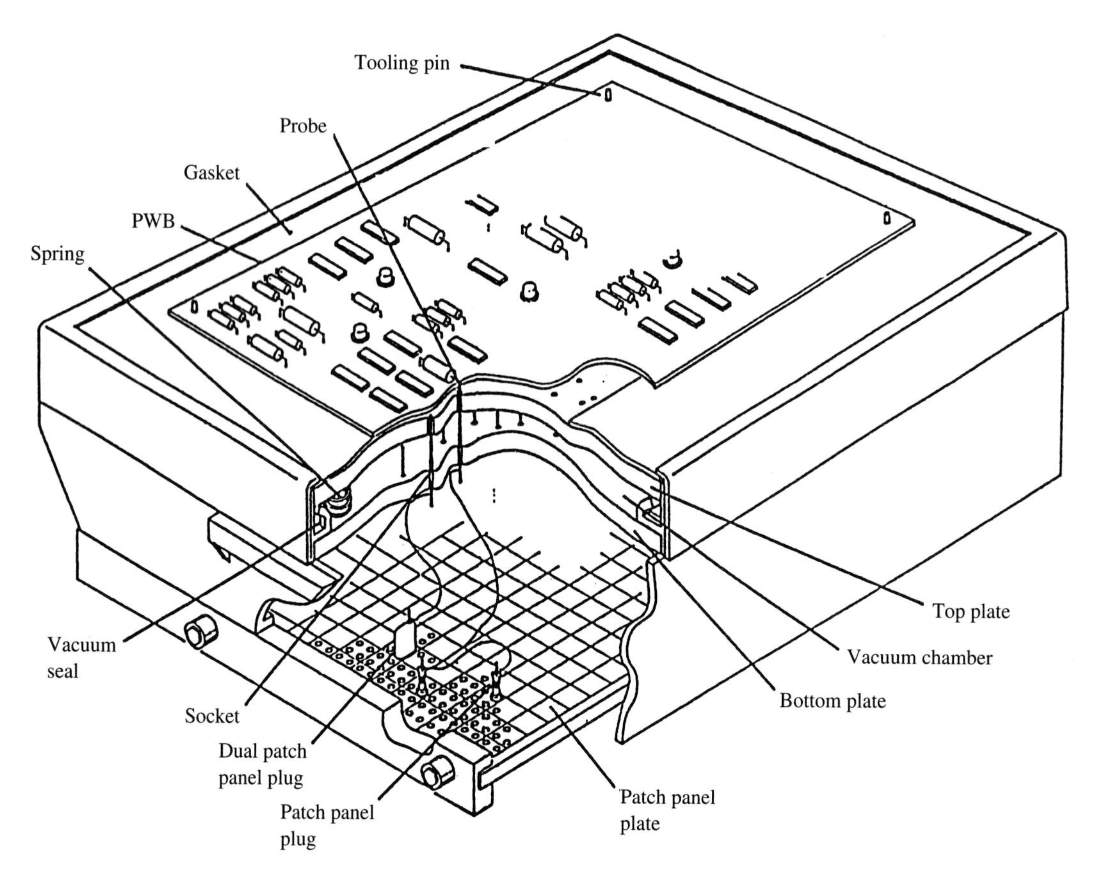

Figure 12-13 Bed-of-nails test fixture.

(*Source:* Ken Parker, *Integrated Design and Test: Using CAE Tools for ATE Programming,* no. 788 (Piscataway, N.J.: Computer Society Press, Institute of Electrical and Electronic Engineers, 1987).)

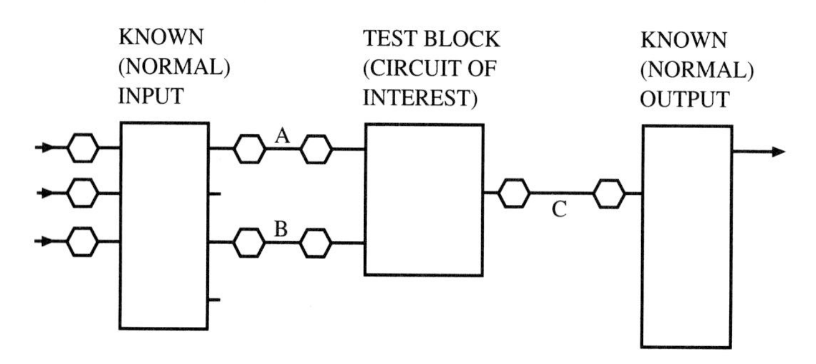

Figure 12-14 An example of backdriving.

With both approaches, the method seeks to isolate the component being tested from the other components in the circuit.

Digital guarding is illustrated in Fig. 12-15. In this illustration, the leftmost part of the circuit is driven in a manner to place nodes A, B, and C in a steady-state logic 1 or 0 condition. These points isolate the circuit of interest from the UUT or guard the circuit from external actions. Test signals are applied to points A and B, and the resulting actions are observed at point C. The test signals applied at points A and B constitute four possible test vectors. The truth table for the device showing these vectors and the corresponding outputs are also given in Fig. 12-15.

In-circuit test programs utilize many preprogrammed test routines. Example routines include those specifically programmed to test memory chips, resistors of specified values, and microprocessor chips.

Each PWB design requires a unique corresponding bed-of-nails fixture. Example bed-of-nails fixtures for two different PWB configurations are illustrated in Fig. 12-16.

Surface-mounted devices (SMDs) can be tested using bed-of-nails fixtures, but the cost is high due to the very small probe contacts that are required on the bed-of-nails fixture. Fixture costs are also higher for designs utilizing both sides of the board for components.

SMT may not necessarily increase testing cost. PWBs can often be designed with test access points spaced far enough apart to allow larger probe contacts to be used. The probe size is not directly related to the lead spacing on the SMDs, since the leads themselves are never probed directly. All access is at through-hole leads or test pads. Only on extremely dense PWBs must smaller probes be used.

Two-sided PWBs can also be tested on conventional bed-of-nails fixtures if designed

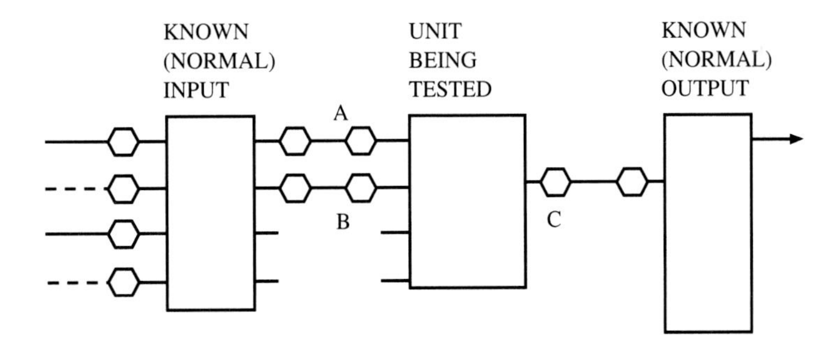

TEST MODE

INPUT		OUTPUT
A	B	C
0	0	0
0	1	1
1	0	1
1	1	1

Figure 12-15 An example of guarding.

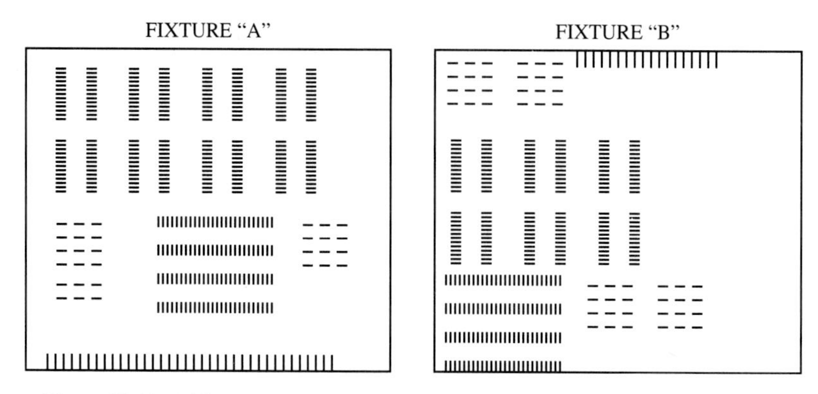

Figure 12-16 Different bed-of-nails fixtures.

such that all nodes are accessible from one side. In the rare case that this cannot be done, a complex double access fixture must be used which contacts both sides of the PWB at the same time. Two separate bed-of-nails fixtures are generally not used. All circuit nodes must be accessed simultaneously in order to detect all possible shorted nodes.

12.4.3 Functional Testing

Functional testing uses the PWB edge connector for test interfacing (Fig. 12-17) as opposed to the bed-of-nails test fixture. Functional testing generally checks for failures such as invalid system inputs or outputs, improper timing, bus failures, and interface problems. With functional testing, the ATE simulates the entire electrical interface that the PWB is likely to experience in actual operation.

Functional testing inputs are generally system commands or vector sets. Corresponding outputs are generally of two types. The first is a stored response, which is a known or expected bit pattern on a specified output pin. The second is called signature analysis. With signature analysis, an output bit pattern is compressed algorithmically to form a bit pattern or signature.

There are two types of functional tests (static and dynamic). Static tests do not check for timing-related faults. Rather, checks are made for specific logic levels. Dynamic

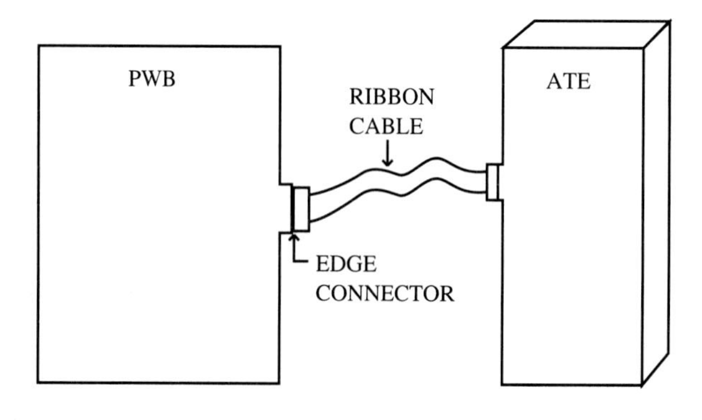

Figure 12-17 Functional testing interface.

testing detects timing-related faults. Tests may be conducted at different operational speeds. Generally, dynamic testing provides more extensive test coverage (detects a higher percentage of failures) than does static testing. With dynamic testing, some degree of signature analysis may be used. Dynamic testing is also more likely to detect soft failures, intermittent problems, or condition-specific faults.

Functional testing may use the concept of guided probing to complete the failure identification process. With guided probing, specific sequences of fault indications guide the test technician to probe additional locations on the UUT to identify the cause of the failure. Guided probing is more expensive than conventional in-circuit testing because of much higher ATE and test programming costs, and because of the manual labor component involved.

Figure 12-18 illustrates a commercially available functional circuit board tester. The oscilloscope on this equipment is used for both timing analysis and probing. When probing is used to supplement functional testing, the resultant procedure is a combination of both in-circuit and functional testing.

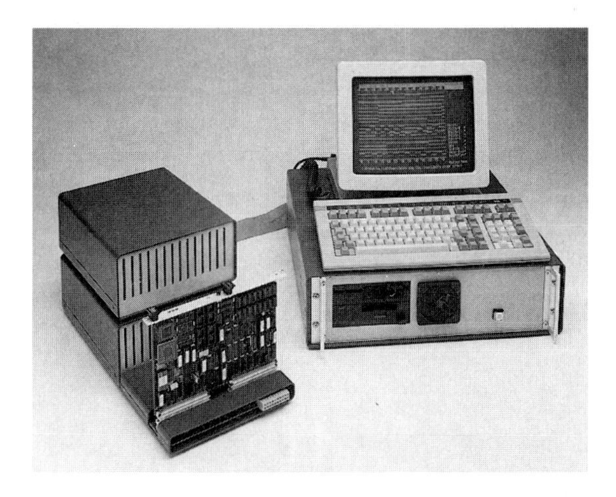

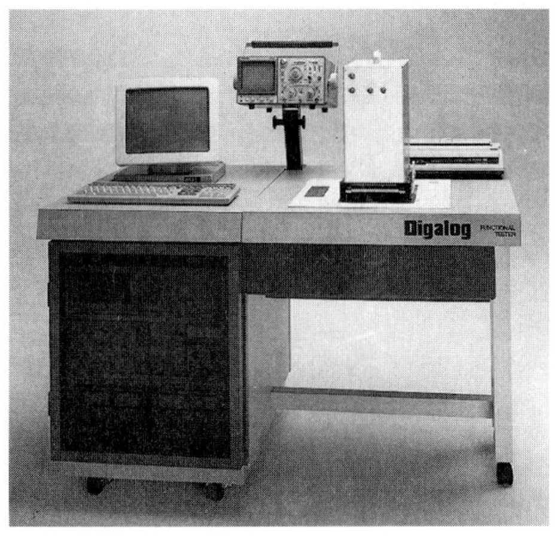

Figure 12-18 Functional circuit board tester. (Reprinted courtesy Digilog Systems, Inc., New Berlin, Wis.)

12.4.4 Systems Testing

Four categories of systems testing exist. The first is dedicated testing. With this method, specialized ATE is designed specifically for use with one and only one product or UUT. Closely related to dedicated testing is the concept of captive testing. In captive testing, all interfaces to the UUT are provided by the corresponding ATE.

A third approach is called in-circuit emulation. With this method, a microprocessor or PWB is replaced with a test microprocessor or PWB. This test device is inserted into the UUT. The test device effectively controls the system. It is necessary for the test device to control the UUT so that functions are relayed back to the ATE controller and measurement systems. The concept of signature analysis can also be used at the systems level. A unique signature can be formed after a significant number of outputs are generated by the UUT. Signature analysis at the systems test level generally offers good speed of testing but a comparatively poor level of failure isolation.

Another approach is called comparison testing and is illustrated in Fig. 12-19. With this method, a pattern generator is interfaced to both the UUT and a known good system. The UUT and the known good system are interfaced in parallel. If the UUT is good, it will produce the same results as the reference system. The exclusive OR gate shown in the illustration will then produce an output consisting only of zeros. Nonzero outputs indicate that one or more failures are present in the UUT. Readers desiring more information on testing techniques are encouraged to consult the reference material listed at the end of this chapter [1,6,7,9,15,17,18,20, and 22].

12.5 DESIGN FOR TESTABILITY

Design for testability (DFT) has evolved through increased circuit complexity resulting from miniaturization. Miniaturization has fostered increased circuit density, resulting in increased failure rates per individual PWB. The goal of DFT is to design electronic

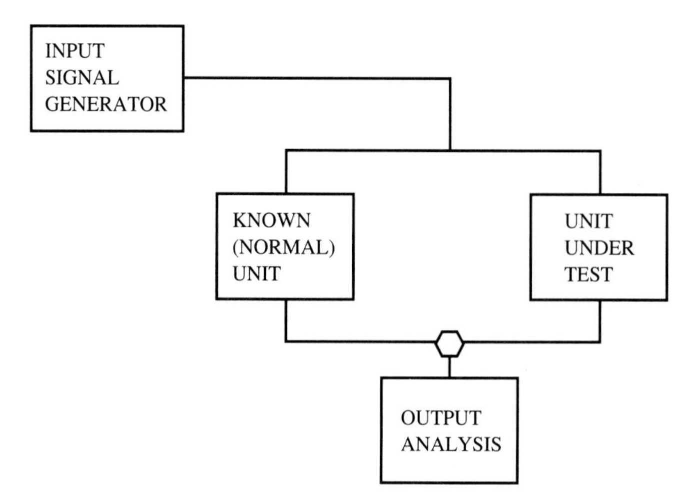

Figure 12-19 Comparison testing.

products to facilitate the testing process. DFT seeks to reduce the costs of testing, troubleshooting, and technician labor. These cost reductions are accomplished at the expense of increased costs of design and manufacturing. However, when DFT is effective, an overall reduction in life-cycle cost results.

Torino[13] defines testability as the ease with which test programs can be created and executed, as well as the ease with which faults may be isolated and diagnosed on defective units. Testability seeks to provide maximum test efficiency and economy from the product design through manufacturing and field maintenance. Testability leads to lower costs of programming, testing, and diagnosis, and to increased test coverage. Testability benefits include the following:

- Shorter time to market (design and manufacturing cycle)
- Lower production test costs
- Reduction of test bottlenecks
- Shorter debugging, troubleshooting, and fault diagnosis times
- Lower field service costs

12.5.1 Self-Testing

There are two ways in which DFT is commonly achieved: through better access for test equipment and probes and through **self-testing.** Self-test is also known as **built-in-self-test** (BIST) **built-in-test** (BIT), or **built-in-test equipment** (BITE). BIT generally means that the product includes the capability to generate test patterns, measure the corresponding test results, and report a failed test.

Providing better circuit access for testing entails the following:

- Making the circuits more controllable (easier to backdrive and isolate)
- Making the circuits more observable (increasing the ease with which test results may be observed)
- Partitioning the circuit (facilitating the troubleshooting process by breaking up the circuit into small regions)

The implementation of BIT generally requires the use of one or more microprocessors. PWBs equipped with microprocessors may have the capability of testing not only themselves but other boards located in the UUT. An example of BIT implementation is shown in Fig. 12-20.

The PWB in Fig. 12-20 contains both a microprocessor and a test chip. The test chip contains the test program for the PWB, eliminating the need for external ATE since the test chip can directly communicate with the circuits and microprocessor. Communication may take place across a unique test bus or across the normal bus. The test chip will control the PWB operation by interrupting the microprocessor and running test sequences.[14]

[13]Torino, "Testability Practices Today,' p. 37.

[14]Parker, *Integrating Design and Test,* np.

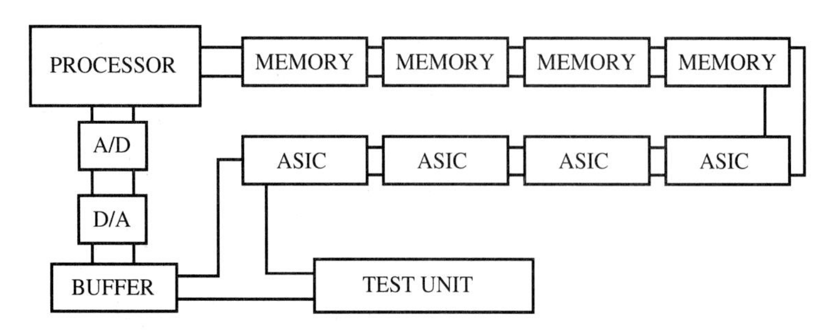

Figure 12-20 An example of BIT implementation.

A more commonly used method of BIT is to have self-test code reside in the microprocessor's program memory and have the microprocessor execute the test sequence upon reset or on commands.

12.5.2 Partitioning

The concept of partitioning basically applies a decomposition (divide-and-conquer) approach to the concept of testing.[15] The smallest circuits are tested first, one at a time. These circuits are then combined and tested as a group. An example of circuit test partitioning is illustrated in Fig. 12-21.

[15]S. Bhatt, F. Ching, and A. Rosenberg, "Partitioning Circuits for Improved Testability," *Advanced Research in VLSI* (Leirson, 1986), pp. 91–107.

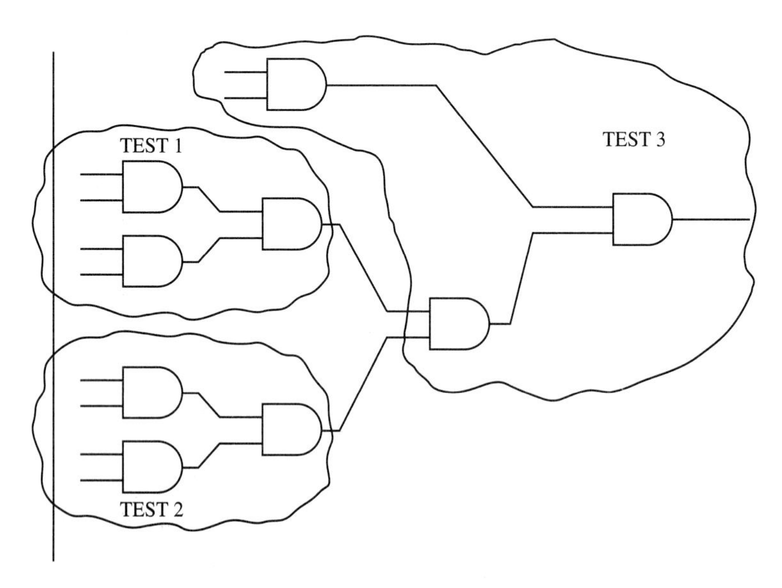

Figure 12-21 An example of test partitioning.

12.5.3 Controllability

The method of **controllability** uses AND gates such that an accessible test circuit can drive the circuit being tested as needed. Refer to Fig. 12-22. With the modified circuit, each input can be tested easily since the AND gate allows the tester to isolate that input. The AND gate requires all inputs to be at logic high in order for an output to occur. Therefore, if one input is a test point, the output can be held off by setting the test input line low. The other inputs to the circuit of interest may then be cycled through the test pattern.

12.5.4 Observability

The technique of **observability** seeks to provide one or more points at the output of a circuit such that the test system may easily measure and/or identify the output value. In this approach, a test connector, with pins connected to deeply embedded circuit points of interest, is added to the PWB. Alternately, test signals can be routed to unused pins on connectors already resident on the PWB.

12.5.5 Grouping

Grouping (illustrated in Fig. 12-23) seeks to place similar functions or devices together such that one type of test may be successfully used to check all components in the group. Circuit grouping also facilitates the test partitioning process (Sec. 12.5.2). Another DFT technique entails breaking feedback loops such that each part of the loop may be tested individually. The test points allow specific sections of the loop to be held steady while other sections are cycled through a test.

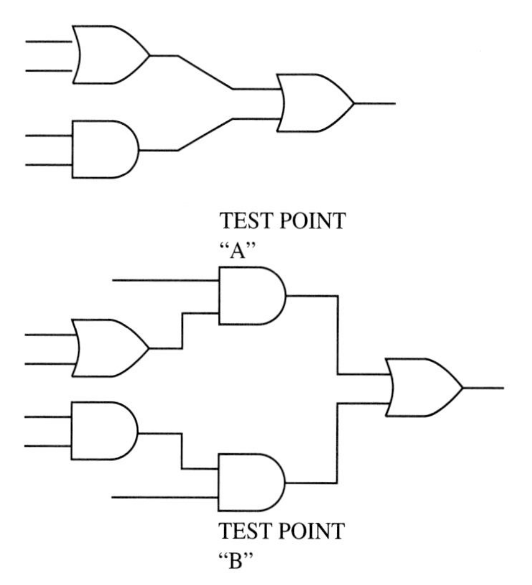

Figure 12-22 Circuit controllability.

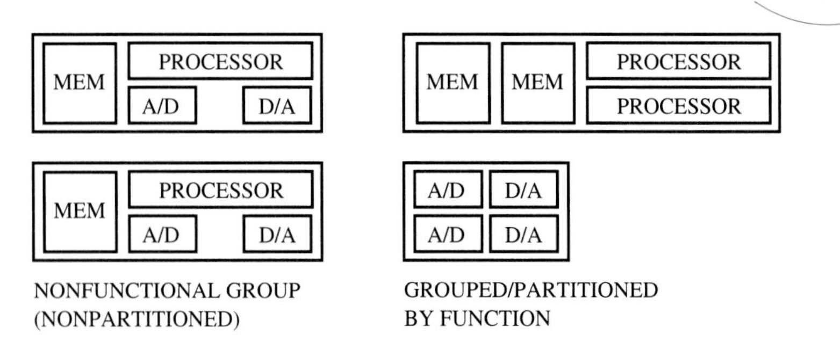

Figure 12-23 An example of grouping.

12.5.6 Clock Control

Many circuits have operating clock speeds that are so fast that the process of testing is either significantly complicated or impossible. The concept of clock control is a DFT method used to provide the capability to control or slow down the circuit clock to facilitate and observe test results. Test points allow the test equipment to insert its own test speed clocks.

12.5.7 Breaking Up a Sequential Process

Another form of circuit partitioning consists of breaking up a sequential process into individual actions. The goal is to break up a dependent electronic test into independent tests. Often, the technique may be accomplished with the insertion of circuitry to effectively isolate sequential stages of a larger circuit.

In many cases, a circuit has an on-board controller. For test purposes, it is often necessary to disable this controller such that the test equipment may control the system. The control lines that the test must control involve the clock output, enable, read, write, reset, and ready, among others. When the test circuit has access to the lines, two test program processes may occur. First, the ongoing operations are interrupted and a test vector is input. Second, the test program can interrupt the operations and read the result of the test vector.

12.5.8 Mechanical DFT Techniques

A number of other DFT techniques are mechanical in nature. These entail physical design considerations including, but not limited to, the following:

- Easy access (slides and stops on racks)
- Visual indicators for self-detected failures (LEDs)
- Placement of test points at the edge of the board
- Orientation of all components in the same direction
- The physical separation of test points from tall components

12.5.9 Advanced DFT Concepts

One advanced DFT concept is that of **logic-sensitive scan design** (LSSD), in which shift register latches are used to facilitate internal circuit control. These latches contain critical test points. An example of LSSD circuit architecture is shown in Fig. 12-24. The shift register latches link the test bus to multiple internal test points, collecting test data and feeding the data to the test system at the appropriate time. All of these DFT concepts are means of adding test circuitry to the design such that the internal control and observation points are maximized while the external or physical access points are minimized.

Boundary Scan (B-S) is an advanced testability approach documented in IEEE Standard 1149.1.[16] The term boundary scan comes from the ability of the circuit to scan the boundary of each chip. This technique requires the PWB design to use B-S versions of chips, which have four additional pins and additional internal circuitry. Test-clock and mode-select signals are routed to all B-S chips, but the serial data in/out pins are daisy chained with the path normally beginning and ending at edge connector pins.

[16]C. M. Maunder and R. C. Tulloss, "An Introduction to the Boundary Scan Standard: ANSI/IEEE Std 1149.1," *Journal of Electric Testing: Theory and Applications"* (1991), np.

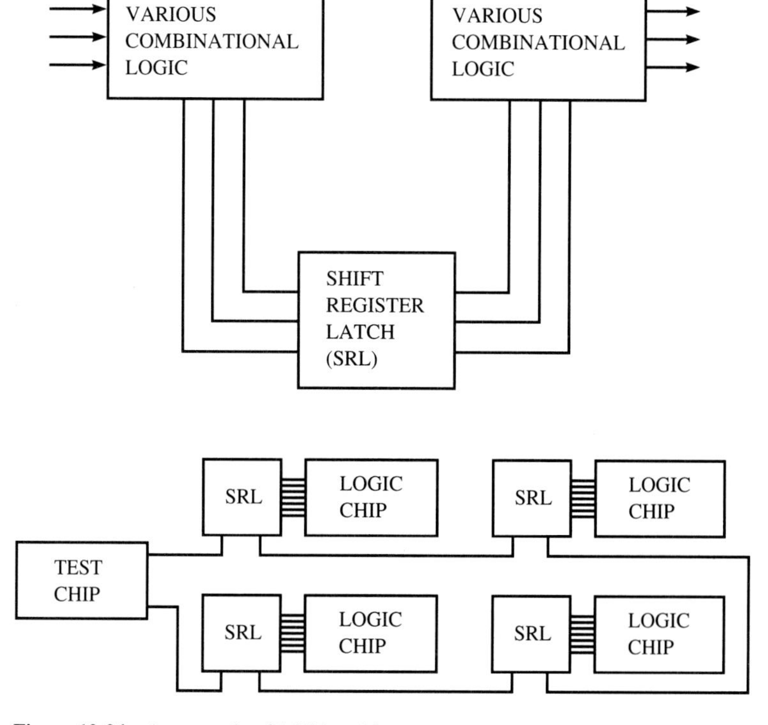

Figure 12-24 An example of LSSD architecture.

Inserted between the chip pin and the chip internal logic are boundary-scan cells or buffers. The buffers act as test points at each pin. B-S allows interconnects (shorts and opens) and some chip testing to be done without using bed-of-nails test fixturing.

12.5.10 References on DFT Concepts

Readers desiring more information on basic DFT techniques should consult the references listed at chapter's end [1, 2, 3, 13, 21, 22]. Additional information on advanced DFT concepts is presented in references [1, 2, 21, and 22].

12.6 SUMMARY

This chapter has sought to overview concepts of electronic testing. Testing is likely to become a topic of ever-increasing importance in the context of electronics manufacturing. As circuit sophistication and miniaturization continue to evolve, the testing portions of the manufacturing process will become more important and account for a greater percentage of product cost.

KEY TERMS

Automatic test equipment (ATE)
Backdriving
Bed-of-nails (interface)
Boundary scan (B-S)
Built-in-self-test (BIST)
Built-in-test (BIT)
Built-in-test equipment (BITE)
Controllability
Design for testability (DFT)
Driver circuits
Guarding
High potential (HIPOT) tests
Hot mockup
Input stimulation
Logic-sensitive scan design (LSSD)
Observability
Self-test
Signal comparison
Test coverage
Test yield
Unit under test (UUT)

EXERCISES

1. What are the three objectives of electronic testing?
2. What is the difference between a continuity test and a HIPOT test?
3. At what level of testing do most defects which cause test failures occur?
4. What is a test vector?
5. How much technology should be procured when choosing an ATE?
6. A two-station test network has 50 units incoming. Station one has a yield of 91 percent and a testing time of .50 minutes per unit. Station two has a yield of 80 percent and a testing time of .30 minutes per unit. What is the number of cycles for station one and station two and the total time to process the 50 units through the test network?

7. What is the difference between a cascade network and a network with feedback?
8. Why is the escalation factor used in the prediction of test yields?
9. In designing for testability, what is the concept of grouping?
10. List a few mechanical DFT techniques.
11. Contrast boundary scan from logic-sensive scan design.

BIBLIOGRAPHY

1. AGARWAL, VISHWANI, *BIST at Your Fingertips: Handbook.* Princeton, N.J.: AT&T, June 1987.
2. BENNETTS, R. G., *Design of Testable Logic Circuits.* New York: Addison-Wesley Co., 1984.
3. BHATT, S., F. CHUNG, AND A. ROSENBERG, "Partitioning Circuits for Improved Testability," *Advanced Research in VLSI,* Leirson, 1986.
4. BYERS, T. J., *Electronic Test Equipment.* New York: McGraw-Hill Book Company, 1987.
5. FUJIWARA, H., *Logic Testing and Design for Testability.* Cambridge, Mass: MIT Press, 1985.
6. HROUNDAS, GEORGE, "Bare Board Testing," in *Printed Circuits Handbook.* New York: McGraw-Hill Book Company, 1988.
7. KEAR, F., "Testing of Printed Circuit Assemblies," *Printed Circuit Assembly Manufacturing.* New York: Marcel Dekker, Inc., 1987.
8. LAHORE, HENRY, *Artificial Intelligence Applications to Testability.* Griffis Air Force Base, N.Y.: Rome Air Development Center, 1984.
9. LANGLEY, FRANK J., "Testing in Assembly," in *Printed Circuits Handbook,* ed. C. Coombs. New York: McGraw-Hill Book Company, 1988.
10. LUETZOW, R. H., *Interfacing Test Circuits with Single-board Computers.* Blue Ridge Summit, Pa.: TAB Books, Inc., 1983.
11. MALSTROM, E., AND R. SHELL, "Determining Expected Assembly Cycles for Electronic Test Networks," *Proceedings of the Midwest Regional Conference,* American Institute for Decision Sciences, Cincinnati, Ohio, May 1978.
12. MALSTROM, E., AND R. MCNEELY, "Empirical Prediction of Yields and Required Times in Electronic Testing," *Proceedings of the 29th Annual AIIE Conference and Convention,* Toronto, May 1978.
13. MARKSTEIN, HOWARD, "Making SMT Assemblies Testable," *Electronic Packaging & Production* (October 1986), pp. 60-64.
14. PABST, STEPHEN, "Elements of VLSI Production Test Economics," *Proceedings, IEEE International Test Conference,* 1987, pp. 982–86.
15. PARKER, KEN, *Integrating Design and Test: Using CAE Tools for ATE Programming,* no. 788. Piscataway, N.J.: Computer Society Press, Institute of Electrical and Electronic Engineers, 1987.
16. PARKER, KEN, "Testability: Barriers to Acceptance," *Design and Test of Computers.* Piscataway, N.J.: IEEE Computer Society Press, 1986.
17. PRIEST, JOHN, *Engineering Design for Producibility and Reliability.* New York: Marcel Dekker, Inc., 1988.
18. PYNN, CRAIG, *Strategies for Electronics Test.* New York: McGraw-Hill Book Company, 1986.
19. PYNN, CRAIG, *The Low-Cost Board Test Handbook.* Zehntel Publishing Company, 1987.
20. STOVER, ALLAN, *ATE: Automatic Test Equipment.* New York: McGraw-Hill Book Company, 1984.

21. TORINO, JON, *Testability Practices Today: Seminar/Workshop.* Campbell, Calif.: Logic Solutions Technology, Inc., 1987.

22. UNGAR, LOUIS, *The Complete ATE Book: Course Notes.* Reseda, Calif.: Test Engineering Solutions, Inc., 1982.

23. WALTON, PAGE, "Testing Surface Mounted Boards," *Electronic Packaging and Production* (February 1987) pp. 8–15.

13 Environmental Stress Screening

13.1 INTRODUCTION

Chapter 11 emphasized the importance of environmental stresses in the design and testing of electronic products. Chapter 12 introduced the methods of functional testing. In this chapter, we discuss the application of principles from Chapters 11 and 12 to environmental stress screening of electronic products for elimination of defects in the factory, prior to customer use.

Section 13.2 defines the concepts, objectives, and terminology of environmental stress screening. Section 13.3 identifies defect failure mechanisms and the types of screens that reveal the defects. In Sec. 13.4 we address the issue of screen placement: At what stage(s) of product assembly should screening be used? Procedures for random vibration and temperature cycling screens are summarized in Sec. 13.5.

13.2 CONCEPT AND TERMINOLOGY

Defects are inherent or induced weaknesses or flaws in a product due to substandard materials or faulty processes. A **patent defect** can be defined as a condition which does not meet specification but is readily detectable by an inspection or test procedure without the need for stress screens. Patent defects represent the majority of the defect population in electronic equipment. A smaller population of **latent defects** cannot be detected by conventional means. If undetected, latent defects will appear as early or premature failures in the operating environment. Latent defects cannot be detected until they are transformed into patent defects by environmental stress applied over time.

The intent of **environmental stress screening** (ESS) is to avoid premature field failures by applying in the factory excessive stresses that accelerate the defect failure mechanisms, without damaging the units and thus shortening the useful life. Utilizing ESS to complement the conventional system of test and repair increases production cost. However, this increase in initial cost can be small in comparison to the reduction in warranty, maintenance, and replacement and other costs. In most cases, the support costs of not screening increase more rapidly than does the initial cost of screening.

We define a screen as an operation applying environmental stress to a product, whereas a test is an operation performed to verify product function in accordance with specifications. The terms **test** and **screen** are sometimes used interchangeably in commercial practice. However, some engineers prefer to speak of **environmental stress testing** (EST) rather than ESS to emphasize that screening out defects is not sufficient. The ultimate goal should be identification of problems and corrective actions to eliminate the causes. Engineers in the defense industry make a different distinction between screening and testing. In defense electronics the precipitation of flaws to the point of failure is the desired product of screening, whereas a test is intended to demonstrate a contractual level of reliability and the ideal outcome of the test is zero failures. Also, a screen can involve 100 percent of all manufactured units, whereas a test usually involves only a small sample of products.

The effectiveness of a screen is measured in terms of **screening strength,** or the fraction of defects revealed as follows:

$$\textit{Screening Strength} = \frac{\textit{Defects Revealed}}{\textit{Incoming Defects}} \tag{13-1}$$

Short of damaging the product, the greater the strength of the screen, the better the product reliability as perceived by the customer.

Figure 13-1 illustrates the impact of ESS on the bathtub curve, for a hypothetical finite population of 700 units. Recall from Sec. 11.5.1 that a bathtub curve is a hazard

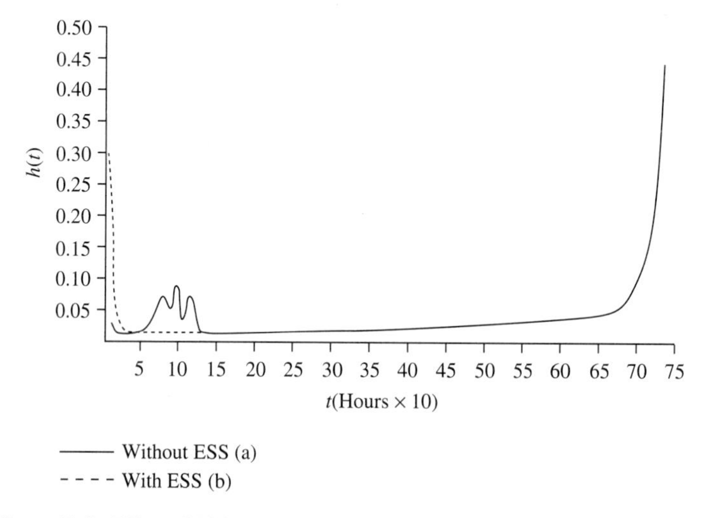

Figure 13-1 Effect of ESS on the bathtub curve.

function for time to first failure and can be constructed, using Eq. 11-34, after all units of the population have experienced first failure. Figure 13-1 contains plots of two hypothetical bathtub curves for the product. Hazard function (a) is a baseline case for benign burn-in. A hump in the hazard function in the range of lifetimes from 60 to 120 hours suggests an abnormally high hazard rate due to a latent defect that was not detectable in the factory but emerged under environmental stresses in field use. Integration of the reliability function for the baseline case, using Eq. 11-31, reveals a mean time to failure of 32 hours.

Hazard function (b) depicts the effect of precipitating these failures in the factory under an environmental stress screen. Assuming all failures in ESS can be reworked in the factory, the repaired units should eventually fail in the field by the same mechanisms as the defect-free units (that is, by random events where stress exceeds design strength, or by wearout where strength deteriorates below the design-level stresses). The MTTF for the case of ESS and rework is over 40 hours. Perhaps more important, the early-life field reliability, wherein customer opinions are often formed, is substantially improved.

ESS typically utilizes random vibration and/or thermal cycling stresses. The screening apparatus consists of a shaker table (multiaxis vibration source) and/or a temperature cycling chamber. Vibration and thermal screens are then conducted either simultaneously or sequentially. Thermal stress levels are typically at temperatures between $-50°C$ and $+100°C$. Vibration screens accelerate the test article at a wide range of frequencies with an average acceleration of up to 6 g. When properly applied, the screens will precipitate failure while not degrading or compromising the product life. Other processes (including high temperature burn-in, cold soak, and power on-off cycling) may be included in the screening.

13.3 SCREEN SELECTION

Table 13-1 summarizes the options of thermal, vibration, and combined screening and the types of latent defects stimulated. In general, vibration screens are more effective than are thermal screens for detecting loose contacts, debris, mounting problems, inadequate strain relief, and loose hardware. Thermal screens are generally more effective for detecting part parameter drift, contamination, poor bonding, defective conformal coating, and improper crimp or mating defects. Many defects are related to soldering problems, and solder defects are susceptible to the temperature changes seen in thermal screening. Other defects are stimulated by expansion and contraction resulting from temperature changes. Some defects are revealed by a combination of thermal and vibration screens.

13.4 SCREEN PLACEMENT

Stress screening can be performed at the part, intermediate (PWB or module), or system manufacturing levels. Stress screen planning involves:

- Defining the yield requirements
- Estimating the number of defects

TABLE 13-1 SCREENING OPTIONS AND DEFECTS STIMULATED

Screening method	Defects revealed
Thermal cycling	IC wire bond defects Thin film separation Cracked die Cracked metallization at oxid steps Open between metallized layers Electromigration Component parameter drift PWB opens/shorts Contact degradation Component incorrectly installed Wrong component Hermetic seal failure Chemical contamination Corrosion Defective harness termination Improper crimp Improper insulation
Vibration cycling	Particle contamination Chafed, pinched, or loose wires Defective crystals Mixer assemblies Adjacent boards rubbing Two components shorting Poorly bonded component Inadequately secured high-mass parts Mechanical flaw Foreign object debris Loose contacts
Thermal and vibration cycling	Defective solder joints Loose hardware Defective components Fasteners Broken component PWB etch defective

- Determining the types of defects present
- Determining required screening strength
- Selecting and placing screens
- Estimating cost elements

It is desirable to eliminate infant failures as early as possible in the manufacturing process. Screening at lower assembly levels is efficient. At lower levels of the product structure, smaller masses are screened. Consequently, for temperature cycling, a greater rate of temperature change is more easily achieved; and for vibration screening, mounting and fixturing of the test article are also simplified. To detect and eliminate most of the intrinsic part defects, screening is often recommended at the part level.

Screening at lower levels eliminates defects before further value is added by subsequent assembly processes. Rework also tends to be simpler at lower assembly levels. Removing and replacing a defective part at the board level is more expensive than eliminating defective parts when detected in a part-level screen. Table 13-2 reports experience from the defense electronics industry; the data is also representative of the relative cost factors in commercial electronics.

Conversely, screening at higher levels of assembly (for example, at the PWB level) has the advantage that several parts and interconnections are screened simultaneously. Some defects are more easily detected by higher assembly-level screens. Examples include drift measurements and marginal propagation delay problems. Defects such as cold solder joints and connector contact defects can be detected only at higher assembly levels.

If the number of latent defects to be precipitated is relatively small, a single screen placed at the lowest level of assembly is most effective. If the number of defects is large, placing screens at two or more levels of assembly, with stronger screens at lower levels, may be necessary. System-level ESS is not common since the cost and complexity of system-level screens is high.

The ESS process should include functional testing. The tests may be run throughout the screen or intermittently. The following factors should be considered in the test strategy:

- *Type of defect:* If the predominant defect is expected to be a weak interconnection, which is transformed to an open circuit with any cycling, then a postscreen test is adequate. If the predominant defect is only detectable at specific temperatures or vibration levels, then the testing must be done under those stresses. An example would be a connection that opens or shorts intermittently due to expansion or contraction. It is prudent to functionally test before screening to establish whether failures are latent defects precipitated by the screen or patent defects induced by the screen.
- *Economics:* The fixtures and associated test equipment to house assemblies, apply power, provide stimuli, and monitor assembly performance can be costly. The trade-

TABLE 13-2 REPAIR COSTS AT PROGRESSIVE ASSEMBLY LEVELS

Component assembly level	Cost multiple
Part	1
Assembly	30
Unit	250
System	500
Field	5000

Source: A. E. Fiorentino and A. E. Saair, "Planning of Production Reliability Screening Programs," *IEEE Transactions on Reliability*, R-32. no. 3 (1983), pp. 247–51.

off is between the cost of fixtures and instrumentation versus the cost of discovering the defects at a later time.

Testability was discussed in Chapter 12 where test coverage was defined as a measure of the percent of possible defects that a test is capable of detecting, as follows:

$$\textit{Test Coverage} = \frac{\textit{Defects Detected}}{\textit{Total Defects}} \tag{13-2}$$

At the PWB level, test coverage ranges from about 85 percent for production line go–no go tests to 90 percent for in-circuit tests and 95 percent for sophisticated automatic tests. For complex products, the test coverage can be difficult to predict analytically, since the set of possible defects is difficult to define. Consequently, test coverage is generally measured on a relative basis. For example, if a test indicates a unit is free of defects, but another test indicates 10 percent defects, then test coverage of the first test relative to the second test is 90 percent.

Figure 13-2 illustrates stress screening at one stage in a manufacturing process. The **process yield** is the ratio of good products to the total number produced as follows:

$$\textit{Process Yield} = \frac{\textit{Total Products} - \textit{Fallout from All Tests}}{\textit{Total Products}} \tag{13-3}$$

Both good and defective units enter the screen. The screening should remove a large fraction of the defective units, as determined by the screening strength. However, since the functional test coverage is imperfect, a fraction of the defects are not detected. The small fraction of remaining defects, not stimulated by the screening process or detected by the functional test, are then passed on to the customer.

13.5 SCREENING PROCEDURES

We will discuss screening procedures for the two primary methods: vibration and thermal cycling.

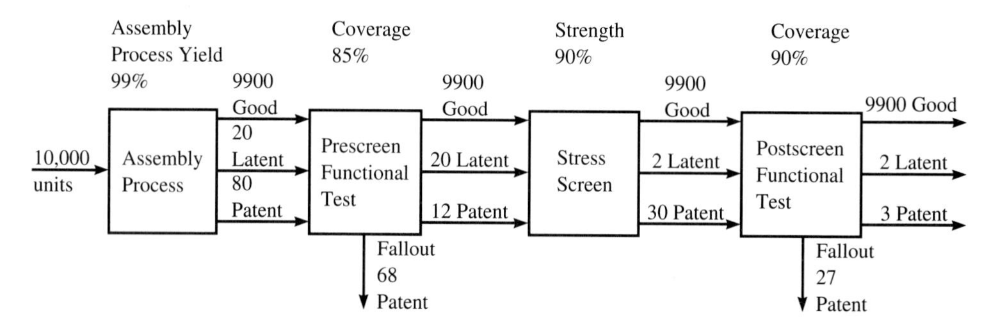

Figure 13-2 ESS in the process flow.

13.5.1 Vibration Stress Screening

Electronic products operate in environments of **random vibration** at various energies. Vibration energy passes through the assembly according to the mechanical property of **transmissibility,** which defines whether the assembly will absorb the vibration energy or pass the energy on to adjoining structures. The transmissibility is a function of the elasticity and damping of the materials from which the assembly is made.

The elastic properties of the assembly also define the natural frequency for each element of the assembly. When the assembly is forced at its natural frequency, it absorbs more of the vibration energy than it would at other frequencies. This condition is known as resonance. For each element of the assembly, there is a range of frequencies (a bandwidth) about the natural frequency at which the assembly absorbs similar energies.

When the assembly absorbs vibration energy, it expends the energy by moving some distance. The amount of movement is called displacement, and the energy expended is called the displacement energy. If the displacement energy exceeds the material bonding strength, fracture occurs.

The displacement energy is a function of both the elasticity of the assembly and the strength of the external force. The energy of the external force is stated in terms of its **power spectral density** (PSD). The PSD units are g^2/Hz (acceleration squared per cycle per second). The PSD can also be expressed as the root-mean-square of the g-acceleration, or grms. The PSD for a typical random vibration screen is shown in Fig. 13-3.

The vibration screen should be designed such that the power spectral density of the screen stimulates defects to detectable failures without producing damaging displacement energy. Screening vibration levels may be significantly higher than the field vibration level, but usually for a much shorter period of time. For example, a typical military screen requires an input force of 6 grms for 10 minutes in each of the three axes. The resulting PSD at the test article is 0.04 to 0.045 g^2/Hz. Commercial standards have not been established for random vibration screens.

Random vibration is common in some use environments for electronic products, and therefore some engineers contend that a screen should produce true random vibration.

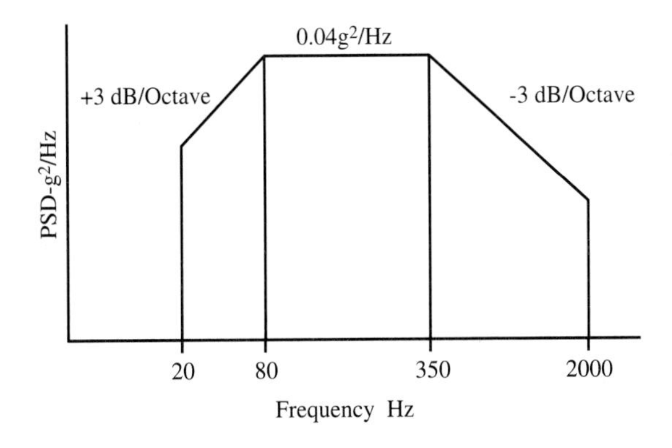

Figure 13-3 Power spectral density for a typical vibration screen.

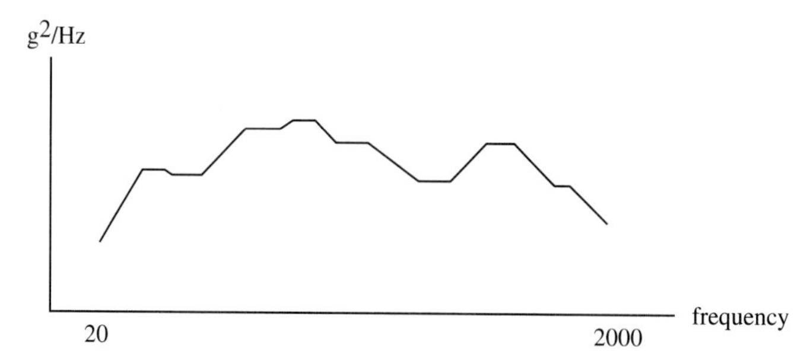

Figure 13-4 Notched screen.

Other engineers believe that vibration screens need only approximate multiaxis random vibration. Pseudo-random vibration is easier to implement in a factory screen.

Notching refers to removing portions of the frequency range that could damage a product if applied at certain energies. An appropriate screen (for example, the screen in Fig. 13-4) would apply all unnotched frequencies in a certain bandwidth and not be too cyclic or repetitive. The vibration does not have to be purely random. The primary objective is to excite the resonant frequencies of the assembly and reveal the defects rather than to reproduce the user environment.

The screening spectrum is generated with electrodynamic or pneumatic shakers. A typical electrodynamic shaker setup (Fig. 13-5) consists of a vibration exciter, test fixture, test specimen, control accelerometers, charge amplifiers, digital control and analysis system, and a power amplifier. The specifications for the required vibration environment are programmed into the digital control system, which uses a feedback control loop to assure the test article experiences the specified vibration spectrum.

When the test begins, the control system emits a drive signal to the power amplifier. The amplifier relays an amplified signal to the shaker. As the shaker begins to move, the accelerometer detects motion and sends a signal to the control system via the charge amplifier. The control system compares this feedback to the desired motion and adjusts the drive signal accordingly. Thus, the drive signal is dependent upon the response of the test article as measured by an accelerometer attached to the mounting fixture.

In order to select a vibration screen, engineers must evaluate the subject assembly to determine the appropriate vibration type and level of excitation. If large, massive articles are screened, low levels of vibration may be effective. In most cases, assembly vibration characteristics may be estimated based on similarity to other assemblies and finite element

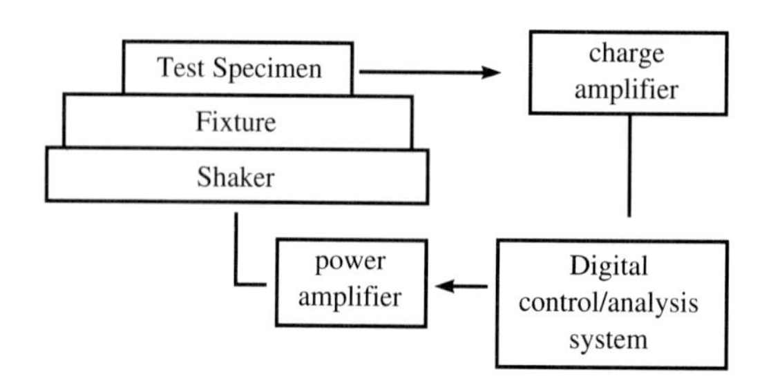

Figure 13-5 Vibration screening system.

modeling and simulation. However, if the dynamic response characteristics are not known, a vibration survey should be conducted to establish proper frequency spectrum and acceleration level. It is desirable to evaluate the screen behavior in each axis.

Two factors to consider in determining the desirability of a PWB vibration screen are the size and stiffness of the board. In general, larger boards will flex more and precipitate latent defects such as cracked metal runs, cold solder joints, and embedded conductive debris. Smaller boards, particularly if conformally coated, are usually quite stiff and thus do not benefit greatly from vibration screening.

13.5.2 Thermal Stress Screening

Thermal stress screening (TSS) is an assembly-level electronics manufacturing process that stimulates latent defects by subjecting the product to **temperature cycling** between extremes beyond the design limits. Figure 13-6 shows an example thermal screening **temperature profile.**

The following terms are used in relation to thermal stress screening:

- *Maximum Temperature:* The maximum air temperature to which the assembly will be exposed.
- *Minimum Temperature:* The minimum air temperature to which the assembly will be exposed.
- *Temperature Range:* The difference between the maximum and minimum applied air temperatures.
- *Temperature Delta:* The absolute value of the difference between the chamber air temperature at which the equipment is being screened and 25°C. (*Note:* In industry practice, the terms temperature range and temperature delta are sometimes used interchangeably to mean maximum temperature minus minimum temperature).
- *Rate of Change of Temperature:* Screening strength tends to increase with higher rates of temperature change up to a limit. The maximum rate of change is dependent on the thermal chamber used, the rate of airflow, and the thermal mass of the items to be screened. The larger the mass, the lower the overall rate of temperature change.
- *Number of Cycles:* The number of transitions between temperature extremes, divided by two. Up to twenty thermal cycles are used for a typical assembly-level screen.

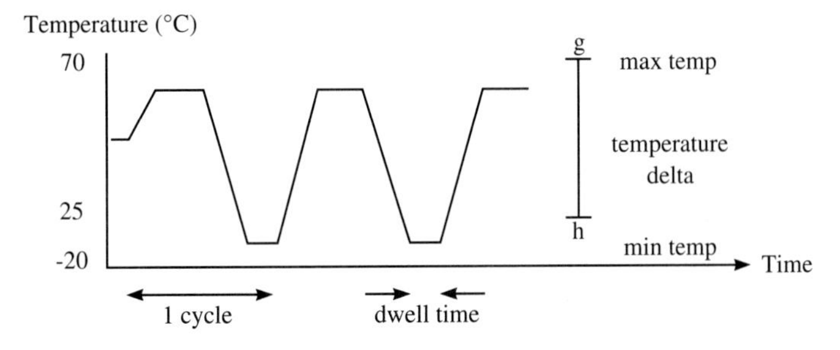

Figure 13-6 Thermal cycling temperature profile.

- *Dwell:* The dwell is the time period for which the temperature is maintained at an extreme temperature. The dwell time is designed to allow the thermal mass to stabilize at the chamber air temperature. If the chamber air changes temperature too quickly, the thermal inertia of the test setup may result in a lower screening strength.

The diverse materials used to fabricate an electronic device have different thermal coefficients of expansion. Thermal stresses are thereby imposed on circuit materials as temperature changes and materials expand and contract at differential rates. Temperature cycling induces mechanical stresses, which activate fatigue mechanisms in the microstructures of metals, polymers, and dielectrics. Flaws such as microcracks or voids tend to accelerate fatigue failure. Thus, thermal stress ages a defective product at a rate greater than for a good product.

Thermal cycling with power applied to the product will have a synergistic effect on the stress at conductor cracks. If a crack exists in a conductive element, the contraction and expansion will cause the crack to grow. The crack growth can be accelerated by applying power. The crack will limit the cross-sectional area in which the electrons may flow. A resulting electrical effect known as electromigration will alter the electrical characteristics of the material. Electrically induced heating at the cracks adds to the thermal stress.

It is important to monitor processes (that is, test the circuit function) during TSS. This monitoring provides information on the phases of the cycle where the failures occur and the number of cycles to failure, and helps in both failure diagnosis and in optimization of the screening procedure.

The ability of an assembly to withstand the temperature extremes is a function of time. Consider a hypothetical piece of equipment that will be damaged if the environment exceeds specifications. The specified extreme temperatures were based on long duration times. Since TSS duration is short, the screening temperature extremes may exceed specified extremes without damaging the product. In Fig. 13-7 the point where temperature will cause damage is plotted as a function of time. Screening below the damage threshold is safe for the product.

The transition and dwell times should be determined by analysis, depending upon the thermal mass properties of the product. The recommended rate of temperature change

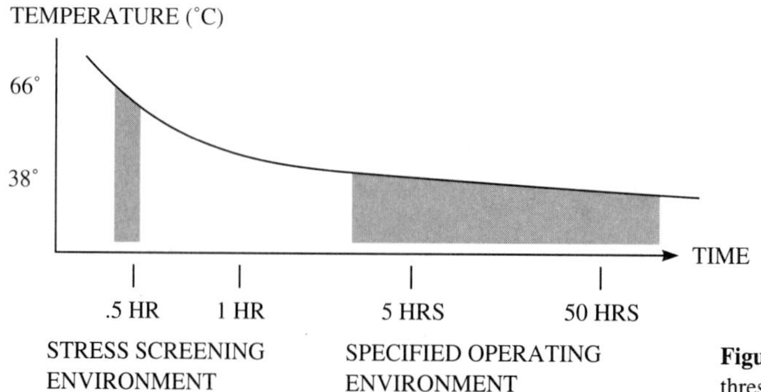

Figure 13-7 Thermal cycling damage threshold.

is between ½°C and 20°C per minute with dwell times between 1 and 10 minutes. Power on/off cycling during the screen is recommended as it adds to the stress on defects. Power should be off during hot to cold transitions to increase the rate of temperature change and decrease cycle time.

The volume of defects stimulated by the screen typically follows a Pareto curve, with the first cycle revealing the largest group of defects and diminishing returns thereafter. The number of thermal cycles employed in the screen is dependent upon board complexity and part density. The necessary number of cycles is determined based on a trade-off between screening cost versus repair and replace costs.

The assembly temperature requirements are defined by the following criteria:

- The weakest component defines the assembly limits. The assembly high-temperature limit is set by the part which is the first to be damaged by high temperature. A similar approach defines the low-temperature limit.
- Use component failure analysis data to determine the temperature rate of change that will stimulate the expected defects. A combination of temperature cycling profiles may be necessary to stimulate all the defects.
- Use screen results as feedback information to determine if more or less extreme temperatures could be used while maintaining expected screening strength.

In situations where the equipment must be rescreened after further assembly, there are some differences in the thermal profile. Higher levels of product assembly have greater mass and therefore the dwell times may be long. The desired temperature rate of change may not be attainable, and the number of temperature cycles may have to be reduced to maintain short throughput times for products in the manufacturing process. Typically, 4 to 12 cycles are used. Equipment-level screens generally include some level of functional testing during temperature cycling, and the test instrumentation adds mass and complexity to the test setup.

13.6 SUMMARY

Thermal, vibration, and other types of environmental stress screens have been used widely in military electronics applications for several years. ESS is becoming increasingly utilized in commercial electronics as well. This chapter has provided a brief overview of the concepts and terminology in ESS, with emphasis on random vibration and thermal cycling screens. ESS is a complex technology, requiring an interdisciplinary approach, and there is a rapidly expanding body of knowledge available for reference by the potential user.

KEY TERMS

Environmental stress screening (ESS)
Environmental stress testing (EST)
Latent defect
Notching
Patent defect
Power spectral density (PSD)

Process yield
Random vibration
Screen
Screening strength
Temperature cycling
Temperature profile
Temperature range (delta)
Test
Thermal stress screening (TSS)
Transmissibility

EXERCISES

1. Discuss the concept of environmental stress screen placement concerning manufacturing level, cost to repair, and test efficiency.
2. Define notching in the context of vibration screening.
3. Explain the distinction between latent and patent defects.
4. Discuss the cost trade-offs to be considered in screen design.
5. What considerations determine the choice of assembly level(s) at which screening is performed?
6. Why is random vibration screening more effective on large PWBs than on small boards?
7. Many manufacturers in the commercial electronics industry prefer thermal cycling over random vibration screening. Provide an engineering rationale for this preference.
8. Explain how electromigration and power on/off cycling accelerate aging of defects.
9. A certain assembly process has a yield of 95 percent. If 20,000 units are assembled, determine the fallout and the number of defective units sent to customers, assuming the following information:

Prescreening test coverage	80%
Screening strength	90%
Postscreening test coverage	95%

10. Describe how you would go about designing a screen for a new product, to precipitate latent defects while not damaging the product.

BIBLIOGRAPHY

1. Dimarogonas, Andrew, and Sam Haddad, *Vibration for Engineers.* Englewood Cliffs, N.J.: Prentice Hall, Inc., 1992.
2. *Environmental Stress Screening Guidelines for Assemblies,* Mount Prospect, Ill.: Institute of Environmental Sciences, 1990.
3. Jensen, Finn, and N. E. Petersen, *Burn-in.* New York: Wiley-Interscience, 1982.
4. Saari, A. E., S. J. VanDenBerg, and J. E. Angus, *Environmental Stress Screening,* RADC-TR-86-149. Griffis Air Force Base, N.Y.: Rome Air Development Center, 1986.
5. Thomas, Lindon C., *Heat Transfer.* Englewood Cliffs, N.J.: Prentice Hall, Inc., 1992.

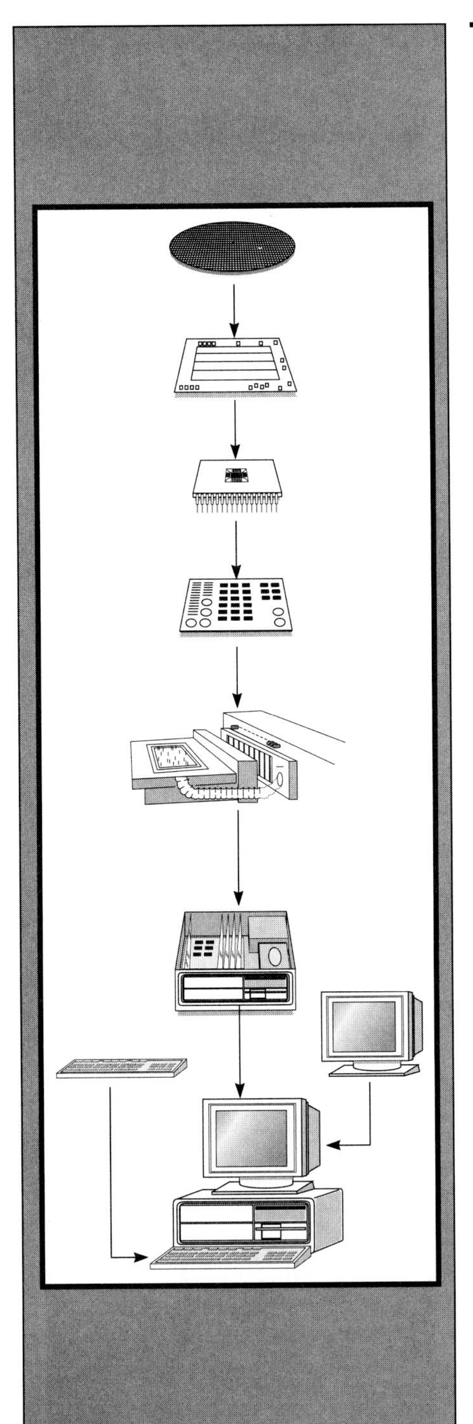

PART FIVE
MANUFACTURING SYSTEMS

14 Facilities and Materials Handling

14.1 INTRODUCTION

The facilities and material handling systems have a major effect on the productivity of an electronics manufacturing process. In this chapter, we discuss the principles of design for facilities and material handling. Section 14.2 discusses facilities planning and design, including the topic of layout, which was introduced in Chapter 2. Section 14.3 presents the principles of materials handling among areas in the plant and discusses the types of material-handling equipment used in the electronics industry. In Section 14.4 we discuss ergonomics, design of modular workstations, and manual materials handling within the workstation.

Section 14.5 introduces the theory and control of electrostatic discharge (ESD). ESD has a significant impact on product quality (yield rates) and is primarily controlled through design of the facilities and the material handling systems. Section 14.6 deals with the stringent requirements for environmental control often seen in electronics manufacturing processes. Clean rooms require specialized equipment and procedures and also impact the design and cost of production facilities. Finally, Sec. 14.7 presents a case study on innovative concepts for the flexible electronics assembly process.

14.2 FACILITIES

Facilities planning includes both facilities location and facilities design. In this chapter, we introduce some basic concepts of location and focus on design. Facilities design includes structural engineering, as might be performed by civil and architectural engineers. Our main interest in electronics manufacturing is in the topics of facilities design and layout and material handling systems.

14.2.1 Location

The decision of where to locate a manufacturing facility significantly affects the cost and service with which a company can supply its customers. The decision is based on quantitative factors, such as transportation costs and property taxes, as well as qualitative factors, such as politics and the quality of life for employees. Operations research techniques aid in determining the optimum location of a facility, based on logistical costs for procuring raw materials and supplying customers. However, the qualitative factors are often decisive in facility location decisions.

Because a facility location is an expensive, long-term commitment, it should be approached as a major strategic decision. The location decision should be well integrated with the overall business plan and should consider the following factors:

- Make versus buy
- Off-shore versus on-shore
- Size of facility
- Geographic centralization
- Ecological impact
- Product and process technology
- Capital budget

The trend toward off-shore assembly operations is significant and is primarily driven by competitive pressures. Electronics manufacturers have located plants in Mexico, South America, and the Pacific Rim countries to benefit from low-cost labor. Large companies such as Texas Instruments and IBM and smaller companies such as Seagate and Western Digital have located some plants on off-shore sites (Taiwan, Singapore, Mexico, Brazil, and Italy, to name a few). However, the costs of transportation and adaptation to the laws and cultures of these countries can offset much of the labor cost savings. Consequently, many companies have made the commitment to locate all or a major part of operations in the United States. This decision can be influenced by the availability of technically skilled professionals and automation technology. In some cases, such as the aerospace/defense industry, on-shore location is mandated by national security considerations. There has also been a trend for multinational companies based in Asia and Europe to locate plants in the United States for better access to markets.

The make-versus-buy decision can also influence facility location. If we choose to rely on outside sources, then location in the proximity of suppliers is important. The JIT methods of production discussed in Chapter 15 depend heavily on close proximity to outside suppliers. Some companies in the United States do not compete in the mass consumer electronics markets dominated by Asian suppliers but rather focus more on the industrial electronics sector. These companies supply customized, integrated solutions (including hardware, software, and technical support), rely on outside suppliers, and concentrate their efforts on the integration tasks. While the assembly and integration operations are located in the U.S., many of the component suppliers are located off-shore and were chosen because of low cost and high quality.

The long-range facility plan should include the following:

- Facility mission statement
- Basic growth plan
- Master site plan[1]

The mission statement should be brief, specific, and consistent with the overall business strategy of the firm. The objectives and priorities should be determined by upper management before the planning process begins. The growth plan outlines the strategies from among the options of relocation, consolidation, decentralization, or expansion. The master site plan usually follows one of three patterns in the electronics industry (Fig. 14-1). The zone concept allows an expansion area (zone) for each major function and is therefore consistent with the functional layout approach discussed in Chapter 2. Texas Instruments has built several plants on a spine pattern (Fig. 14-2), so that additional modules can be added to zones as new functions are needed. The block concept provides expansion area around each building and is appropriate for campus-style site plans. The duplicate concept simply replicates an existing facility design to a new location on the site.

14.2.2 Layout

Layout refers to the spatial arrangement of physical resources (facilities, equipment) that are used to create the product or deliver the service. The layout design and the material handling approach vitally impact the overall productivity of an operation. Consequently, a thorough and systematic approach is needed in layout design. Apple identified several tasks that should be performed in roughly the following sequence:

- Collect data
- Analyze data
- Design the productive process
- Plan the material flow pattern
- Consider the general material plan
- Determine equipment requirements
- Plan individual workstations
- Select specific material handling equipment
- Coordinate groups of related operations
- Design activity interrelationships
- Determine storage requirements
- Plan service and auxiliary activities
- Determine space requirements

[1]H. Lee Hales, "Time Has Come for Long-Range Planning of Facilities Strategies in the Electronics Industry," *Industrial Engineering,* 17, 4 (1985), pp. 29–39.

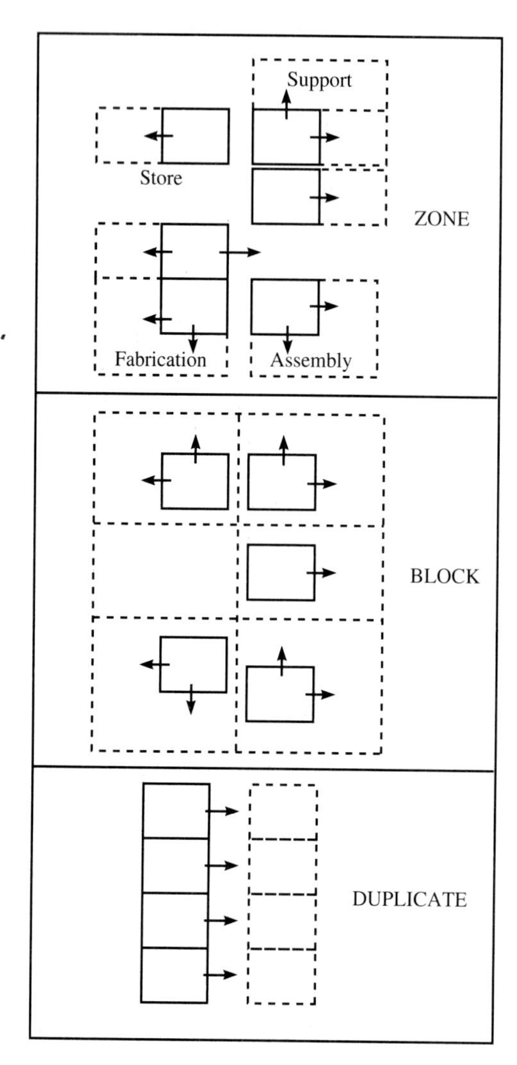

Figure 14-1 Facility site plans.

(*Source:* Adapted from Richard Muther and Lee Hales, *Systematic Planning of Industrial Facilities,* vol. 1 and 2 (Kansas City, Mo: Management and Industrial Research Publications Inc., 1980), p. 20–4.).

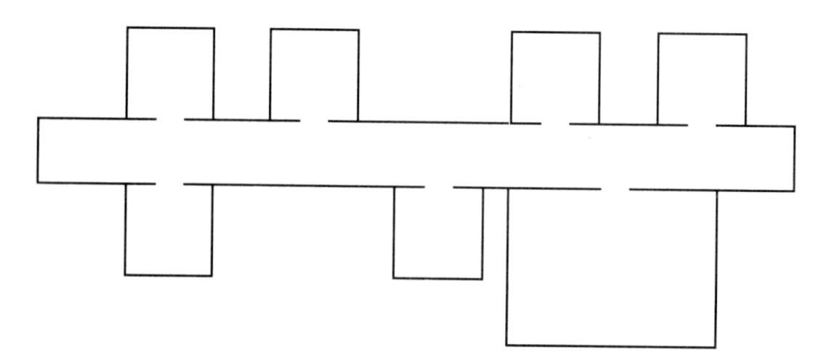

Figure 14-2 Spine-type modular facilities.

- Allocate activities to space
- Consider building types
- Construct master layout
- Evaluate, adjust, and check the layout with the appropriate persons
- Obtain approvals
- Install the layout
- Follow up on the implementation of the layout[2]

Systematic layout planning (SLP) is a methodology developed by Muther. SLP considers both material flow data (from-to chart) and subjective relationships (activity relationship chart) among functions (departments). The activity relationship chart documents closeness ratings for adjacency of pairs of departments, ranging from absolutely necessary to undesirable. A graphical model called the relationship diagram depicts the strength of relationships among departments. Space requirements are overlaid on the relationship diagram to form a space relationship diagram, which is then shaped into alternative block layouts fitting rectangular space.[3]

Figure 14-3 illustrates the four main types of layouts:

- *Fixed:* The product is built in one place, such as on a workbench in the laboratory. All materials, tools, and skills are brought to the location. The **fixed layout** is typical in the construction industry.
- *Functional:* In the **functional layout,** equipment is arranged by the process or function performed. This layout is very flexible and therefore consistent with the needs of a batch shop where many different products are made in low volumes.
- *Product:* The **product layout** arranges equipment in an efficient straight-line flow optimal for a high-volume product.
- *Group:* The **group layout** blends the flexibility of a process layout with the efficiency of a product layout. This approach is made possible by identifying a product family (group) with common processing requirements. The grouping creates the greater volume needed to justify dedicating a work center to a specific process. This analytical technique of defining product families is called group technology. The U-shaped arrangement allows for some flexibility in the sequence in which operations are performed.

Figure 14-4 summarizes the general types of layouts suited to product-process combinations. A specific application might blend or differ from the layout recommendations in Fig. 14-4. Also, a specific plant may include a mixture of layout designs for a variety of product-process combinations. Fixed and functional layouts are not optimized for a single product and thus may be described as process-focused. Conversely, the group and product layouts tend to be specifically designed for a product or family of products.

[2]James M. Apple, *Plant Layout and Material Handling,* 3rd ed. (New York: John Wiley & Sons, 1977), pp. 26–32.

[3]Richard Muther, *Systematic Layout Planning,* 2nd ed. (Boston: Cahners Books, 1973), p. 2-2.

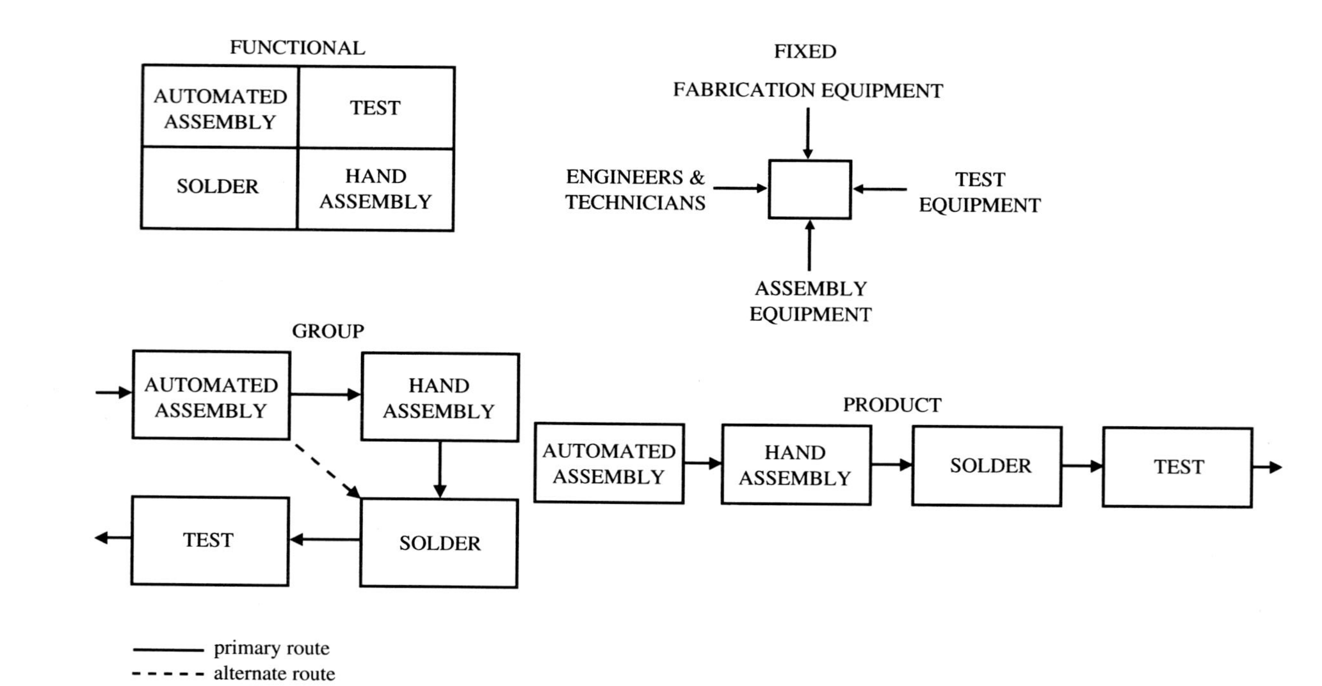

Figure 14-3 Layout types.

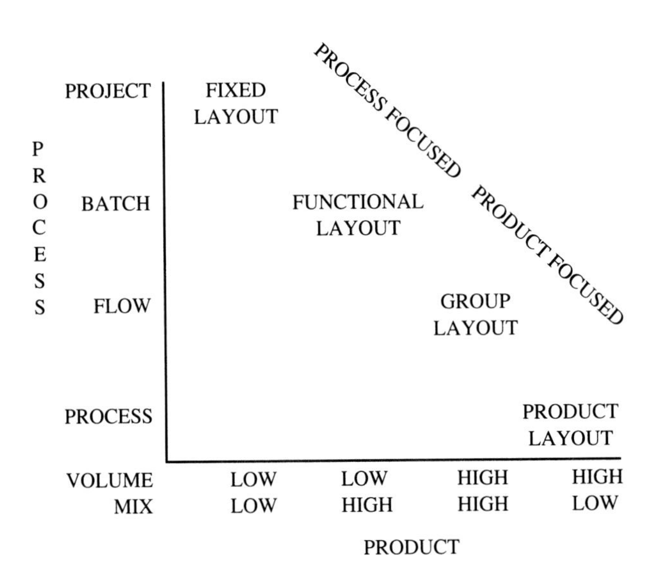

Figure 14-4 Layout alternatives for products and processes.

The fixed layout is used in the laboratory or model shop. It is also common in rework and maintenance facilities where electronic products are repaired. The advantages include:

- Minimal movement of material
- Good teamwork and job enrichment
- High flexibility

The fixed layout also has several disadvantages:

- Low utilization of space and equipment
- High skill requirements
- Complicated scheduling and control on large projects

The functional layout is very common in electronics manufacturing, but often it is not the most appropriate layout for the product mix and processes. Often when we analyze the situation we find that a substantial proportion of the products share common processes and operation sequence, such that a product or group layout is more appropriate. The functional layout has certain advantages which account for its wide application as follows:

- High flexibility
- High utilization of labor and equipment
- Low-cost, general-purpose equipment may be used

However, there are several critical disadvantages:

- Heavy material handling activity
- Accumulation of work-in-process inventory

- Excessive throughput (lead) times
- High skill requirements
- Complicated production planning/control

Product layouts are common in PWB assembly operations where volume is high and the product mix is not diverse. Machines are arranged in a straight-line flow, and product movement is mechanized on conveyor tracks connecting the machines. Figure 14-5 is an example for through-hole technology.

The product layout has the following advantages:

- Efficiency
- Smooth, straight-line flow
- Lower skill requirements

The disadvantages include the following:

- Inflexibility
- Risk of obsolescence
- High capital investment
- Slowest machine (station) paces the production rate
- Equipment breakdowns stop the line

The group layout is increasing in popularity, particularly with electronics companies organizing operations into cellular teams and implementing JIT techniques. Figure 14-6 shows the group layout for a PWB assembly shop in a high-mix/low-volume environment. In the PWB shop, many different board types are produced in low volume, but all boards go through the same processes and in the same sequence. Consequently, a group-oriented layout can be used to take advantage of the common process routing and form a process cell that is efficient for the family but is also flexible.

In Fig. 14-6 there is a standard flow, for most products in the family, among machines arranged in a U-shaped layout. Because of the low production volumes, handling between machines is performed manually in batches. Because of the U-shaped arrangement, alternate routings (for products that do not exactly follow the standard process sequence of the family) are quite convenient. Also, personnel staffing can be varied in response to

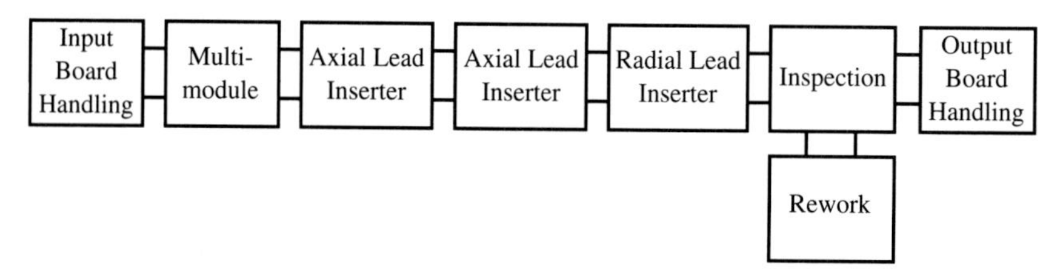

Figure 14-5 PWB assembly line for through-hole insertion.

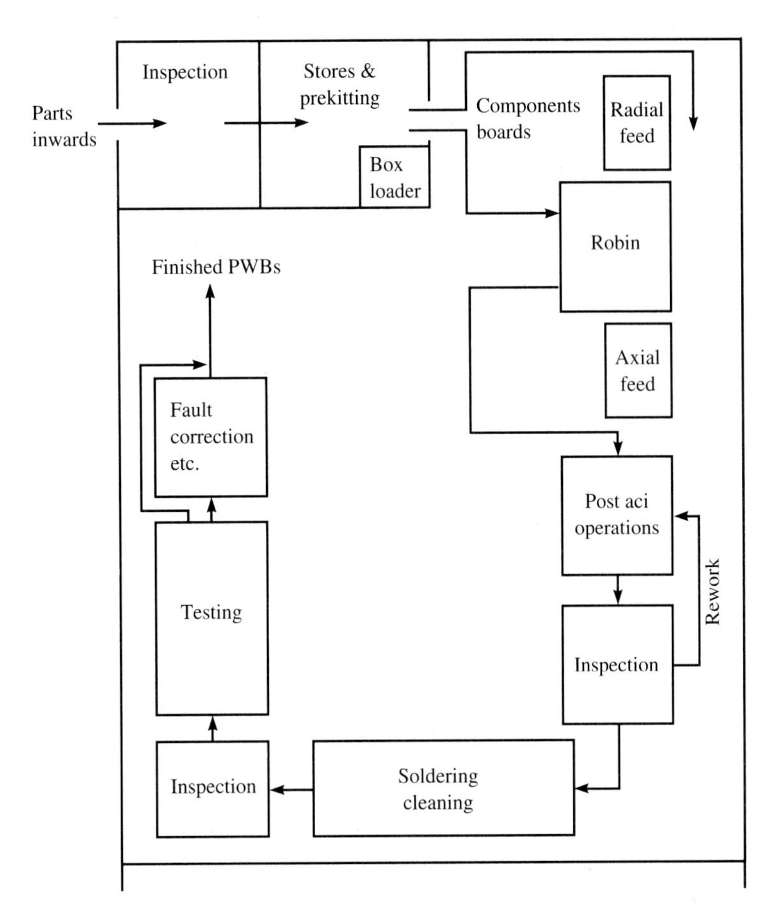

Figure 14-6 Group layout for PWB assembly.

(*Source:* P. J. W. Noble, *Printed Circuit Board Assembly,* (New York: John Wiley & Sons, 1989), p. 70.)

changing customer demands. Although the U-shaped layout offers several advantages, facility constraints—such as the locations of utility drops (electrical, shop air, pure water, computer network ports)—can limit the use of U-shaped layouts. In some cases, the layout flexibility within an area served by a utility grid may be more important than the process routing flexibility of a U-shaped layout.

In summary, the group layout blends the advantages of both the functional and product layouts, namely:

- High flexibility
- Low-cost, general-purpose equipment may be used
- Good utilization of labor and equipment
- Smooth material flows over short distances

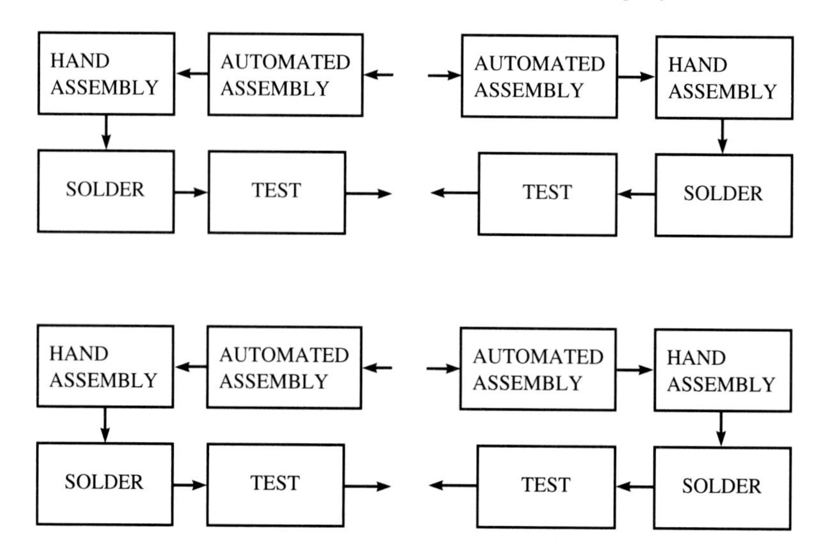

Figure 14-7 Layout of groups cells.

- Teamwork and job enlargement

The disadvantages of the group layout are few, but include the following:

- Requirements for versatile, skilled labor
- Dependence on production control to balance flows

The flexibility of the group layout can be further exploited by forming an array of group cells, as shown in Fig. 14-7. The input and output points are all located along a common flow path, thereby minimizing the material handling among cells. If material handling is done manually, then one person could handle the input and output for at least two adjacent cells. The common flow path is also convenient for mechanized handling modes such as a conveyor. The flexible manufacturing system (FMS) is a special case of the group layout. Group technology is used to define a family of products that are high-value or high-volume in nature, therefore justifying the capital cost of automation. The FMS is characterized by highly automated processes (machines) linked by systems of flexible, automated material handling.

14.3 MATERIALS HANDLING

The *Materials Handling Handbook* states: "Materials handling is a system or combination of methods, facilities, labor, and equipment for moving, packaging, and storing of materials to meet specific objectives.[4]"

[4]Raymond A. Kulwiec, ed., *Materials Handling Handbook* (New York: John Wiley & Sons, 1981), p. 4.

14.3.1 Principles

The Material Handling Institute (MHI) has summarized many years of experience into a body of twenty material handling principles:[5]

- *Orientation:* Study the system relationships thoroughly prior to preliminary planning in order to identify existing methods and problems and physical and economic constraints, and to establish future requirements and goals.
- *Planning:* Establish a plan to include basic requirements, desirable options, and the consideration of contingencies for all material handling and storage activities.
- *Systems:* Integrate those handling and storage activities which are economically viable into a coordinated system of operation including receiving, inspection, storage, production, assembly, packaging, warehousing, shipping, and transportation.
- *Unit Load:* Handle product in as large a unit load as is practical.
- *Space Utilization:* Make effective utilization of all cubic space.
- *Standardization:* Standardize handling methods and equipment wherever possible.
- *Ergonomics:* Recognize human capabilities and limitations by designing material handling equipment and procedures for effective interaction with the people using the system.
- *Energy:* Include energy consumption of the material handling systems and material handling procedures when making comparisons or preparing economic justification.
- *Ecology:* Minimize adverse effects on the environment when selecting material handling equipment and procedures.
- *Mechanization:* Mechanize the handling process where feasible to increase efficiency and economy in the handling of materials.
- *Flexibility:* Use methods and equipment that can perform a variety of tasks under a variety of operating conditions.
- *Simplification:* Simplify handling by eliminating, reducing, or combining unnecessary movements and/or equipment.
- *Gravity:* Utilize gravity to move material wherever possible, while respecting limitations concerning safety, product damage, and loss.
- *Safety:* Provide safe material handling equipment and methods which follow existing safety codes and regulations in addition to accrued experience.
- *Computerization:* Consider computerization in material handling and storage systems, when circumstances warrant, for improved material and information control.
- *System Flow:* Integrate data flow with the physical material flow in handling and storage.
- *Layout:* Prepare an operational sequence and equipment layout for all viable system solutions, then select that system which best integrates efficiency and effectiveness.

[5]*The Twenty Principles of Material Handling* (Charlotte, N.C.: College, Industry Council on Material Handling Education), np.

- *Cost:* Compare the economic justification of alternate solutions in equipment and methods on the basis of economic effectiveness as measured by expense per unit handled.
- *Maintenance:* Prepare a plan for preventive maintenance and scheduled repairs on all material handling equipment.
- *Obsolescence:* Prepare a long-range and economically sound policy for replacement of obsolete equipment and methods with special consideration to after-tax life-cycle costs.

14.3.2 Unit Loads

The unit load is foundational to all good materials handling design. We can define the **unit load** as that quantity of material to be moved, handled, or stored. The unit load should be designed to

- Minimize handling
- Assemble materials for economy of handling/storage
- Assemble materials as soon as possible
- Maintain materials as long as possible
- Utilize cubic space

Traditionally, the unit load concept has emphasized handling in as large a quantity as possible, as reflected in the MHI twenty principles. While it is true that handling in a larger load tends to save trips and thus reduce material handling, we must be cautious to avoid excessive inventory. The JIT approach to production advocates frequent movements of small quantities, to reduce the average inventory level.

In modern manufacturing, our goal is to determine the optimum unit load, considering such factors as

- Product
- Process
- Customers
- Suppliers
- Carriers (transportation modes)
- Containers
- Packaging
- Handling and storage equipment

Major progress has been made in the design of unit load handling and storage containers in the electronics industry. The trend toward JIT has accelerated the development of optimum unit loads. The containers achieve the unit load principle, while also conforming to the following important principles:

- Systems integration
- Flexibility
- Modularity
- Standardization

To observe the systems principle we look at how the unit load functions throughout the operation and how it can benefit overall systems integration. To achieve flexibility we design the unit loads so that the maximum array of materials can be accommodated. Standardization reduces costs and simplifies operations by limiting the number of different containers to the minimum required.

The containers should also be modular. Modularity implies that the containers might be nestable (for compact storage), stackable (without crushing of contents), and compatibly sized (for example, four small containers fit neatly into one larger-sized container). In electronics manufacturing, the containers should also aid in the control of electrostatic discharge, as discussed in Sec. 14.5.

Several electronics manufacturers have adopted modular container systems that meet the objectives described. For example, Texas Instruments (TI) has long been a leader in the design of unit load handling and storage systems. In the early 1970s, TI made the decision to design an entire integrated system and worked with the Herman Miller furniture company to develop designs evolving into the unit load containers, storage equipment, and handling devices shown in Fig. 14-8. There are four main sizes of **tote pans,** with the same footprint (allowing the use of one standard lid), but different depths. Smaller bins are designed to fit inside the totes in integer multiple numbers. Storage lockers and carts are designed around the standard tote pans. The tote pans have lips for hanging on the carts. The same standard shelves that are used on the workcenter furnishings also fit on the carts.

In the late 1980s, TI developed a new generation of tote pans. The new design was needed to better support automation and bar code interfacing and for better ESD protection. Among the design criteria were the following:

- *Dimensions:*
  - Various sizes
  - Modularity
  - Compatibility with standard pallet 48 inch (1.2 m) $\times$ 40 inch (1 m)
  - Compatibility with IC tube packaging
- *Material:*
  - ESD protection
  - Strength
  - Chemical resistivity
  - Color coding
  - Weight
  - Cleaning
  - Fire rating

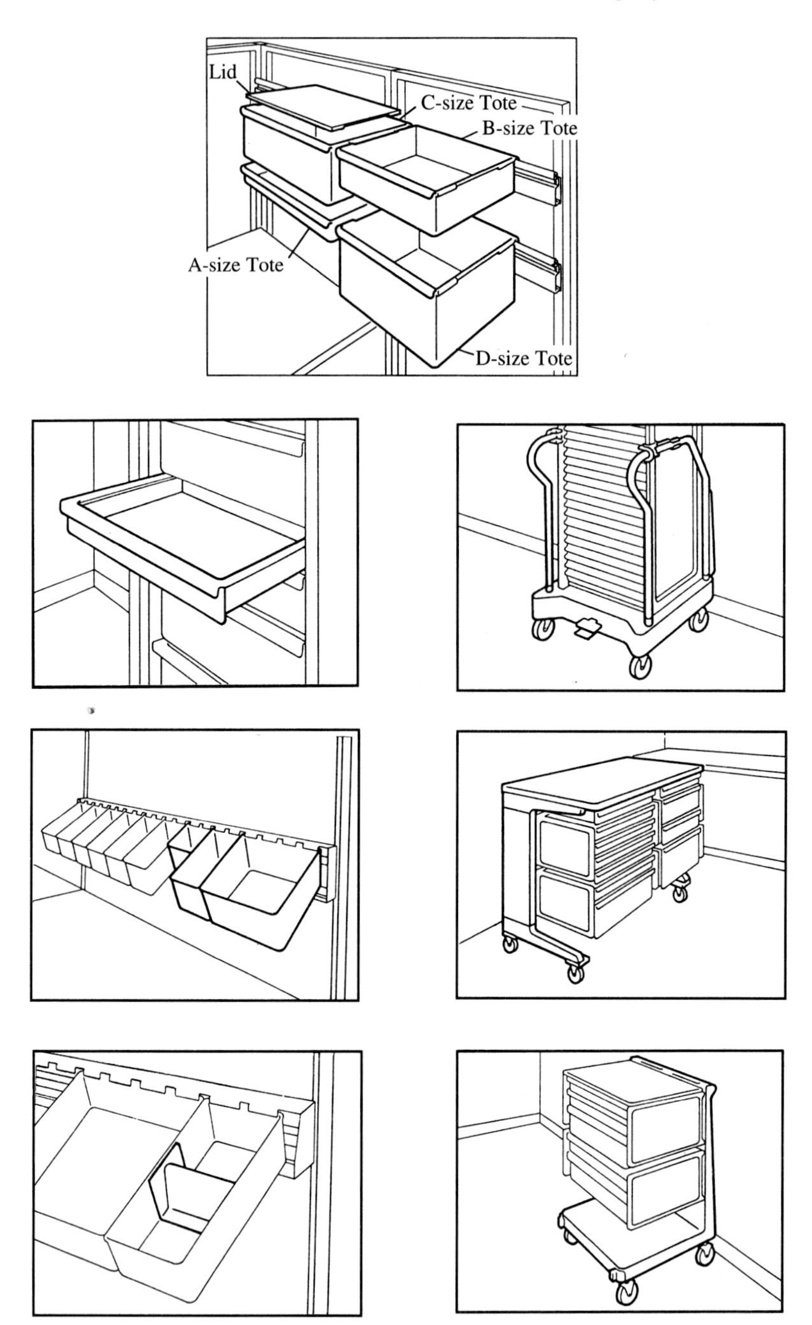

Figure 14-8 Unit load storage and handling system.

(*Source:* Courtesy of Herman-Miller, Minneapolis, Minn., and Symbiote, Inc., Holland, Mich.)

- *Features:*
  - Transportability
  - Use of cubic space
  - Hangability
  - Bar code application
  - Robotic handling
  - Manual handling
  - Safety
  - Stackability
- *Other:*
  - Kitting
  - PWB handling
  - Accommodation of paperwork
  - Compartmentalization

The new TI tote has outside dimensions of 24 inches × 16 inches × 6 inches (approximately 61 cm × 41 cm × 15 cm) and can be hung from any of the four sides. The tote has internal ribbing on 0.2 inch (0.5 cm) centers providing rigidity and compartmentalization through the insertion of dividers. Tote capacity is increased by adding optional 2 inch (5 cm) tall collars to make the depth 18 inches (46 cm). The tote can carry 95 percent of the IC tubes in the industry. All components of the tote are conductive for ESD control, and paperwork is enclosed in a built-in pouch on the outside of the tote.[6]

14.3.3 Equipment

There are two main classifications of material handling equipment (MHE): discrete product and bulk. Bulk material handling is used in industries dealing with primary resources (agriculture, mining, forestry) and is not discussed here. The electronics industry makes use of most MHE designed for handling of discrete products. The following paragraphs discuss:

- Storage equipment
- Powered industrial trucks
- Conveyors
- Automatic guided vehicle systems
- Automatic storage and retrieval systems

Figure 14-9 illustrates a wide array of storage equipment. Bins, shelving, drawers, and flow racks are designed for manual storage and retrieval. Pallet racks and cantilever racks are designed for storage and retrieval by forklift trucks and at lower levels by hand.

The **carousel** rack is another mode of storage used widely in electronics manufacturing. Figure 14-10 shows single-level horizontal carousels and multilevel horizontal carou-

[6]A. Nager, P. Pole, D. Smith, "Material Handling Tote Developed to Answer the Needs of Electronics Manufacturing," *Industrial Engineering,* 21, 2 (1989), pp. 32–33.

Figure 14-9 Storage equipment.

(*Source:* Raymond A. Kulwiec, *Basics of Material Handling* (Pittsburgh, Pa: Material Handling Institute, 1981), pp. 16–17.)

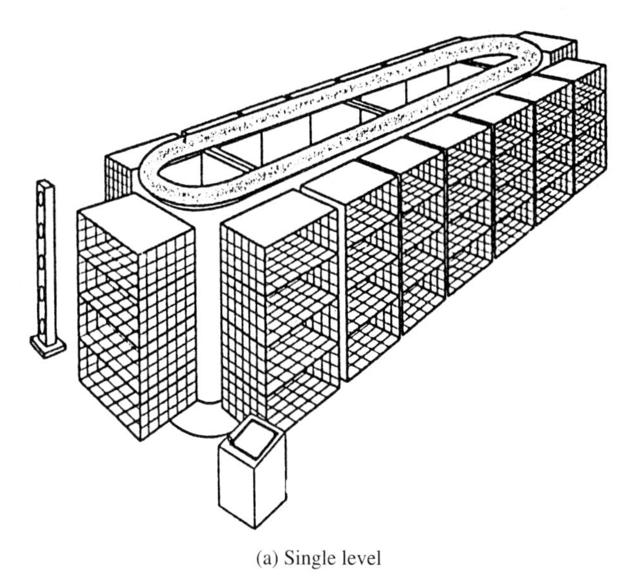

(a) Single level

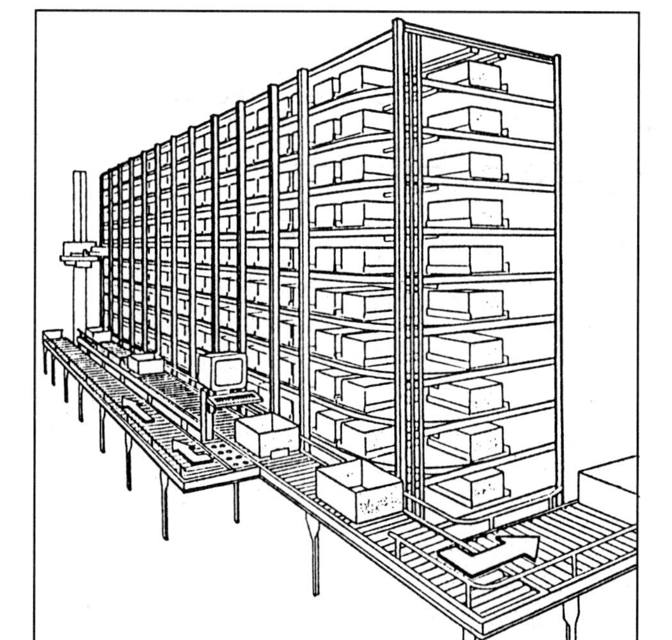

(b) Multilevel

Figure 14-10 Carousels.

(*Source:* Courtesy of Stanley-Vidmar Company, Cincinnati, Ohio.)

sels (rotary racks). These devices are used for storage of small-parts and work-in-process (WIP) inventory and also for temporary storage of finished electronic products undergoing burn-in testing. Vertical carousels (Fig. 14-11) provide excellent protection and security for high-value small parts.

Common powered industrial trucks are shown in Fig. 14-12. Industrial trucks are powered by either battery or internal combustion engine. Most applications in electronics manufacturing are indoors and thus require storage battery-powered electric trucks to avoid exhaust fumes.

Conveyors are also used widely in electronics manufacturing. Figure 14-13 shows the types of conveyors used for handling of general material. Wheel conveyors are unpowered and rely on gravity or manual motive force. Roller conveyors are used to transport tote boxes and cartons. Belt conveyors are used for handling of materials in many industries, including electronics, but they can cause electrostatic discharge damage to electronic products. Overhead trolley conveyors are used for WIP handling, including movement of materials through furnaces or painting operations. Tow conveyors pull carts along the floor by means of a chain drive embedded in the floor, but they are being replaced in many applications by automatic guided vehicles which are more flexible.

Specialized, light-duty, horizontal conveyors provide for material handling between machines in an assembly line. These conveyors include two parallel tracks of continuously moving chain. Either the board edges are supported directly on the chain tracks, or the boards ride on captive pallets. Figure 14-14 shows examples of PWB conveyance.

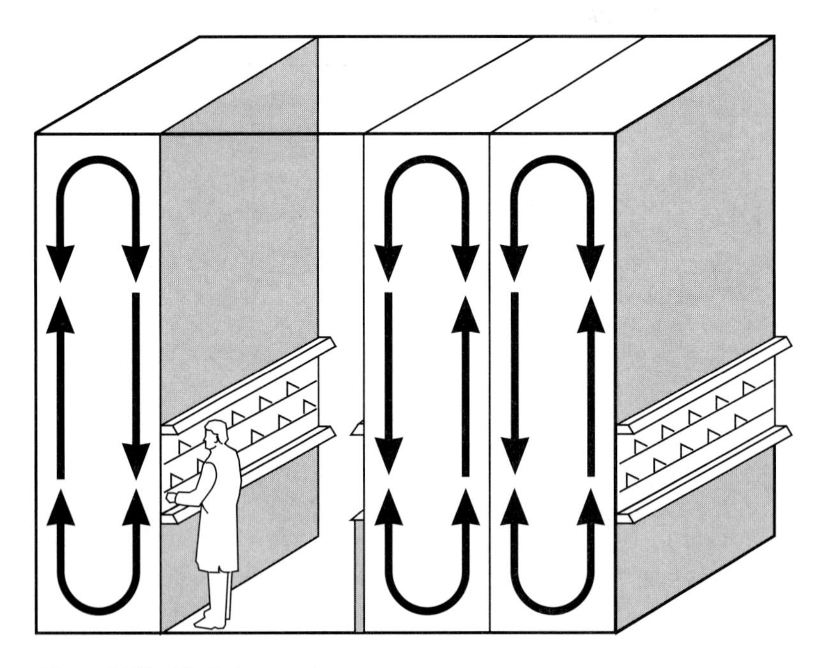

Figure 14-11 Vertical carousel.

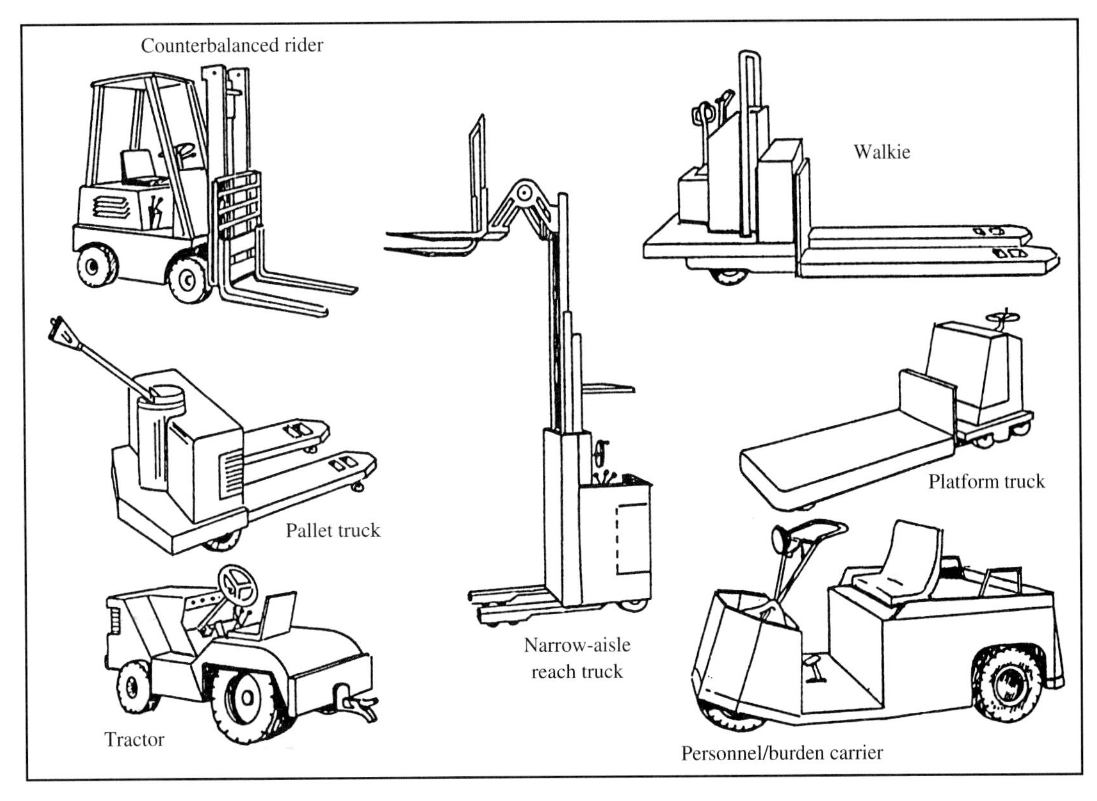

Figure 14-12 Powered industrial trucks.

(*Source:* Kulwiec, *Basics of Material Handling,* p. 14.)

The **automatic guided vehicle system** (AGVS) consists of driverless, automated carts operating on a path, which may be either an electromagnetic wire embedded in the floor or a chemical or retroreflective line on the floor. Figure 14-15(a) illustrates the AGVS for handling of palletized unit loads. The pallet load AGVS is generally used in warehousing and employs wire guidance. The AGVS in part (b) of Fig. 14-15 is a typical light-duty AGVS used for moving tote boxes in an electronics manufacturing plant. The light-duty AGVS may use wire guidance (as shown) or optical sensing for guidance on chemical or retroreflective tape paths. Figure 14-16 is a typical path layout in manufacturing for an AGVS.

Automatic storage and retrieval systems (AS/RS) are primarily used in the electronics industry for warehousing of purchased materials and components for kitting operations to support assembly and for storage of finished goods. Prior to the 1980s, AS/RS were also used for WIP. There are limited applications of AS/RS for WIP, but with the trend to JIT and inventory reduction, the AS/RS is increasingly difficult to cost justify for WIP. Electronics companies prefer to decentralize the WIP and thereby expose it as

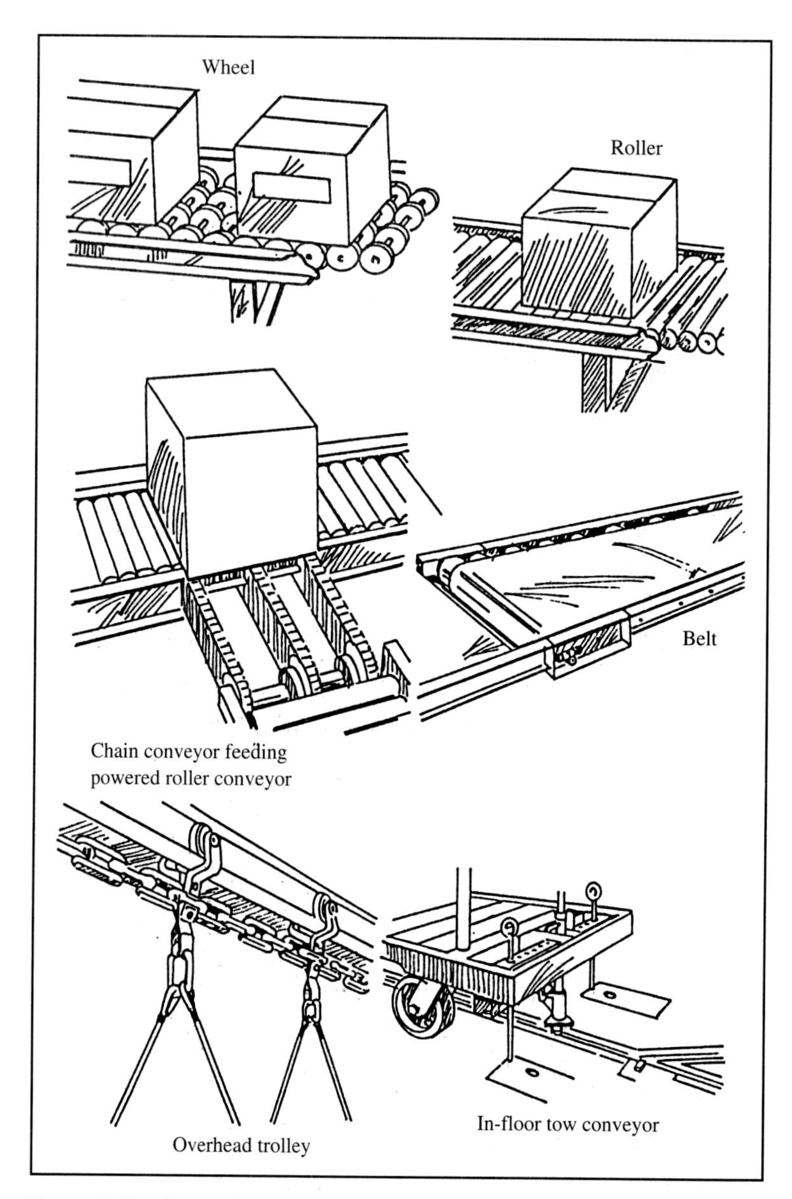

Figure 14-13 Conveyors.

(*Source:* Kulwiec, *Basics of Material Handling,* p. 13.)

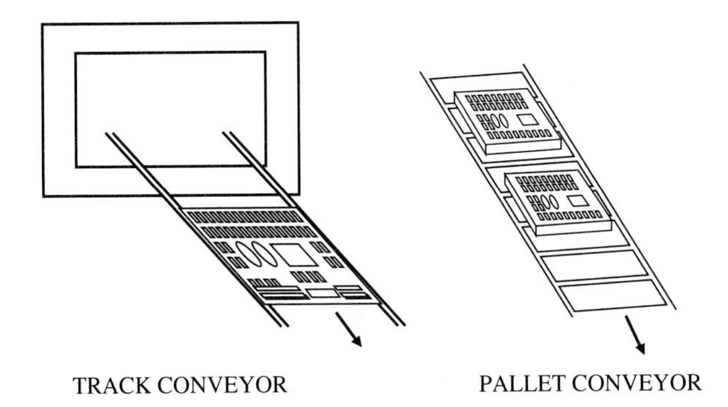

Figure 14-14 PWB conveyance.

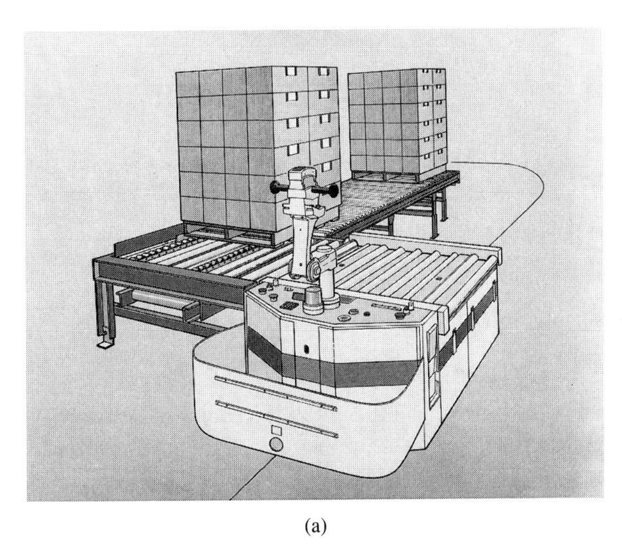

(a)

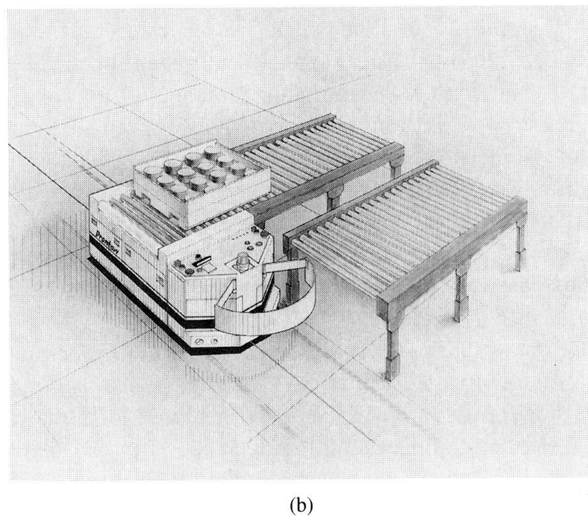

(b)

Figure 14-15 Automatic guided vehicles.

(*Source:* Courtesy of Control Engineering Company, Affiliate of Jervis B. Webb Company.)

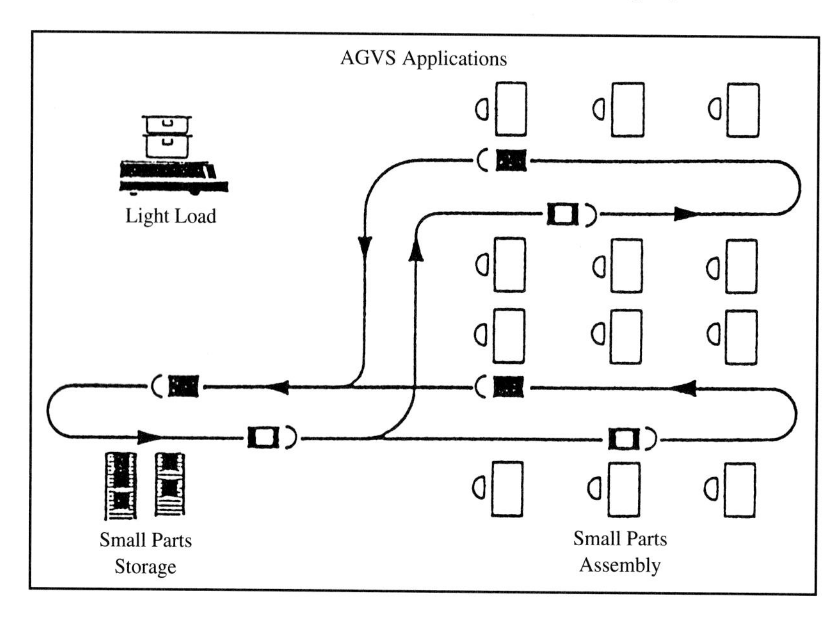

Figure 14-16 AGV path in manufacturing.

(*Source:* Gary A. Koff, "A Primer on AGVS," *Machine and Tool Blue Book,* June 1985, p. 57.)

a resource to be managed rather than hiding it in a centralized area. Small AS/RS called use-point-managers are located at strategic points on the shop floor for focused storage of purchased components. There are two major types of AS/RS: unit load and miniload.

The unit load AS/RS employs automated stacker cranes, operating in captive aisles between high-rise rack structures to store and retrieve palletized loads. Figure 14-17 shows a unit load AS/RS with rack-supported structure and single-mast stacker crane.

The miniload AS/RS is used for small-parts storage. The system operates under computer control, while pickers stand at end-of-aisle stations to handle parts in and out of captive trays (Fig. 14-18). These systems are useful for support of kitting operations, in which the parts needed for an assembly are collected together into kits, then dispatched into assembly. Miniload AS/RS is also used for warehousing of service parts.

Transporters are material handling systems used in the electronics industry for handling WIP material in assembly operations. The mode of transportation is either a conveyor or an AGVS. There are two main classes of transporters: open-loop and closed-loop.

The open-loop transporter is a synchronous, machine-paced system for progressive assembly in mass production (product-focused operation). The pace of the line is determined by the cycle time of the slowest operation on the line. The closed-loop transporter is used in process-focused operations and utilizes a buffer inventory to absorb the differences in processing rates of different operations. In manual assembly environments, closed-loop transporters are referred to as operator-paced systems, since each assembler can operate at a different rate. One operation can be disrupted without stopping all production;

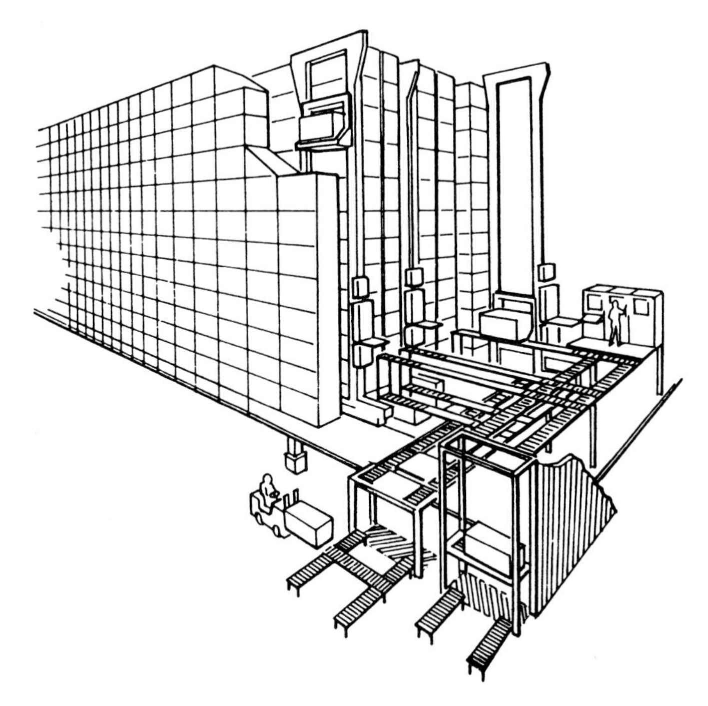

Figure 14-17 Unit load AS/RS.

(*Source:* Kulwiec, *Basics of Material Handling,* p. 46.)

Figure 14-18 Miniload AS/RS.

(*Source:* Kulwiec, *Basics of Material Handling,* p. 44.)

however, there is a penalty of excessive inventory in the buffer storage. With the increasing emphasis on minimizing inventories, closed-loop transporters are becoming less popular.

The open-loop transporters may use either a conveyor or a fleet of light-duty AGVs. The closed-loop transporter generally employs a series of belt conveyors arranged in a loop. Tote boxes containing WIP, in various states of completion, circulate on the transporter loop. The assembly and test operations are located at workstations along the conveyor loop. At least one station is an input/output point for interfacing with buffer storage area(s). Buffer storage can be a series of gravity flow racks (Fig. 14-9) designed to hold the tote boxes and staffed by material handling personnel. Carousel storage systems (Fig. 14-10) are also used, with I/O performed either manually or by a robotic insert/extract machine. The transporter used by Hewlett-Packard for assembly of HP 9000 computers (Fig. 2-13) included a carousel storage system. There are also AS/RS systems, called tote stackers, available, which handle general-purpose tote boxes rather than captive trays.

14.4 ERGONOMICS AND WORKSTATIONS

The use of automation is expanding in electronics manufacturing. Some processes, such as SMT assembly, involve repetitive operations on miniaturized products and require high-quality levels beyond human capacity. However, electronics plants produce a mix of products, some of which require manual handling and assembly. Therefore, it is essential to do a good job of designing the manual processes to achieve overall productivity objectives. Experience has shown that well-designed manual processes are sometimes cost-competitive with automation, while adding flexibility. The prudent approach to process improvement is to simplify the process as much as possible, then apply cost-effective automation. However, successful automation requires that the product be designed for automated processes. Chapter 10 discussed design for automated assembly. This section introduces design principles for manual workstations, based on human capabilities and limitations.

14.4.1 Ergonomics

Ergonomics is the science of human work, and it provides insights into the capabilities and limitations of human operators. Ergonomics research provides knowledge about the distribution of physical capabilities (such as height, reach, lifting capability) in subgroups of the population. These data are compiled in data bases called anthropometric tables. Ergonomics also establishes standards for environmental factors such as illumination, noise, and ventilation.

Most assembly, test, and inspection tasks in electronics manufacturing are performed in the seated position. However, it is desirable to provide the operator the flexibility to alternate between seated and standing positions to reduce fatigue and boredom. In fact, flexibility is a major requirement for any work environment, to accommodate the individual differences among operators.

Figure 14-19 illustrates the proper design for a seated work space. The work surfaces should be at a height which is comfortable for most of the population and should be adjustable, if possible. The engineer designing a workstation should consult recent and complete anthropometric tables and if possible adapt for the specific workers selected to perform the task, considering the work surface height, fixtures, and workpieces.

The work surface should be arranged so that parts, tools, and equipment are comfortably located within the reach of the operator, as shown in Fig. 14-20. Parts bins and chutes are used to locate parts in an arc within the assembler's reach envelope. Tools should be designed and located to avoid awkward holding.

Good ergonomic seating incorporates the following design principles:

- Adjustable lower lumbar support in the seat back
- Waterfall edge on seat front to promote blood flow
- Forward/backward seat adjustment/movement
- Recessed seat to give evenly distributed support
- Pneumatic height adjustment
- Five-star base for stability

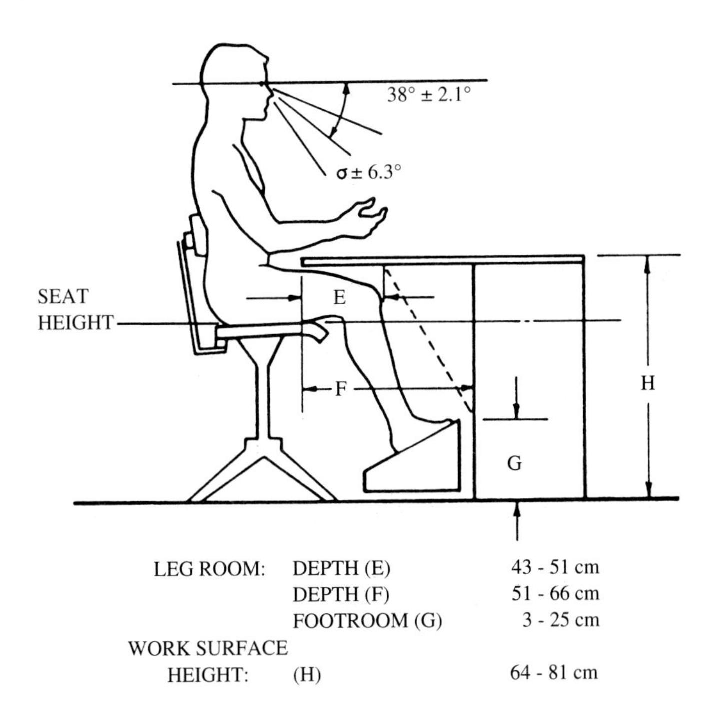

LEG ROOM:	DEPTH (E)	43 - 51 cm
	DEPTH (F)	51 - 66 cm
	FOOTROOM (G)	3 - 25 cm
WORK SURFACE HEIGHT:	(H)	64 - 81 cm

Figure 14-19 Seated operator work space.

(*Source:* Vern Putz-Anderson, *Cumulative Trauma Disorder: A Manual for Musculoskeletal Diseases of the Upper Limbs* (Bristol, Pa: Taylor & Frances Ltd, 1988), p. 98.)

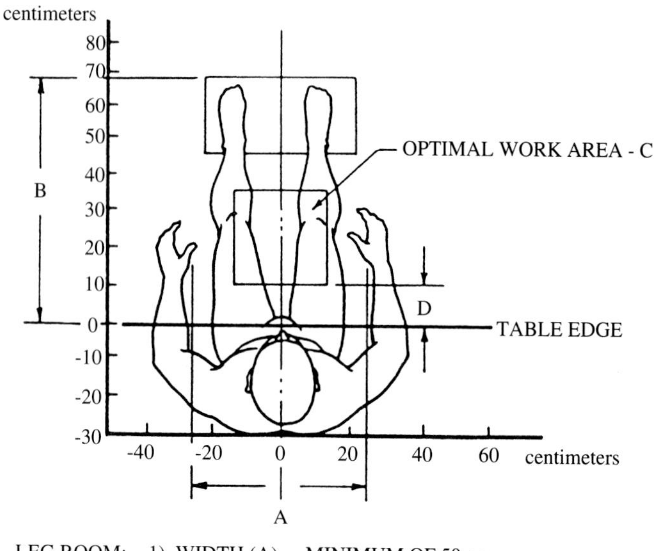

LEG ROOM:	1) WIDTH (A)	MINIMUM OF 50 cm
	2) DEPTH (B)	65 cm, IF LIMITED
OPTIMUM WORK AREA:	3) (C)	APPROXIMATELY 25 x 25 cm
	4) (D)	10 cm FROM EDGE OF WORK SURFACE

Figure 14-20 Seated operator reach envelope.

(*Source:* Putz-Anderson, *Cumulative Trauma Disorder,* p. 90.)

The work surface and seating should promote good posture, with

- Wrists straight
- Elbows down
- Minimal moments on the spine
- Minimal twisting and bending
- Adjustable chairs and work surfaces[7]

Adequate illumination and glare control are also essential for assembly of electronic products. Lighting is particularly important where fine details are critical as in placement of fine-pitch components and in inspection tasks. Both direct (task) and indirect (area) lighting should be provided, and nonglare work surfaces should be selected. Handbooks such as MIL-STD-1472 provide standards for task illumination.

A comfortable ambient temperature is also important. If the temperature is too warm, alertness tends to decrease with resulting decrease in productivity and quality and increase in accidents. Ventilation should be provided to remove the vapors that result from soldering and the use of chemicals such as cleaning solvents.

[7]Ray Pukanic and Donald L. Morelli, "A System Approach to Ergonomically Sound Design for Electronics Assembly/Test Stations," *Industrial Engineering,* 17, 7 (1985), 44–53.

If the workstation is improperly designed, problems will develop. Typical symptoms include:

- Poor morale
- Low productivity
- High rates of defects and scrap
- Absenteeism
- Lost-time injuries
- Increases in worker compensation claims

These problems can be avoided by the use of industrial engineering principles.

Task analysis (the process of scientifically designing work) utilizes information collected by

- Interviewing employees
- Surveying employees by questionnaire
- Observing operations
- Analyzing production records
- Analyzing injury claims

The most effective means of identifying the causes and effects of improper workstation design is to talk to the employees. Most employees are eager to improve their work life and will have greater commitment to any corrective actions in which they are involved in the early stages. When observing operations, the engineer should look for poor posture and identify the causes. For example, scope work in inspection tasks may force an operator to lean forward and hold the neck in a tense, motionless position for extended periods of time. This improper posture leads to neck and shoulder pain. Employees may modify the workplace and/or the task to relieve strain and fatigue. These workplace modifications indicate the problems and possible solutions. Other problem indicators include

- Chairs or work surfaces at improper heights
- Poor lighting or ventilation
- Noise
- Hand tools forcing awkward holding
- Tasks requiring twisting
- Tasks requiring a large range of motion
- Use of hands instead of fixtures to hold workpieces

Cumulative trauma disorder (CTD) problems build up slowly over an extended period of months or even years. An employee may work for several years before the symptoms eventually appear. Similarly, an employee may work for some time with no problems, until the task, workstation, or pace of production is changed; then the symptoms appear.

CTDs are soft tissue problems that tend to be caused by repeated exertions and body movements such as are common in high-volume repetitive assembly in the electronics industry. Assembly tasks often involve repetitive gripping, pushing, or reaching, and these motions may be repeated as often as 25,000 times in an 8-hour shift. Employees suffering from CTD complain of cramping pain, inflammation, and aches. The symptoms usually appear in nerves, tendons, tendon sheaths, or muscles. The occupational risk factors cited with CTD are

- Repetitiveness
- Forcefulness
- Mechanical stress
- Awkward posture
- Vibration
- Low temperature

In the electronics industry carpal tunnel syndrome (CTS) is an important CTD problem. The carpal tunnel is a passageway in the wrist structure, between the wrist bones and the carpal ligament (Fig. 14-21). The tendons that operate the fingers pass through this tunnel. The carpal tunnel is a confined space, and the repetitive flexion and extension of the hand and fingers in some assembly tasks causes inflammation as the tendons slide back and forth in the narrow carpal tunnel. Eventually the inflammation and swelling can

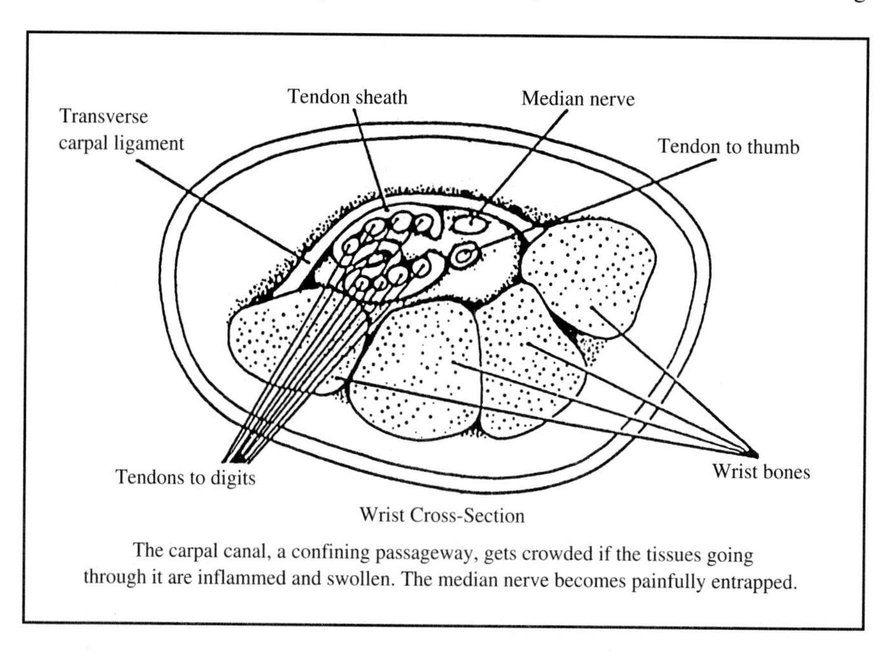

The carpal canal, a confining passageway, gets crowded if the tissues going through it are inflammed and swollen. The median nerve becomes painfully entrapped.

Figure 14-21 Wrist structure and carpal tunnel syndrome.

(*Source:* Courtesy of Occupational Medicine Consultants, Minneapolis, Minn.)

become debilitating. In some cases, employees have been occupationally disabled for several months, and it is not unusual for a CTS case to cost an employer over $20,000 in worker compensation claims.

Power tools are used widely in the electronics industry, particularly for inserting fasteners in chassis assemblies. Operators often hold these tools improperly. Also, they are usually electric or pneumatic, and the vibration aggravates the carpal tunnel syndrome. Figure 14-22(a) illustrates proper and improper holding of powered drivers. Figure 14-22(b) shows some good and improper orientations for holding fixtures.

14.4.2 Workstations

The workstation must be designed to accommodate the ergonomic factors described in Section 14.4.1 and to integrate with the unit loads and storage/handling equipment.

To determine the space requirements for a workstation, we first analyze the process, using tools such as the flow process chart (Fig. 2-11) discussed in Chapter 2. We must define the tools and equipment and estimate the personnel staffing to be located in the workstation. Sufficient space must be provided for a minimum quantity of work-in-process inventory, including parts and supplies to support near-term operations. The optimum space allocation depends on a trade-off between the flexibility and space utilization principles. A small workstation saves on facilities cost and places all processes in close reach. However, a more spacious workstation provides for greater flexibility to rearrange the work area as products and processes change.

Modular furniture systems are popular for design of workstations in the electronics industry. These systems provide a good balance among the following material handling principles:

- Space utilization
- Standardization
- Flexibility
- Unit load
- Ergonomics

Modular workstations are assembled from a variety of standard components, including partitions, work surfaces, and accessories. All furnishings must be durable against impact damage and abrasion and resistant against stains. The partitions may be of material with acoustic deadening to reduce noise, or transparent to allow a more open appearance and better supervision. Work surfaces must be nonglare and conductive. The surfaces should also be resistant to the heat of soldering. Utility outlets and task lighting may be built into the modular components. Alternatively, dense utility grids (with utility drops arranged in a matrix pattern) and adequate ambient lighting can promote layout flexibility.

Accessories are available to enhance productivity, for example:

- Equipment carousels
- Parts bins

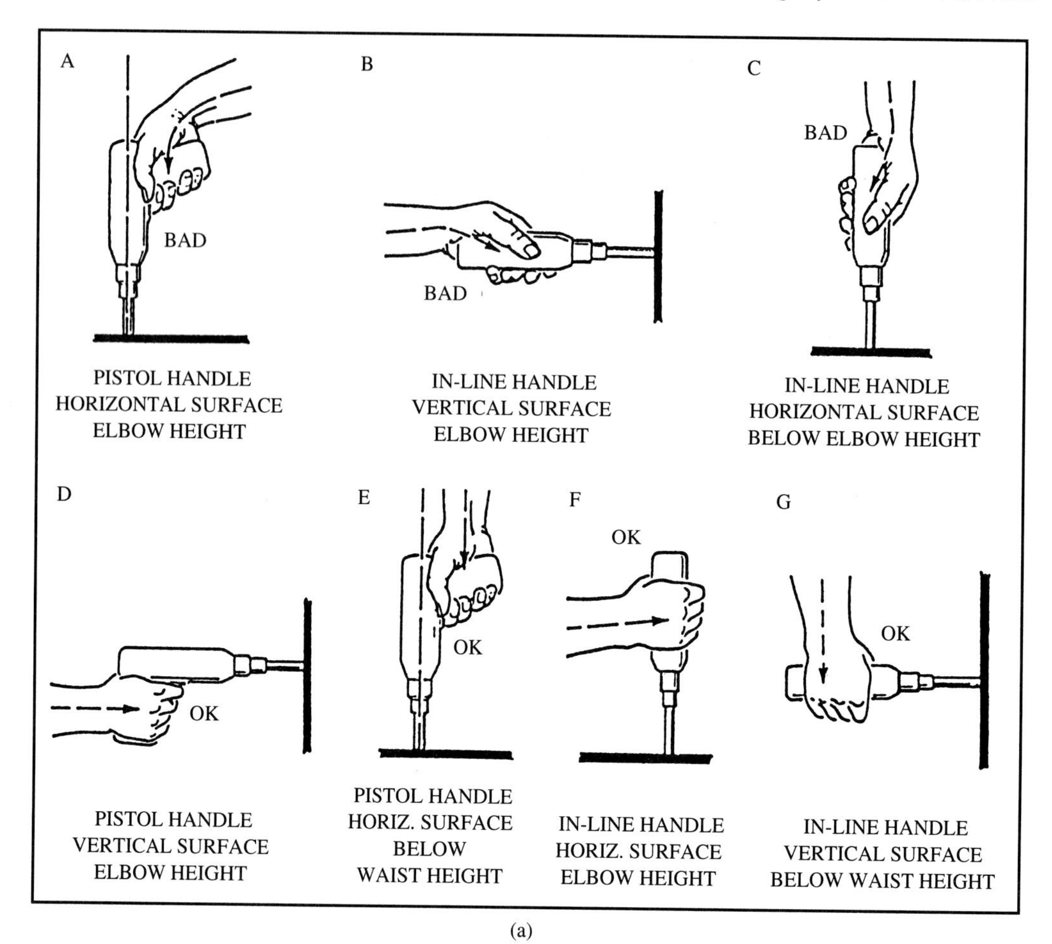

(a)

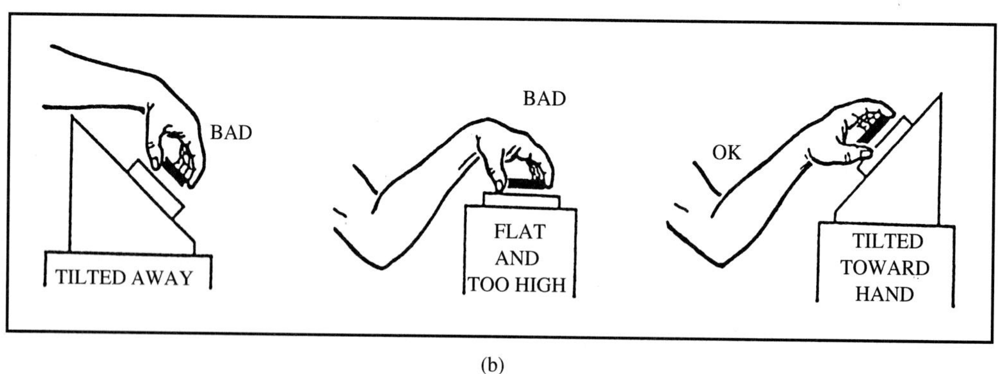

(b)

Figure 14-22 Proper wrist postures.

(*Source:* Putz-Anderson, *Cumulative Trauma Disorder,* pp. 100, 106.)

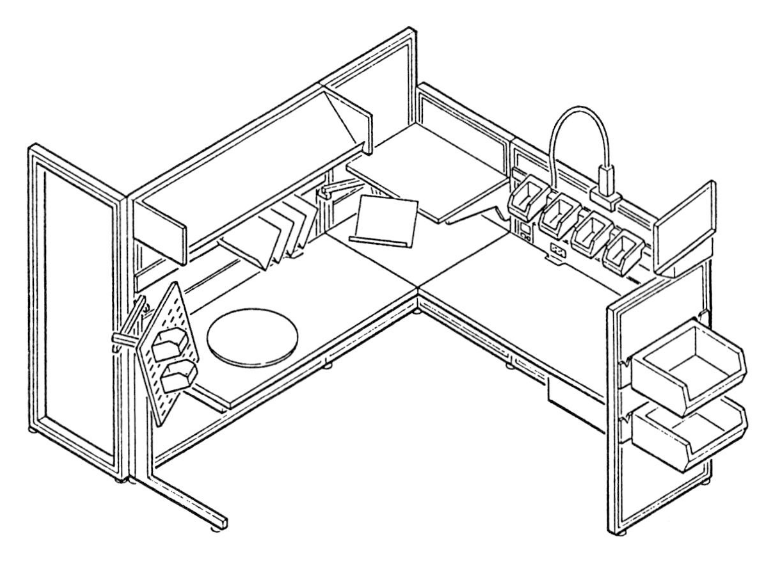

Figure 14-23 Modular workstation.

(*Source:* "Environmental Modular Workstation," U.S. Navy Electronics Manufacturing Productivity Facility, EMPF0003, August 1988, p. 5.)

- Shelves
- Paper flow racks
- Tool balances
- Tote hangers

Figure 14-23 pictures a sample workstation developed by the Naval Electronics Manufacturing Productivity Facility (EMPF) and Panel Concepts/ETS Division.

14.5 ELECTROSTATIC DISCHARGE CONTROL

Control of electrostatic discharges is essential to ensure quality of electronic products, and it is an integral criterion for design of material handling systems and workstations.

14.5.1 Fundamentals

We observe **electrostatic discharge** (ESD) when we walk across new carpet and touch a door knob. We may see a spark and feel a shock. What we have experienced is called **triboelectric charging** and electrostatic discharge. These phenomena occur naturally and cannot be eliminated. The goal in electronic manufacturing is to control ESD and minimize its effects on products.

In the discussion of ESD, we will use the following definitions:

- *Static Electricity:* Electrical charge (excess or deficiency of electrons) at rest (also called **electrostatic charge**).
- *Triboelectric charging:* Accumulation of electrostatic charges (positive and negative ions) on the surfaces of two materials (one or both are insulators) when brought into contact and separated.
- *Electrostatic discharge:* Sudden transfer of electrons between two bodies at different static potentials.

In order to understand these phenomena of electrostatic discharge, we must look at the atomic level. At the center of the atom is a nucleus containing positively charged protons. Negatively charged electrons circle about the nucleus. The atom is electrically balanced so long as it maintains an equal number of electrons and protons. If an electron attains a specified energy, it will leave the atom (provided there is some nearby atom to which it may jump). The old atom becomes positively charged and is called a positive ion. The electron makes its new atom negatively charged—a negative ion. These charged ions collect on surfaces such as human hands, work surfaces, and the leads of components. The process is illustrated in Fig. 14-24.

When a human hand rubs against a vinyl surface, the friction energizes electrons. The electrical characteristics of the hand and vinyl are such that the electrons migrate to the vinyl, making it negatively charged. The hand leaves the vinyl and has a strong positive charge that attracts electrons in any nearby object.

There are two primary models for ESD failure. The human-body model (HBM) describes the situation where a person builds up electrostatic charge (such as by walking on carpet), then transfers the charge to a workpiece, such as an integrated circuit. The charged-device model (CDM) assumes a charge on the lead frame of a device and other conductive paths that is quickly discharged through one pin to ground. This case arises when a DIP builds up electrostatic charge on its surface sliding down a tube feeder. The first corner pin to contact the assembly machine provides a path for ESD.

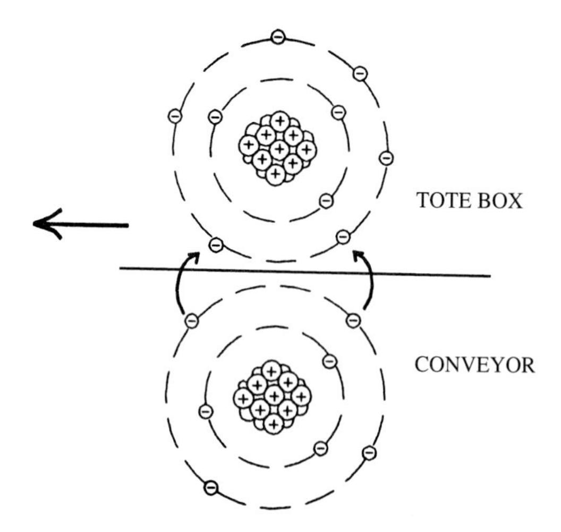

Figure 14-24 Atomic view of ion formation.

Induction is a magnetic phenomenon allowing ESD to occur without direct contact. When we see a magnet move iron pieces without touching them, we have witnessed induction. When a strong electrical charge creates a magnetic field, induction may result. Induction can be either a charging or discharging process, depending on the vantage point. When the magnetic attraction becomes strong enough, the electrons will jump between two objects.

Table 14-1 shows the triboelectric series. The materials at the top of the table can

TABLE 14-1 TRIBOELECTRIC SERIES

Materials	Polarity (+ or −)
Asbestos	Acquires a more positive charge
Acetate	
Glass	
Human hair	
Nylon	
Wool	
Fur	
Lead	
Silk	
Aluminum	
Paper	
Polyurethane	
Cotton	
Wood	
Steel	
Sealing Wax	
Hard Rubber	
Acetate Fiber	
MYLAR*	
Epoxy Glass	
Nickel, Copper, Silver	
UV Resist	
Brass, Stainless Steel	
Synthetic Rubber	
Acrylic	
Polystyrene Foam	
Polyurethane Foam	
SARAN†	
Polyester	
Polyethylene	
Polypropylene	
PVC (vinyl)	
TEFLON*	
Silicone Rubber	Acquires a more negative charge

*Trademark of E. I. Du Pont de Nemours & Co. Inc.

†Trademark of Dow Chemical U.S.A.

Source: Theodore B. Dangelmayer, *ESD Program Management* (New York: Van Nostrand Reinhold, 1990), p. 40.

TABLE 14-2 ELECTRONIC DEVICE CLASSIFICATION

Device class	Static threshold
Unprotected MOS Junction FETs LSI Microcircuits & Hybrids Ultra-high & microwave devices	< 1,000 volts
Schottky diodes Precision resistor networks TTL and high-speed ECL Protected LSI	1,000 to 4,000 volts
Piezoelectric crystals Resistor chips Low-power chopper transistors Silicon transistors Small signal diodes (< 1 watt), excluding zener diodes	4,000 to 15,000 volts

hold a positive charge most easily, while the materials at the bottom hold a negative charge most easily. Cotton is considered a neutral or middle point in the series. The further separated two materials are in the series, the greater the charge generated between the two when in contact.

The ability of a material to hold a charge is related to its conductivity. Materials are classified by their surface conductivity, whereas electronic devices are classified by their susceptibility to voltage (Table 14-2).

The susceptibility of a device to ESD is based on its ability to withstand voltage differences. So how does ESD damage the device? ESD will produce a voltage at one or more pins. Inside the device this voltage will drive a proportionally strong charge into the circuit (Fig. 14-25). If this charge is strong enough, it will puncture a hole through the circuit's transistors. A large hole will cause immediate device failure. A smaller hole

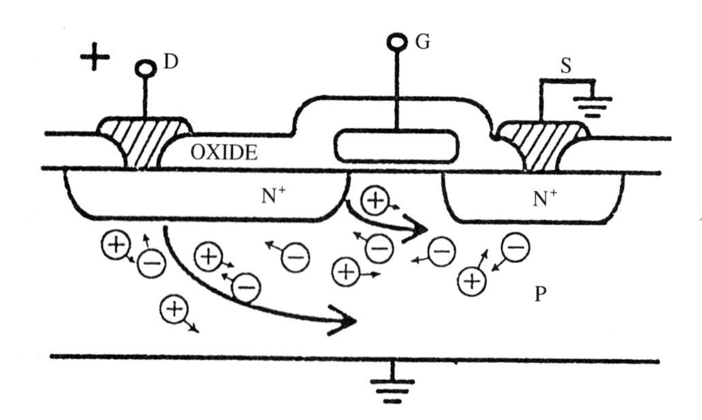

Figure 14-25 Generic ESD damage to a transistor.

(*Source:* Timothy J. Maloney, "Contact Injection: A Major Cause of ESD Failure in Integrated Circuits," *EOS/ESD Symposium Proceedings* (1986), p. 167.)

is an example of a latent defect, as defined in Chapter 13, and although not immediately detectable, it will eventually alter the operating characteristics and cause the device to fail.

Production facilities with minimal or no ESD control have found that up to 90 percent of the infant mortality was related to ESD. Failure analyses for several military products found that about 70 percent of all failures were ESD related. Under proper ESD control, these rates may be reduced to 5 percent.

Proper ESD control requires a workforce educated concerning ESD and carrying out ESD-controlled material handling. In order to demonstrate such handling, we will first discuss ESD controls and then look at an ESD-controlled production process.

14.5.2 Controls

Since ESD results from stored charges, one way to defeat ESD is to prevent charges from accumulating. This strategy, called the grounding principle, requires a ground path at any point where friction may occur. Another way to defeat ESD is to use ESD protective materials. The U.S. Department of Defense has classified ESD protective materials into three groups: conductive, static dissipative, and antistatic. If the static-sensitive device has developed a surface charge, then ESD may occur by contacting a conductive or static dissipative material. A conductive material has a surface resistivity of less than 10^5 ohms/cm^2. Materials such as metals, bulk conductive plastics, wire impregnated materials, and conductive laminates are conductors. If the surface resistivity is between 10^5 and 10^9 ohms/cm^2, then the material is classified as static dissipative.

An antistatic material could be defined as a material with a surface resistivity in the range of 10^9 to 10^{14} ohms/cm^2. Example antistatic materials according to this definition are melamine laminates, high-resistance conductive plastics, cotton, wood, and paper products. It is easiest to think of an antistatic material as being any material with a surface conductivity that discourages triboelectric charging. Such surfaces are achieved by using soap-based additives. These are the same additives used in laundry detergents to remove static cling. In the aerospace/defense industry, these additives are sometimes called **topical antistats.** A more generic name for these antistatic materials is **hygroscopic oils.** These chemicals are generally sprayed on the surface and initially saturate the material, then work towards the surface over time. At the surface, hygroscopic oils are effective for a period of time, but they are eventually rubbed off of the protected surface.

A hygroscopic antistatic material uses moisture from the air to alter the surface conductivity to a point where friction cannot generate a charge. Nonhygroscopic materials would use a lubricant to discourage friction but are not used as widely as are hygroscopic treatments.

Another approach to preventing ESD damage is to design devices to withstand ESD. Protection diodes, field oxide transistors, and resistors can be used to protect the device from damage. The key to these designs is that the device must respond to the transient fast enough to redirect the charge before it affects the protected device. This approach is mostly applied at the board level by grounding the devices. However, this approach is ineffective unless the board is grounded.

Shielding (the use of a faraday cage, as shown in Fig. 14-26) is the only way to

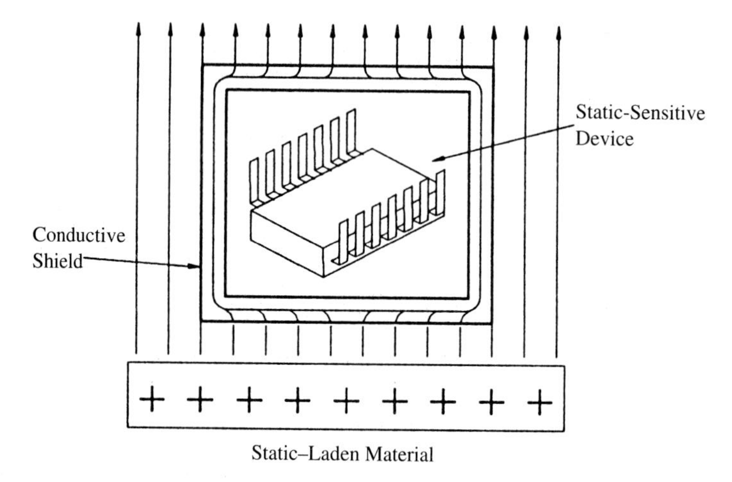

Figure 14-26 The faraday cage.

approach 100 percent protection from ESD. The faraday cage completely surrounds an object with conductive material and passes any charge around the object without affecting the charge on the protected object.

A shielded, antistatic bag is a good example of a faraday cage. This bag consists of three layers. The outer layers are made of antistatic material, and the inner layer is a conductive material. The idea is to minimize the surface charge both inside and outside the bag, while shielding the contents from induction. There are several different types of antistatic bags available on the market, but not all are faraday cages.

Where faraday cages are infeasible and antistatic or grounding practices are inadequate, ionizers must be used. An ionizer is a device that broadcasts positive and negative ions out into the workplace. These ions come in contact with and neutralize the developing charges. Ionizers are effective but only over small areas. Beyond a radius of one meter the ions become too dispersed to be effective in neutralizing charges.

14.5.3 Material Handling

In properly ESD-controlled material handling, at any point where friction can occur there is a conductive path. Any object that may touch the static-sensitive device is either made of antistatic material or is grounded. Figure 14-27 illustrates the possible components of an ESD-controlled workstation: ionizer, conductive bench top (connected to ground), a conductive floor and/or mat, conductive seat cover, conductive apron, and grounding straps on one or both wrists and one or both feet. A bottle of antistatic spray might be located nearby. Many common nonconductive objects would also be restricted from the workstation (for example, styrofoam cups, cellophane tape, and vinyl shop carriers containing workorders and other shop paperwork, envelopes, and notebooks).

Material handling of a static-sensitive device usually requires a faraday cage, which may be a shipping container, tote box, or antistatic bag. However, storage and transportation of electronic devices requires ESD controls beyond the faraday cage. There are two reasons why the faraday assumption is weak:

- Antistatic tote boxes lose their antistatic properties.
- Tote boxes do not completely surround their contents.

At any point where the device is moved from one faraday cage to another, the workstation controls must be used.

The production facility itself must be designed to reduce charge accumulation. The floors must be well grounded. The flooring must not be synthetic, plastic coated, carpeted, wooden with lacquered finish, or completely painted. Araldite is a recommended floor material. The floors should be periodically washed with aqueous glycerine. Any handrails must be conductive and grounded. Wooden structures are not sufficient conductors for use as ground paths. If the facility is in an area with low humidity, then humidifiers and/or ionizers should be used. When humidity drops below 65 percent, the surface conductivities may not develop. Below 40 percent humidity, topical antistats are ineffective.

There are some manufacturing processes that generate static electricity very close to sensitive devices. Examples are wave soldering, board trimming, and any cutting action such as routing. These processes may require ionizers in addition to grounding and antistatic devices. There are also specific corrective actions on manufacturing processes. For example, solder-suckers must use metallized exteriors, and when using cleaning boards, an antistatic cleaner should be used.

Manufacturers' guidelines for ESD control specify techniques for different levels

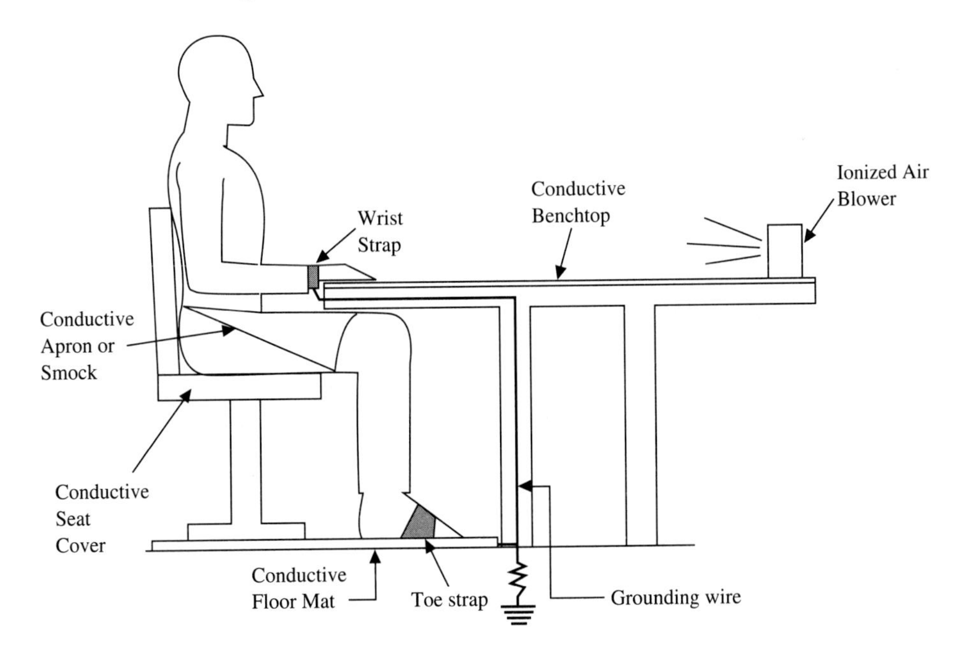

Figure 14-27 ESD-controlled workstation.

of device sensitivity, in terms of threshold voltages. The guidelines also specify control techniques as optional, required, or not required in typical functional areas, such as storage, kitting, assembly, test, and rework.

ESD-controlled storage usually requires antistatic bags inside antistatic totes. ESD-protective racks are located away from induction sources such as high-voltage equipment.

Movement of static-sensitive devices requires that the transporting system be grounded as much as possible. The shop floor should be conductive and at ground potential. Any movable transportation vehicle (for example, a cart, forklift truck, or AGVS) will need a grounding drag-chain or other grounding device. Once the transporting system is grounded, the device being transported must not touch a conductive object. This is accomplished by keeping the device in an antistatic bag or tote box.

Standard belt-driven conveyors are not recommended, as they are inherently static-laden. Manufacturers have found strong surface charges on antistatic tote boxes coming off conveyors. A transporter belt can easily develop an 80 kV charge (in contrast to a human who develops 20 kV walking on carpet). If a conveyor must be used, the belt should be made of a conductive material (typically carbon-based). The bearings and framework must be grounded and lubricated with admixtures of graphite or molybdenum sulfide. A conveyor system will need to be frequently tested for charge accumulation. In general, all handling controls should be routinely audited and tested to ensure they are providing the proper ground paths.

The shipping container and packing material must also be made of antistatic materials. Static-safe packaging is clearly labeled with symbols such as the industry-standard Static Awareness Symbol in Fig. 14-28. The industry tends to use a pink-tinted, almost transparent, coloring scheme for antistatic packages. The Electronics Industries Association has developed static-safe standards for shipping containers.

A common shipping approach for leaded components is to press the leads into an antistatic carbon-base pad. The pad is then attached to an antistatic plastic box, or placed inside an antistatic bag. The box or bag is placed in a shipping container and packed with antistatic materials. Figure 14-29 shows an example of a shipping box containing tubes of ESD-sensitive DIP devices.

ESD controls are only effective if the employees understand the need for and know how to apply the controls. Employees need ESD training if they are in any way involved

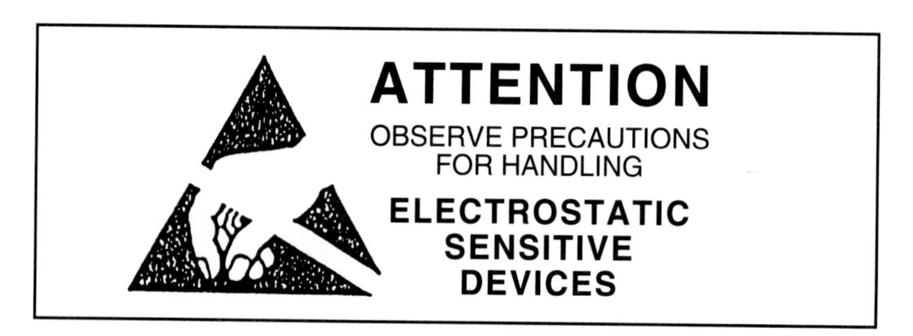

Figure 14-28 Industry standard symbol for ESD-sensitive devices.

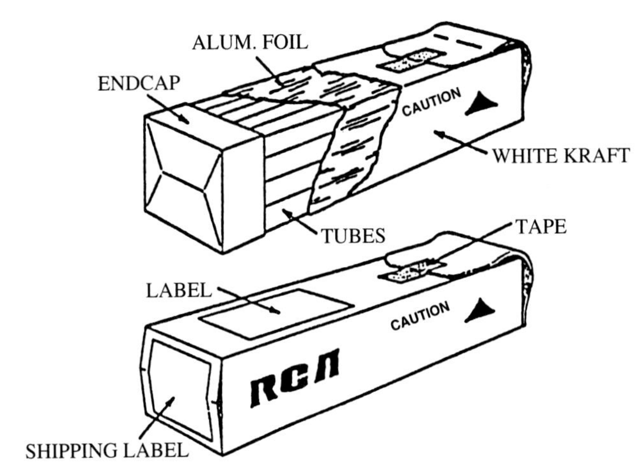

Figure 14-29 Example shipping container for tubes.

(*Source:* A. F. Murello and L. R. Avery, "Study of Antistatically Coated Shipping Tubes Using Static Decay and Triboelectric Tests," *EOS/ESD Symposium Proceedings* (1986), p. 158. Courtesy of RCA Corp.)

with ESD-sensitive devices. The following list of topics has been suggested for an in-house ESD training program:[8]

- *Static electricity:*
  Methods of charge generation
  Induction charging
  Electrostatic field
  Sources of charges
  Distribution and dissipation of charges
- *ESD-sensitive parts:*
  Part susceptibility to ESD
  Sensitivity classifications
  ESD failure modes/mechanisms
  Examples of failed parts
  Equipment-level failures
- *ESD control techniques:*
  Design considerations
  ESD protective circuitry
  ESD protective materials and equipment
  ESD protective packaging and labeling
  Handling precautions and procedures
  Implementation requirements and guidelines.

In many electronics plants, one individual will be designated as responsible for ESD training and audits. During the ESD control audit, this person will test the ESD controls to ensure the production process will not damage products. Figure 14-30 is a typical format for auditing the proper use of ESD controls at a workstation.

[8]Spyros Vrachrnas, "Guidelines for Performing ESD Control Program Evaluation," *Evaluation Engineering,* June 1988, pp. 100–106.

ESD AUDIT FORM S/N: ______________________

Audit Date: ______________ Area: ______________________
Audit Time: ______________ Number of Personnel: ______________
Auditor: ______________

_______ Verify Previous Corrective Action Requirements

_______ New Corrective Action Requirements

_______ Wearing wrist straps _______ # Nonconforming
_______ Wrist strap test _______ # Tested _______ # Nonconforming
_______ Wearing smocks _______ # Nonconforming
_______ Ground workbench _______ # Tested _______ # Nonconforming
_______ Proper use of static bags, boxes
_______ Proper storage of sensitive parts
_______ Static generating material in work area
_______ Floor mat usage required
_______ Heel strap usage required
_______ Air ionizer required

Comments: ____

Corrective Action Response

Person Receiving Audit Report: ______________________ Date: ______________

Correction action taken for any items checked in the upper portion of form:

Signed: ______________________ Date: ______________
Please respond by: ______________________

Figure 14-30 Example ESD audit form.

14.6 ENVIRONMENTAL CONTROL SYSTEMS

Some production processes in the electronics industry require stringent environmental controls, either at specific workstations or throughout a general work area. Environmental controls may be required on the air, water, and gasses.

14.6.1 Clean Rooms

Control of airborne contamination is a critical issue in manufacturing today, particularly for semiconductors, printed wiring boards, hard-disk drives, aerospace systems, medical equipment, and pharmaceuticals.

A particle diameter of 1 μm may cause the read/write head of a hard-disk drive to crash. In the semiconductor industry, particles as small as 10 nm can have an impact on device performance and/or yield if deposited in a critical area. With the increasing trends of large-scale integration and mechatronics, there will be continual challenge to eliminate ever-smaller particles of contamination from the work place.

It is only through an exceptionally well-controlled local environment that the presence of such contaminants can be prevented. This environment is called the clean room. The clean room prevents contamination from the outside ambient atmosphere and is designed so that particles generated internal to the clean room work space are eliminated.

Of all the manufacturing industries, the semiconductor industry has one of the most difficult challenges in contamination control. Path dimensions in microelectronics are frequently less than 0.5 μm, requiring control of contaminants less than 0.2 μm in size. Our discussion of clean room methods, accessories, and specifications will be based on the requirements of IC wafer fabrication, unless otherwise stated.

14.6.2 Nature and Origins of Contamination

Contamination is classified in four general categories:

- *Particulate:* Dust, chips, fibers, etc.
- *Chemical:* Gases, liquids, oily films, etc.
- *Biological:* Bacteria, virus, fungi, spores, etc.
- *Energy or Changes of State:* Light contaminates film, magnetic fields contaminate iron or nickel, radiation affects living cells, etc.

Section 14.5 discussed a form of energy-based contamination: ESD. This section will be concerned with the first three categories of contamination and the methods by which they are controlled. In the typical local atmosphere there are thousands of particles that go unseen and unnoticed. In a rural area under normal atmospheric conditions, a count of 40,000 particles (of $\geq$ 0.5 μm diameter) per cubic foot (0.03 m^3) is not unusual. In metropolitan areas, contamination counts of up to 1.5 million particles per cubic foot (0.03 m^3) are normal. General industrial areas will have up to 1 million particles above 1.0 μm, with a normal range of distribution up to 600 μm. Table 14-3 lists some typical sources and sizes of industrial contamination.

In addition to the external, atmospheric contaminants which have been mentioned, ordinary activities generate contamination internal to the clean room. See Table 14-4 for some typical sources. The number of personnel that are allowed in the clean room at any given time is also of importance in controlling airborne contamination. Each additional person in the clean room generates further contamination, as shown in Table 14-5.

TABLE 14-3 SOURCES OF INDUSTRIAL CONTAMINATION

Source	Particle size range, μm
Combustion products	
Power generating plants	0.5 to 50
Refineries	0.5 to 50
Commercial transportation (including private autos)	0.1 to 10
Heating plants	0.1 to 1200
Exhaust from chemical processing plants	2 to 10
Construction	
Erection of new buildings	1 to 50
Demolition of buildings	1 to 100
Construction and repair of streets and roads	1 to 100
General	
Mining and quarries	1 to 500
Cement plants, foundries, steel mills	0.5 to 1000
Domestic coal smoke	0.01 to 5

Source: NASA, 1969.

There are three types of contamination transfer:

- *Airborne contamination:* The contamination can be of any size, shape, or nature and is transferred by air currents or by the trajectory of a particle during a work operation.
- *Direct contact:* Particulate, biological, or chemical matter carried from one subject to another by direct contact of the subjects.
- *Fluids:* Transfer of contaminant to product by way of cleaning solutions, lubricants, and propellants.

There are six major areas that must be addressed in the control of clean room contamination:

TABLE 14-4 TYPICAL SOURCES OF PARTICLES

Activity	Particle size, μm
Rubbing ordinary painted surface	90
Sliding metal surfaces (nonlubricated0	75
Crumpling or folding paper	65
Rubbing an epoxy painted surface	40
Seating screws	30
Belt drive	30
Writing with ballpoint pen on ordinary paper	20
Handling passivated metals such as fastening materials	10
Vinyl fitting abraded by a wrench	8
Rubbing the skin	4

Source: NASA, 1969.

TABLE 14-5 INCREASE OF CONTAMINATION LEVELS BY PERSONNEL

Activity	Times increase over ambient levels (particles, 0.2 to 50 μm)
Personnel movement	
Gathering together 4 to 5 people at one location	1.5 to 3
Normal walking	1.2 to 2
Sitting quietly	1 to 1.2
Laminar flow workstation with hands inside	1.01
Laminar flow workstation—no activity	None
Personnel protective clothing (synthetic fibers)	
Brushing sleeve of uniform	1.5 to 3
Stamping on floor without shoe covering	10 to 50
Stamping on floor with shoe covering	1.5 to 3
Removing handkerchief from pocket	3 to 10
Personnel *per se*	
Normal breath	None
Breath of smoker up to 20 min after smoking	2 to 5
Sneezing	5 to 20
Rubbing skin on hands and face	1 to 2

Source: NASA, 1969.

- Facility design
- Airflow
- Equipment
- Ultrapure water and gases
- Personnel training
- Maintenance and record keeping

These six topics, and the options involved in each, will be discussed individually in the following sections.

14.6.3 Facility Design

Clean room facilities are classed according to the number and size of particles per cubic foot (0.03 m^3) of clean room space that can be detected with current sampling methods. The levels of these classes are based on statistical distribution for average particle sizes. For statistical purposes, the particle counts are taken during periods of normal work activity and at locations where the air approaches the work area. Clean room classifications are set according to U.S. Federal Standard 209c. For example:

- *CLASS 100:* This class allows a maximum concentration of 100 particles of diameter $\geq$ 0.5 μm per cubic foot (0.03 m^3) and not more than one particle 4.0 μm or larger.

- *CLASS 10,000:* This class allows a particle count per cubic foot (0.03 m^3) of air not to exceed 10,000 particles 0.5 μm and larger, 65 particles 5.0 μm and larger, and not more than 1 particle 35 μm and larger.
- *CLASS 100,000:* This class allows a particle count per cubic foot (0.03 m^3) of air not to exceed 100,000 particles 0.5 μm and larger, 700 particles 5.0 μm and larger, and not more than 1 particle 100 μm and larger.

14.6.4 Airflow

Before air can enter a clean room, it must go through a three-stage filtering process. The first stage is a roughing filter made of glass wool or a similar material. This stage will stop the larger contaminants of size 10 μm or greater. The second stage may be an electrostatic filter or fiberglass, and it will generally have an efficiency rating from 35 to 85 percent for particles above 0.3 μm.

The final stage **High Efficiency Particulate Air (HEPA) filter** uses a medium of dry ultrafine fibers (usually less than 1 μm in diameter) that have been formed into thin porous sheets. These sheets are fan-folded or pleated to render maximum surface area for filtering the air and are bonded to a rigid frame with an adhesive to prevent leakage around the edges.

HEPA filters will provide a minimum airflow capacity of 500 cubic feet (14 m^3) per minute and have a minimum efficiency rating of 99.97 percent for 0.3 μm particles, at airflows of 20 and 100 percent of the rated flow capacity of the filter assembly. These filters are fragile and demand great care in handling during monitoring, installing, removing, and repairing. HEPA filters cannot be cleaned of contaminants but rather must be replaced.

The air in a clean room is constantly moving and being recirculated through the HEPA filter banks. Since the air in a clean room is much cleaner than outside air, only the minimal amount of new air is added as necessary to maintain a positive pressure with respect to the areas surrounding the clean room. This positive pressure prevents particle migration into the clean room when people or products pass through the air locks to enter or leave the area.

There are three basic approaches in designing the airflow for clean rooms, and each approach has its advantages and disadvantages. The three types of airflow are conventional (nonlaminar), horizontal laminar, and vertical laminar.

Conventional flow clean rooms are characterized by nonlaminar airflow. The airflow is in random patterns, rather than straight, sheetlike patterns. There are various arrangements for the air inlet and exhaust grilles for this type of clean room, as shown in Fig. 14-31. The conventional flow system offers the following advantages:

- Production line work-flow patterns are not critical and are simple to lay out.
- Filters and air handling are less complex and are easy to maintain.
- Size is more flexible and easy to expand.
- Construction and operation are least expensive.

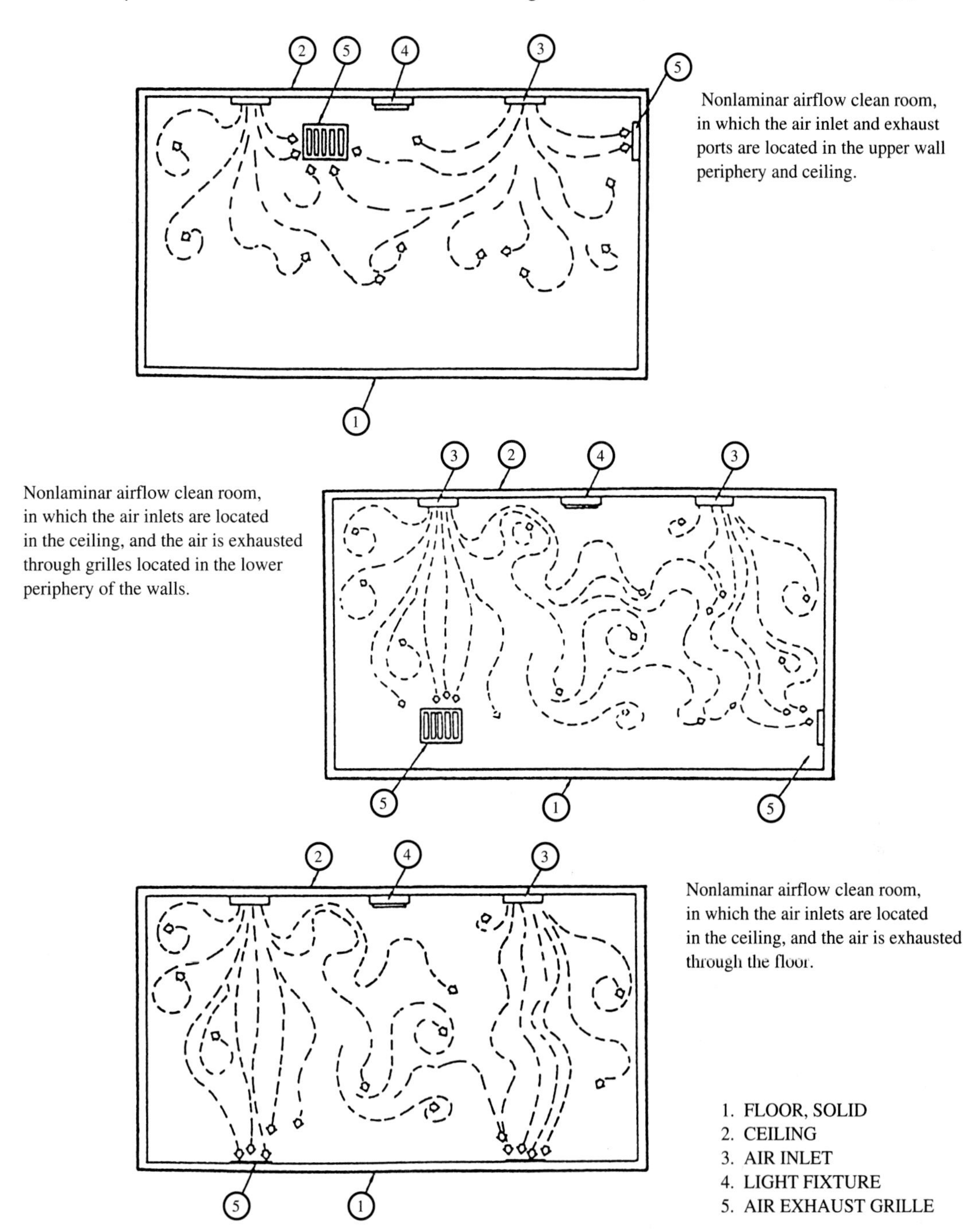

Figure 14-31 Random airflow patterns.

(*Source:* NASA, 1969.)

The primary disadvantages of the conventional flow system are as follows:

- Recovery from contaminated condition is slow.
- Changes of air are few (20 to 25 per hour).
- Frequent janitorial service is required.

Laminar airflow is defined as airflow in which the entire body of air within a confined area moves with uniform velocity along parallel lines. This is a much more practical and efficient method of maintaining low particle count in a controlled environment than is the conventional flow system. The laminar flow concept offers the choice of horizontal or vertical laminar flow, and a wide range of flow rates may be used to suit various product needs. Laboratory tests indicate that vertical laminar airflow rooms should not be operated below 65 linear feet per minute (0.33 m/sec). Horizontal facilities require air velocities from 100 feet per minute (0.51 m/sec) for installations up to 25 feet (7.6 m) in length, to speeds ranging from 125 to 140 feet per minute (0.64 to 0.71 m/sec) for facilities of greater length.

The advantages of the laminar airflow system are as follows:

- Rapid recovery from a contaminated condition
- Deposition and resuspension of particles is minimal
- Reduced maintenance is possible (airborne contamination produced internally is exhausted from the system, so vacuuming the area is unnecessary)
- Class 1 is achievable.
- Air showers, double door airlocks, and large dressing rooms are usually unnecessary.

Disadvantages of the laminar airflow system include the following:

- The higher air velocity requires larger fans, motors, and ducts.
- Ideal uniform velocity profiles are difficult to attain with personnel, equipment, and movement in the room.
- Cost per unit of floor area is considerably higher than conventional airflow types.

In the **horizontal laminar flow** (HLF) room, the HEPA filter bank covers one entire wall, and the opposite wall is a grille-covered exhaust area (Fig. 14-32). The air moves horizontally from inlet wall to outlet wall.

An HLF is considered a zoned space. The station nearest the HEPA filter bank may be class 100, but air quality degrades toward the outlet because particles from one station tend to migrate downstream to other stations.

In the **vertical laminar flow** (VLF) room, the HEPA filter bank covers the entire ceiling and the floor grating is the exhaust area (Fig. 14-33). This system is extremely efficient in the removal of airborne particles that are generated internal to the clean room, and it will operate well within the class 10 level.

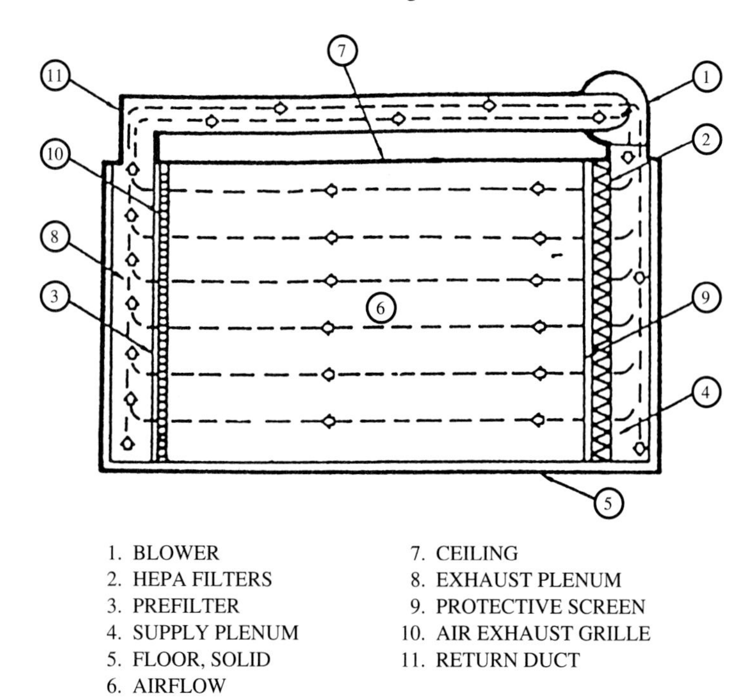

Figure 14-32 Horizontal laminar flow clean room.

(*Source:* NASA, 1969.)

14.6.5 Equipment

Clean room equipment includes tools, furniture, personnel garments, and accessories. Tools selected for clean rooms should be chosen to minimize particulate dispersion during use. Moving parts should be shrouded with vinyl boots or other devices to contain excessive contamination. Friction points should be equipped with DuPont Teflon™, nylon, or other self-lubricating bearings. Stainless steel ball-bearing hinges should be used for doors and openings. All exhaust air should be dumped outside the clean room.

Laminar flow workbenches are used to provide a clean environment in a small area, rather than requiring the entire work area to meet a stringent standard. For example, a laminar flow workbench may provide a class 100 work space in a class 10,000 clean room. Like the clean room itself, a laminar flow workbench may be of either the horizontal (a) or vertical (b) flow variety (Fig. 14-34).

Clean room furniture is chosen to minimize cleaning, maintenance, and particle dispersion. In general, there should be a minimum of sharp edges and corners. Ledges and horizontal cracks should be avoided. The work surface should be of laminated plastic, and it should be resistant to heat, moisture, and abrasion. If stainless steel is used, it should be treated to reduce reflections and glare.

Clean room garments, known as bunny suits, act as a personal filter to prevent human-generated contamination from damaging the product being manufactured. Clean

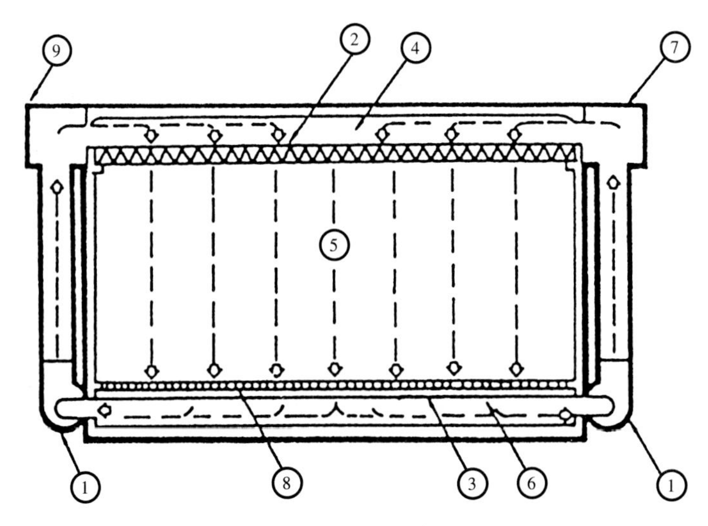

1. BLOWER
2. HEPA FILTERS
3. PREFILTER
4. SUPPLY PLENUM
5. AIRFLOW
6. EXHAUST PLENUM
7. SUBPLENUM
8. FLOOR, GRATED OR PERFORATE

Figure 14-33 Vertical laminar flow clean room.

(*Source:* NASA, 1969.)

room garments when used properly can minimize the contaminating effects of humans working in the clean room without themselves contributing to the airborne particle count. Well-designed garments must act as effective particle barriers, be static dissipative, resist chemical attack, wear, and abrasion, and still be ergonomically acceptable.

Federal Standard 209 addresses clean room and workstation requirements, but makes no stipulations for garments. The Institute of Environmental Sciences (IES) has developed recommendations for clean room apparel: "Garments Required in Cleanrooms and Controlled Environmental Areas," IES-RP-CC-003-87-T (or RP3).

Most static-dissipative fabrics are made of polyester, which is low linting and is interwoven with a grid of conductive carbon filament (about 1 percent by weight). The grid pattern will dissipate the electrostatic charge over the entire garment at low enough levels that the charge further dissipates into the atmosphere by corona discharge. Some garment manufacturers also include an ESD snap in the coat that connects through the wrist strap to the grounding cord, while others use a static-dissipative boot sole, which essentially connects the garment to a grounded clean room floor.

The type of garment that is required to be worn in a clean room is dependent on the class of the clean room. A list of requirements for different classes of clean rooms is shown in Table. 14-6. Many class 1 clean rooms require that operators wear masks that exhaust expired air through portable HEPA canisters.

Due to their relatively low cost, clean room accessories are usually given very little attention. However, in clean room design and practice, details are vitally important.

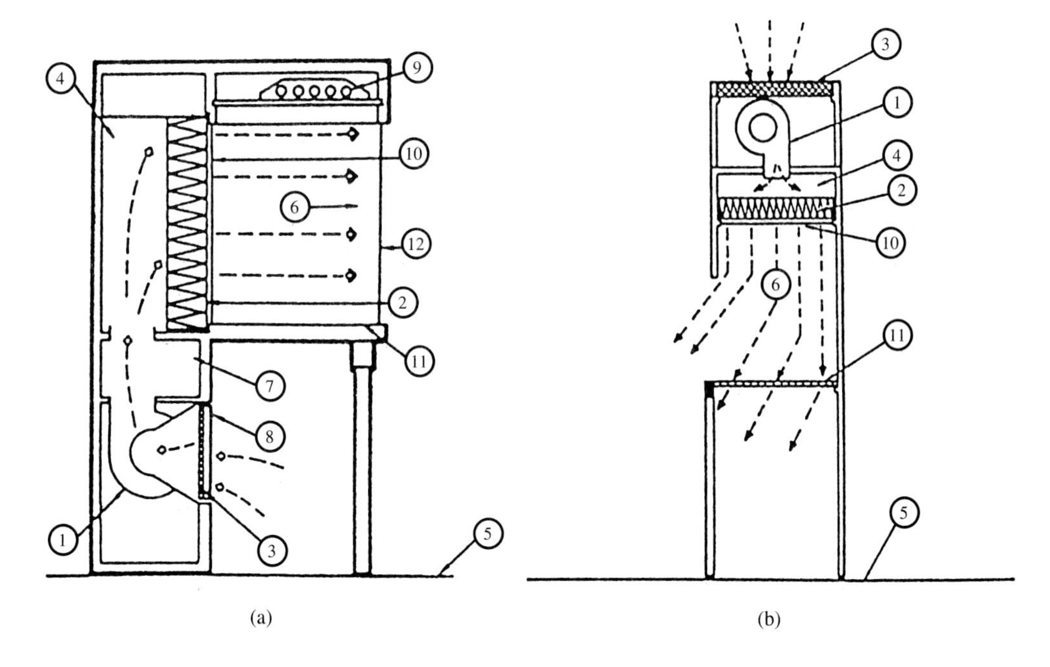

1. BLOWER
2. HEPA FILTERS
3. PREFILTER
4. SUPPLY PLENUM
5. FLOOR, SOLID
6. AIRFLOW
7. SUBPLENUM
8. AIR INLET
9. LIGHT FIXTURE
10. PROTECTIVE SCREEN
11. WORK SURFACE
12. END PANEL

Figure 14-34 Laminar flow clean benches.

(*Source:* NASA, 1969.)

TABLE 14-6 RECOMMENDED GARMENT USAGE FOR VARIOUS CLEAN ROOM CLASSES

	Class 10	Class 100	Class 1000	Class 10,000
Body cover	Coverall	Coverall	Coverall	Coverall or frock
Head cover	Complete facial closure	Facial closure	Hood, beard cover	Hood, beard cover
Foot cover	Full cover boots	Boots	Footwear	Footwear
Hand cover	Gloves	Gloves	Gloves	Optional gloves
Frequency of change	Each entry	Daily	Daily	Daily

Source: David A. Toy, "Clean Room Garments: Performance and Comfort," *Semiconductor International*, November 1989, p. 93.

Accessories (such as paper, pens, forms for record keeping, tacky mats, wipers, sponges, swabs and solvents for cleaning, and packaging materials) are necessary to the operation of a clean room but are also potential sources of contamination. Recycling concerns are also beginning to impact this facet of clean room operation.

Paper can be one of the largest sources of particles in the clean room. One answer to this problem is a paper that is latex impregnated. The latex binds the random fibers in the paper and reduces particle contamination by up to 1600 times more than standard bond paper. Companies are moving toward paperless clean rooms by using bar codes and scanners to keep track of a product as it moves through the various processes.

Adhesive floor mats are used at the clean room entrance to help minimize tracked-in dirt from feet, casters, and wheels. Two basic types are used. One type is cleaned and reused, while the other type is a stack of disposable sheets that are peeled off when contaminated. Several steps must be taken on the mat to remove most of the particles from the clean room boots.

14.6.6 Ultrapure Water and Gases

Clean room manufacturing procedures, especially in IC wafer fabrication, often require the use of ultrapure water and ultrapure gases.

Water is used in almost all phases of IC fabrication. The traditional concern when producing semiconductor-grade water has been the content of inorganics (ions such as sodium and potassium). Resistivity (megohm-cm) is a measure of the presence of dissolved ions.

The trend is for manufacturers to specify a broad range of water-borne contaminants. The Semiconductor Equipment and Materials Institute (SEMI) guidelines suggest that ionic levels from 0.1 to 0.002 ppb, depending on the ion species, are attainable.[9] Dissolved ions are removed from water by reverse osmosis (RO), deionization (DI), and sometimes by electrodialysis reversal (EDR).

Bacteria are controlled through a combination of filtration, sterilization, high pipe velocities, and point-of-use filtration. Point-of-use filters can become a source of bacterial contamination if they are not frequently sterilized or replaced.

There is a relationship between dissolved oxygen (DO) levels in ultrapure water and the growth of natural oxides. It has been confirmed that a reduction in DO levels can significantly retard the formation of natural oxides. Dissolved oxygen can be reduced to about 40 ppb by vacuum deaeration, but the cleaning tank cannot then be exposed to the open atmosphere or reabsorption will occur. A technique called ammonium reduction can reduce DO to between 3 and 5 ppb.

Many compressed gases are also used in IC fabrication. Commonly used gases include hydrogen, oxygen, nitrogen, compressed air, diborane (used as a source of boron to dope silicon), argon, and helium. The wafer yield will be directly related to the purity of these gases.

Potential contaminants in gases can include particles, chemical residues in solid,

[9] Sandra Leavitt, "Optimizing Clean Room Efficiency," *Semiconductor International,* May 1985, pp. 25–27.

liquid, or vapor form, and moisture. The contaminants may come from the gas as supplied, or they may come from the gas storage/delivery system itself.

Methods of preventing contamination of products in the clean room due to gas impurities include:

- Use of nonlubricating type compressors to eliminate oil contamination
- HEPA filters for retention of particles to 0.3 μm
- Point-of-use membrane filters for retention of particles to 0.1 μm
- Continuous gas flows in delivery pipes to avoid gas stagnation in any part of the line

14.6.7 Personnel Training

The training of clean room personnel has proven to be one of the most important aspects of maintaining a quality environment in the clean room. Maintenance people must understand the importance of preventive maintenance and must learn why their tools must be virtually sterilized. They must realize that if a filter system requires preventive maintenance once a week and it is not properly maintained, a class 10 area may become class 100.

Training must also include proper gowning methods and the proper sequence that should be followed to enter the clean room. Before entering the building or entrance to the gowning room, workers may be required to stick their feet into a shoe scrubber that will clean the bottom, top, and sides of their street shoes. Personnel are then allowed to enter the gowning area and put on the bunny suit, clean shoes, booties, hair net, and gloves.

Most companies forbid the use of cosmetics, gum, and food in the clean room. Often the use of a special lotion or moisturizing cream is required, to reduce skin flaking and chafing.

Sometimes the gowning room is a vertical laminar flow room. After gowning, the employee may have to walk through an air shower and then across an adhesive mat to enter the clean room. Each stage prior to clean room entry will be successively more positive in air pressure to prevent particle migration from one stage to the next.

14.6.8 Maintenance and Record Keeping

The maintenance of a clean room is divided into two categories: preventive and janitorial. A good preventive maintenance program can practically eliminate unscheduled shutdowns. Preventive maintenance is an ongoing operation and is particularly dependent upon proper record keeping.

Detailed clean room records must be kept to ensure the opportunity to recognize deficiencies in contamination control. These records should include continuous measurement of pressure differentials, temperature, humidity, particulate distribution, and frequency of personnel entry into the room.

Schedules for janitorial maintenance will vary from company to company. Some companies perform janitorial maintenance on every shift, while others choose to do it on a daily or weekly basis. The cleaning sequence is from ceiling to walls to floors and proceeds from the area most sensitive to contamination to the area that is least sensitive.

Janitorial cleaning should be carried out using mops and wipes that are designed for clean room use. Special cleaning agents like deionized water and isopropyl alcohol should be chosen for their low-particulate, low-residue qualities.

14.7 CASE STUDY

Throughout this chapter, Texas Instruments has been cited as an industry leader in facilities and material handling system design. We have noted that TI pioneered the modular design of manufacturing facilities, based on the spine configuration (Fig. 14-2). TI has also developed innovative applications of modular furniture, equipment, and material handling devices (Fig. 14-8) and a system of standard ESD-controlled tote boxes (Sec. 14.3.2).

14.7.1 Planning Methodology

TI has formalized a methodology of factory planning for world-class manufacturing. As shown in Fig. 14-35, the planning process incorporates functions similar to those recommended by Apple [1]. There are four major phases:

- Proposal support
- Upfront planning
- Factory design
- Installation

Consistent with the concurrent engineering approach, TI factory planning specialists become involved in preliminary planning for new products. In the upfront planning phase, the location and production concept are defined and the facilities project is scheduled. A major expenditure authorization request (MEAR) not only justifies the need for capital equipment but also serves as a tool to plan and evaluate the company's major capital investment projects on an annual basis. Factory design results in the detailed layout and material handling and storage system as well as the operational concepts for the production area. The facility planning process continues through installation and validation. Subsequently, as the product mix, volumes, and operational requirements change, factory modifications and upgrades are planned and executed.

14.7.2 Multiproduct Flexible Factory

TI's Defense Systems and Electronics Group (DSEG) has developed a flexible-factory concept for a changing defense industry that will require high-mix, low-volume electronics assembly. The multiproduct flexible factory is intended to process a variety of products in the start-up, growth, and decline phases of the product life cycle, and products manufactured intermittently (with breaks in the production run).

To address high-mix, low-volume production requirements, TI set four objectives for the multiproduct flexible factory:

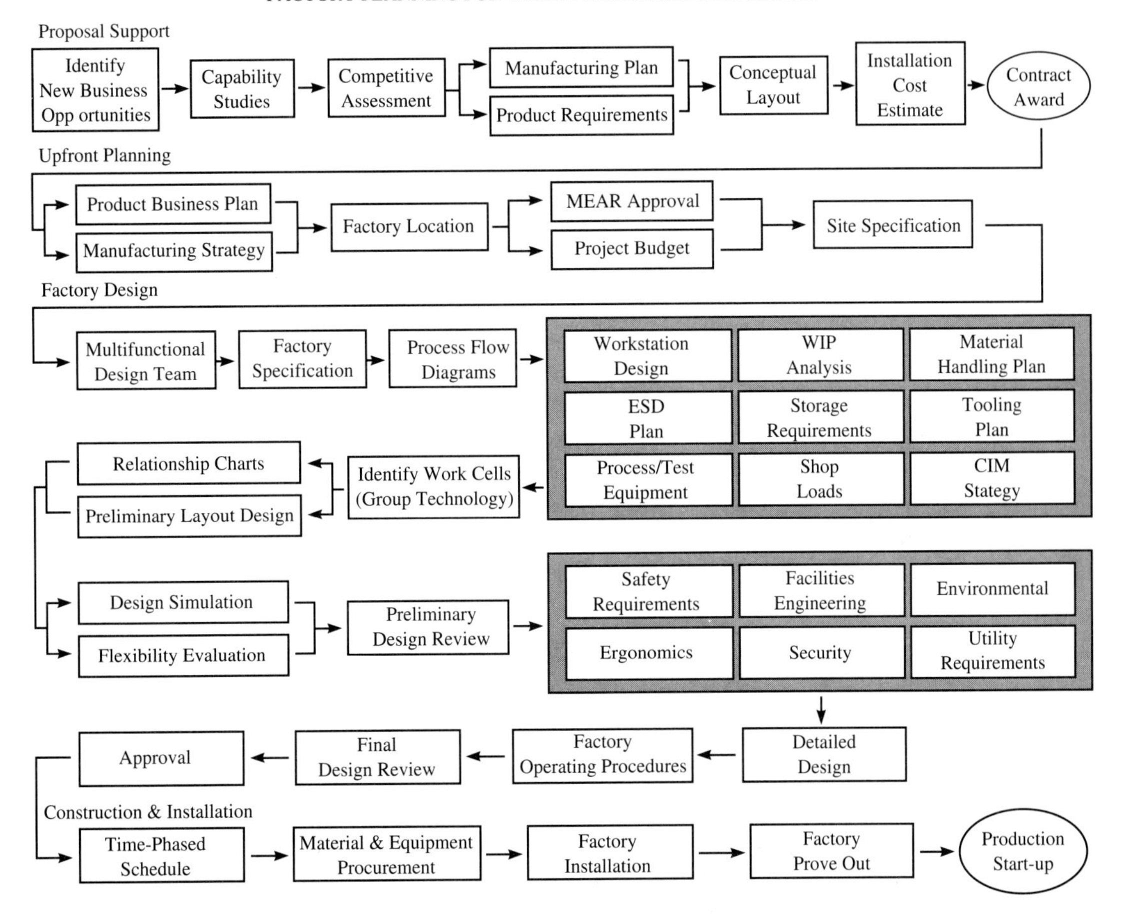

Figure 14-35 Factory planning process at Texas Instruments.

(*Source:* Courtesy of Texas Instruments Incorporated.)

- Workstation and equipment flexibility
- Minimization of setups and setup time
- Reduction of work-in-process inventory
- Integration of support functions

In order to accomplish these objectives, the facility design approach has included the following features:

- Cross-functional facilities design team
- Cellular production
- Self-directed work teams

- Interchangeable, cross-trained labor force
- Flexible utilities featuring quick disconnects
- ESD-controlled modular, portable workstations evaluated by an operator team
- Ergonomically designed universal carts utilizing interchangeable fixtures

The cellular production approach allows for a focused area for each product. A flexible layout is maintained by minimizing the number of inflexible resources (for example, immoveable machines), positioning the necessary inflexible resources around the perimeter of the work area, and keeping the area inside the perimeter open. Resources (furniture and equipment) are not lagged to the floor unless absolutely necessary. A product cell is located in an area including the required inflexible resources at the perimeter and the necessary amount of adjacent open space. Cells are partitioned in a highly flexible manner using low-profile furniture partitions, natural aisle patterns, portable storage racks, and tape-marked boundaries on the floor.

As products move through their life cycles, the work cells can grow, contract, or consolidate. When a product's volume declines, the size of its area shrinks accordingly, and the vacated space is reallocated to other products.

Self-directed work teams develop their own layout designs with the assistance of functional experts. Assembling operators perform many of the production control, test, and inspection functions. The assemblers are highly skilled and cross-trained to perform a variety of technical functions and/or work on a variety of products.

The workcenter furnishings are restricted to low profiles to preserve visibility and promote team communication within and across cells (Fig. 14-36). The workcenter furnishings are modular, but with the aid of specially designed jacks (Fig. 14-37) they can be moved to another location without the disassembly and reassembly typical with conventional modular furniture (see Fig. 14-23). Workstation furnishings have built-in house

Figure 14-36 Flexible workcenter furnishings.

(*Source:* Courtesy of Texas Instruments Incorporated.)

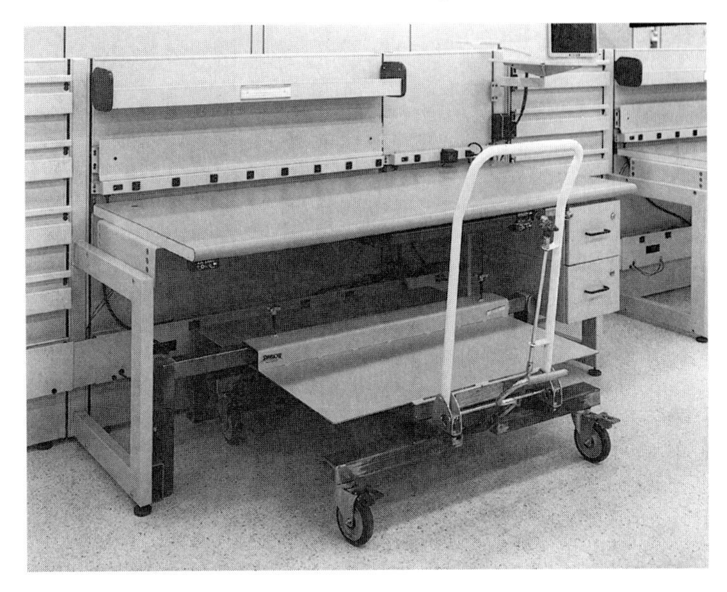

Figure 14-37 Jack for relocation of workcenter furnishings.

(*Source:* Courtesy of Texas Instruments Incorporated.)

power and compressed air (vacuum and low-pressure nitrogen are optional) with quick disconnects (Fig. 14-38). ESD control is designed into the facility, with conductive flooring, continuous electronic ESD monitors, and static dissipative laminate applied to workstation surfaces.

A pair of workstations is shown in Fig. 14-39. Adjustable-height work surfaces are 72 inches (1.8 m) long by 30 inches (0.76 m) deep. The standard workstation includes a

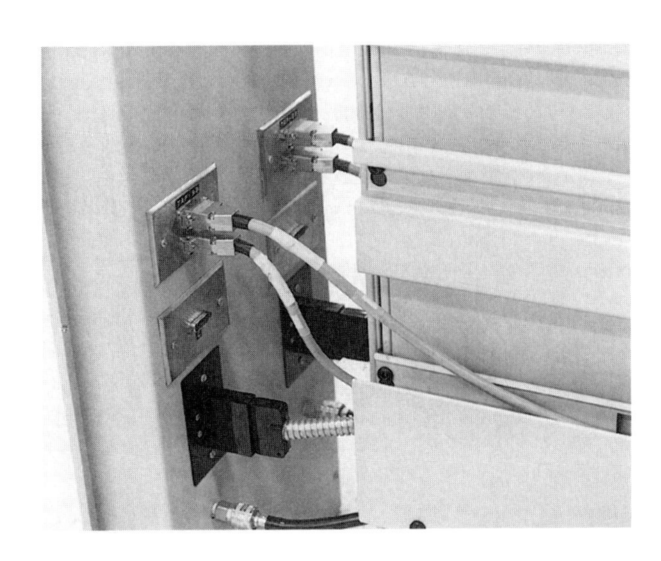

Figure 14-38 Quick disconnects.

(*Source:* Courtesy of Texas Instruments Incorporated.)

Figure 14-39 Workstations (adjustable-height work surface, tote rails, ergonomic chair).

(*Source:* Courtesy of Texas Instruments Incorporated.)

microshelf, part bin rail, tote rails for mounting of tote boxes adjacent to the worksurface (Fig. 14-39), storage drawers, document holder, and pencil drawer. For paperless applications, a monitor platform is substituted for the document holder and a keyboard drawer for the pencil drawer. Ambient illumination of 100 foot-candles is supplied at the work surface from ceiling fixtures. Task lighting is not used except where absolutely required, as in an inspection process. Elimination of task lighting for assembly increases the layout flexibility and reduces the maintenance tasks of changing bulbs and cleaning surfaces. Ergonomically designed chairs have five-star bases and five different positioning adjustments.

The standard unit load throughout the multiproduct flexible factory is the tote box described in Sec. 14.3.2. Special carts have been designed for handling and storage of the tote boxes. The L-cart (Fig. 14-40) holds up to four totes on trays with slide rails for accessibility to all levels. Tote pedestals (Fig. 14-41) are provided for workcenters requiring additional tote storage. The pedestals contain up to three totes and fit underneath the work surface of the workstation (Fig. 14-42). Factory personnel can work directly off of either the carts or the tote pedestals. Limitations on the number of carts and pedestals per workstation enforce inventory control by restricting the amount of WIP inventory storage.

TI's multiproduct flexible factory clearly exhibits several of the material handling principles defined in Sec. 14.3.1:

- Orientation
- Planning
- Systems
- Unit load
- Space utilization
- Standardization

Figure 14-40 L-cart.

(*Source:* Courtesy of Texas Instruments Incorporated.)

- Ergonomics
- Flexibility
- Simplification
- Safety
- Computerization
- Maintenance

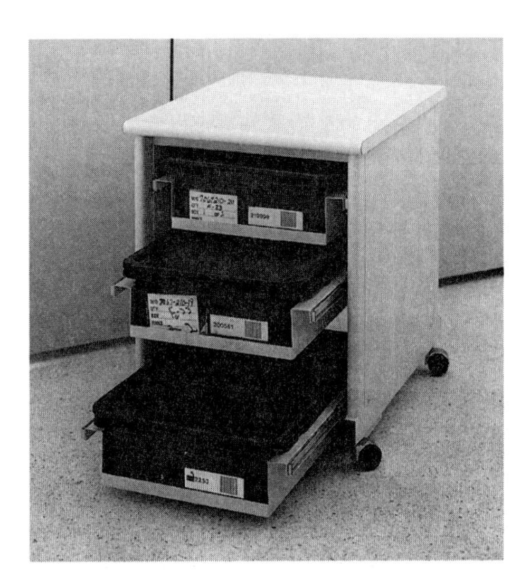

Figure 14-41 Tote pedestal.

(*Source:* Courtesy of Texas Instruments Incorporated.)

Figure 14-42 Tote pedestal stowed beneath work surface.

(*Source:* Courtesy of Texas Instruments Incorporated.)

14.8 SUMMARY

This chapter has discussed the major elements of facilities design and material handling in the electronics industry. Facilities location and choice of layout concept are strategic decisions, profoundly affecting the productivity and competitiveness of the electronics manufacturer. The appropriate layout concept depends upon the product-process mixture.

The layout and the material handling system should be designed concurrently, according to the principles of material handling, with the objective of minimizing the amount of material storage and handling. Definition of the unit load is perhaps the first step in design of layout and material handling systems. In the electronics industry, the ESD-controlled tote box is the typical unit load. A wide variety of handling and storage equipment is available.

There has been a long-term trend toward mechanization and automation in the electronics industry. Automated material handling (for example, via AGVS and AS/RS) has paralleled the automation of fabrication and assembly processes. However, with the emphasis on flexibility, inventory reduction, and frequent production in small lots, the human element is receiving renewed emphasis in production. Principles of ergonomics and workstation design promote the productivity and well-being of personnel.

The environment in the production facility impacts not only the productivity of workers, but also the quality of the product. Electrostatic discharge (ESD) and contamination (air-, water-, and gas-borne) are major sources of quality and reliability problems for electronics products. This chapter has provided principles of facility design and material handling practice to control ESD and contamination.

A case study demonstrated an innovative integration of the facilities design and material handling principles presented in this chapter. High-mix, low-volume environments are increasingly prevalent in the electronics industry, and accordingly Texas Instruments has chosen a factory planning strategy emphasizing flexibility.

KEY TERMS

Automatic guided vehicle system (AGVS)
Automatic storage and retrieved system (AS/RS)
Carousel
Electrostatic discharge (ESD)
Fixed layout
Functional layout
Group layout
High-efficiency particulate air (HEPA) filter
Horizontal laminar flow (HLF)
Hygroscopic oil
Laminar airflow
Product layout
Static electricity (electrostatic charge)
Systematic layout planning
Topical antistat
Tote pan
Transporters
Triboelectric charging (series)
Unit load
Vertical laminar flow (VLF)

EXERCISES

1. Relate layouts to processes and product mix.
2. What are the advantages and disadvantages of each of the four layouts?
3. What are the 20 principles of material handling? Visit an electronics manufacturing facility, and try to identify how it does or does not follow the principles of material handling.
4. What is carpal tunnel syndrome (CTS), and how do power tools aggravate the illness?
5. Why is ESD considered a serious problem in electronic manufacturing?
6. What are the types of electrostatic discharge controls, and how do they work?
7. A company uses a belt conveyor to transport electronic devices. What precautions should be taken?
8. What is the disadvantage of horizontal laminar flow in relation to vertical laminar flow?
9. What is the most important aspect of maintaining a high-quality environment in a clean room, and why?
10. Describe how the Texas Instruments multiproduct flexible factory design follows the material handling principles.
11. Visit an electronic assembly plant, and determine the types of layout concepts and ESD and environmental controls used.

BIBLIOGRAPHY

1. JAMES M. APPLE, *Plant Layout and Material Handling* (3rd ed.), New York: John Wiley & Sons, 1977.
2. BARNES, RALPH M., *Motion and Time Study: Design and Measurement of Work,* New York: John Wiley & Sons, 1980.
3. DANGELMAYER, G. THEODORE, *ESD Program Management,* New York: Van Nostrand Reinhold, 1990.
4. FRANCIS, RICHARD L., LEON F. MCGINNIS, AND JOHN A. WHITE, *Facility Layout and Location: An Analytical Approach* (2nd ed.), Englewood Cliffs, N.J.: Prentice Hall, Inc., 1992.

5. KULWIEC, RAYMOND A., ed., *Materials Handling Handbook,* New York: John Wiley & Sons.
6. PULAT, MUSTAFA, *Fundamentals of Industrial Ergonomics,* Englewood Cliffs, N.J.: Prentice Hall, Inc., 1992.
7. TOMPKINS, JAMES A., AND JOHN A. WHITE, *Facilities Planning,* New York, N.Y.: John Wiley & Sons, 1984.

15 Production and Inventory Control

15.1 INTRODUCTION

This chapter addresses the planning and control of production and inventory. It is important to note that entire books have been written about many of the topics that will be covered in this chapter. The chapter therefore cannot provide complete stand-alone coverage of many of the topics that will be addressed. The approach that has been used is to provide an overview of a wide range of topics related to production planning and inventory control. Where more specific information is desired, the reader is referred to a number of additional references in each of the chapter sections. A detailed list of supplementary reading material appears at the end of this chapter.

The chapter begins by overviewing time series forecasting methods (Sec. 15.2). Forecasting is necessary to accurately anticipate future demand for products and goods. Steps in the purchasing process are overviewed next (Sec. 15.3).

Significant coverage is devoted to inventory control and lot-sizing methods. Both explosion-based (Sec. 15.4–15.7) and reorder-point (Sec. 15.8) inventory systems are discussed in detail. Within explosion-based systems, the topics of master scheduling, material requirements planning (MRP), and lot-sizing heuristics are addressed. Several reorder-point lot-sizing methods are overviewed.

Just-in-time (JIT) inventory methods are described in Sec. 15.9. Case studies documenting examples of successful JIT implementation are overviewed. Section 15.10 is

Much of this chapter has been excerpted from a chapter by Malstrom entitled, "Planning and Control of Manufacturing Systems." This chapter appears in the 4th edition of the *Industrial Engineering Handbook,* edited by William K. Hodson and published by McGraw-Hill Book Company. The adaption of this handbook chapter for this textbook has been completed with the written permission of the publisher.

devoted to the topic of scheduling. MRP scheduling is addressed in Sec. 15.11 as related to the topic of capacity planning. Traditional single and multiple machine scheduling rules are also overviewed in Sec. 15.12. The concluding sections of the chapter address the topics of dispatching (15.13) and expediting (15.14).

15.2 FORECASTING METHODS

Before production may be either planned or scheduled, customer demand must be assessed over future time periods. This forecast is usually the basis for a scheduled plan of production regardless of whether reorder-point, explosion-based, or just-in-time inventory systems are being used. This section overviews traditional time-series and curve-fitting forecasting methods.

In forecasting sales or demand levels, a variety of time-series methods are commonly used. The term **time series** infers that the past history of demand levels will influence the future behavior of this demand over time. A variety of curve-fitting and time-series forecasting techniques are described in the sections that follow.

15.2.1 Ordinary Least Squares Regression

Ordinary least squares (OLS) regression was introduced in Chapter 11 as one of the seven quality tools; it is a curve-fitting method which applies when a linear trend is present in data. Assume a scatterplot of historical data. A first-order equation of the form

$$Y_i = a + bX_i, \qquad i = 1, 2, \ldots, n \tag{15-1}$$

is fitted through the data. The intercept, a, and slope, b, of Eq. (15-1) are selected by the regression procedure to minimize the sum of the squares of the distances from each data point in the scatterplot to the straight line represented by the fitted equation. The values for both the intercept and slope are determined by simultaneously solving a set of least squares normal equations. The minimization of the sum of squares is illustrated in Fig. 15-1.

The quality of the fit of the straight line through the data is generally measured by the coefficient of determination, r^2. This parameter varies between 0 and 1, and it measures the percentage of scatterplot data variation accounted for by the fitted equation. An r^2 value of 1.0 indicates that all data variation is accounted for by the fit, and it is equivalent to saying that all of the data points in the scatterplot fall exactly on the fitted straight line. An r^2 value of 0 indicates that none of the data variation is accounted for by the first-order fit.

In practice, r^2 values in excess of 90 percent usually indicate a fit of good quality. Values of 50 percent or less suggest a poor fit and indicate that linearity was not present in the data. Scatterplots indicative of both high and low r^2 values are depicted in Fig. 11-8 and Fig. 15-2.

Once a fitted line has been defined for the scatterplot, the forecasting procedure is straightforward. The fitted line is used to extrapolate demand values for future time periods, as illustrated in Fig. 15-3.

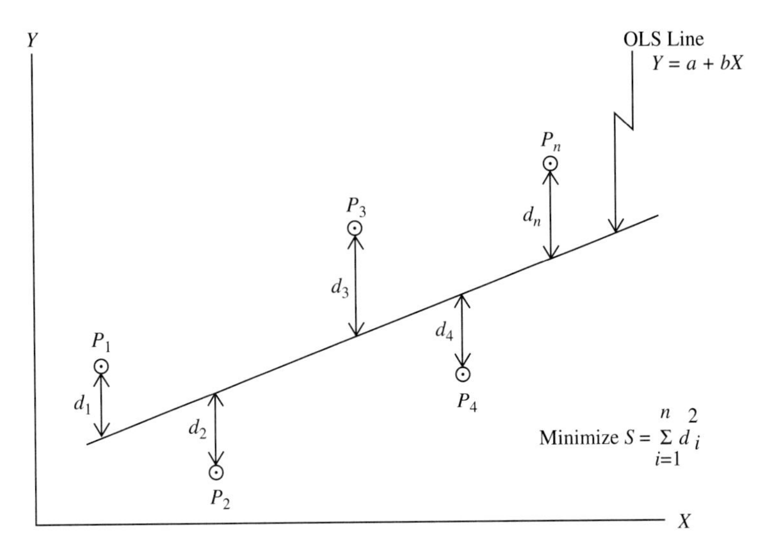

Figure 15-1 OLS minimization of the sum of squares.

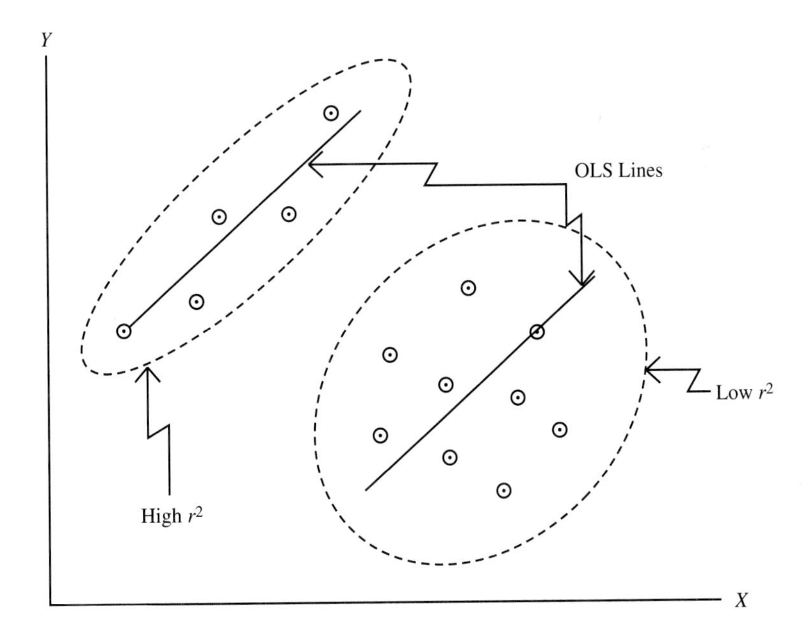

Figure 15-2 High and low r^2 values.

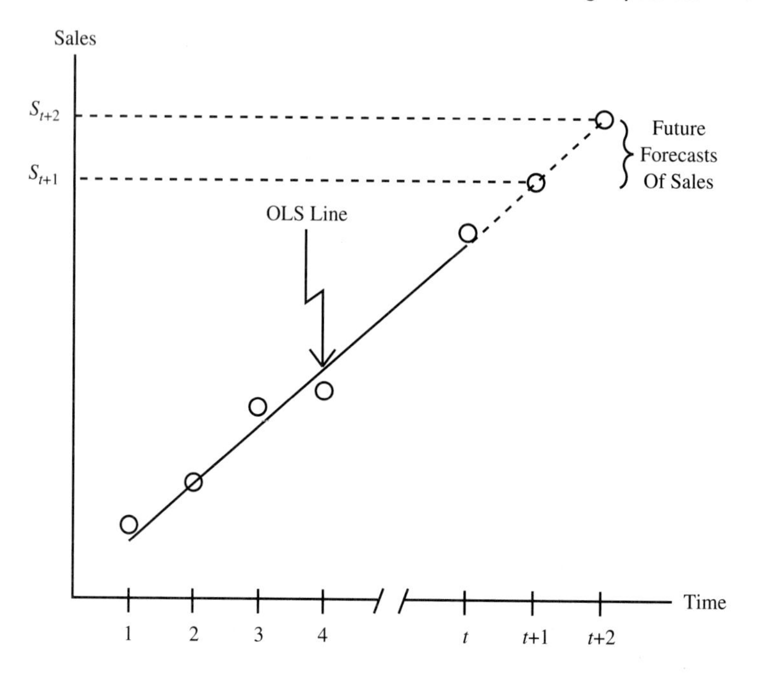

Figure 15-3 Extrapolation of OLS line.

The OLS procedure is not confined to making linear or first-order fits. The method may be adapted to two types of exponential equations, as follows:

$$Y_i = ab_i^X \qquad i = 1, 2, \ldots, n \tag{15-2}$$

$$Y_i = aX_i^b \qquad i = 1, 2, \ldots, n \tag{15-3}$$

The OLS procedure is adapted by taking logarithms of Eqs. (15-2) and (15-3) to convert the equations to first order. The OLS procedure is then modified by substituting logarithms of these values into the OLS normal equations.

15.2.2 Other Curve-Fitting Methods

For scatterplots more irregular in form, other polynomial curve fits may be used. Orthogonal polynomials may be used to fit an *n*th-order polynomial through historical data. The equation is of the form

$$Y_i = a + bX_i + cX_i^2 + dX_i^3 + \cdots + mX_i^n \tag{15-4}$$

Caution should be used when using *n*th-order polynomials as a forecasting tool. It is possible to use this method to obtain a perfect curve fit if a polynomial order equal to the number of data points in the scatterplot is used. This perfect curve fit may lead to unrealistic expectations on the part of the user in terms of the accuracy of the forecasts that may be obtained by extrapolating the fitted curve to forecast future demand.

15.2.3 Moving Averages

A variety of time-series forecasting methods exist. The **moving average forecast** assumes that the forecast for the next period is an average of previous historical sales values. With a simple moving average, the forecast for the next period is merely the average of the historical actual sales values for the n preceding periods:

$$F_{t+1} = A_t = (X_t + X_{t-1} + \cdots + X_{t-n+2} + X_{t-n+1})/n \qquad (15\text{-}5)$$

where

A_t = Average over n periods through period t
F_{t+1} = Forecast for period $t + 1$
X_t = Sales or demand for period t

The term, n, reflects the number of periods in the moving average. For example, consider a four-period moving average. If it is now period 14, the forecast for period 15 would be the average of demands for periods 11, 12, 13, and 14.

A simple moving average assigns equal weights to all past data points comprising the forecast. With a weighted moving average, different weights are assigned to all past data:

$$F_{t+1} = A_t = w_t X_t + w_{t-1} X_{t-1} + \cdots + w_{t-n+1} X_{t-n+1} \qquad (15\text{-}6)$$

where

$$\sum_{i=t-n+1}^{t} w_i = 1$$

and,

w_i = Weighting factor corresponding to period i

For most applications, weights are generally selected to place more weight on recent data. As an example, suppose it is now period 22. A four-period weighted moving average might be written as

$$F_{23} = A_{22} = .4X_{22} + .3X_{21} + .2X_{20} + .1X_{19}$$

15.2.4 Exponential Smoothing

Exponential smoothing is a time-series forecasting method that effectively includes all past data. The technique is similar to the weighted moving average except that all previous demand values are contained in each forecast. A parameter called the smoothing constant may be adjusted to place more or less emphasis on more recent demand values.

Consider an actual demand $X_t = 120$. The corresponding forecast for period t, F_t, is known to be 100. Since the forecast is lower than the actual demand, it is apparent that the forecast for the next period, $t + 1$, should be higher. The question is, by how much? The exponential smoothing equation is defined as

$$New\ Forecast = Old\ Forecast + \alpha\ (Present\ Demand - Old\ Forecast) \quad (15\text{-}7)$$

where

α = The smoothing constant, $0 \leq \alpha \leq 1$

Substituting previously defined notation, we obtain

$$\begin{aligned} F_{t+1} &= F_t + \alpha(X_t - F_t) \\ &= F_t + \alpha X_t - \alpha F_t \\ &= \alpha X_t + (1-\alpha)F_t \end{aligned} \quad (15\text{-}8)$$

where

F_t = Forecast for period t
F_{t+1} = Forecast for period $t + 1$
X_t = Actual demand during period t

Adjusting the smoothing constant allows the user to vary the extent to which the forecasting model is responsive to data fluctuations. The use of smoothing constants near 1.0 results in forecasts that place more weight on recent data, thus causing the forecasts to be responsive to data fluctuations in recent historical demand values. The use of lower smoothing constants (generally less than 0.7) places more weight on older data, thus tending to dampen out the effects of irregularities in demand data.

15.2.5 Model Selection and Validation

Which forecasting method to use is not always an obvious decision. A good way to select the best method is to simulate how different techniques would have actually worked in practice when compared to one another. This may be accomplished in the following manner.

Model validation requires several periods of actual demand or sales data, usually over a period of two or more years. This historical data base is then partitioned into two sections. The first section is used to develop the forecasting model. Models are developed on the first section of the data base without the use or knowledge of any of the data values in the second section.

Forecasts are then made using the model. Forecast values are compared, period by period, with the actual demand values known to have occurred in the second section of the data base. Deviations between forecasted and actual values are tabulated for each forecasting method evaluated. The best method is the one that yields the smallest sum of absolute deviations between forecasted and actual values.

Both curve-fitting and time-series forecasting methods offer distinct advantages. Time-series methods are usually easier to employ and are responsive to nonarithmetic changes in data periods. Time-series methods are constrained in that the forecast is usually limited to one period into the future.

Curve-fitting methods usually require more work initially. While not responsive to sudden changes in demand data, curve fitting offers the advantage of extrapolation, permitting forecasts to be made several periods into the future. The material in this section is intended to provide only an overview of forecasting techniques and methods. Readers desiring more detailed information on this subject are urged to consult references [2, 3, 10, 12, 13, 19, and 24].

15.3 THE PROCUREMENT PROCESS

One of the initial steps in the production process is that of procuring purchased parts. The make/purchase decision is not always straightforward. Purchased parts are by definition those not manufactured in-house.[1,2] Occasions exist when parts that can be manufactured internally are also purchased from the outside. Under normal conditions, the least expensive alternative is selected.

Exceptions to the least-cost procurement alternative do exist. It may be preferable to make parts in-house that can be procured at a lower cost from an outside vendor. One situation is where fill workload is needed in the plant to keep from laying off employees who will be needed for other production tasks in the near future. Another situation is where the organization has a short **lead time** (time interval from issuance of order to receipt of material) requirement and can produce the parts internally faster than it can procure the parts from an external vendor.

Several steps exist in the procurement process that make the purchase of component parts an expensive procedure. These steps are overviewed in the sections that follow.

15.3.1 Purchase Order Requisitions

The first step is the initiation of a purchase order requisition. The **requisition** is usually completed by the person who wishes the material to be procured. The purchaser selects the type of part or material, its specifications, and a desired vendor. The purchaser usually estimates the cost of the part(s) being ordered. Finally, an account number to which the cost of the order is to be charged is also supplied. Purchase order requisitions normally originate in the manufacturing departments of most production organizations.

15.3.2 Solicitation of Bids

The requisition is next forwarded to the purchasing department. If the cost of the order is high enough ($500 to $700 or more), the purchasing department will contact a variety of vendors to obtain bids on the order. In some cases, the lowest bid is selected. The part requested on the purchase order requisition may be replaced by a lower-cost equivalent part or material. However, there is a trend to form stronger relations with fewer vendors, considering other factors, such as quality, in addition to price.

[1]Eric M. Malstrom, ed. *Manufacturing Cost Engineering Handbook,* (New York: Marcel Dekker, Inc., 1984), p. 94.

[2]Eric M. Malstrom, *What Every Engineer Should Know About Manufacturing Cost Estimating.* (New York: Marcel Dekker, Inc., 1981), p. 53.

15.3.3 Purchase Orders

The purchasing department next initiates a **purchase order** (PO), which is forwarded to the vendor from whom the parts will be procured. The PO is a contractual document indicating the organization's intent to procure a specified quantity of a part or material at a negotiated unit cost. The intent to purchase is usually contingent upon the vendor's ability to deliver parts of an acceptable quality level by a specified date.

15.3.4 Inspection of Incoming Parts

After the order is placed, follow-up procedures are initiated, particularly if the order is not delivered on time. Upon receipt, the parts may be inspected to ensure that they conform to defined quality levels. If the order does not conform to these levels, it is rejected and sent back to the vendor. A follow-up order for replacement parts is sometimes initiated. Alternately, the order may be canceled and a new order placed with a different vendor.

15.3.5 Issuing of Purchased Parts

After the order has been successfully inspected, it is either placed into warehouse stock or issued directly to the production floor for use in a manufacturing operation.

15.3.6 Cost of the Procurement Process

The cost of completing each of these steps is significant. In large organizations the cost can range from $60 to in excess of $300 per order. Many organizations have adopted JIT inventory procedures which repetitively place successive orders for parts with the same set of preferred vendors. In such cases **blanket purchase orders** (BPOs) are generated. BPOs set up in advance agreements on cost, order quantity, quality levels, and delivery schedules. A significant initial cost is incurred in setting up a BPO with a preferred vendor, but this cost is offset by significantly reduced costs of placing successive orders for the same part.

15.4 INVENTORY SYSTEMS AND LOT-SIZING METHODS

The inventory system must address two questions with regard to parts that are either made or purchased: when should the order be placed, and how many parts should be ordered (**lot size**)?

Reorder-point inventory systems are used for parts whose demands are known to be independent of one another. Retail department stores, grocery stores, and automobile parts stores are examples of organizations that would use reorder-point inventory systems.

In a manufacturing environment, assembly relationships usually exist between a final assembly that is shipped to the customer and all of its component parts. If the demand for final assemblies is known, then this demand defines corresponding demands for all component parts in the final assembly. Reorder-point inventory systems do not lend themselves to these types of production situations. Since the demands for end items and their components are functionally related, availability of component parts at the time they are needed in the manufacturing process cannot be ensured by reorder-point systems. **Explosion-based inventory systems** are therefore used to address the question of lot sizes.

Explosion-based inventory systems use bills of materials, master schedules, and lot-sizing heuristics. Many of these heuristics are based on reorder-point methods. The following sections overview both the explosion-based and reorder-point methods.

15.5 MASTER SCHEDULING

Explosion-based inventory systems rely on requirements planning. **Requirements planning** may be defined as the management of raw materials, components, and subassemblies to ensure timely availability in sufficient quantity to satisfy the requirements for end products.

End products are scheduled to be produced in accordance with the master schedule. The **master schedule** is a forecast, by time period, of the anticipated demand for production end items. Many of the forecasting methods described in the preceding sections can be applied to project this demand.

Vollman, Berry, and Whybark[3] define master scheduling as the anticipated build schedule for manufactured end products. The master schedule is not the specific result of a sales forecast. It is a statement of scheduled production that is likely to satisfy anticipated demand.

Sales forecasts may be regarded as critical inputs in determining master schedules. However, the master schedule also takes into account both limitations in factory capacity and the need to utilize such capacity as fully as possible.[4]

The master schedule indirectly determines the demands and related procurement schedules for all production components contained in the end items being produced. For example, a production schedule of 100 automobiles per month infers the need for prior procurement of 500 tires per month (four tires per car plus one spare). This procurement needs to be completed in advance of the car being assembled so the tires are available to mount on the car at the time it is built. An explosion-based inventory system known as material requirements planning (MRP) facilitates the determination of procurement and production of production components. This system is the subject of a later section in this chapter.

15.5.1 Time-Phased Records

Vollman, Berry, and Whybark,[5] define a variety of master scheduling techniques. One of the more detailed approaches is the **time-phased record,** in which cumulative production

[3]Thomas E. Vollman, William L. Berry, and D. Clay Whybark, *Manufacturing Planning and Control Systems,* 2nd ed. (Homewood, Ill.: Richard D. Irwin Inc., 1988), p. 297.

[4]Vollman et al., *Manufacturing Planning and Control Systems,* pp. 297–98.

[5]Vollman et al, *Manufacturing Planning and Control Systems,* pp. 302–305.

and the cumulative sales forecasts are plotted over a specified planning horizon. Actual and forecasted sales are compared with one another.

With this method, a backlog of orders may exist. In other words, the demand for end items of production may exceed the supply available in any period. The master schedule, forecasted sales, and actual sales are compared with one another. This comparison permits end items to be committed for shipment to customers in future time periods in the planning horizon.

15.5.2 A Numerical Example

Consider the table shown in Fig. 15-4. A 12-month planning horizon applies. It is now January 1st. An on-hand balance in the amount of 40 units has been carried over from the preceding month.

The entries in the table are computed in the following manner. Orders in the amount of 5 units per month have been promised for each of the first 4 months. These orders must be satisfied in addition to the sales forecast. For January the total demand is the 10 units from the sales forecast plus the 5 units previously promised. This leaves $40 - 15 = 25$ available units at the end of January. Since 15 total units have been promised for February, March, and April, only $25 - 15 = 10$ units are available to promise at the end of January.

The quantity available at the end of any period can be determined from the relationship

$$A_i = A_{i-1} + MPS_i - F_i - O_{pi} \tag{15-9}$$

where

A_i = Stock available at the end of period i
MPS_i = Quantity scheduled for production in period i by the master schedule
F_i = Forecasted demand for period i
O_{pi} = Order quantity of units previously promised for delivery during period i

The amount available to promise in any period is defined by

$$ATP_i = A_i - \sum_{j-i+1}^{n} O_{pj} \tag{15-10}$$

	J	F	M	A	M	J	J	A	S	O	N	D
Forecast	10	10	10	10	10	20	20	20	20	20	20	20
Orders	5	5	5	5								
Available	25	10	55	40	30	10	50	30	10	50	30	10
Avail. to promise	10	0	50				50			50		
MPS			60				60			60		

On hand 1/1 = 40

Figure 15-4 Time-phased record.

where

ATP_i = Amount available to promise in period i

Orders are generated on the master schedule in the following manner. A total of 10 units is carried forward from the end of February. The total demand for March is 15 units, 10 units from the sales forecast and 5 units that have been previously promised. The 10 units available are not sufficient to satisfy this demand. The master schedule therefore calls for an additional 60 end items to be available by the beginning of March. Applying Eq. (15-9) yields the amount available at the end of this month (45 units). Eq. (15-10) yields a value of 50 units available to promise during March.

All other entries in the table are determined in this manner. The reader should note that the MPS quantity of 60 is arbitrary. Methods for determining actual lot sizes will be discussed in a later section. Readers desiring more detailed information on the subject of master scheduling should consult references [13, 18, and 25] at the end of this chapter.

15.6 MATERIAL REQUIREMENTS PLANNING

MRP is the most popular explosion-based inventory system. The most substantive treatment of MRP in early technical literature has been given by Orlicky [18]. MRP examines the assembly relationships between the component parts of an end item being produced. These relationships are used to generate both production and purchase schedules for made and purchased parts. These schedules ensure that sufficient components and subassemblies will be produced at the right time and in the right quantities to satisfy the forecasted demand for end items.

15.6.1 Part Explosion Diagrams

A **part explosion diagram** is also called a **bill of materials.** The diagram indicates the what-goes-into-what relationship of a manufactured end item. Each discrete part or subassembly is indicated by a separate node in the diagram. A sample node is illustrated in Fig. 15-5. As indicated in this illustration, the upper half of the node contains the part number of the component or assembly. The lower left portion of the node indicates the number of the part or subassembly required at the next higher assembly level. Finally, the lower right portion of the node indicates whether the part is to be made or purchased.

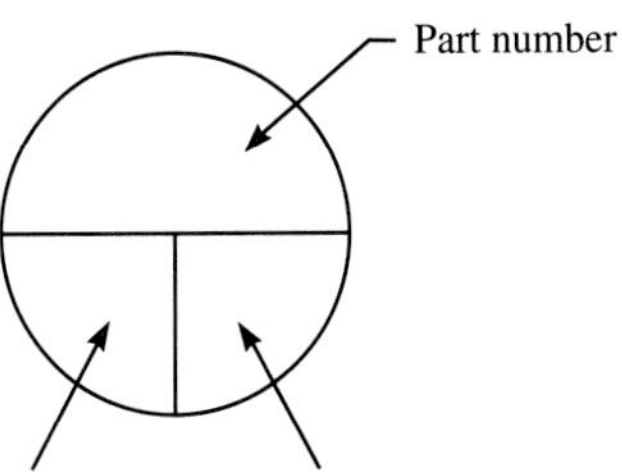

Figure 15-5 Part explosion diagram node.

(*Source:* Reprinted from Eric M. Malstrom, *What Every Engineer Should Know About Manufacturing Cost Estimating* (New York: Marcel Dekker, Inc., 1981).)

A sample part explosion diagram for personal computers is illustrated in Fig. 15-6. The diagram illustrates the component parts for a microcomputer to be fabricated by an electronics manufacturing facility. The reader should note that four distinct assembly levels exist, and that the nodes corresponding to each assembly level are arranged in columns on the diagram. Required make or purchase lead times are indicated to the right of each node, on the connecting assembly links of the diagram.

15.6.2 Generation of MRP Tables

The MRP logic is best illustrated with a detailed numerical example. All production schedules are a function of time. The time periods used in generation of these schedules are called time buckets. In practice, most commonly used time buckets are either in weeks or months.

Each production schedule for a component or assembly consists of a table with four sets of entries. These include the following:

- *Gross Requirements (GR):* The amount of the parts or components required to satisfy the master schedule in any time bucket.
- *Scheduled Receipts (SR):* An order of quantity Q units scheduled to arrive at the beginning of the time bucket. This order was placed LT time buckets ago where LT is the lead time for the part or component.
- *On Hand (OH):* The **on-hand balance** of the part or component that remains at the end of the time bucket. The on-hand balance for any period, t, is given by

$$OH_t = OH_{t-1} + SR_t - GR_t \tag{15-11}$$

where:

OH_t = Parts on hand at the end of period t
SR_t = Scheduled receipts that are to arrive at the beginning of period t
GR_t = Gross requirements to be satisfied in period t

- *Planned Orders (PO):* An order of size Q initiated during period (time bucket) t. This order will arrive at the beginning of period $(t + LT)$ as a **scheduled receipt.**

Suppose the master schedule for the end item (a ZPD 2000 computer) shown in Fig. 15-6 is as shown in Table 15-1. The MRP logic may be applied in the following manner. From Table 15-1 we see that 150 end items must be available to ship at the beginning of period 20. From Fig. 15-6, it is apparent that the lead time for end items (part # 076) is one week. This time allows for the final assembly of part # 076 which consists of parts 143, 137, and 129.

15.6.3 MRP Table for End Item

We begin by determining the **gross requirements.** Since a schedule for end items is desired, the Gross Requirements row of the table is merely the master schedule from Table 15-1. The result is shown in Table 15-2.

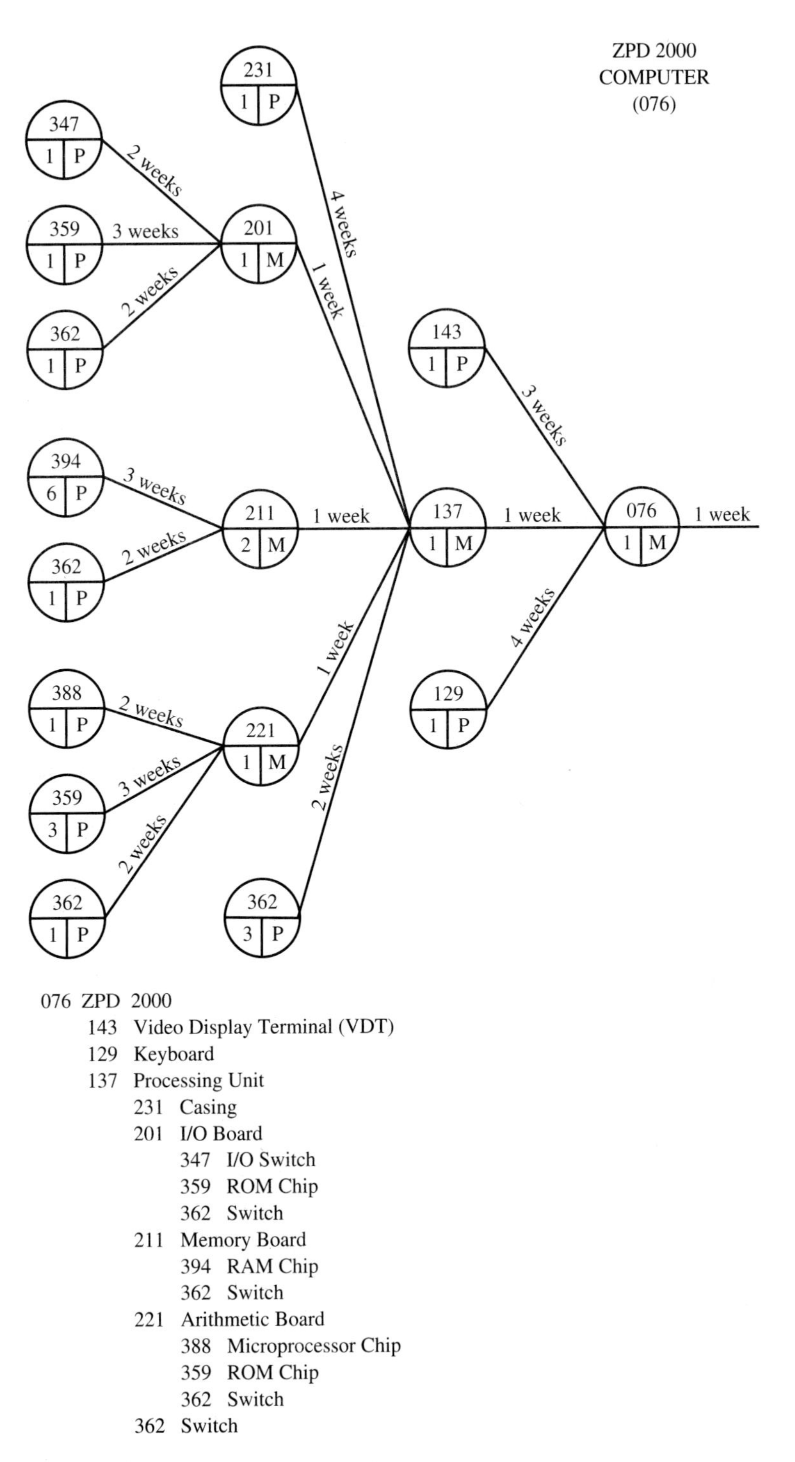

Figure 15-6 Example part explosion diagram.

TABLE 15-1 MASTER SCHEDULE FOR ZPD 2000 COMPUTERS

Period	20	21	22	23	24	25	26	27	28	29	30
Demand	150	220	200	200	250	260	160	220	220	200	180

TABLE 15-2 MRP TABLE FOR ZPD 2000 COMPUTERS PART NUMBER 076

Period	19	20	21	22	23	24	25	26	27	28	29	30
Gross Req.		150	220	200	200	250	260	160	220	220	200	180
Sch. Rec.			220	200	200	250	260	160	220	220	200	180
On Hand	150	0	0	0	0	0	0	0	0	0	0	0
Planned Ord.		220	200	200	250	260	160	220	220	200	180	

We will assume that a separate order is placed for each period's demand. (This lot-sizing decision is arbitrary, and specific MRP lot-sizing methods are presented in Sec. 15.7). The order schedule then becomes the **Planned Order** row of the table. Since the required lead time for the part is one week and a separate order is placed for each period, the Planned Order row is identical to the Gross Requirements row but starts one week earlier in time. (This example assumes a previous balance of 150 end items from period 19.)

Refer again to Table 15-2. The Scheduled Receipts row depicts the arrival of each order one week after it is placed. Since the scheduled receipts exactly equal the gross requirements for each period, all end-of-period on-hand balances are zero.

15.6.4 MRP Tables for Level-Two Parts

Consider now the generation of a procurement schedule for the video display, part #143. From Fig. 15-6 it is apparent that the video display is purchased and requires a lead time of three weeks.

We begin by determining the gross requirements. For any component part, the gross requirements may be determined by considering the relationship between parent and component nodes in the part explosion diagram. A component node is any node that goes into a node at the next higher assembly level in the part explosion diagram. The node that the component node goes into is called the parent node.

Refer to Fig. 15-6. Part # 076 is a parent node for the component nodes corresponding to part numbers 143, 137, and 129. Similarly, part # 137 is a parent node for parts 231, 201, 211, 221, and 362, and so on. The gross requirements for any component node may be defined as a function of the planned orders of the parent node in accordance with

$$GR_c = PO_p \times Q_g \qquad (15\text{-}12)$$

where:

GR_c = Gross requirements of the component node
PO_p = Planned orders of the parent node
Q_g = Goes-into quantity between the component node and the parent node

TABLE 15-3 MRP TABLE FOR VIDEO DISPLAY PART NUMBER 143

Period	19	20	21	22	23	24	25	26	27	28	29	30
Gross Req.		220	200	200	250	260	160	220	220	200	180	
Sch. Rec.				500		500		500			500	
On Hand	500	280	80	380	130	370	210	490	270	70	390	
Planned Ord.	500		500		500			500				

From Fig. 15-6 the goes-into quantity between part # 076 and part # 143 is one. This means that one video display is required at the next higher assembly level. The gross requirements for the video display are then the planned orders for part # 076, taken from Table 15-2, times one, as shown in Table 15-3.

Suppose a total of 500 video displays are in stock at the end of period 19. The on-hand balances at the end of each period may be obtained by subtracting the gross requirements for each period from the inital stock balance, as shown in Table 15-3. A balance of 80 video displays is projected for the end of period 21, which is not sufficient to satisfy the scheduled demand for period 22 (200 units).

MRP does not allow part shortages to occur. It is therefore necessary to schedule an order to arrive at the beginning of period 22. Suppose this order size is 500. (Again, the selection of this lot size is arbitrary and is used for example purposes only.) The purchase lead time for video displays is three weeks. To arrive in period 22, the order would have to be placed in period 19. A planned order in this amount is therefore shown for this period.

Applying Eq. (15-11) yields the remaining on-hand balances. The on-hand balance at the end of period 23 is 130. This is not sufficient to satisfy the demand for period 24 (260 units). It is thus necessary to schedule another order to arrive at the beginning of period 24 to prevent a shortage from occurring. To arrive in period 24, this order must be placed in period 21. An order size of 500 units is again used.

Applying this logic to the remaining entries in this table results in additional orders that must be scheduled to arrive in periods 26 and 29. These orders are placed in periods 23 and 26, respectively.

The time-phased planning procedure may be used to generate production or procurement schedules for all of the remaining nodes in Fig. 15-6. The MRP table for the keyboard is Table 15-4. From Fig. 15-6, it is obvious that only one keyboard is required for each end item. The Gross Requirements row of this table is thus the Planned Orders row from the MRP table for the end item (Table 15-2). The required purchase lead time for the

TABLE 15-4 MRP TABLE FOR KEYBOARD PART NUMBER 129

Period	19	20	21	22	23	24	25	26	27	28	29	30
Gross Req.		220	200	200	250	260	160	220	220	200	180	
Sch. Rec.					500		500		500			
On Hand	700	480	280	80	330	70	410	190	470	270	90	
Planned Ord.	500		500		500							

TABLE 15-5 MRP TABLE FOR PROCESSING UNIT PART NUMBER 137

Period	19	20	21	22	23	24	25	26	27	28	29	30
Gross Req.		220	200	200	250	260	160	220	220	200	180	
Sch. Rec.				200	250	260	160	220	220	200	180	
On Hand	420	200	0	0	0	0	0	0	0	0	0	
Planned Ord.			200	250	260	160	220	220	200	180		

keyboard is four weeks. An on-hand balance of 700 keyboards is assumed at the end of period 19. An order size of 500 is again used. The order sizes are selected to keep inventories low and support approximately two weeks of production.

The MRP table for the processing unit, part # 137, is Table 15-5. From Fig. 15-6, we see that only one processing unit is present in each end item. The gross requirements again are the planned orders for part # 076 (Table 15-2). The processing unit is a costly part that consists of a number of subassemblies and purchased parts. We therefore desire to keep inventory levels low. Orders are placed each week to support the required demand levels. From Fig. 15-6, we note that the required lead time is one week. Assuming an initial inventory balance of 420 units at the end of period 19 yields the table values shown in Table 15-5.

15.6.5 MRP Tables for Level-Three Parts

It is next necessary to generate the MRP tables for the nodes in level three of Fig. 15-6. Five separate parts go into the processing unit, part # 137. These include the casing (part # 231), an I/O board (part # 201), a memory board (part # 211), an arithmetic board (part # 221), and a switch (part # 362).

The MRP table for the casing is shown in Table 15-6. From Fig. 15-6, we know that only one casing is required for the processing unit, and the lead time is four weeks. The gross requirements are thus the planned orders for part # 137 (Table 15-5) times one. No initial balance of stock exists, and an order size of 500 is again used. This order size supports about two weeks of production.

The MRP table for the I/O board, part # 201, is Table 15-7. Only one board is required at the next assembly level, and the lead time is one week. The gross requirements are again the planned orders for part # 137 times one. No initial stock balance exists. An order size of 500 is again assumed (enough to support about two weeks of production).

Part # 137 requires two memory boards, part # 211. The gross requirements for

TABLE 15-6 MRP TABLE FOR CASING PART NUMBER 231

Period	17	18	19	20	21	22	23	24	25	26	27	28
Gross Req.					200	250	260	160	220	220	200	180
Sch. Rec.					500		500		500		500	
On Hand					300	50	290	130	410	190	490	310
Planned Ord.	500		500		500		500					

TABLE 15-7 MRP TABLE FOR I/O BOARD PART NUMBER 201

Period	17	18	19	20	21	22	23	24	25	26	27	28
Gross Req.					200	250	260	160	220	220	200	180
Sch. Rec.					500		500		500		500	
On Hand				0	300	50	290	130	410	190	490	310
Planned Ord.				500		500		500		500		

TABLE 15-8 MRP TABLE FOR MEMORY BOARD PART NUMBER 211

Period	17	18	19	20	21	22	23	24	25	26	27	28
Gross Req.					400	500	520	320	440	440	400	360
Sch. Rec.					1000		1000		1000		1000	
On Hand				0	600	100	580	260	820	380	980	620
Planned Ord.				1000		1000		1000		1000		

part # 211 are thus the planned orders for part # 137 times two. No initial stock balance is assumed, and an order size of 1,000 is used. The MRP table for part # 211 is Table 15-8.

One arithmetic board, part # 221 is required by part # 137. The lead time noted from Fig. 15-6 is one week. No initial stock balance is assumed, and an order size of 500 is used. The resulting MRP table is Table 15-9.

The last part in level three of the part explosion diagram is a switch, part # 362. Three switches are required in each processing unit in level two. The gross requirements for this part are thus the planned orders for part # 137 times three. Part # 362 has a purchase lead time of two weeks. The MRP table for this part is shown in Table 15-10. No initial stock balance is assumed, and an order size of 1,500 is used.

TABLE 15-9 MRP TABLE FOR ARITHMETIC BOARD PART NUMBER 221

Period	17	18	19	20	21	22	23	24	25	26	27	28
Gross Req.					200	250	260	160	220	220	200	180
Sch. Rec.					500		500		500		500	
On Hand				0	300	50	290	130	410	190	490	310
Planned Ord.				500		500		500		500		

TABLE 15-10 MRP TABLE FOR SWITCH PART NUMBER 362

Period	17	18	19	20	21	22	23	24	25	26	27	28
Gross Req.					600	750	780	480	660	660	600	540
Sch. Rec.					1500		1500		1500		1500	
On Hand			0	0	900	150	870	390	1230	570	1470	930
Planned Ord.			1500		1500		1500		1500			

TABLE 15-11 MRP TABLE FOR I/O SWITCH PART NUMBER 347

Period	17	18	19	20	21	22	23	24	25	26	27	28
Gross Req.				500		500		500		500		
Sch. Rec.				500		500		500		500		
On Hand				0	0	0	0	0	0	0		
Planned Ord.		500		500		500		500				

15.6.6 MRP Tables for Level-Four Parts

The final step is to generate the MRP tables for all nodes in level four of Fig. 15-6. The I/O board, part # 201, has three components. These include an I/O switch (part # 347), a ROM chip (part # 359), and a switch (part # 362). All three parts are purchased, and only one of each is required at the next assembly level.

Part # 347 has a lead time of two weeks. The gross requirements for this part are the planned orders for part # 201 (Table 15-7) times one. The MRP table for this part is Table 15-11. No initial stock balance exists. An order size of 500 is again used (enough to support about two weeks of production).

Part # 359 has a lead time of three weeks. The gross requirements are again the planned orders for part # 201 times one. No initial stock balance is assumed, and an order size of 500 is used. The resulting MRP table is Table 15-12.

Part # 362 has a lead time of two weeks. The gross requirements are again the planned orders for part # 201 times one. The MRP table is Table 15-13. As with the other components contained in the I/O board, no initial stock balance exists, and an order size of 500 is used.

The memory board, part # 211, consists of two components. These include a RAM chip (part # 394) and a switch (part # 362). Both components are purchased.

Each memory board requires six RAM chips. The gross requirements for this part

TABLE 15-12 MRP TABLE FOR ROM CHIP PART NUMBER 359

Period	17	18	19	20	21	22	23	24	25	26	27	28
Gross Req.				500		500		500		500		
Sch. Rec.				500		500		500		500		
On Hand				0	0	0	0	0	0	0		
Planned Ord.	500		500		500		500					

TABLE 15-13 MRP TABLE FOR SWITCH PART NUMBER 362

Period	17	18	19	20	21	22	23	24	25	26	27	28
Gross Req.				500		500		500		500		
Sch. Rec.				500		500		500		500		
On Hand		0	0	0	0	0	0	0	0	0		
Planned Ord.		500		500		500		500				

TABLE 15-14 MRP TABLE FOR RAM CHIP PART NUMBER 394

Period	17	18	19	20	21	22	23	24	25	26	27	28
Gross Req.				6000		6000		6000		6000		
Sch. Rec.				6000		6000		6000		6000		
On Hand	0	0	0	0	0	0	0	0	0	0		
Planned Ord.	6000		6000		6000		6000					

are the planned orders for the memory board (from Table 15-8) times six. RAM chips require a purchase lead time of three weeks. The MRP table for the RAM chips is Table 15-14. An order size of 6,000 is used (enough to support two weeks of production). No initial stock balance of this part is assumed in this table.

The remaining component of the memory board is the switch, part # 362. Each board requires only one switch, which has a purchase lead time of two weeks. The gross requirements for this part are the planned orders of the memory board times one. The MRP table is Table 15-15. No initial stock balance is assumed, and an order size of 1,000 is used.

The arithmetic board has three components. These include a microprocessor chip (part # 388), a ROM chip (part # 359), and a switch (part # 362). One each of part numbers 388 and 362 are used in the arithmetic board. Three ROM chips, part number 359, are required.

The microprocessor chip has a purchase lead time of two weeks. The MRP table for this part is Table 15-16. The gross requirements are the planned orders for the arithmetic board (from Table 15-9) times one. No initial stock balance exists, and an order size of 500 is used.

The ROM chip has a purchase lead time of three weeks. The gross requirements are the planned orders for part # 221, the arithmetic board, times three. No initial stock

TABLE 15-15 MRP TABLE FOR SWITCH PART NUMBER 362

Period	17	18	19	20	21	22	23	24	25	26	27	28
Gross Req.				1000		1000		1000		1000		
Sch. Rec.				1000		1000		1000		1000		
On Hand		0	0	0	0	0	0	0	0	0		
Planned Ord.		1000		1000		1000		1000				

TABLE 15-16 MRP TABLE FOR MICROPROCESSOR CHIP PART NUMBER 388

Period	17	18	19	20	21	22	23	24	25	26	27	28
Gross Req.				500		500		500		500		
Sch. Rec.				500		500		500		500		
On Hand		0	0	0	0	0	0	0	0	0		
Planned Ord.		500		500		500		500				

TABLE 15-17 MRP TABLE FOR ROM CHIP PART NUMBER 359

Period	17	18	19	20	21	22	23	24	25	26	27	28
Gross Req.				1500		1500		1500		1500		
Sch. Rec.				1500		1500		1500		1500		
On Hand	0	0	0	0	0	0	0	0	0	0		
Planned Ord.	1500		1500		1500		1500					

balance is assumed, and an order size of 1,500 is used. The resulting MRP table is Table 15-17.

The final component is a switch, part # 362. Its purchase lead time is two weeks. The gross requirements are the planned orders for the arithmetic board times one. The MRP table is Table 15-18. No initial inventory exists, and an order size of 500 is used.

15.6.7 Composite MRP Tables

By now the reader may have noted that some parts appear in more than one position in the part explosion diagram of Fig. 15-6. The switch, part # 362, appears in four separate locations in Fig. 15-6. Separate MRP tables have been generated for each of these nodes and appear as Tables 15-10, 15-13, 15-15, and 15-18. In practice, orders for each of these nodes would not be placed separately. Instead, the separate demands (gross requirements) for each node would be totaled. A new order policy would be developed to satisfy the aggregate demand for all four nodes.

Table 15-19 is an aggregate MRP table for part # 362. The gross requirements from Tables 15-10, 15-13, 15-15, and 15-18 have been totaled in this table. An order size of 3,300 placed every two weeks is sufficient to satisfy the aggregate demand.

One other part appears more than once in the part explosion diagram of Fig. 15-6.

TABLE 15-18 MRP TABLE FOR SWITCH PART NUMBER 362

Period	17	18	19	20	21	22	23	24	25	26	27	28
Gross Req.				500		500		500		500		
Sch. Rec.				500		500		500		500		
On Hand		0	0	0	0	0	0	0	0	0		
Planned Ord.		500		500		500		500				

TABLE 15-19 COMPOSITE MRP TABLE FOR SWITCH PART NUMBER 362

Period	17	18	19	20	21	22	23	24	25	26	27	28
Gross Req.				2000	600	2750	780	2480	660	2660	600	540
Sch. Rec.				3300		3300		3300		3300		
On Hand		0		1300	700	1250	470	1290	630	1270	670	130
Planned Ord.		3300		3300		3300		3300				

TABLE 15-20 COMPOSITE MRP TABLE FOR ROM CHIP PART NUMBER 359

Period	17	18	19	20	21	22	23	24	25	26	27	28
Gross Req.				2000		2000		2000		2000		
Sch. Rec.				2000		2000		2000		2000		
On Hand	0	0	0	0	0	0	0	0	0	0		
Planned Ord.	2000		2000		2000		2000					

Part # 359, the ROM chip, appears in two separate locations in the part explosion structure. Separate MRP tables for each location appear as Tables 15-12 and 15-17.

Table 15-20 shows an aggregate MRP table for part # 359. The gross requirements from Tables 15-13 and 15-18 have been totaled in Table 15-21. An order size of 2,000 placed every other week is sufficient to satisfy the aggregate demand for this part.

15.6.8 References for Further Reading

The previous example is intended to present the reader with an overview of MRP inventory logic. Readers desiring a more detailed description of MRP should consult references 2, 3, 5, 8, 10, 13, 18, 25, and 27 at the end of this chapter.

15.7 MRP LOT-SIZING HEURISTICS

Lot sizing in an MRP environment is equivalent to determining how many periods of gross requirements to combine into a planned order. The lot sizes used in the preceding example were arbitrarily selected to keep inventory levels low. This procedure tends to minimize the cost of keeping purchased parts, finished goods, and work-in-process inventory in stock. The following subsections present MRP lot-sizing heuristics. These heuristics can be used to determine MRP order sizes. The order sizes tend to minimize the total costs of **ordering** (setups and orders) and the costs of **carrying** inventory in stock.

When more than one level of the part explosion diagram is considered, it is usually not possible to prove the optimality of MRP lot-sizing methods. MRP lot-sizing heuristics are based on reorder-point lot-sizing methods. These techniques are addressed in Sec. 15.8.

The effectiveness of MRP lot-sizing methods has been thoroughly investigated by simulating different types of rules with a variety of part explosion structures and demand patterns. The following subsections overview some typical MRP lot-sizing methods. The comparative effectiveness of these rules in terms of total annual inventory cost is also described.

15.7.1 Lot-for-Lot Heuristic

The **lot-for-lot** (LFL) heuristic specifies that a separate order is placed for each period or time bucket. No periods of demand are combined. The order size equals the gross requirement for the period in question leaving a constant on-hand balance of zero.

The LFL method typically has high order costs, since separate orders are placed for each period with a nonzero demand. Inventory carrying costs are minimized by this approach, since the stock is always used in the period in which it arrives. The LFL method is similar to the JIT order philosophy (Sec. 15.9) in that it tends to minimize inventories.

15.7.2 Economic Order Quantity Heuristic

The **economic order quantity** (EOQ) heuristic applies the economic order quantity logic of reorder-point inventory systems (Sec. 15.8). The EOQ approach attempts to select the lot size that minimizes the sum of both ordering and carrying costs. The EOQ method assumes that demand from period to period is relatively constant; it is based on Eq. (15-13) in Sec. 15.8. Each time it is necessary to place an order, Eq. (15-13) is used to determine the order size. If the order size happens to be less than the gross requirement for the period in question, the order size is increased to a level just large enough to prevent shortages from occurring. When an initial stock balance exists, the annual demand, R, in Eq. (15-13) is reduced by the amount of the initial stock balance. Carrying costs for the initial inventory are added to the total annual inventory cost.

15.7.3 Periodic Order Quantity Heuristic

The **periodic order quantity** (POQ) heuristic uses the EOQ logic to determine the optimal time interval between orders. An order is then initiated for a quantity just large enough to cover the demand that is scheduled to occur over this time interval. Time periods in the interval are totaled in such a way that no order is scheduled for receipt during a period that has zero demand, avoiding incurring unnecessary carrying cost. The POQ method responds well to demand patterns with wide fluctuations.

15.7.4 Least Unit Cost Algorithm

The **least unit cost** (LUC) algorithm computes for various order sizes the cost per unit chargeable to orders, setup, and storage. An order size is selected that minimizes the total cost per unit.

15.7.5 Least Total Cost Algorithm

It may be shown that the cost minimum corresponding to the optimal order size of Eq. (15-13) occurs at the point where the annual order costs and the annual carrying costs are equal. The **least total cost** (LTC) algorithm analyzes the gross requirements over a specified planning horizon. Various order quantities are evaluated. An order quantity is selected that makes the resulting order and carrying costs most closely equal one another.

15.7.6 Part-Period Balancing Algorithm

The **part-period balancing** (PPB) algorithm is very similar to the LTC approach to lot sizing. The primary difference between the two methods is an adjustment (look ahead/look back) routine. This feature prevents inventory intended to cover peak period demands from being carried in stock for long periods of time. This PPB approach also helps prevent orders from being keyed to periods with low requirements.

15.7.7 Silver-Meal Algorithm

The **Silver-Meal** (SM) **algorithm** is computationally more robust than the methods previously described. The method is based on selecting the order quantity so as to minimize the cost per unit time over the time periods during which the order quantity lasts, under the assumption that all inventory needed during a period must be available at the beginning of that period. This assumption of stock availability also holds for all of the previously described algorithms and heuristics.

15.7.8 Wagner-Whitin Algorithm

The **Wagner-Whitin** (WW) **algorithm** uses an optimizing procedure that is based on a dynamic programming model. It evaluates all possible combinations of orders to cover requirements in each period of the planning horizon. The WW objective is to arrive at an optimal ordering strategy for the entire requirements schedule.

The algorithm does minimize the total cost of setup and carrying inventory, but only for the assembly level of the part being considered and only for the planning horizon. The algorithm has the disadvantage of a high computational burden due to its mathematical complexity.

15.7.9 Comparative Performance of Heuristics

The comparitive performance of lot-sizing heuristics has been studied in detail. Results of key studies have been documented by Choi, Malstrom, and Classen,[6,7] Choi, Malstrom, and Tsai,[8] and Taylor and Malstrom.[9]

Digital simulation has been used to simulate the heuristics under a variety of part explosion and demand conditions. More recent studies have evaluated larger part explosion product structures with increasing amounts of actual manufacturing data as inputs.

Of the heuristics previously described, the consistent best performer has been the POQ rule. Other rules that have performed well include the LTC and LUC rules. Marginal rules on the basis of performance have included the EOQ, WW, and SM methods. The LFL rule has been consistently the worst performing heuristic in all evaluations. Total annual inventory costs associated with this method are three to twenty times more expensive

[6]Richard H. Choi, Eric M. Malstrom, and Ronald L. Classen, "Computer Simulation of Lot Sizing Alternatives in Three Stage Multi Echelon Inventory Systems," *Journal of Operations Management,* 4, 3 (May 1984), pp. 259–77.

[7]Richard H. Choi, Eric M. Malstrom, and Ronald L. Classen, "Evaluation of Lot Sizing Alternatives in Multi-Echelon Inventory Systems," *Proceedings of the Fall Systems Conference,* Institute of Industrial Engineers, Washington, D.C., December 1981, np.

[8]Richard H. Choi, Eric M. Malstrom, and Russell D. Tsai, "An Extended Simulaton of MRP Lot Sizing Alternatives in Multi-Echelon Inventory Systems," *Production and Inventory Management,* 29, 4, (1988) pp. 4-10.

[9]R. Bruce Taylor and Eric M. Malstrom, "Simulation of MRP Lot Sizing Heuristic," (Unpublished Research Report, Department of Industrial Engineering, University of Arkansas, March 1990).

than the best performing rules in the variety of simulation studies that have been conducted.[10,11,12,13]

The feature that has consistently distinguished good rules from the ones that are not cost effective has been the order policy structured for end item products. Those rules that trigger frequent separate orders for end items incur annual setup or order costs that are extremely high. The increase in these costs is not completely offset by the corresponding lower carrying costs that are obtained.

This conclusion has some interesting implications for JIT inventory systems. JIT order policies are similar to the LFL heuristic. This method has historically been the least effective of the rules evaluated. Malstrom[14] has determined that setup or order costs must be reduced to levels equal to 1/100 or less of the corresponding carrying cost for each node before the performance of the LFL heuristic begins to significantly improve relative to other lot-sizing methods.

It is questionable whether this reduction in both setup and order costs is always attainable when JIT policies have been implemented. Burney, Malstrom, and Parker[15] and Malstrom[16] have stated that such potential increases in order costs have the potential to negate many of the possible savings attainable with JIT policies. JIT inventory procedures are discussed in greater detail in Sec. 15.9.

15.8 REORDER-POINT INVENTORY SYSTEMS

Unlike explosion-based inventory systems previously described, reorder-point (ROP) systems do not consider assembly relationships depicted by the parts explosion diagram. ROP systems are used for separate parts whose demands are known to be functionally independent of one another. Some spare parts as well as inventories in grocery stores and other retail outlets are example applications for ROP inventory systems.

A variety of ROP lot-sizing methods exist. Most of the mathematically straightforward models are based on a number of restrictive assumptions. Many of these assumptions are not true in practice. As these assumptions are relaxed, the computational complexity of the lot-sizing models increases significantly. The assumptions are summarized as follows:

[10]Choi et al., "Computer Simulation of Lot Sizing Alternatives," pp. 259–77.

[11]Choi et al., "Evaluation of Lot Sizing Alternatives," np.

[12]Choi, et al., "An Extended Simulaton of MRP Lot Sizing Alternatives," pp. 4–10.

[13]Taylor and Malstrom, "Simulation of MRP Lot Sizing Heuristic," np.

[14]Eric M. Malstrom, "Set Up Cost Reduction Requirements for JIT Lot Sizing," Summary of Class Project Reports for IE 541, Advanced Production Control, Department of Industrial Engineering, Iowa State University, 1986, np.

[15]M. A. Burney, Eric M. Malstrom, and Sandra C. Parker, "A Computer Assisted Cost Assessment of Just-In-Time Inventory Systems," *Proceedings of the Spring Annual Conference,* Institute of Industrial Engineers, San Francisco, May 1990, pp. 420–25.

[16]Eric M. Malstrom, "Assessing the True Cost Savings with Just-In-Time Inventory Systems," *Proceedings of the Fall Annual Conference,* Institute of Industrial Engineers, St. Louis, November 1988, pp. 141–46.

- Annual demand is constant and is known exactly.
- Orders are received instantly.
- Lead time is known and is constant.
- Order costs are known and are independent of order size.
- Purchase price is constant. Price may vary with the order size.
- Storage capacity is available to store up to one year's demand of an item.

Entire texts have been written on lot-sizing models. It is therefore not feasible to cover all of them in detail in this section. The approach used will be to summarize popular methods in increasing order of mathematical complexity. The assumptions associated with each method will be summarized. Mathematical derivations of each approach will not be presented. However, lot-sizing formulas will be included, where appropriate, to assist the reader in selecting the appropriate method.

15.8.1 Notation

In describing lot-sizing notation, it is necessary to address the concept of inventory cycles. This is best accomplished by reviewing the inventory stock level of a given part over time, as illustrated in Fig. 15-7. The first inventory cycle begins by assuming that an order in the amount of Q units has just been received. A constant demand is assumed, so the stock level is depleted at a linear rate.

Initially, stock shortages are assumed not to occur. A second order is placed when the stock level reaches Q_{RO} units, the **reorder point.** This value defines the part's lead time, LT, since a new order must arrive exactly when the stock level for the part reaches zero. The maximum stock level is Q units; the minimum level is zero. It follows that the average stock level, $\bar{I}$, during the inventory cycle time, t, is $Q/2$ units.

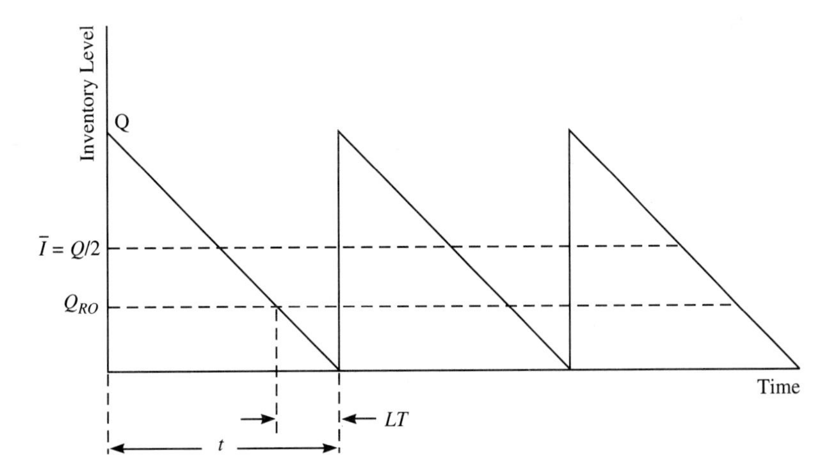

Figure 15-7 Inventory cycles and notation.

A standarized set of notation for lot sizing has yet to be developed. Commonly used notation in many texts is similar to the following:

TIC = Total inventory cost
TIC_0 = Optimal or minimum TIC for a given lot size
Q = Lot size or order quantity
Q_0 = Optimum lot size corresponding to TIC_0
R = Annual demand in units per year
C_H = Holding cost in dollars per unit-year
C_P = Order cost in dollars per order
C_S = Shortage cost in dollars per unit short-year
Q_{RO} = Reorder point in units
LT = Lead time
B = Buffer or safety stock level
I = Inventory level
S = Sales price in dollars per unit

15.8.2 Classical EOQ Model

The classical economic order quantity (EOQ) model was first developed by Harris[17] in 1915. All of the restrictive assumptions listed in the preceding subsection apply for the EOQ model. In addition, part shortages are not allowed.

The model determines that order quantity which minimizes the sum of annual order costs and annual inventory carrying costs for the part being ordered. The optimal order quantity is

$$Q_0 = \sqrt{\frac{2RC_p}{C_H}} \tag{15-13}$$

The corresponding minimum total annual inventory cost is as follows:

$$TIC_0 = \sqrt{2RC_pC_H} \tag{15-14}$$

15.8.3 EOQs with Shortages

It is possible to adapt the EOQ model to allow it to address situations where stock shortages occur. Consider the inventory pattern shown in Fig. 15-8. In this illustration, the maximum inventory balance during any cycle is I_{max}. The period of positive inventory balance is t_1. During period t_2, a shortage in the amount of $Q - I_{max}$ units accrues. An order of size Q is needed to restore the inventory to its previous level of I_{max}. Q is the order size. Of this total, $Q - I_{max}$ units are temporarily unsupplied (**back-ordered**). The optimal order size and corresponding minimum inventory cost are as follows:

[17]F. Harris, *Operations and Cost* (Chicago, Ill: Factory Management Series, Shaw, 1915), pp. 48–52.

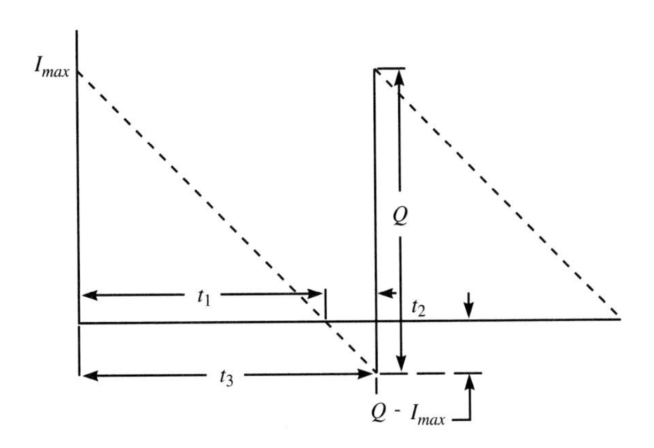

Figure 15-8 EOQ with shortages.

$$Q_0 = \sqrt{\frac{2RC_p}{C_H}} \times \sqrt{\frac{C_H + C_s}{C_s}} \tag{15-15}$$

$$TIC_0 = \sqrt{2RC_pC_H} \times \sqrt{\frac{C_s}{C_H + C_s}} \tag{15-16}$$

Readers should be advised that inventory cost Eqs. (15-14) and (15-16) are valid only when $Q = Q_0$.

15.8.4 EOQ with Price Breaks

The EOQ methodology may be modified to apply in situations where price breaks occur. Generally, vendors will offer products at discounted prices when larger orders are placed. For this model, it is necessary to define a new carrying cost parameter, F_H, which defines holding costs as a fixed percentage of the annual inventory value of the part being stocked. The optimal lot size and total annual inventory costs for this model are

$$Q_0 = \sqrt{\frac{2RC_p}{SF_H}} \tag{15-17}$$

$$TIC_0 = \frac{C_pR}{Q} + SR + SF_H \times \frac{Q}{2} \tag{15-18}$$

Equations (15-17) and (15-18) are applied in the following manner to solve for the optimal lot size. For an order situation with price breaks, each price must have a specific quantity interval. The quantity intervals may not overlap. The price per unit must decrease as the order quantity intervals increase in size.

Equation (15-17) is used to solve for Q for all values of S that apply for the quantity intervals in question. For each computation, the user must check to ensure that the value of Q obtained falls within the quantity interval for which the value of S used in the computation applies. Equation (15-18) is used to compute the total inventory cost associated with the quantity interval.

If Eq. (15-17) yields a value of Q lower than the lowest value of the quantity interval, the value of Q is not used in the computation. Instead, the lowest value of Q in the quantity interval for which S applies is selected and substituted in Eq. (15-18) to obtain the total inventory cost.

If the value of Q is greater than the largest value in the quantity interval, the value of Q from Eq. (15-17) is again not used. Instead, the largest value of Q in the quantity interval for which S applies is selected and substituted in Eq. (15-18) to obtain the total inventory cost.

The calculations of Eqs. (15-17) and (15-18) are performed for all different values of S and their corresponding quantity intervals. An inventory cost associated with each value of S is determined. The optimal order policy is that quantity (and value of S) that has the smallest total inventory cost.

15.8.5 Economic Production Quantity Model

The economic production quantity (EPQ) model applies the EOQ logic to parts that are made, as opposed to those purchased from an outside vendor. The production situation is depicted in Fig. 15-9. A part is produced internally at the rate of p units per day for a period of t_p days. If the daily demand for the part is r units per day, then the inventory balance increases by $(p - r)$ units for each day of production.

At the end of the production period there exists an inventory balance of $t_p(p - r)$ units. This stock level is depleted at the rate of r units per day for the remainder of the inventory cycle. When the stock balance reaches zero, production of the part is again initiated, and the inventory cycle repeats. The optimal order size and corresponding inventory cost are given by Eqs. (15-19) and (15-20):

$$Q_0 = \sqrt{\frac{2RC_p}{C_H(1 - r/p)}} \tag{15-19}$$

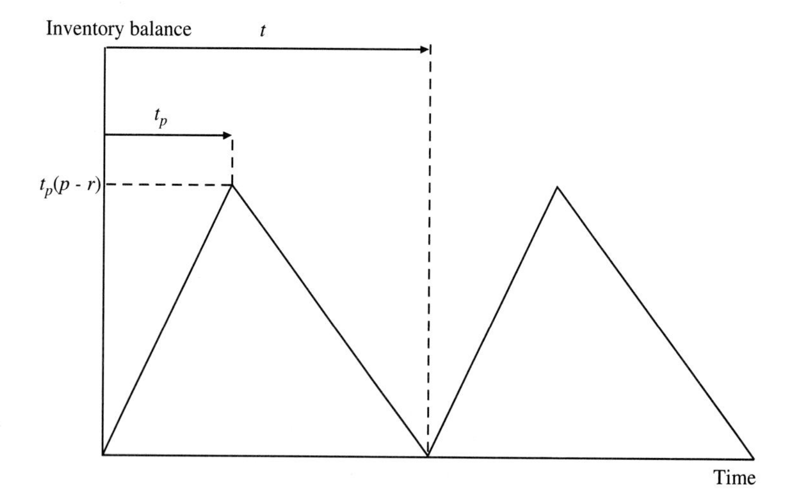

Figure 15-9 Economic production quantity stock levels.

$$TIC_0 = \sqrt{2RC_pC_H(1 - r/p)} \tag{15-20}$$

As before, inventory cost Eq. (15-20) is valid only when $Q = Q_0$.

15.8.6 Variable Demand, Constant Lead-Time Models

The preceding inventory models have all assumed that the demand for the product is constant both during the lead time and the total inventory cycle. This assumption of constant demand is rarely true in practice. Consider the situation shown in Fig. 15-10. The demand from the beginning of each order cycle occurs at some average rate, $\overline{D}$. While the stock is shown to be depleted at a constant rate during the inventory cycle, it will actually vary in accordance with some statistical distribution until the reorder point, Q_{RO}, is reached.

For analysis purposes, it is not necessary to know the demand variation prior to the time Q_{RO} is reached. The approach concentrates on determining demand variation during the lead time, *LT*. The lead time in this case is assumed to be constant.

The variance in the demand during the lead time is accounted for by carrying a **buffer** or **safety stock.** If the lead-time demand continues at the average rate, the stock balance will be depleted exactly to zero by the time the next order arrives. The buffer stock is carried to satisfy lead-time demand up to a rate of D_{max} units per day.

The level of D_{max} selected in determining the buffer stock level, *B*, determines the service level associated with the order policy. The **service level** is that percentage of the time during any order cycle that a stockout will not occur. The higher the service level, the higher the level of buffer stock.

Variable demand, constant lead-time models are of two types: back-order and lost sales. **Back-order** models assume that when a shortage occurs, the product can be back-

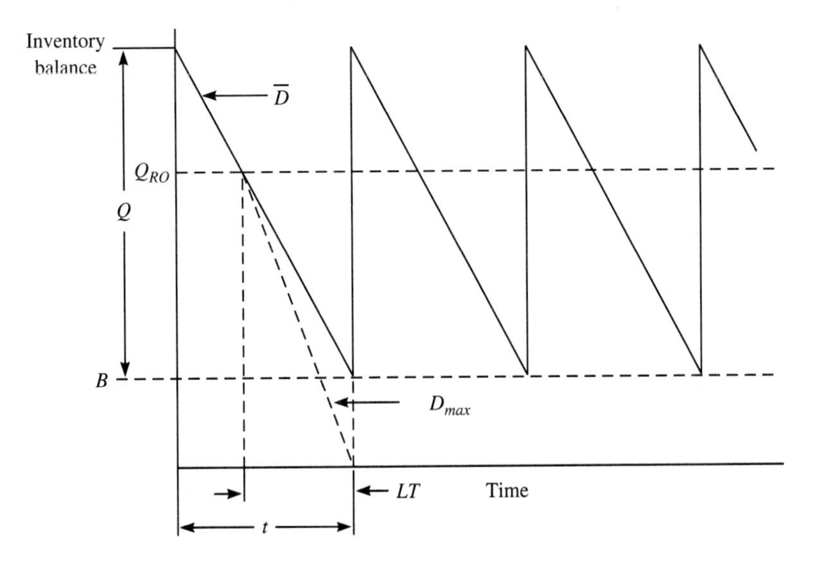

Figure 15-10 Variable demand, constant lead-time stock levels.

ordered, thus satisfying the demand at a later date. Lost sales models assume that when a shortage occurs, the demand for the units short is permanently lost.

The optimal lot size is again the one that minimizes total costs. In this case ordering costs and carrying costs again exist. However, there are now additional carrying costs associated with the buffer stock. The cost of back-ordering parts or lost sales also results when shortages occur during any inventory cycle.

Most models presented in the literature derive solutions corresponding to situations where the lead-time demand is known to vary in accordance with normal, Poisson, or exponential distributions. Product demand in practice rarely varies in accordance with these types of distributions. Discrete probability distributions are therefore recommended for these types of inventory situations. Readers desiring more information on this type of inventory model should consult the references that appear in the bibiliography.

15.8.7 Constant Demand, Variable Lead-Time Models

The constant demand, variable lead-time model addresses the exact opposite of the situation described in Sec. 15.8.6. Figure 15-10 again applies, modified to account for demand that is constant at r units per day and lead time that varies in accordance with a known statistical distribution.

The demand that occurs during the lead time still must be determined. The concepts of buffer stock, back-order costs, and costs of lost sales still apply, as does the concept of service levels. Discrete probability distributions are recommended for use in describing lead-time variation. A model accommodating discrete probability distributions has been presented by Riggs.[18] This model has been significantly refined by Lee, Malstrom, Vardeman, and Petersen[19] to address true average inventory levels when stockouts occur. Readers desiring more information on this type of model should consult these references.

15.8.8 Variable Demand, Variable Lead-Time Models

Variable demand, variable lead-time models impose the fewest restrictive analysis assumptions but are also the most complicated set of inventory models. In this analysis situation, both the demand during the lead time and the lead time itself are allowed to vary. The concepts introduced in the preceding two sections still apply. The problem now becomes one of constructing a joint probability distribution in terms of both demand and lead time. This joint distribution will describe the lead-time demand. Discrete probability distributions are also recommended to describe both demand and lead-time variation.

[18]James L. Riggs, *Production Systems: Planning, Analysis, and Control,* 2nd ed. (New York: John Wiley & Sons, 1976), pp. 450–56.

[19]Ted S. Lee, Eric M. Malstrom, Stephen B. Vardeman, and Volker P. Petersen, "On the Refinement of the Variable Lead Time, Constant Demand Lot Sizing Model: The Effect of True Average Inventory Level on the Traditional Solution," *International Journal of Production Research,* 27, 5 (1989), pp. 883–99.

15.8.9 Reorder-Point Model References

Many of the models described in this section have been exerpted from Buffa and Miller.[20] However, there are a variety of newer texts that also describe these models in greater detail. Interested readers desiring more information on this subject should consult references 2, 3, 10, 12, 13, 19, 25, and 27 at the end of this chapter.

15.9 JUST-IN-TIME INVENTORY SYSTEMS

Just-in-time (JIT) inventory systems are known by a variety of names and terms. These include material as needed (MAN), minimum inventory production systems (MIPS), stockless production, continuous flow manufacturing (CFM), kanban, pull system, and others. JIT has as its goal the elimination of waste, and consequently it represents a philosophy of manufacturing management that is broader than simply an inventory system. Waste is generally defined as anything other than the absolute minimum resources of material, machines, and labor required to add value to the product being produced.

15.9.1 JIT Benefits

The JIT inventory techniques are most effective in situations of low-mix, high-volume repetitive demand at a reasonably constant rate. In most cases, JIT results in significant reductions of all forms of inventory. Such forms include inventories of purchased parts, subassemblies, work-in-process (WIP), and finished goods. Such inventory reductions are accomplished through improved methods not only of purchasing, but also the scheduling of production.

JIT requires significant modifications to traditional methods by which parts are procured. Preferred suppliers are selected for each purchased part. Special purchase arrangements are contractually structured to provide for small orders. These orders are delivered at exact times as required by the user's production schedule and in quantities small enough to be used in very small time periods.

Daily and hourly deliveries of purchased parts are not uncommon in JIT systems. Vendors contractually agree to deliver parts that conform to prescribed quality levels, thereby eliminating the need for the purchaser to inspect incoming parts. The arrival time of such deliveries is extremely important. If the parts arrive too early, the purchaser must carry additional inventory; conversely, if they arrive too late, part shortages occur that can stop scheduled production.

Manufacturers consuming parts on a JIT basis may pay increased unit costs to have parts delivered in this manner. While the start-up costs of structuring the purchase agreement can be significant, the follow-on costs of procuring individual lots of parts on a frequent periodic basis can be reduced to near zero levels. Not having to inspect incoming

[20] E. S. Buffa and Jeffery G. Miller, *Production Inventory Control Systems,* 2nd ed. (Homewood, Ill.: Richard D. Irwin, Inc., 1975), pp. 105–154.

parts can result in increased product quality, reduced inspection costs, and reduced lead time.

Fabricated parts are produced just in time so as to minimize WIP inventory and stockpiles of finished goods. Work is pulled through the process steps of fabrication in response to customer demand. In some JIT systems the need for production is signaled by a simple card called a **kanban.** The JIT philosophy forces manufacturers to solve production bottlenecks and design problems that were previously not visible because of surplus inventory.

A number of organizations have successfully implemented JIT procedures resulting in significant cost savings. Readers desiring a more detailed overview of JIT policies, procedures, and benefits should consult references 4, 11, 15, 16, 17, 20, 21, 22, 23 and 26 at the end of this chapter.

15.9.2 Cost Effectiveness of JIT Systems

The benefits of JIT are realized after some significant investments of effort associated with JIT implementation. The cost of structuring blanket purchase arrangements with a variety of preferred suppliers can be significant. Large costs may also be associated with sophisticated tool design procedures to reduce setup costs to near-zero levels.

Reductions in setup costs are absolutely necessary if a lot-for-lot (LFL) order policy is to be applied, as discussed in Sec. 15.7.9. According to Malstrom,[21] setup costs must be reduced to at least 1/100 of the corresponding carrying costs for the part being produced. If this reduction is not possible, the setup costs associated with an order policy requiring frequent small lots may be significant enough to negate the benefits associated with JIT policies.

Software to assess the cost effectiveness of JIT has been developed by Burney, Malstrom, and Parker.[22] Written in the C language, the software utilizes long sequences of pop-up screens. The screens contain sequences of tutorials that step the user through a detailed cost assessment procedure.

The procedure begins by helping the user to compile a detailed estimate of costs associated with the inventory system currently in use. JIT implementation costs are next estimated. Costs of blanket purchase agreements, setup reduction, personnel training, and lot sizing are all separately estimated. The user is also guided in estimating the costs of facilities modifications required by JIT.

JIT benefits are next assessed. The net change in inventory costs is estimated. The user is guided in ways to quantify the cost savings associated with better product quality and improved customer delivery. Readers desiring a more detailed description of the developed software should consult reference 4.

[21]Malstrom, "Set Up Cost Reduction Requirements for JIT Lot Sizing," np.

[22]Burney et al., "A Computer Assisted Cost Assessment of Just-In-Time Inventory Systems," pp. 420–25.

15.10 SCHEDULING

This section addresses two different types of scheduling. At the macro level, master schedules (Sec. 15.5) must consider the capacity of both the plant and the individual work cells. The master schedule must be continuously adjusted to match workloads with the capacities of machines, facilities, and available personnel. This goal is accomplished through the process of capacity planning.

At the micro level, the scheduling problem becomes one of determining a priority sequence for competing jobs or orders awaiting processing by a single machine or group of production facilities. The sections that follow individually address these topics.

15.11 CAPACITY PLANNING

Capacity planning is a method by which the master schedule is adjusted to balance the due dates of jobs or orders against the capacity of the plant and its individual work cells and facilities.

Capacity planning is perhaps best illustrated with the use of an example. Consider a hypothetical workcenter with one machine which is staffed by one worker. Let us suppose that a one-shift operation applies. Consider the ten-week production schedule shown in Fig. 15-11. For simplicity, assume that each part requires one hour of processing time on the machine.

From Fig. 15-11 it is apparent that there is not enough work to fully occupy the machine and its worker during weeks 20, 21, 28, and 29. The demand in weeks 22, 23, 26, and 27 can be satisfied with the use of overtime. The demand in weeks 24 and 25 cannot be satisfied even if the worker completes six 12-hour days (a total of 72 hours).

The schedule may be adjusted as shown in Fig. 15-12. Suppose 25 units each from weeks 24 and 25 are moved to weeks 20 and 21. The schedule for weeks 20 through 27 inclusive may now be satisfied with the use of 10 hours per week of overtime. While no full workload exists for weeks 28 and 29, it is likely that additional orders will arrive in the next several weeks to fully utilize the production facility during these time periods.

In periods of work underload, the capacity planning procedure seeks to move orders back in time to match workload levels with existing capacities. In periods of work overload, orders are moved forward in time to reduce workload levels. When such schedule

Week	20	21	22	23	24	25	26	27	28	29
Demand	25	25	50	50	75	75	50	50	25	25

Figure 15-11 Sample production schedule.

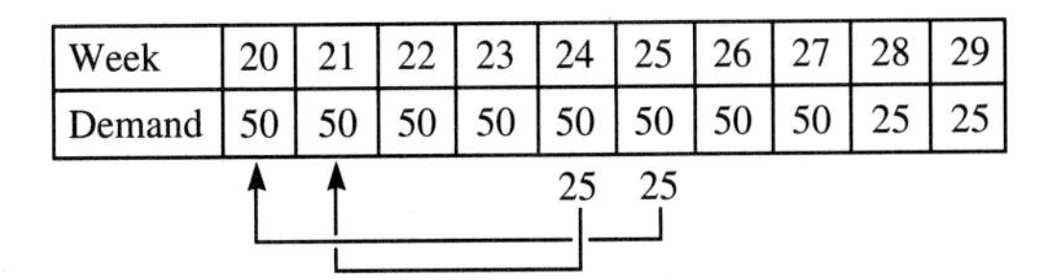

Week	20	21	22	23	24	25	26	27	28	29
Demand	50	50	50	50	50	50	50	50	25	25

Figure 15-12 Schedule changes to smooth production.

adjustments are not possible, it is necessary to hire additional people, add shifts, or lay off personnel (undesirable alternatives because of the extra costs incurred).

Vollman, Berry, and Whybark[23] have described three separate types of capacity planning methods. The methods differ in the amount of production data used to afford increasing levels of detail in assessing workload levels. These methods are separately described in the following subsections.

15.11.1 Capacity Planning Using Overall Factors

Capacity planning using overall factors (CPOF) is a relatively simple approach that results in a rough-cut capacity plan. The inputs come from the master schedule rather than from the MRP tables associated with individual parts in the bill of materials. Workload levels are derived from performance standards or historical data for end products only. Fabrication times for components included in the end item are embedded in these totals. The end item forecasts and component processing data are used to derive workload levels. The CPOF method does not consider the time shift associated with the lead times for all component parts in the end item.

15.11.2 Capacity Bills

Capacity bills provide a more direct linkage between different end products being produced and the respective capacities required by these different end items in various workcenters. The method is responsive to changes in product mix of the end items produced. Additional data are required to use this approach. Lot sizes for each end product and their respective components must be known. Setup and run times for each lot must be defined for each workcenter in which processing is required.

15.11.3 Resource Profiles

The **resource profile** approach further refines the capacity bills procedure, by considering the lead-time requirements associated with each node in the part explosion diagram. All data for the capacity bill method is used, but it is defined to occur in the specific period during which the work on a specific part or subassembly is scheduled to take place. The resource profile is the most detailed (and time-consuming) of the three approaches that have been described.

More detailed information on each of these three capacity planning methods is presented in reference 25. The descriptions of each method are illustrated with detailed numerical examples.

15.12 MACHINE SCHEDULING METHODS

A variety of methods exist for the scheduling of jobs or orders within a given work cell. For most rules a notation of the form *n/m/C* applies. In this notation, the term *n* denotes the number of jobs or orders that are to be scheduled. The term *m* refers to the number

[23]Vollman et al., *Manufacturing Planning and Control Systems*, pp. 119–130.

of machines within the work cell. Finally, the term C refers to the objective or criterion addressed by the developed schedule.

The typical scheduling objective is to deliver or complete the orders by the due dates required by the customer, and it is accomplished by minimizing the average or maximum lateness for a sequence of jobs or orders. Another common objective is to minimize the elapsed time that the order or job is in process within the work cell, which is equivalent to minimizing the average or maximum flow time for a sequence of jobs.

The mathematics involved in proving that job sequences derived from specific rules satisfy specific scheduling criteria are complex. Most early work in analyzing scheduling methodologies therefore focused on work cells consisting of only one or two machines. A review of this early work is the topic of the subsections that follow.

15.12.1 Shortest Processing Time Rule

The **shortest processing time** (SPT) rule schedules jobs across a machine or set of production facilities in order of increasing processing times. For n jobs that are sequenced across a single machine, it may be proved that the SPT rule minimizes the mean flow time for all jobs. Flow time refers to the sum of the times the job spends in queue plus its processing time.

The primary disadvantage of the SPT rule is that jobs with long processing times usually are delayed in reaching the front of the queue. They are therefore often completed long after the due date required by the production schedule. This problem is addressed by using a truncated form of the SPT rule which forces jobs with long processing times to the front of the queue after they have awaited processing a specified length of time.

15.12.2 Due Date Rule

The **due date** (DDATE) rule sequences jobs across a machine or set of production facilities in ascending order of date by which the order or job is due to be completed. Those jobs with the earliest due dates are worked on first. For n jobs and one machine, it may be proved that the DDATE rule minimizes the maximum lateness for the sequence of jobs that are scheduled.

15.12.3 Slack Time Rule

The **slack time** rule (SLACK) sequences jobs across a machine or set of production facilities in order of increasing slack time. Slack time is the difference been a job's due date and its processing time. For any job i in a sequence of n jobs, the slack time is defined as

$$t_i = d_i - p_i \tag{15-21}$$

where:

t_i = Slack time for job i
d_i = Due date for job i
p_i = Processing time for job i

For the SLACK rule, jobs with minimal slack have the greatest risk of being late and are therefore placed first in the scheduling sequence. For n jobs and 1 machine, it may be proved that the SLACK rule maximizes the minimum lateness for the sequence of jobs or orders that are scheduled.

15.12.4 Multiple Machine Rules

Many scheduling applications involve the use of more than one machine. Conway, Maxwell, and Miller[24] overview two methods which address scheduling problems for n jobs on 2 machines ($n/2$) or 2 jobs on m machines ($2/m$). Johnson's algorithm is applicable for $n/2$ scheduling problems. According to Johnson's algorithm, in an optimal schedule job i preceeds job j if the minimal processing time for job i on either machine 1 or machine 2 is less than or equal to the minimal processing time for job j on either machine 1 or machine 2. Application of this procedure will yield a sequence that will minimize the maximum flow time of all n jobs across the two machines. The authors also illustrate a graphical scheduling procedure for a $2/m$ scheduling problem. The goal is to minimize the maximum flow time for two jobs across a set of m machines. Times at which both jobs will need a given machine are depicted on a two-dimensional graph as conflict areas. Scheduling paths are illustrated which pass around these regions, and which attempt to maximize the amount of time that both jobs receive simultaneous processing. Readers desiring more information on either of these methods should consult references [1] and [9].

15.12.5 First Come, First Served, and Random Scheduling

These two methods are equivalent to doing no scheduling at all. In the first come, first served (FCFS) method, jobs are processed in the order in which they arrive at the machine or facility. With the random (RANDOM) method, a completely arbitrary job sequence is randomly selected. The value of these methods comes from comparing them to other scheduling rules and heuristics. The FCFS and RANDOM rules serve as comparison benchmarks to show how much improvement can be obtained through use of other scheduling methodologies.

15.12.6 The RAND Simulation Studies

The RAND simulations were performed in the 1960s by the Rand Corporation. They have been described in detail by Conway, Maxwell, and Miller.[25] These studies are significant in that they are among the first large-scale digital simulation studies that analyzed a variety of scheduling rules in a multiple-machine environment.

Much of the RAND evaluation focused on an $n/9$ scheduling environment. A variety of evaluation criteria were defined, including average number of jobs in queue, work

[24] Richard W. Conway, William L. Maxwell, and Louis W. Miller, *Theory of Scheduling* (Reading, Mass.: Addison-Wesley Publishing Company, 1967), p. 287.

[25] Conway et al., *Theory of Scheduling,* p. 287.

hours remaining, work hours completed, average flow time, average job tardiness, and fraction of jobs tardy.

Job due dates were generated in four different ways. These included a constant multiple of the job's processing time, a date proportional to the number of operations in the job, a constant due date for all jobs, and due dates that were randomly assigned.

A variety of scheduling rules were analyzed, including SPT, DDATE, SLACK, RANDOM, and FCFS. Additional rules included those based on the amount of work in queue, the amount of work remaining, and the number of job operations remaining, and those that prorated both due dates and slack time among a job's operations.

The SPT rule was consistently among the best performers for all evaluation criteria. SPT-scheduled jobs were found to have the smallest average flow times. The SPT rule also performed best in terms of average tardiness and the number of jobs tardy.

The results of the RAND simulations have been confirmed in a number of subsequent simulation analyses. Because of the excessive lateness of SPT sequenced jobs with large processing times, a truncated version of the SPT rule is generally recommended for use. Readers desiring additional information on the RAND simulations and scheduling rules in general should consult references 1, 6, 7, and 9.

15.13 DISPATCHING

Dispatching is a very important facet of the production control process. **Dispatching** involves the movement of parts, components, subassemblies, and end items so that they arrive at the appropriate workcenter, exactly at the time they are needed in the production process. An alternate name for this procedure is **shop floor control.**

Three types of material are usually moved in the dispatching process. The first is the movement of a partially completed part or subassembly to the appropriate workcenter. The second is the movement of raw materials or components that are to be added at a particular process operation. The third is the movement of tooling, fixtures, gages, and inspection equipment to the workcenter.

Not all parts survive all steps in the manufacturing process. For example, in metal-working a part that has a dimension that is too large may often be reworked. In contrast, a part that has too much metal removed resulting in a small dimension is often scrapped.

A common procedure is for process planners to start greater quantities than are required at the beginning of the process sequence. Suppose historical scrap rates for a given set of processes are 10 percent. If 50 parts are required at the end of the sequence, a start quantity of 55 parts might be specified to begin the first process operation.

Those parts that can be reworked must be moved back through those workcenters required to perform the necessary rework operations. Parts completing the rework process must be rejoined with the remaining parts of the order.

When the scrap rate for a set of processes is higher than expected, the start quantity may not be sufficient to satisfy the order requirements. Supplemental orders of sufficient quantity to account for excessive scrap rates may have to be placed. Moving these supplemental orders to and from the necessary workcenters is also part of the dispatching process.

15.14 EXPEDITING

Greene[26] defines **expediting** as "the process of pushing shop orders that have fallen behind schedule." Expediting also encompasses the follow-up of orders for purchased parts that have not arrived on time.

The production scheduling process is dynamic. Based on data extracted from the initial master schedule, it is possible to establish required dates for both made and purchased parts through MRP and the product explosion structure (bill of materials). Unfortunately, these dates are often subject to change. Capacity planning both compresses and stretches the master schedules for different end items. The result is a necessity to change previously established requirement dates for both purchased parts and fabricated subassemblies.

Other production difficulties result in the need for expediting. These include labor problems, equipment failures, and unanticipated rework or scrap. In JIT environments, stretched schedules may result in delayed requirement dates. Expediting in these cases may entail delaying the receipt dates for those parts and assemblies affected by the revised schedule.

15.15 SUMMARY

This chapter has sought to overview a number of principles and techniques of production planning and control. A wide variety of topics have been addressed. Several books have been written about many of the major topics that have been addressed.

The approach used has been to provide overviews of all subjects. Where appropriate, numerical examples and mathematical notation have been used to illustrate concepts and procedures. None of the sections in this chapter is intended to provide comprehensive coverage on any topic. References have been provided throughout the chapter that provide literature sources containing additional information on each major subject. Readers desiring additional information are urged to consult these references. A detailed bibliography of suggested additional reading is provided at the end of this chapter.

KEY TERMS

Back-order
Bill of materials
Blanket purchase orders (BPO)
Buffer (safety) stock
Capacity bill
Capacity planning using overall factors
Carrying cost
Dispatching
Due date
Economic order quantity (EOQ)
Expediting
Explosion-based inventory systems
Exponential smoothing
Gross requirement (GR)
Kanban
Lead time

[26] James H. Greene, *Operations Management: Productivity and Profit* (Reston, Va.: Reston Publishing Company, 1984), pp. 571–72.

Least total cost (LTC)
Least unit cost (LUC)
Lot-for-lot (LFL)
Lot size
Master schedule
Moving average forecast
On-hand balance
Ordering cost
Part explosion diagram
Part period balancing (PPB)
Periodic order quantity (POQ)
Planned orders
Purchase order (PO)
Reorder point
Reorder-point inventory systems
Requirements planning
Requisition
Resource profile
Scheduled receipt
Service level
Shop floor control
Shortest processing time (SPT)
Silver-Meal (SM) algorithm
Slack time
Time-phased record
Time series
Wagner-Whitin (WW) algorithm

EXERCISES

1. List the methods used for forecasting demand, and indicate when these methods are applicable.
2. Describe the advantages and disadvantages between the curve-fitting and time-series forecasting methods.
3. Briefly describe the steps in the procurement process.
4. What are the basic types of inventory systems, and when are they used?
5. List the MRP lot-sizing methods. Which method is consistently the best performer? Why?
6. What are the assumptions associated with some of the more mathematically straightforward reorder-point lot-sizing models?
7. What are the benefits of a JIT system?
8. Briefly describe different machine scheduling rules. What rule would you recommend for general-purpose use? Why?
9. Define dispatching and expediting.
10. An inventory application exists whose characteristics satisfy economic order quantity (EOQ) assumptions. It costs \$50.00 each time an order for this part is placed. The part's carrying cost is \$100 per unit-year. The shortage cost is \$1,000 per unit short-year. If the annual demand for the part is 10,000 units, what is the optimal order size? What is the corresponding annual inventory cost?
11. An internally produced part has an annual demand of 10,000 parts per year and can be produced at the rate of 50 units per day. The carrying cost is \$100 per unit-year. Set-up time to produce a batch of parts costs \$200 per order. If a work year has 250 days, what is the optimal order size? What is the corresponding annual inventory cost?
12. A company wishes to produce a high-pass filter assembly whose part explosion diagram is shown in Fig. 15-13. Part lead times are shown on the links. The master schedule for the high-pass filter is as follows:

Period	10	11	12	13	14	15	16	17	18	19	20
Demand	100	150	150	200	250	200	175	150	150	100	100

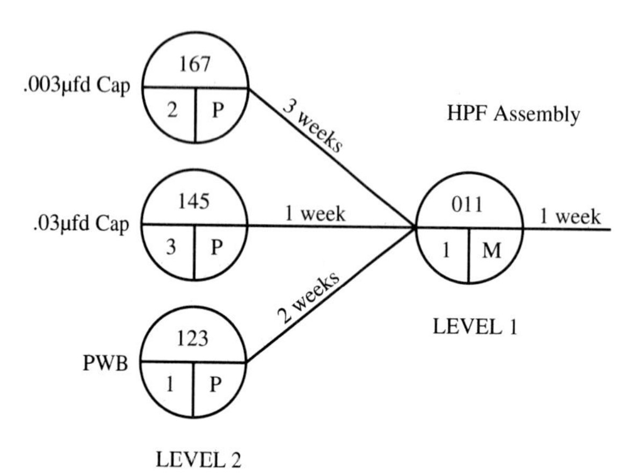

Figure 15-13 High-pass filter part explosion diagram.

Assume a separate order is placed for each period's demand. Complete the missing entries on the MRP tables that follow.

MRP TABLE FOR HPF ASSEMBLY PART NUMBER 011

Period	9	10	11	12	13	14	15	16	17	18	19	20
Gross Req.												
Sch. Rec.			100	100		300	200	150		125	100	100
On Hand	200		50		0	50		25	25		0	0
Planned Ord.			100		300	200	150		125	100	100	

MRP TABLE FOR .003 μFD CAP. PART NUMBER 167

Period	9	10	11	12	13	14	15	16	17	18	19	20
Gross Req.												
Sch. Rec.				400		500	400		250	300	100	
On Hand	400	200		0	0	100		0	0		0	
Planned Ord.		600	500	400		250	300	100				

MRP TABLE FOR .03 μFD CAP. PART NUMBER 145

Period	9	10	11	12	13	14	15	16	17	18	19	20
Gross Req.												
Sch. Rec.			400	600	900		300	400	375			
On Hand	300		100		100	200	50		0	300	0	
Planned Ord.		400	600		700	300	400		600			

MRP TABLE FOR PRINTED WIRING BOARD PART NUMBER 123

Period	9	10	11	12	13	14	15	16	17	18	19	20
Gross Req.												
Sch. Rec.				200		100		100	100		75	
On Hand	200			0	100		150	100				
Planned Ord.			400	100	300		100		75			

BIBLIOGRAPHY

1. BAKER, KENNETH R., *Introduction to Sequencing and Scheduling,* New York: John Wiley & Sons, 1974.
2. BANKS, JERRY, AND W. J. FABRYCKY, *Procurement and Inventory Systems Analysis,* Englewood Cliffs N. J.: Prentice Hall, Inc., 1987.
3. BEDWORTH, DAVID D., AND JAMES E. BAILEY, *Integrated Production Control Systems: Management, Analysis and Design,* 2nd ed., New York: John Wiley & Sons, 1987.
4. BURNEY, M. A., ERIC M. MALSTROM, AND SANDRA C. PARKER, "A Computer Assisted Assessment of Just-in-Time Inventory Systems," *Proceedings of the Spring Annual Conference,* Institute of Industrial Engineers, San Francisco, May 1990.
5. CARBOY, JAMES. D., GEORGE FOO, L. P. JONES, L. E. KINNEY, AND D. C. KRUPKA, "Striving for Excellence at the Denver Works: A Summary," *AT&T Technical Journal,* 69, no. 4 (July/August 1990).
6. CHOI, RICHARD H., AND ERIC M. MALSTROM, "Evaluation of Work Scheduling Rules in a Flexible Manufacturing System Using a Physical Simulator," *Journal of Manufacturing Systems,* 7, no. 1 (1988).
7. CHOI, RICHARD H., AND ERIC M. MALSTROM, "Physical Simulation of Work Scheduling Rules in a Flexible Manufacturing System," *Proceedings of the 8th Annual Conference on Computers and Industrial Engineering,* University of Central Florida, Orlando, March 1986.
8. COHEN, MELVIN I., AND L. C. SEIFERT, "Striving for Manufacturing Excellence," *AT&T Technical Journal,* 69, no. 4 (July/August 1990).
9. CONWAY, RICHARD W., WILLIAM L. MAXWELL, AND LOUIS W. MILLER, *Theory of Scheduling,* Reading, Mass.: Addison-Wesley Publishing Company, 1967.
10. EVANS, JAMES R., D. R. ANDERSON, D. J. SWEENEY, AND T. A. WILLIAMS, *Applied Production and Operations Management,* 2nd ed., St. Paul, Minn.: West Publishing Company, 1987.
11. GORDON, JAY, "JBL: In-Plant Harmony Helps Darth Vader Speak in Final Volume," *Distribution,* 85, no. 8 (1986).
12. GREENE, JAMES H., *Operations Management: Productivity and Profit,* Reston, Va.: Reston Publishing Company, 1984.
13. KRAJEWSKI, LEE, AND LARRY P. RITZMAN, *Operations Management: Strategy and Analysis,* 2nd ed., Reading, Mass.: Addison-Wesley Publishing Company, 1990.
14. MALSTROM, ERIC M., *What Every Engineer Should Know About Manufacturing Cost Estimating,* New York: Marcel Dekker, Inc. 1981.
15. MALSTROM, ERIC M., "Assessing the True Cost Savings Associated with Just-in-Time Inventory Systems," *Proceedings of the Fall Annual Conference,* Institute of Industrial Engineers, St. Louis, November 1988.
16. MARTIN-VEGA, L. A., M. PIPPIN, E. GORDON, AND R. BURCHMAN, "Applying Just-in-Time in a Wafer Fab: A Case Study," *IEEE Transactions in Semiconductor Manufacturing,"* 2, no. 1 (1989).
17. NEIL, GEORGE, AND JIM O'HARA, "The Introduction of JIT into High Mix Electronics Manufacturing Environment," *International Journal of Operations and Production Management,* 7, no. 4 (1987).
18. ORLICKY, JOSEPH, *Material Requirements Planning,* New York: McGraw-Hill Book Company, 1975.

19. REINFELD, NYLES V., *Production and Inventory Control,* Reston, Va.: Reston Publishing Company, 1982.

20. SCHONBERGER, RICHARD J., *World Class Manufacturing,* London: The Free Press, Collier MacMillan Publishers, 1986.

21. SCHONBERGER, RICHARD J., *Japanese Manufacturing Techniques: Nine Hidden Lessons in Simplicity,* London: The Free Press, Collier MacMillan Publishers, 1982.

22. SCHONBERGER, RICHARD J., "Some Observations on the Advantages and Implementation Issues of JIT Production Systems," *Journal of Operations Management,* 3, no. 1 (November 1982).

23. SEPEHRI, M., AND RICHARD C. WALLEIGH, "HP Division Programs Reduce Cycle Times, Set Stage for Ongoing Process Improvements," *Industrial Engineering,* 18, no. 3 (1986).

24. THOMOPOULOS, NICK T., *Applied Forecasting Methods,* Englewood Cliffs, N.J.: Prentice Hall, Inc., 1980.

25. VOLLMAN, THOMAS E., WILLIAM L. BERRY, AND D. CLAY WHYBARK, *Manufacturing Planning and Control Systems,* 2nd ed., Homewood, Ill.: Richard D. Irwin, Inc., 1988.

26. VOSS, C. A., *Just-in-Time Manufacturing,* United Kingdom: IFS Publications, Ltd., 1987.

27. WIGHT, OLIVER W., *Production Inventory Management in the Computer Age,* Boston, Mass.: CBI Publishing Company, 1974.

16 Production Economics

16.1 INTRODUCTION

This chapter addresses the topic of production economics. Different types of production costs are initially overviewed in Secs. 16.2–16.6. Economic aspects of manufacturing automation are next addressed in Sec. 16.7. Special attention is devoted to those conditions that must exist to make manufacturing automation an economically viable alternative. A review of engineering economic analysis is given in Sec. 16.8. Section 16.9 presents a case study in economic evaluation of process automation. The chapter concludes with a description of learning or product improvement curves (Sec. 16.10). Concepts of cost/quantity relationships are overviewed in relation to both production efficiency and standard times required to produce specific assemblies and subassemblies.

16.2 TYPES OF PRODUCTION COSTS

A variety of different types of production costs exist. These costs are described in the sections that follow.

16.2.1 Possible Pricing Situations

Malstrom [5] has described a variety of factors that can affect the way an organization sets the retail price for a specific product. For some situations, the price is set to be

Portions of this chapter are adapted from the following references, with the permissions of the publishers: MALSTROM, ERIC M., *What Every Engineer Should Know About Manufacturing Cost Estimating.* New York, New York: Marcel Dekker, Inc., 1981.

MALSTROM, ERIC M., AND RICHARD L. SHELL, "A Review of Product Improvement Curves," *Manufacturing Engineering,* 82, no. 5, 1979 (Courtesy of the Society of Manufacturing Engineers).

competitive with similar products on the market. If a variety of similar products are being manufactured by competitors, the organization will have little control over the retail price that is charged for the product. This price will be determined by supply, demand, and consumer preferences.

Alternately, the organization may opt to set high prices to attract customers desiring a prestige item. This philosophy capitalizes on the fact that some consumers equate product excellence with the price paid for the specific product. The price may also be set by adding as profit margin a fixed percentage of the cost to produce the product. The profit margin plus the product cost then comprises the selling price. Products produced under military contracts are often priced in this manner, referred to as cost-plus contracting.

Products that are patented or made by a secret (proprietary) process may have no competition on the open market. In these cases, the producing organization may opt to set the price as high as the market will tolerate. A good example of this pricing strategy evolved with the early Polaroid Land camera that was first introduced in the 1950s.

Another scenario involves the setting of a low initial price usually associated with a new product with which consumers have not had prior experience. As the product gains acceptance, the price is raised. Consumers realize they may not want to do without the product after having become exposed to it. An example of this pricing scenario would be introduction of cable television and subsequent rate hikes as viewers became reluctant to do without this service.

16.2.2 Direct Labor

Direct labor is the hands-on effort in the production process and adds value to the product. Typical direct labor activities include machining, assembly, inspection, electronic/mechanical testing, and troubleshooting. Examples of labor activities that are not normally considered direct include:

- Dispatching and movement of production parts (Chaps. 14 and 15)
- Shop supervision
- Preparation of cost estimates
- Production scheduling activities (Chap. 15)

16.2.3 Direct Material

Direct material cost is defined as the cost of all components and raw materials included in the product. Examples of materials that are not considered direct include:

- Raw material purchased for production tooling
- Coolants, solder, and similar supplies consumed by the production process
- Material required for end product packaging during shipment

16.2.4 Relationships Between Component Costs

Niebel and Draper [8] have defined relationships between costs associated with manufacturing. The **prime cost** of a product is defined as the sum of its direct labor and direct material:

$$\textit{Prime Cost} = \textit{Direct Labor} + \textit{Direct Material} \qquad (16\text{-}1)$$

Factory expenses are defined as the total of all costs for rent, heat, electricity, water, expendable factory supplies, and indirect labor. **Factory cost** is defined as the sum of factory expenses and prime costs:

$$\textit{Factory Cost} = \textit{Factory Expenses} + \textit{Prime Cost} \qquad (16\text{-}2)$$

General expenses are defined as the costs of design, engineering, purchasing, office salaries/supplies, and depreciation. **Manufacturing cost** then becomes the sum of general expenses and factory costs:

$$\textit{Manufacturing Cost} = \textit{General Expenses} + \textit{Factory Costs} \qquad (16\text{-}3)$$

Sales expenses are defined as all costs incurred in selling and delivering the product. These costs also include advertising expenses, allowances for bad debts, shipping costs, and commissions for sales personnel. **Total costs** to manufacture the product then become the sum of sales expenses and the manufacturing costs:

$$\textit{Total Cost} = \textit{Sales Expenses} + \textit{Manufacturing Costs} \qquad (16\text{-}4)$$

The **selling price** of the product then is the sum of the total cost and profit margin:

$$\textit{Selling Price} = \textit{Total Costs} + \textit{Profit} \qquad (16\text{-}5)$$

The composition and relationships of costs and sales price are depicted in Fig. 16-1.

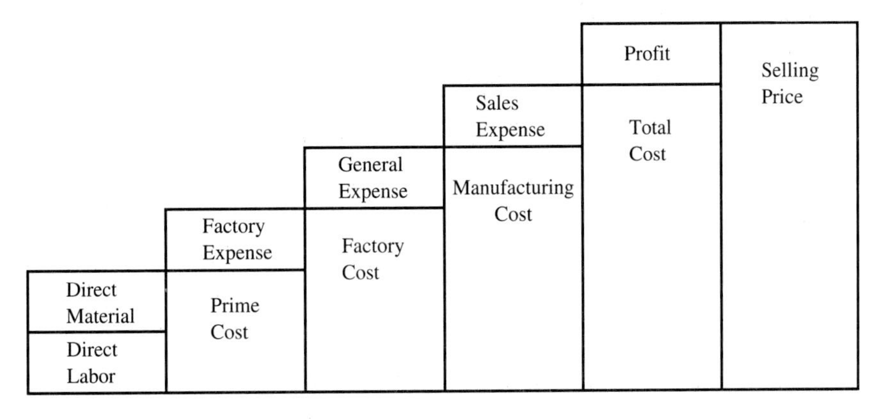

Figure 16-1 Interrelationships between product cost components.

(*Source:* Adapted from Benjamin W. Niebel and Alan B. Draper, *Product Design and Process Engineering* (New York: McGraw-Hill Book Company, 1974).)

16.3 INDIRECT LABOR AND ORGANIZATIONAL SIZE

Indirect labor cost is defined as the cost of all labor efforts that cannot be directly associated with the manufacture of a product. Examples include manufacturing supervision, production scheduling, cost estimating, and salaries for personnel who work in the purchasing department.

Malstrom [5] has observed that the cost of indirect labor increases sharply with an organization's overall size. This phenomenon is best illustrated with an example. Consider a small organization with only 10 employees. An organizational chart for this company is illustrated in Fig. 16-2. If this organization is managed by an eleventh person who serves as the shop manager, the ratio of direct to indirect employees is 10:1.

Suppose this organization is compared with a larger company whose organizational chart is depicted in Fig. 16-3. For purposes of consistency in comparison, only the

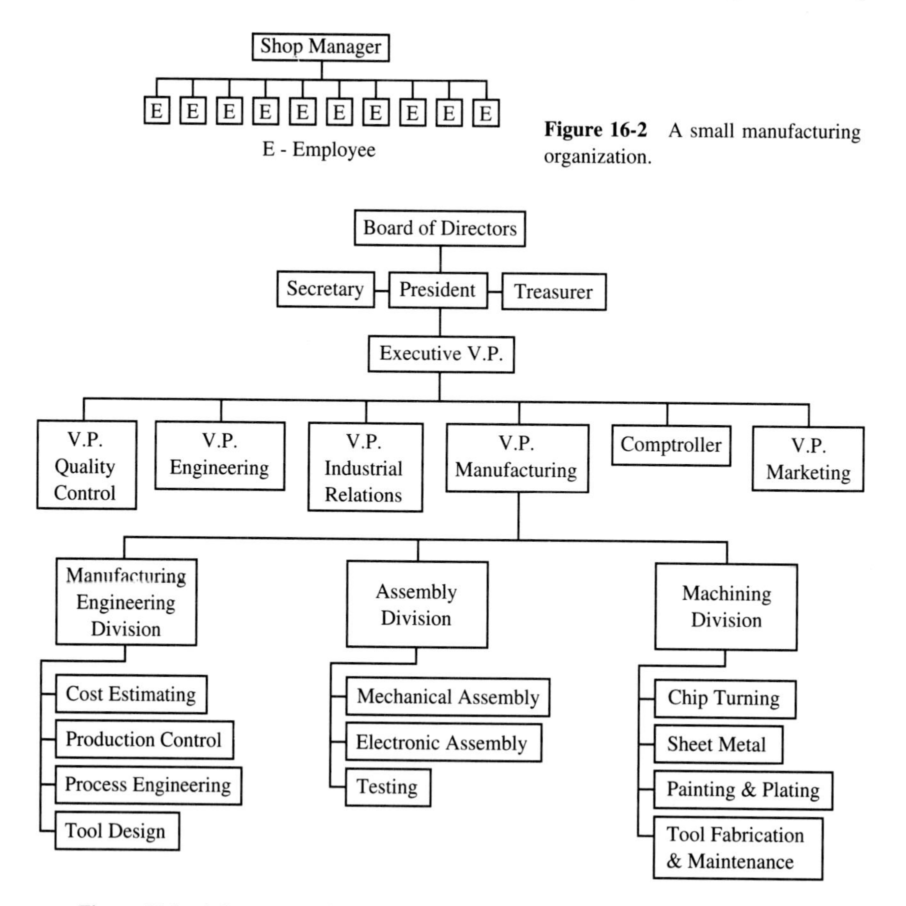

Figure 16-2 A small manufacturing organization.

Figure 16-3 A larger manufacturing organization.

(*Source:* Excerpted from Malstrom, *What Every Engineer Should Know About Manufacturing Cost Estimating,* p. 10.)

Manufacturing Department in Fig. 16-3 will be compared with the smaller shop having 10 employees. For additional consistency, it will be assumed that each branch supervisor in Fig. 16-3 oversees the actions of 10 employees. This will provide for a supervisory span of control identical to that shown in Fig. 16-2.

The direct labor branches in Fig. 16-3 are

- Chip turning
- Sheet metal
- Painting and plating
- Tool fabrication and maintenance
- Mechanical assembly
- Electronic assembly
- Testing
- Process engineering (?)
- Tool design (?)

In this list of direct labor branches, question marks have been placed adjacent to the process engineering and tool design branches. In some organizations these functions are considered to be indirect labor, but to keep this comparison conservative, they will be treated as direct labor functions. The total direct workers in the Manufacturing Department can now be summarized as follows:

$$\text{Total Direct Workers} = \frac{10 \text{ Workers}}{\text{Branch}} \times 9 \text{ Branches} = 90$$

In the Manufacturing Engineering Division, the cost estimating and production control branches are normally considered to be indirect labor. The total number of indirect workers in the Manufacturing Department may now be calculated as follows:

$$\text{Total Indirect Workers} = \frac{10 \text{ Workers}}{\text{Branch}} \times 2 \text{ Branches} + 3 \text{ Division Managers} + 11 \text{ Branch Supervisors} = 34$$

Thus, there is a ratio of 2.65 to 1 direct to indirect employees as follows:

$$\frac{\text{Direct Employees}}{\text{Indirect Employees}} = \frac{90}{34} = 2.65$$

No clerical support has been included in this example. Clerical support exists in larger organizations and would be costed as indirect labor. Since the ratio of direct to indirect employees has decreased from 10:1 to 2.65:1, it is apparent from this example that indirect costs increase significantly with organizational size. Indirect labor has a significant impact on the hourly rate of indirect cost per direct labor hour that a larger organization must charge.

16.4 FIXED AND VARIABLE COSTS

Manufacturing costs may be classified based on how the cost varies with the production quantity being produced. There are three main categories of manufacturing costs (Fig.16-4):

- *Fixed costs:* Those costs that are independent of production quantity, including set-up costs, costs to program component insertion/placement and automatic testing equipment, and tooling costs.
- *Variable costs:* Those costs that vary with the production quantity on a per-unit basis, including direct material and direct labor activities completed on specific production units in the processes of machining, inspection, assembly, and testing.
- *Semifixed costs:* Those costs that are dependent on production quantity and vary with specific groups of units that are produced. Examples include costs for periodic recalibration of machines and costs required to perform scheduled maintenance [5].

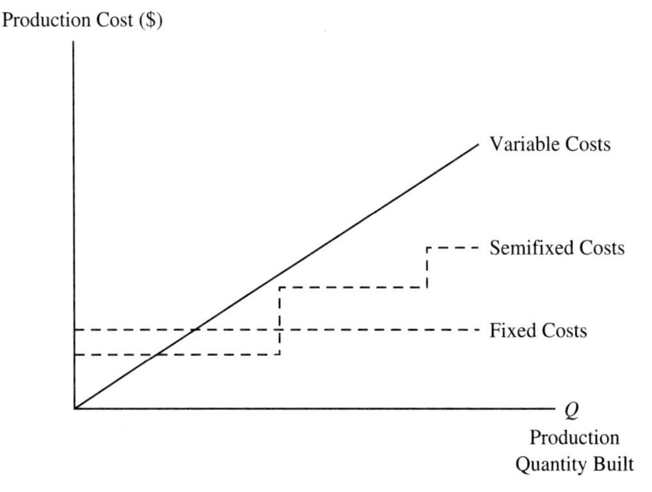

Figure 16-4 Other types of shop costs.

(*Source:* Malstrom, *What Every Engineer Should Know,* p. 19.)

16.5 BREAK-EVEN ANALYSIS

Break-even analysis has been described by Malstrom [5]; it addresses the question of determining what production quantity is required to recover initial fixed costs associated with the start-up of production.

16.5.1 Linear Break-even Analysis

An example involving linear break-even analysis will be considered first. The following notation applies.

Let:

Q = Number of units built or sold

P = Sales price per unit
$I = Q \times P$ = Sales income
F = Fixed costs
V = Variable costs for a production level of Q units
R = Profit at a production level of Q units
C = Total costs

$$C = F + QV \tag{16-6}$$

$$R = I - C \tag{16-7}$$

The relationship between these variables is depicted in Fig. 16-5. The break-even point occurs where total income equals total cost. In the linear break-even example of Fig. 16-5, losses occur to the left of the break-even point, whereas to the right of the break-even point, the organization will incur profits. The break-even point may be expressed mathematically in the following manner:

$$\begin{aligned} \textit{Sales Income} &= \textit{Total Costs} \\ QP &= F + QV \\ Q &= \frac{F}{P - V} \end{aligned} \tag{16-8}$$

Example:

A plant has **fixed costs** in the amount of \$35,000. **Variable costs** are incurred in a production process in the amount of \$57.00 per unit. If the units can be sold for \$75.00 each, how many units must be produced before the plant breaks even?

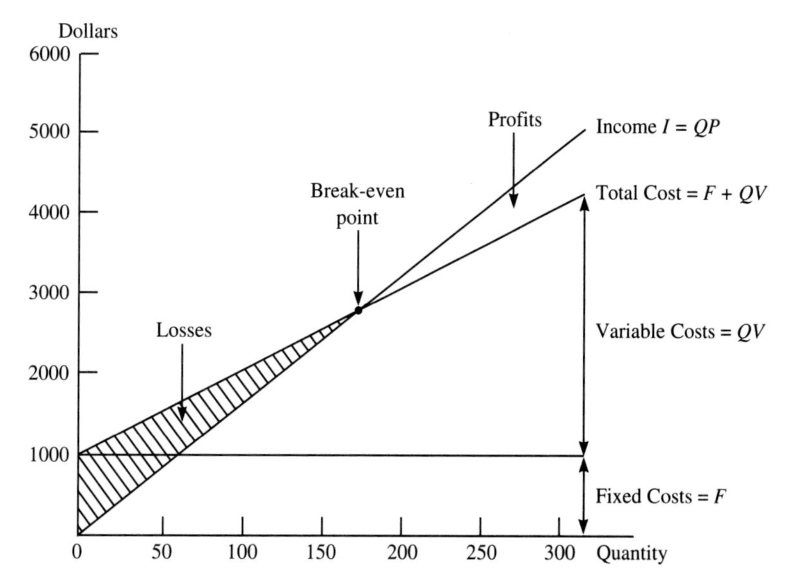

Figure 16-5 Linear break-even chart.

(*Source:* Malstrom, *What Every Engineer Should Know,* p. 20.)

$$Q = \frac{F}{P - V} = \frac{\$35,000}{\$75 - \$57} = 1,945 \text{ units}$$

What is the margin of profit at a production level of 3,000 units?

$$\text{Income} = I = QP = 3,000 \text{ units} \times \$75/\text{unit} = \$225,000$$
$$\text{Costs} = C = F + QV = \$35,000 + 3,000 \text{ units} \times \$57/\text{unit} = \$206,000$$
$$\text{Profit} = R = I - C = \$225,000 - \$206,000 = \$19,000$$

16.5.2 Piecewise-Linear Break-even Analysis

Piecewise-linear break-even analysis allows for the inclusion of **semifixed costs** and price breaks, resulting in a reduced price charged for larger production quantities. Consideration of these factors may result in multiple break-even points in the production process.

Example:

Suppose a process exists which has an initial fixed cost of $1,000.00. Suppose further that a semifixed cost in the amount of $1,500.00 is incurred at unit # 300. The sales price is assumed to be $15.00 per unit for units 1 through 500 inclusive. The price drops to $2.50 per unit for units 501 and up. The piecewise-linear unit break-even chart for this example is presented in Fig. 16-6. From this illustration, it is apparent that three break-even points exist, at quantities of approximately 180, 425, and 575 units, respectively. Over the quantity interval shown, there exist two regions of profits and three regions of financial losses.

The following example illustrates the supplier's profits under price breaks.

Example:

Suppose 125 units are to be produced. The first 75 are to be sold for $20 each. The next 25 are to be sold at a price of $15 per unit. The last 25 are to be sold at $10 each. Fixed costs are known to be $100. A semifixed cost of $75 is incurred at unit 100 and every 100 units

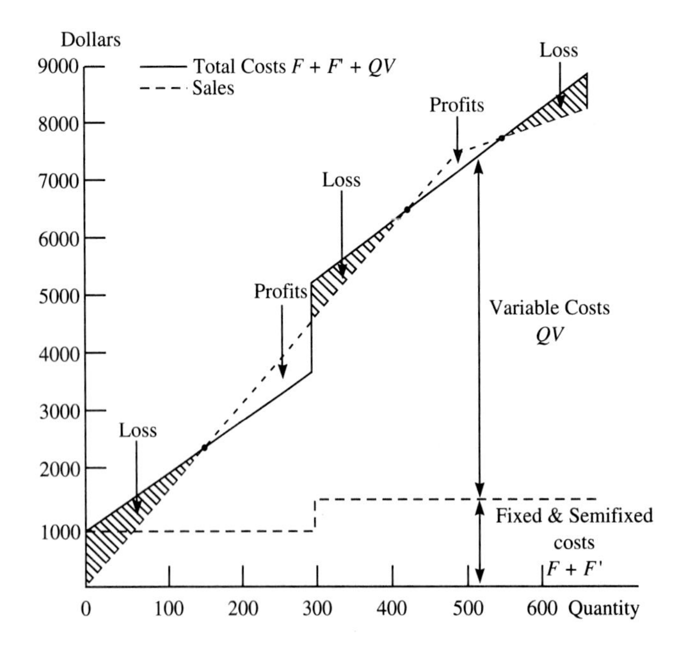

Figure 16-6 Piecewise-linear break-even chart.

(*Source:* Malstrom, *What Every Engineer Should Know,* p. 22.)

thereafter. Variable costs are \$8 per unit for the first 85 units and increase to a level of \$15 per unit thereafter. What is the profit level of this operation if all 125 units are built?

Solution:

$$\text{Total Income} = I = Q_1P_1 + Q_2P_2 + Q_3P_3$$
$$= 75\ (\$20) + 25\ (\$15) + 25\ (\$10)$$
$$= \$1{,}500 + \$375 + \$250 = \$2{,}125$$
$$\text{Total Costs} = C = F + F' + Q_4V_1 + Q_5V_2$$
$$= \$100 + \$75 + 85\ (\$8) + 40\ (\$15)$$
$$= \$100 + \$75 + \$680 + \$600$$
$$= \$1{,}455$$
$$R = I - C = \$2{,}125 - 1{,}455 = \$670$$

16.6 MATERIAL COSTS

Material costs vary both with time and with the quantity ordered. When larger quantities are ordered, price breaks are often received from vendors. In constructing historical purchase data files, prices are seen to increase for the same part over time, primarily due to inflation. For purposes of cost estimating, it is often desirable to be able to historically retrieve data that will determine the estimated price that should be allowed for a production process soon to begin.

The most accurate way to determine such prices is to solicit quotations from vendors. The bid process is time-consuming and only feasible for higher-cost items. For low-value items, historical purchase prices can be retrieved. If the items have been purchased before, a data file is likely to contain the dates at which orders were placed, the quantity ordered, and the unit cost paid per part for each order on file. Historical analysis of purchased component prices is readily computerized. For very low-cost items not previously purchased, the following example illustrates a useful procedure [5].

Example:

The following historical data file contains the quantities and unit prices for five different orders.

Date of Order	Quantity Purchased	Unit Cost
06/01/X0	100	\$4.50
12/15/X0	50	\$4.80
03/01/X1	35	\$5.00
01/15/X2	500	\$4.00
05/01/X2	250	\$4.25

Suppose annual inflation is 12 percent per year. If today's date is 11/1/X2, what is the estimated cost for an order of 137 parts?

$$12\%/\text{year} = 1\% \text{ per month}$$

$$Q = 100 \text{ purchased 29 months ago}$$

$$\text{Cost today} = U/C_{100} = 4.50\,(1.01)^{29} = \$5.98$$

$$Q = 250 \text{ Purchased 6 months ago}$$

$$\text{Cost today} = U/C_{250} = \$4.25\,(1.01)^{6} = \$4.50$$

These order sizes bracket the order quantity in question. Then,

$$U/C_{137} = \$5.98 - \left[\frac{137 - 100}{250 - 100}\right] \times (\$5.98 - \$4.50)$$

$$= \$5.62/\text{unit}$$

$$\text{Order cost} = \$5.62/\text{unit} \times 137 \text{ units} = \$770.00$$

16.7 ECONOMICS OF AUTOMATION

Automated production in the electronics industry can assume a variety of forms (Chaps. 7–9, 12), including, but not limited to, the following:

- Automated component insertion/placement
- Automated electronic testing
- Wire wrapping
- Soldering
- Automated fault diagnostics
- Automated assembly

A variety of costs are associated with manufacturing automation. The first of these is the high initial purchase cost for most automated equipment. There also exist a number of recurring costs associated with automation, including the following:

- Programming costs
- Program verification/checkout costs
- Program storage costs
- Increased complexity of maintenance
- Increased complexity of equipment operation

The indirect nature of many of these costs often causes them to be overlooked by cost analysts. When expensive automated equipment has been purchased, there is often pressure from management to utilize the equipment, even when less expensive conventional methods of production may be available.

When a variety of production tasks are routed through automated equipment, significant bottlenecks can occur if the equipment fails. Also workers may fear losing their jobs as a result of automation and fail to cooperate.

A variety of benefits accompany manufacturing automation, including

- Increased accuracy
- Increased repeatability
- Improved product quality
- Reliable completion of monotonous tasks
- Constant rate of operation and output
- Ability to operate in inhospitable environments
- Ability to complete tasks that cannot be accomplished by manual means
- Creation of new job positions with requirements for higher levels of skill

Higher initial costs are associated with automation than with any other production function. This higher initial investment results in a lower per-unit cost of parts that are produced. Conventional production results in higher per-unit costs but avoids the initial investment associated with the procurement and installation of automatic equipment. The result is that higher production quantities are usually necessary to cost justify automated production.

16.8 ENGINEERING ECONOMIC ANALYSIS

To fully assess the economic impact of investments in production equipment, it is useful to analyze the life-cycle costs and savings using methods of engineering economy.

Engineering economy accounts for the time value of money in assessing the cash flows of a production process. The following sections overview interest factors, present, future, and end-of-period dollar payments, cash flow diagrams, present worth concepts, and equivalent annual cost values.

16.8.1 Cash Flow Diagrams

A **cash flow diagram** is a schematic method of representing payments and receipts of funds over time. A payment of funds is indicated as a downward arrow. Conversely, a receipt of funds is indicated as an upward arrow, as shown in Fig. 16-7. Any payment or receipt located at time zero is designated as *P*. Any future payment or receipt *n* periods in the future is designated by the term *F*. Refer to Figs. 16-8 and 16-9. Any sequence of equal, end-of-period payments or receipts is designated by the term *A* (see Fig. 16-10).

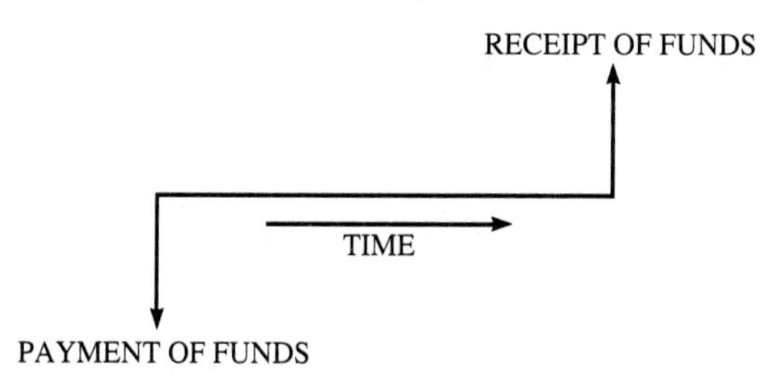

Figure 16-7 Payment and receipt of funds.

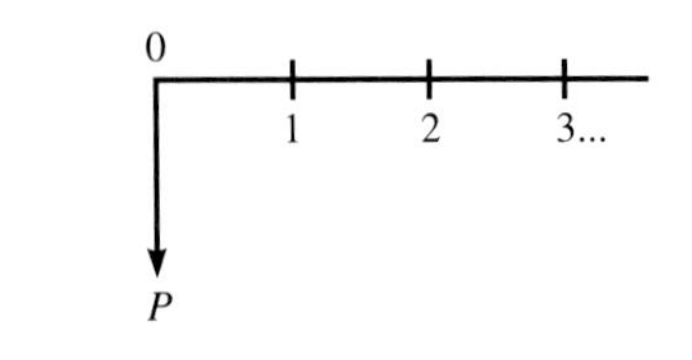

Figure 16-8 Payment at time zero.

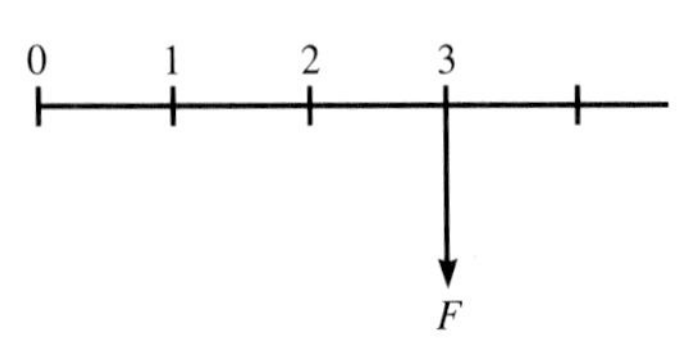

Figure 16-9 Future payment of funds.

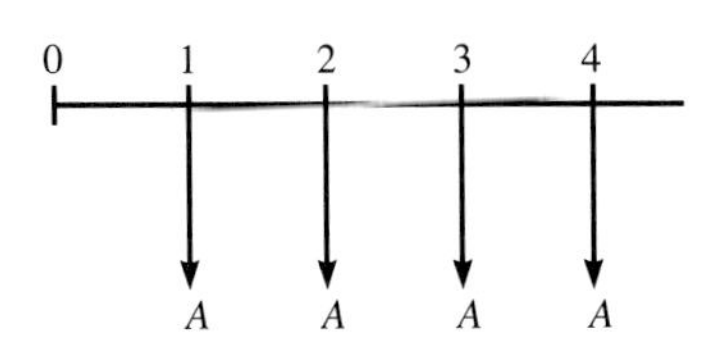

Figure 16-10 End-of-period payments.

16.8.2 Interest Factors

All present and future sums displaced by a specified number of time periods are equivalent at some interest rate, i. Interest factors provide a way to equate and compare time-displaced sums of money. Suppose that a dollar sum in the amount of P is invested at time zero with interest rate i percent. We wish to know what the value of this sum (F) will be n periods from now, as shown in Fig. 16-11. The relationship between F and P is given by Eq. (16-9):

$$F = P(1 + i)^n \tag{16-9}$$

We define $\left(\frac{F}{P}, i\%, n\right) = (1 + i)^n$ as the **single-payment compound amount factor** (SPCAF). The notation is read as F given P at i percent for n periods. It follows that

$$F = P\left(\frac{F}{P}, i\%, n\right) = P(1 + i)^n$$

The reverse situation of this example is when a value, F, is known n periods from now. At an interest rate of i percent, we wish to know the equivalent value of P that

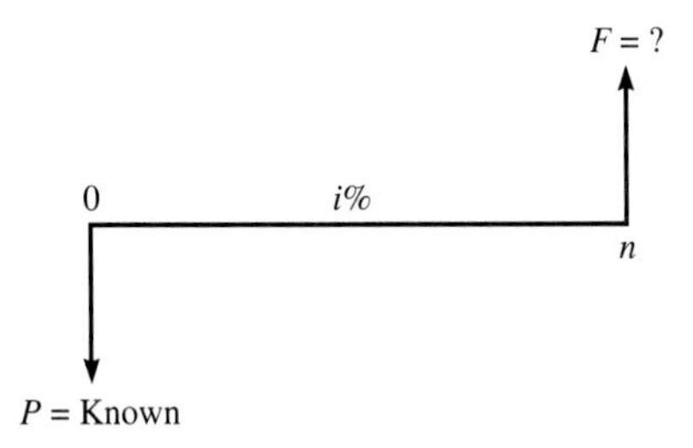

Figure 16-11 Cash flow for single-payment compound amount factor.

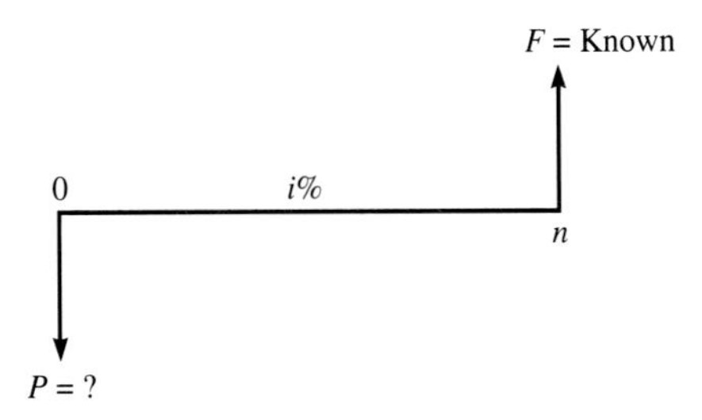

Figure 16-12 Cash flow for single-payment present worth factor.

should have been invested n periods before at i percent to yield a value of F at period n. The relationship between F and P is described as follows:

$$P = \frac{F}{(1 + i)^n} \tag{16-10}$$

We define $\left(\frac{P}{F}, i\%, n\right) = 1/(1 + i)^n$ as the **single-payment present worth factor** (SPPWF). The cash flow diagram for this scenario is shown in Fig. 16-12. It follows that

$$P = F\left(\frac{P}{F}, i\%, n\right) = \frac{F}{(1 + i)^n}$$

The next case involves a present sum, P, and a uniform series of payments, A. Suppose an investment of P dollars is made at time zero. It is desired to withdraw a sum of A dollars at the end of every period for n periods. How large will P have to be in order to allow these end-of-period withdrawals to be made over n periods? The situation is illustrated in Fig. 16-13. The relationship between P and A is given by Eq. (16-11):

$$P = A\frac{(1 + i)^n - 1}{i(1 + i)^n} \tag{16-11}$$

We define

$$\left(\frac{P}{A}, i\%, n\right) = \frac{(1 + i)^n - 1}{i(1 + i)^n}$$

as the **uniform series present worth factor** (USPWF). It follows that

$$P = A\left(\frac{P}{A}, i\%, n\right) = A\,\frac{(1 + i)^n - 1}{i(1 + i)^n}$$

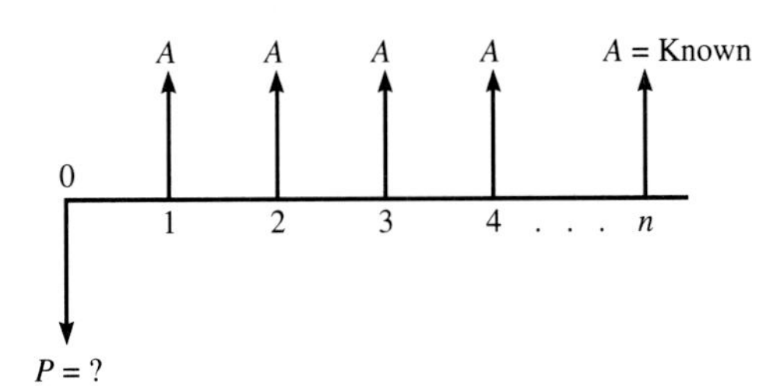

Figure 16-13 Cash flow to determine deposit for fixed withdrawals.

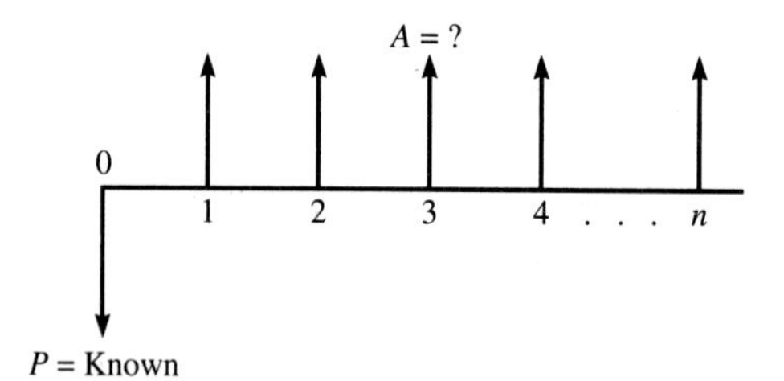

Figure 16-14 Cash flow to determine fixed withdrawal for known deposit.

The reverse situation is as follows. Suppose we deposit a known sum at time 0. We wish to know how much may be withdrawn from the account each period at i percent to exactly deplete the account balance to zero over n periods, as depicted in Fig. 16-14. The relationship between A and P is as follows:

$$A = P\frac{i(1 + i)^n}{(1 + i)^n - 1} \tag{16-12}$$

We define

$$\left(\frac{A}{P}, i\%, n\right) = \frac{i(1 + i)^n}{(1 + i)^n - 1}$$

as the **capital recovery factor** (CRF). It follows that

$$A = P\left(\frac{A}{P}, i\%, n\right) = P\frac{i(1 + i)^n}{(1 + i)^n - 1}$$

Suppose we desire to have a sum of F dollars in an account at the end of n periods as shown in Fig. 16-15. We wish to know how much should be deposited at the end of each period, for n periods, in order for this sum to be available. The relationship between F and A is as follows:

$$A = F\frac{i}{(1 + i)^n - 1} \tag{16-13}$$

We define

$$\left(\frac{A}{F}, i\%, n\right) = \frac{i}{(1 + i)^n - 1}$$

as the **sinking fund deposit factor** (SFDF). It follows that

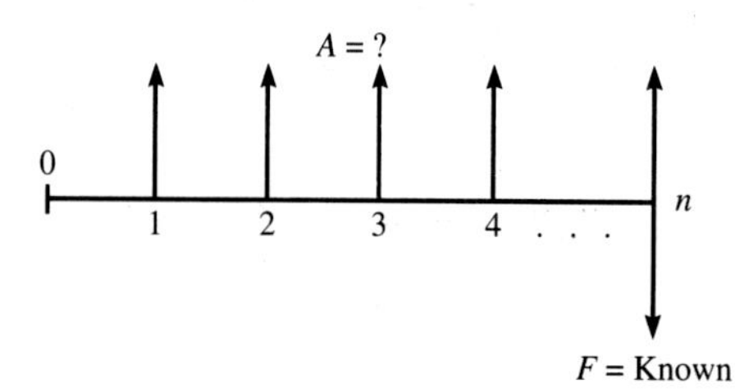

Figure 16-15 Cash flow for sinking fund deposit factor.

$$A = F\left(\frac{A}{F}, i\%, n\right) = F\frac{i}{(1 + i)^n - 1}$$

The reverse situation is where the uniform end-of-period series of payments, A, is known. Suppose A dollars are deposited at the end of each period for n periods. How much will be available to withdraw in a lump sum in period n? This situation is illustrated in the cash flow diagram shown in Fig. 16-16. The relationship between F and A is as follows:

$$F = A\frac{(1 + i)^n - 1}{i} \tag{16-14}$$

We define

$$\left(\frac{F}{A}, i\%, n\right) = \frac{(1 + i)^n - 1}{i}$$

as the **uniform series compound amount factor** (USCAF). It follows that

$$F = A\left(\frac{F}{A}, i\%, n\right) = A\frac{(1 + i)^n - 1}{i}$$

16.8.3 Numerical examples

The following examples illustrate the use of Eqs. (16-9) through (16-14).

Example:

Many people prepare for retirement through a monthly or annual savings program. If \$2,000 is put aside at the end of each year at 10 percent interest, how much will the account be worth at the end of 20 years?

The cash flow diagram for this fund is shown in Fig. 16-17. A is known, and we wish to find F as follows:

$$F = A\left(\frac{F}{A}, 10\%, 20\right) = \$2{,}000\left[\frac{(1 + .10)^{20} - 1}{.10}\right] = \$2{,}000(57.27) = \$114{,}540$$

Example:

A lottery winner recently won \$100,000 and is to be paid \$10,000 per year for 10 years. The winner elects to save all the winnings. If the first payment occurs at the end of year one, find the present value of the accumulated savings at $i = 8$ percent.

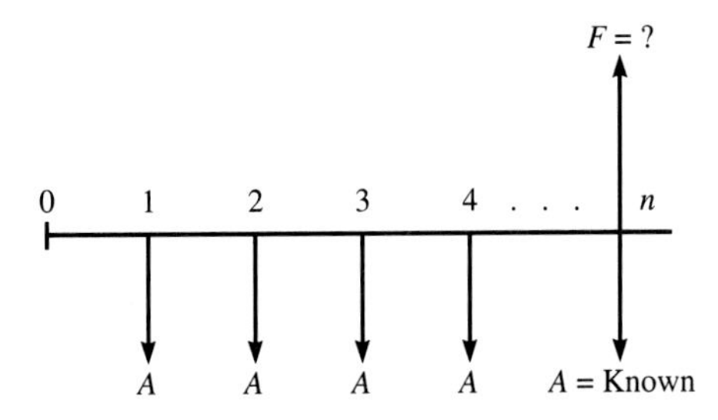

Figure 16-16 Cash flow for uniform series compound amount factor.

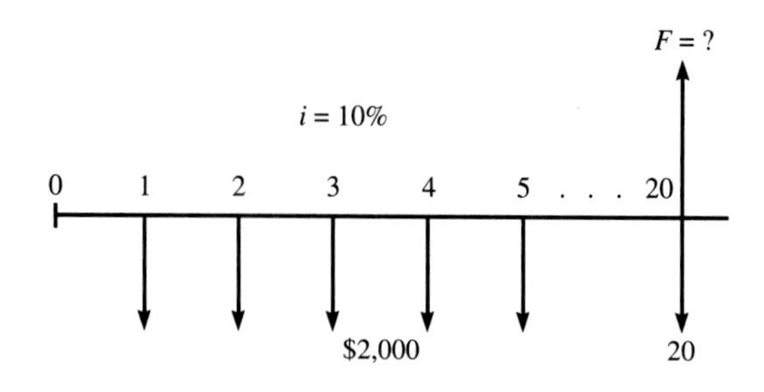

Figure 16-17 Retirement fund cash flow.

The cash flow diagram appears in Fig. 16-18. We know A and wish to find P as follows:

$$P = A\left(\frac{P}{A}, 8\%, 10\right) = \$10{,}000\left[\frac{(1 + .08)^{10} - 1}{.08(1 + .08)^{10}}\right]$$

$$= \$10{,}000\,(6.71) = \$67{,}100$$

Example:

A sophisticated SMT placement machine costs $1 million installed. The estimated salvage value after 10 years is $100,000. If the interest rate is 12 percent, what is the equivalent annual cost of owning the machine over the 10-year period?

The cash flow diagram is illustrated in Fig. 16-19. The solution procedure is as follows:

Step 1: Move $100K back to $t = 0$.

$$F\left(\frac{P}{F}, 12\%, 10\right) = \$100{,}000\left[\frac{1}{(1.12)^{10}}\right] = \$100{,}000(0.322)$$

$$= \$32{,}200$$

Thus

$$P = \$1{,}000{,}000 - \$32{,}200 = \$967{,}800$$

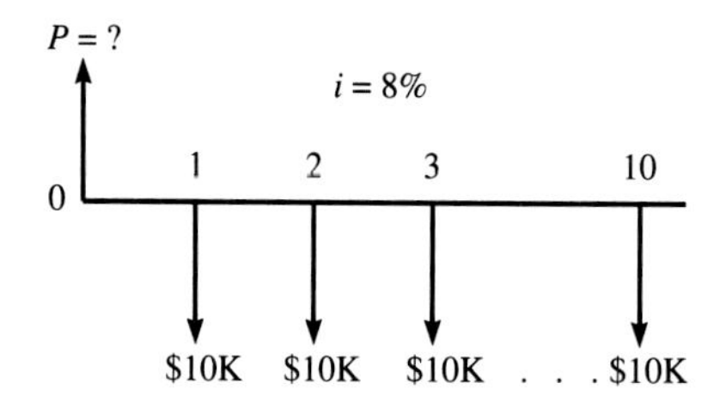

Figure 16-18 Cash flow for lottery winnings.

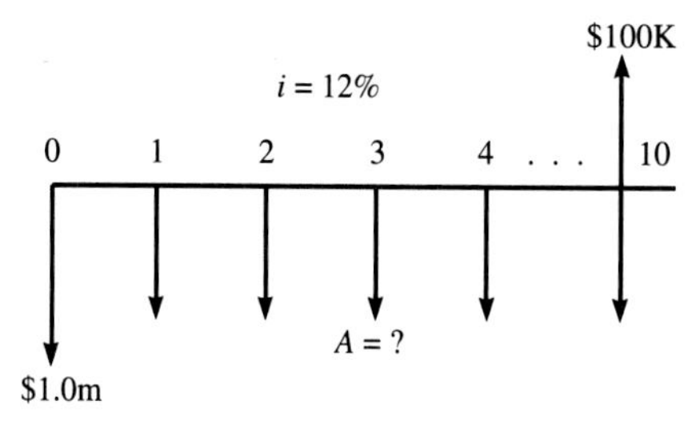

Figure 16-19 Cash flow for component insertion machine.

Step 2: Find the annual equivalent of $967,800 over 10 years.

$$A = P\left(\frac{A}{P}, 12\%, 10\right) = \$967,800\left[\frac{.12(1 + .12)^{10}}{(1 + .12)^{10} - 1}\right] = \$967,800(.177)$$

$$= \$171,301$$

16.9 A CASE STUDY INVOLVING AUTOMATION ECONOMICS

Consider a through-hole component insertion robot purchased for $50,000. The company purchasing the robot has a **minimum acceptable rate of return** (MARR) of 15 percent. It is desired to know how much the robot will have to save in order for the $50,000 purchase to be cost justified. The following assumptions apply:

- Useful life of robot = 20 years
- Salvage value after 20 years = $0
- Benefits = 25 percent of wages
- Factory worker's wages = $15,000 per year
- Maintenance technician's wages = $20,000 per year
- Programmer's wages = $25,000 per year
- Engineer's wages = $40,000 per year
- 50 work weeks per year exist × 40 hours per week = 2,000 hrs per year

A number of nonrecurring costs associated with the robot exist in addition to its purchase price. Suppose it is assumed that 40 hours are required on the part of the factory worker to install the robot. A maintenance technician might also spend the equivalent of 80 hours checking out the system to ensure that it is fully operational. A variety of personnel will require initial training to become familiar with how to operate the new robotic system. Suppose a two-week training course is attended by a factory operator, a maintenance technician, a programmer, and an engineer. Allowing for a 25 percent benefit rate, the cost associated with these personnel is summarized in Sec. 16.9.1.

16.9.1 Personnel Costs

Front-end personnel costs are summarized as follows:

- *Installation:*
  Factory worker (40 hours)

$$\frac{\$15,000 \times 1.25}{2,000 \text{ hrs}} = \$9.37/\text{hr} \times 40 \text{ hrs} = \$375$$

- *Checkout:*

  Maintenance technician (80 hrs)

$$\frac{\$20{,}000 \times 1.25}{2{,}000 \text{ hrs}} = \$12.50/\text{hr} \times 80 \text{ hrs} = \$1{,}000$$

- *Training:*

  Factory operator (2 weeks)

$$80 \text{ hrs} \times \$9.37/\text{hr} = \$750$$

  Maintenance technician (2 weeks)

$$80 \text{ hrs} \times \$12.50/\text{hr} = \$1{,}000$$

  Programmer (2 weeks)

$$\frac{25{,}000 \times 1.25}{2{,}000 \text{ hrs}} = \$15.62/\text{hr} \times 80 \text{ hrs} = \$1{,}250$$

  Engineer (2 weeks)

$$\frac{40{,}000/\text{yr} \times 1.25}{2{,}000 \text{ hrs}} = \$25.00/\text{hr} \times 80 \text{ hrs} = \$2{,}000$$

The preceding totals reflect only labor. If overhead costs are assumed to be equal in magnitude to the labor costs (that is, if overhead is 100 percent of direct labor), then the costs to initially put the robotic system in operation may be summarized as follows:

		Labor	Overhead	Total
Installation		$375	$375	$750
Checkout		1,000	1,000	2,000
Training:	Operator	750	750	1,500
	Maint. Tech	1,000	1,000	2,000
	Programmer	1,250	1,250	2,500
	Engineer	2,000	2,000	4,000
Totals		$6,375	$6,375	$12,750

16.9.2 Batch Costs

Assume that the robot is used for four new production applications each year. Suppose each application requires new fixtures. The fixtures must be both designed and fabricated for each new application. If $50 in raw material is consumed in the fabrication of each fixture, the total cost to design and fabricate fixtures for each application is summarized as follows:

- *Design Time:*

16 hours per application × $25.00/hr = $400 Labor
$400 O/H
$800 Subtotal

- *Fabrication Time:*

$$16 \text{ hours per application} \times \$9.37/\text{hr} = \begin{array}{l} \$150 \text{ Labor} \\ \underline{\$150} \text{ O/H} \\ \$300 \text{ Subtotal} \end{array}$$

- *Raw Material:* $50

Thus the total fabrication cost, including labor and raw material, is $350.

Suppose that programming and checkout costs average 16 hours and 8 hours per application, respectively. These costs may then be tallied as follows:

- *Programming Costs:*

$$16 \text{ hours per application} \times \$15.62/\text{hr} = \begin{array}{l} \$250 \text{ Labor} \\ \underline{\$250} \text{ O/H} \\ \$500 \text{ Subtotal} \end{array}$$

- *Checkout Costs:*

$$8 \text{ hours per application} \times \$15.62/\text{hr} = \begin{array}{l} \$125 \text{ Labor} \\ \underline{\$125} \text{ O/H} \\ \$250 \text{ Subtotal} \end{array}$$

Annual batch costs associated with each of the four different production applications may now be summarized as follows:

Design	$800
Fabrication	350
Programming	500
Checkout	250
	$1,900/Application × 4 Applications/yr = $7,600/yr

16.9.3 Maintenance Costs

Assume that the robot has an up time of 95 percent and runs two shifts, 50 weeks per year. The annual maintenance cost may be calculated in the following manner:

$$\text{Down time} = 5\% \text{ of } 2{,}000 \text{ hrs/shift} \times 2 \text{ shifts} = 200 \text{ hours}$$

$$\text{Maintenance cost} = 200 \text{ hours x } \$12.50/\text{hr} = \begin{array}{l} \$2{,}500 \text{ Labor} \\ \underline{+\$2{,}500} \text{ O/H} \\ \$5{,}000 \text{ Subtotal} \end{array}$$

16.9.4 Operator Costs

Assume that one operator can oversee the operation of five robots. The annual cost of the operator must be considered over two shifts. Assume there is a 15 percent wage increase for second shift workers. Then annual cost to provide an operator over two production shifts for an entire year of 50 work weeks may be determined as follows:

$$\text{Annual Operator Cost} = \frac{15{,}000/\text{yr} \times 1.25 + 15{,}000/\text{yr} \times 1.15 \times 1.25}{5}$$

$$= \frac{\$18{,}750 + \$21{,}562}{5} = \frac{\$40{,}312}{5} = \begin{array}{rl} \$8{,}062 & \text{Labor} \\ +\underline{\$8{,}062} & \text{O/H} \\ \$16{,}124 & \text{Subtotal} \end{array}$$

16.9.5 Utility Costs

Assume that the electrical power required to operate the robot over two shifts during the entire work year is \$200/year.

16.9.6 Summary

The costs over 20 years to operate the robot may be summarized as follows:

- *Nonrecurring Costs:*

Purchase price	\$50,000
Installation, checkout, and training	\$12,750
Total nonrecurring costs	\$62,750

- *Annual Recurring Costs:*

Batch cost	\$ 7,600
Maintenance cost	5,000
Operator cost	16,124
Utilities cost	200
Total annual cost	\$28,924

16.9.7 Engineering Economic Analysis

The cash flow diagram associated with the costs in this example is shown in Fig. 16-20. At an interest rate of 15 percent it is desired to know the equivalent savings that would have to exist in order to justify the purchase of the robot and the cost associated with its installation, checkout, maintenance, and operation. This value may be determined as follows, where PW is the present worth and EUAC is the **equivalent uniform annual cost:**

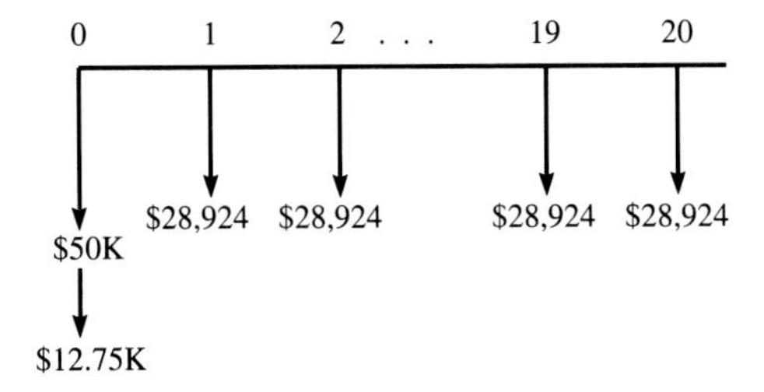

Figure 16-20 Robot costs over 20 years.

$$\text{PW costs} = \$62{,}750 + \$28{,}924\left(\frac{P}{A}, 15\%, 20\right)$$

$$= \$62{,}750 + 28{,}924\left[\frac{(1.15)^{20} - 1}{.15(1.15)^{20}}\right]$$

$$= \$62{,}750 + \$28{,}294(6.25933)$$

$$= \$62{,}750 + \$181{,}045 = \$243{,}795$$

$$\text{EUAC} = \$243{,}795\left(\frac{A}{P}, 15\%, 20\right)$$

$$= \$243{,}795\left[\frac{.15(1.15)^{20}}{(1.15)^{20} - 1}\right]$$

$$= \$243{,}795(0.15976)$$

$$= \$38{,}949/\text{yr}$$

An alternate analysis approach is to let S be the required annual savings to justify the investment. Then

$$\$62{,}750 = -\$28{,}924\left(\frac{P}{A}, 15\%, 20\right) + S\left(\frac{P}{A}, 15\%, 20\right)$$

$$\$62{,}750 = -28{,}924\left[\frac{(1.15)^{20} - 1}{.15(1.15)^{20}}\right] + S\left[\frac{(1.15)^{20} - 1}{.15(1.15)^{20}}\right]$$

$$\$62{,}750 = -\$28{,}924(6.25933) + 6.25933S$$

$$6.25933S = \$62{,}750 + \$181{,}045 = \$243{,}795$$

$$S = \frac{\$243{,}795}{6.25933} = \$38{,}949/\text{yr}$$

The robot must therefore save nearly $40,000 per year in order to be economically justified. This saving is equivalent to replacing two factory workers.

16.10 PRODUCT IMPROVEMENT CURVES

Much of the material in this section has been adapted from references [5] and [7] with the written permission of the publishers.

Product improvement (learning) curves reflect the respective decrease in assembly time as additional units are produced. Product improvement curves are defined by the relationship

$$Y = KX^n \tag{16-15}$$

where

K = Number of hours to build the first unit
X = Cumulative total units built
n = A negative numerical value which determines the percentage by which Y decreases each time X is doubled

The shape of product improvement curves is illustrated in Fig. 16-21. In Eq. (16-15) the exponent, n, determines the percentage by which Y decreases each time X is doubled. For example, with an 80 percent learning curve, the term Y declines to 0.8 of its former value each time the term X is doubled. An 80 percent learning curve is illustrated in Fig. 16-22. With a 60 percent curve the term Y declines to 0.6 of its former value each time the term X is doubled. A 60 percent product improvement curve is illustrated in Fig. 16-23.

The smaller the learning curve percentage, the greater the rate of improvement as additional units are fabricated. Both 80 and 60 percent improvement curves are illustrated for comparison in Fig. 16-24 on a common set of axes with a common starting point.

16.10.1 Types of Product Improvement Curves

There are two types of product improvement curves. The only difference between the two curves is in the way the term, Y, is defined. With **unit curves,** the term Y in Eq. (16-15) is expressed in hours to build the Xth unit. The second form of curve is a **cumulative average curve.** For this type of curve, the term, Y, is expressed as the cumulative average hours to build a quantity of X units.

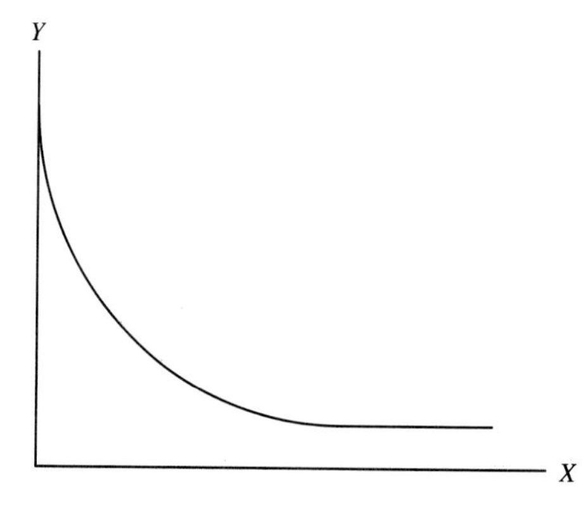

Figure 16-21 Product improvement curve shape.

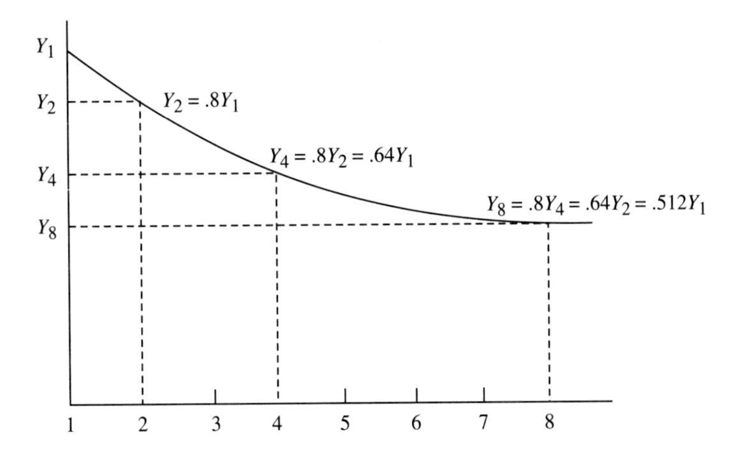

Figure 16-22 An 80 percent improvement curve.

16.10.2 Cumulative Average Curves Derived from Unit Curves

While it is possible to derive a cumulative average learning curve from a unit curve, conversion between the two curve types is not a straightforward process. Consider a 75 percent unit improvement curve where the time required to build the first unit is known to be 100 hours. Learning curve data points may then be generated from the following relationships:

$$\begin{aligned} K &= 100 \text{ hours} \\ n &= -0.415 \text{ for a 75 percent curve} \\ Y &= KX^n = 100X^{-.415} \text{ hrs/unit} \end{aligned} \tag{16-16}$$

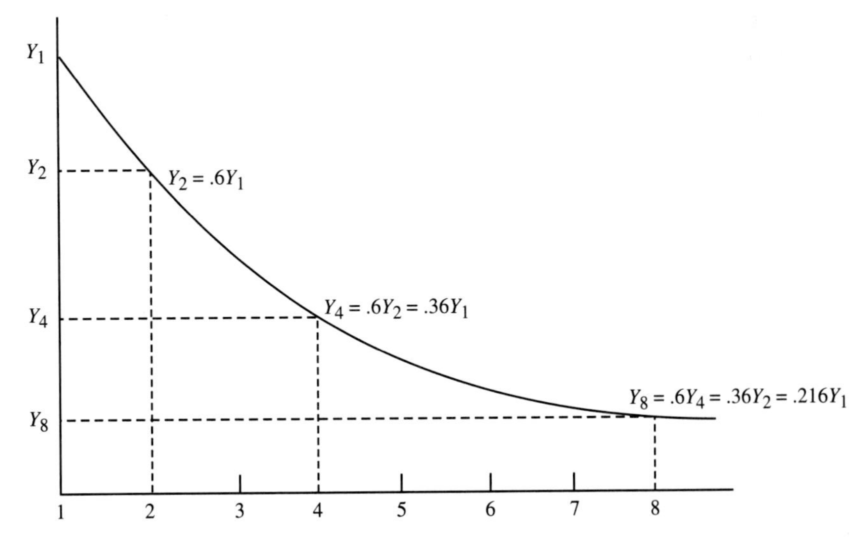

Figure 16-23 A 60 percent improvement curve.

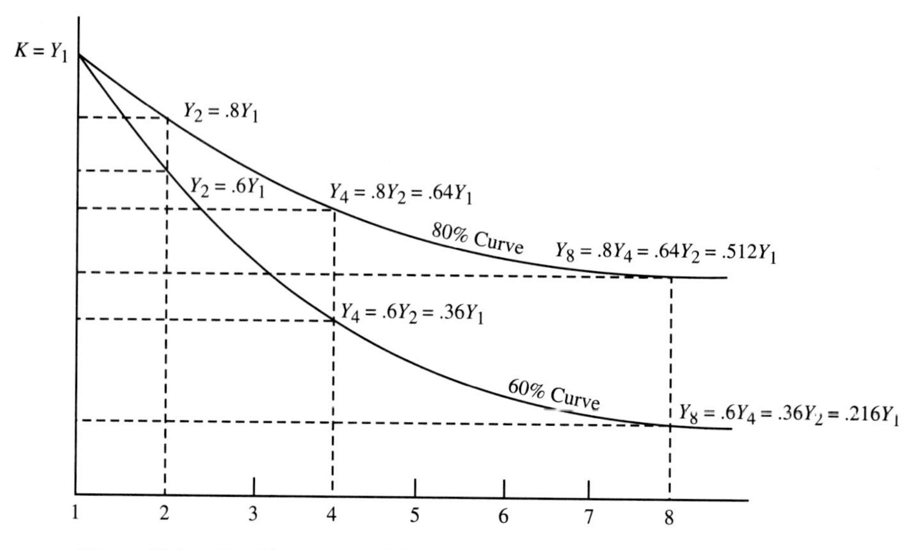

Figure 16-24 The 80 percent and 60 percent improvement curves with a common starting value.

Equation (16-16) may be used to generate corresponding Y values for the first eight units produced. These values are shown in the Table 16-1.

Cumulative average curve data points from these values may be determined in the following way. Consider the data in Table 16-2. The values in the second column of the table have been obtained directly from Eq. (16-16). The values in the third column reflect the total hours required to build a quantity of X units. For example, the second entry in this column is 175 hours, representing the total of the hours required to build unit 1 (100 hours) and unit 2 (75 hours). The values in the fourth column reflect the average time per unit to build a quantity of X units. For example, the second entry in this column is 175 hours/2 units = 87.5 hours per unit. All values in Table 16-2 are obtained in this manner.

TABLE 16-1
CALCULATED VALUES FOR A 75 PERCENT UNIT CURVE

X (units)	Y (hrs/unit)
1	100.00
2	75.00
3	63.69
4	56.25
5	51.28
6	47.54
7	44.59
8	42.18

TABLE 16-2 CUMULATIVE AVERAGE VALUES CALCULATED FROM A 75 PERCENT UNIT CURVE

Total units produced (X)	Time required to produce the Xth unit	Cumulative total hours	Cumulative avg. hours per unit
1	100.00	100.00	100.00
2	75.00	175.00	87.50
3	63.39	238.39	79.46
4	56.25	294.64	73.66
5	51.28	345.92	69.18
6	47.54	393.46	65.57
7	44.59	438.05	62.57
8	42.18	480.23	60.02

Both unit and cumulative average curve data points are illustrated in Fig. 16-25. Note that the values in the second column of Table 16-2 decrease by a factor of 0.75 each time X is doubled. This is not true for the derived cumulative average values in the fourth column. These values decrease each time X is doubled, but not by a constant factor.

16.10.3 Unit Values Derived from Cumulative Average Curves

The procedure in Sec. 16.10.2 may also be applied to derive unit curve data points starting with the cumulative average curve. Equation (16-16) may again be applied to generate the Y values. The only difference is that now Y is in cumulative average hours for a quantity of X units. The equation is

$$Y = KX^n = 100X^{.415}$$

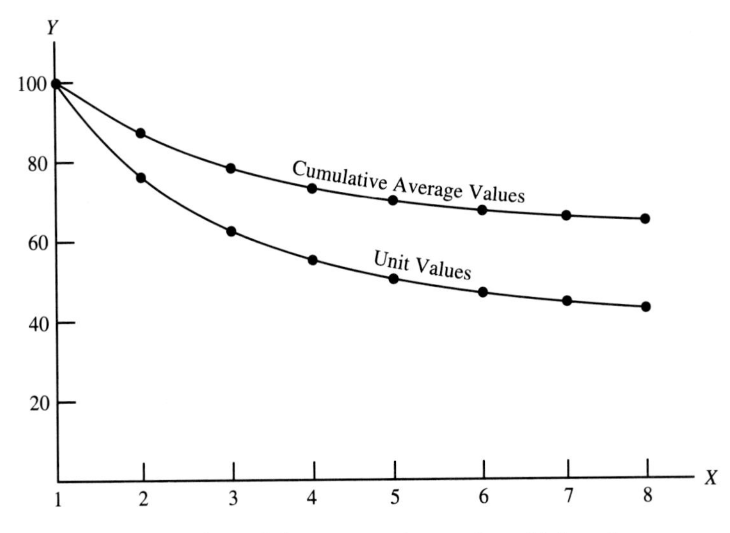

Figure 16-25 Unit and cumulative average values starting with the unit curve.

TABLE 16-3
CALCULATED VALUES FOR A 75 PERCENT CUMULATIVE AVERAGE CURVE

X (units)	Y (hrs/unit)
1	100.00
2	75.00
3	63.69
4	56.25
5	51.28
6	47.54
7	44.59
8	42.18

where the units for Y are in cumulative average hours, for X units. Data points for eight consecutively produced units may now be obtained from Eq. (16-16) and are shown in Table 16-3.

Unit values may be derived from the cumulative average curve data points in the following manner. Refer to Table 16-4. The first two columns are the cumulative average curve data points for the eight units produced. The third column reflects the cumulative total hours to build a quantity of X units and is the product of the first two columns. For example, it takes an average of 75 hours per unit to build a quantity of two units. The total hours required to build two units then becomes $75 \times 2 = 150$. The hours per unit in the last column may now be obtained by subtraction. If 100 hours were required to build the first unit and 150 hours were required to build a total of two units, it follows that the hours required to build unit 2 are $150 - 100 = 50$. Continuing, the cumulative total hours to build three units may be obtained as $63.39 \times 3 = 190.17$. Total hours required to build the third unit then are $190.17 - 150.00 = 40.17$. This procedure is repeated for $X = 1$ through 8.

In Table 16-4, the hours in the second column decrease by a factor of .75 each time

TABLE 16-4 UNIT VALUES CALCULATED FROM A 75 PERCENT CUMULATIVE AVERAGE CURVE

X	Y Avg. hrs/unit	Cum. total hours ($Y \times X$)	Hours per unit
1	100.00	100.00	100.00
2	75.00	150.00	50.00
3	63.39	190.17	40.17
4	56.25	225.00	34.83
5	51.28	256.40	31.40
6	47.54	285.24	28.84
7	44.59	312.13	26.89
8	42.18	337.44	25.31

X is doubled. This is not true for the derived unit values that appear in the fourth column. No constant percentage decrease is associated with these values. Data points for both the cumulative average and derived unit curves are shown in Fig. 16-26.

Both preceding examples illustrate the point that either cumulative average or unit improvement curves may be used in production economic analysis. Unit values may be derived from an initial cumulative average curve. The converse is also true for cumulative average values that are derived from an initial unit curve. It is not possible to associate a constant learning percentage with derived curve data points. If unit curve values are applied, it is therefore not possible to convert these learning curves to an appropriate cumulative average curve that has a constant percentage. The same is true for unit curves that are derived from cumulative average curves.

16.10.4 Relating Curve Percentages to Exponent Values

For an 80 percent learning curve, it does not follow that the exponent n assumes the value of 0.8. For a specific curve percentage, the corresponding exponent value, n, may be derived in the following manner.

For an 80 percent curve, consider two data points:

$$(X_1, Y_1),\ (X_2, Y_2)$$

$$X_2 = 2X_1$$

Then

$$Y_1 = KX_1^n$$

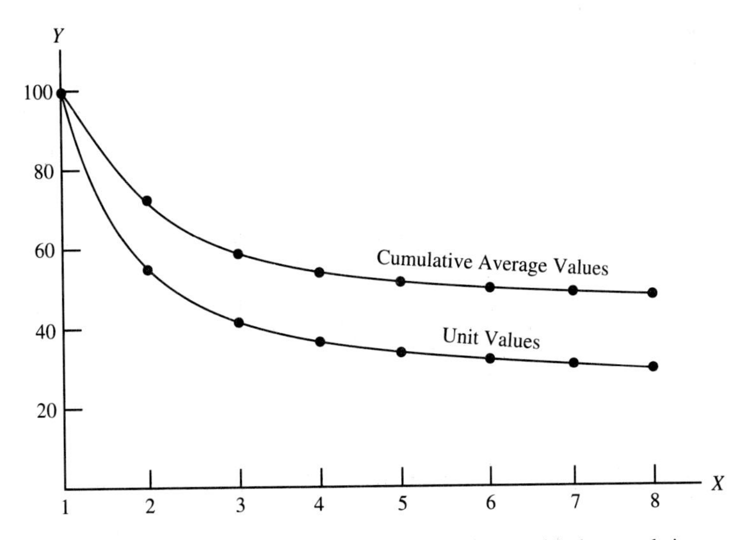

Figure 16-26 Unit and cumulative average values starting with the cumulative average curve.

and

$$Y_2 = KX_2^n$$

and

$$\frac{Y_1}{Y_2} = \frac{KX_1^n}{KX_2^n} = \frac{X_1^n}{X_2^n}\left[\frac{X_1}{X_2}\right]^n$$

If $X_2 = 2X_1$, then $Y_2 = .8Y_1$ and

$$\frac{Y_1}{Y_2} = \frac{Y_1}{.8Y_1} = \left[\frac{X_1}{X_2}\right]^n = 1.25 = \left[\frac{X_1}{2X_1}\right]^n = \left(\frac{1}{2}\right)^n$$

Thus,

$$\log(1.25) = n \log(1/2)$$

$$n = \frac{\log(1.25)}{\log(0.5)}$$

$$= \frac{0.096}{-0.301}$$

$$= -0.322$$

Repeating this procedure for common learning curve percentages yields the values in Table 16-5.

16.10.5 Historical Improvement Curve Construction

To construct improvement curves from actual data, both expended labor hours and cumulative total units built per unit time must be known. Each pair of data points should correspond to a common point in time. It is not unusual to collect improvement curve data at the end of each production week, shift, or day. Cumulative average curves are

TABLE 16-5
EXPONENT VALUES FOR TYPICAL IMPROVEMENT CURVE PERCENTAGES

Curve percentage	n
65	$-.624$
70	$-.515$
75	$-.415$
80	$-.322$
85	$-.234$
90	$-.152$
95	$-.074$

TABLE 16-6 HISTORICAL LABOR AND UNIT COMPLETION DATA

Week	Cumulative total labor hours expended	Cumulative total units complete
1	25	0
2	55	0
3	110	1
4	135	1
5	160	2
6	170	2
7	200	3
8	210	3
9	220	4
10	260	5

recommended for use because of the way in which the data format lends itself to the curve construction procedure.

An example will serve to illustrate how actual improvement curves may be constructed. Consider a 10-week production period during which a total of five units are produced. Suppose the labor expenditures and unit completions follow the schedule shown in Table 16-6.

Cumulative average learning curve data points may be obtained each time an additional total unit is completed at the end of any production week. The determination of the cumulative average values is illustrated in Table 16-7.

As shown in Fig. 16-27, the curves from Table 16-7 fall near but not exactly on the decaying exponential curve that has been sketched between the data points. An actual learning curve may be constructed by the method of least squares regression (Chap. 15). The following method applies:

$$Y = KX^n$$

$$\log Y = \log K + n \log X$$

Taking the logarithms of both the X and Y values in Table 16-7 yields the data in Table 16-8. The logarithms of the data points are plotted in Fig. 16-28. As can be seen from this illustration, the points fall near but not exactly on the straight line that has been illustrated.

TABLE 16-7 CUMULATIVE AVERAGE DATA POINTS

Week	X	Y
3	1	110.0 hrs/1 unit = 110.0 hrs/unit
5	2	160.0 hrs/2 units = 80.0 hrs/unit
7	3	200.0 hrs/3 units = 66.7 hrs/unit
9	4	220.0 hrs/4 units = 55.0 hrs/unit
10	5	260.0 hrs/5 units = 52.0 hrs/unit

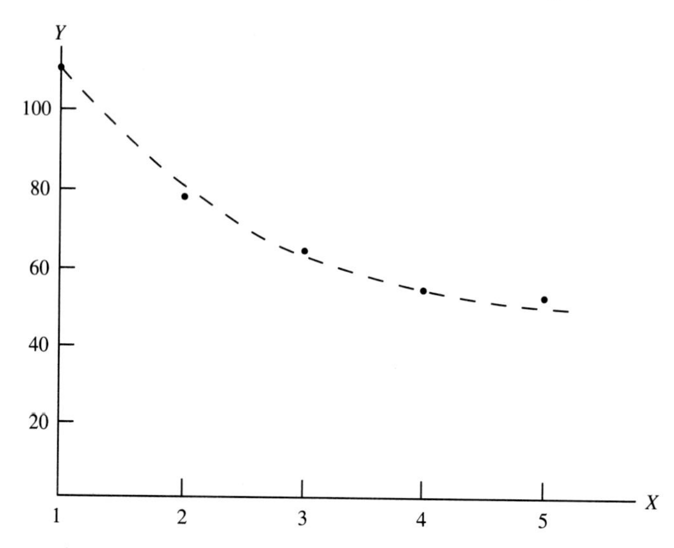

Figure 16-27 Data points from constructed improvement curve.

TABLE 16-8 LOGARITHMS OF DATA POINTS

X	$\text{Log}_{10} X$	Y	$\text{Log}_{10} Y$
1	0.000	110.0	2.041
2	0.301	80.0	1.903
3	0.477	66.7	1.824
4	0.602	55.0	1.740
5	0.699	52.0	1.716

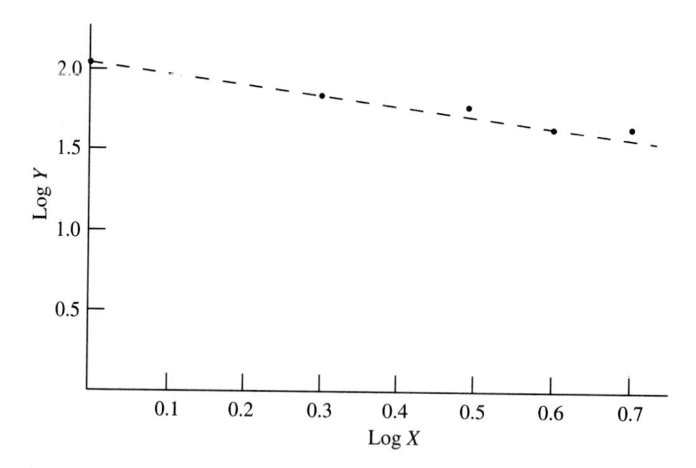

Figure 16-28 Plotted logarithms of (X, Y) values for actual improvement curve.

We want to fit a straight line of the form $Y = a + bx$ through the log-transformed data. The term, a, is the intercept of the derived equation and is equivalent to the log(a) time required to build the first unit on the improvement curve. The slope, b, is equivalent to the exponent n in the learning curve equation. Applying a first-order curve fit by the method of least squares yields the following:

$$\text{Slope} = n = -.4776$$

$$\text{Intercept} = \log(K) = 2.043$$

$$r^2 = .994$$

$$K = \log^{-1}(2.043) = 110.41$$

$$Y = 110.41X^{-.4776}$$

16.10.6 Unit Curve Formulas

Consider a group of consecutively manufactured units. Let:

$$\text{First Unit} = X_f$$

$$\text{Last Unit} = X_e$$

Let the total hours required to build a group of consecutively manufactured units be T_i. Then,

$$T_i = \frac{K}{n+1}[(X_e + .5)^{n+1} - (X_f - .5)^{n+1}] \tag{16-17}$$

The total hours to build X units can be obtained as follows. In the preceding expression let $X_f = 1$. Then

$$T_i = \frac{K}{n+1}[(X_e + .5)^{n+1} - (.5)^{n+1}] \tag{16-18}$$

The average hours required to build X_e units can be expressed as

$$A = \frac{T}{X_e} \tag{16-19}$$

16.10.7 Cumulative Average Curve Formulae

The total hours for X_e units can be written as

$$T = Y \times X_e^n$$
$$T = KX_e^n \times X_e = KX_e^{(1+n)} \tag{16-20}$$

The time to build a group of consecutively manufactured units becomes

$$\begin{aligned} T_i &= Y_eX_e - Y_f(X_f - 1) \\ &= KX_e^n \times X_e - KX_f^n(X_f - 1) \\ &= KX_e^{1+n} - KX_f^n(X_f - 1) \end{aligned} \tag{16-21}$$

where

$$Y_e = \text{Cumulative average hours for } X_e \text{ units}$$
$$Y_f = \text{Cumulative average hours for } X_f \text{ units}$$

The time U_x to build unit X is expressed as

$$U_x = (1 + n)KX^n \quad (16\text{-}22)$$

16.10.8 Curves and Cost Estimate Calculations

Learning curve construction requires knowledge of the values of both parameters K and n. Improvement curve percentages, hence the value of n, may be obtained from historical data. Relating an appropriate value of K, the time to build the first unit, to the standard time is a more complex problem. For example, suppose a quantity of 500 units is to be built. The setup time for this operation is 3 hours. A standard time of 0.5 hours per unit applies for fabrication. Without consideration of improvement curves, the total time to fabricate all 500 units is

$$\text{Total Time} = 3 \text{ Hours} + 500 \text{ Units} \times \frac{(0.5 \text{ hours})}{\text{unit}} = 253 \text{ Hours}$$

Consideration of the effects of learning will yield an appreciably different numerical result. Suppose that the historical improvement percentage associated with this assembly operation is known to be 85 percent. Furthermore, suppose from experience it is known that, on the average, the standard time is reached by the 35th unit on the learning curve. The term, K, may then be found in the following manner:

$$n = .234 \text{ for an 85 percent curve}$$
$$U = (1 + n)KX^n$$
$$u_{35} = 0.50 \text{ hours} = (1 - .234)K(35)^{-.234}$$
$$K = \frac{0.50}{(1 - .234)(35)^{-.234}} = \frac{0.50}{(.766)(.435)}$$
$$= 1.50 \text{ hours}$$

Now that the total value of K is known, the total number of hours required to complete the assembly operation may be calculated. For 500 units,

$$T = KX_e(1 + n) = 1.50\,(500)^{(1-.234)} = 175 \text{ hours}$$
$$\text{Run time} = 175 \text{ hours}$$
$$\text{Setup time} = 3 \text{ hours}$$
$$\text{Total time} = 178 \text{ hours}$$

16.10.9 Product Improvement Example

An assembly process has an improvement rate of 85 percent associated with a cumulative average curve. Five hundred units are to be built. A setup time of 3 hours has been estimated. Standard data predicts the run time to be 0.5 hours per piece. What value of n applies? The value of n in Table 16-5 is found as follows:

$$\text{Let } 2X_1 = X_2 \qquad Y_2 = .85Y_1$$

$$\frac{Y_1}{Y_2} = \frac{KX_1^n}{KX_2^n} = \left[\frac{X_1}{X_2}\right]^n = \left[\frac{1}{2}\right]^n = \frac{1}{.85} = 1.176$$

$$n \log \frac{1}{2} = \log(1.176)$$

$$n = -.234$$

At the completion of production, it is known that 35 hours, in total, were required to complete the last 100 units. If an 85 percent learning rate applies, how many hours were expended to build the first unit?

$$T_{1\text{-}500} = KX^n \times X = K(500)^n \times 500$$

$$= K(500)^{-.234} \times 500 = 116.79K$$

$$T_{1\text{-}400} = KX^n \times X = K(400)^n \times 400$$

$$= K(400)^{-.234} \cdot 400 = 98.44K$$

$$T_{401\text{-}500} = T_{1\text{-}500} - T_{1\text{-}400} = 116.79K - 98.44K = 18.35K$$

$$18.35K = 35 \text{ hours}$$

$$K = 1.907 \text{ hours}$$

How many units were produced before the standard time was reached?

$$U = (1 + n)KX^n$$

$$.5 = (1 - .234)(1.907)X^{-.234}$$

$$X^{-.234} = \frac{.5}{(1 - .234)(1.907)}$$

$$X = \left[\frac{.5}{(0.766)(1.907)}\right]^{-1/.234} = 97.67 \text{ units}$$

How many hours were spent to build the 200th unit?

$$U = (1 + n)KX^n$$

$$U = (1 - .234)(1.907)200^{-.234}$$

$$= (.766)(1.907)(.289) = .422 \text{ hours}$$

How many hours were spent to build units 50–100, inclusive?

$$T_l = KX_e^{1+n} - KX_f^n(X_f - 1)$$
$$= 1.907(100)^{1-.234} - 1.907(50)^{-.234}(49)$$
$$= 1.907(34.04) - 1.907(.400)(49)$$
$$T_{50-100} = 64.91 - 37.37 = 27.54 \text{ hours}$$

16.11 SUMMARY

Production costs are affected by manufacturing methods, organizational size, and quantity produced. These costs form the basis for a manufacturer's pricing and competitive position in national and international markets.

This chapter has introduced manufacturing costs and methods for economic modeling and decision making, including break-even analysis, engineering economic analysis, and product improvement curves. Readers desiring additional information on this topic are urged to consult the references in the bibliography.

KEY TERMS

Break-even analysis
Capital recovery factor (CRF)
Cash flow diagram
Direct Labor Cost
Direct Material Cost
Equivalent uniform annual cost (EUAC)
Factory cost
Fixed cost
Indirect labor cost
Manufacturing cost
Minimum acceptable rate of return (MARR)
Prime cost
Product improvement (learning) curves (unit and cumulative average)
Selling price
Semifixed cost
Single-payment compound amount factor (SPCAF)
Single-payment present worth factor (SPPWF)
Sinking funding deposit factor (SFDF)
Time value of money
Total cost
Uniform series compound amount factor (USCAF)
Uniform series present worth factor (USPWF)
Variable cost

EXERCISES

1. What is indirect labor, and what is its relationship to size of the organization?
2. What is the difference between fixed costs and variable costs?
3. A plant has fixed costs of $100,000. The product has a variable cost of $50/unit and is sold for $100/unit. How many units must be sold to break even?
4. What should be considered in evaluating the alternatives of manual versus automated production?

5. What are the two types of improvement curves?
6. If a process has a 76 percent improvement curve, what is the value of n?
7. An assembly process has an improvement rate of 80 percent. If the time to build the tenth unit is 2 hours, what is the time to build the first unit?
8. An assembly process has an improvement rate of 85 percent. If 28 hours are required to complete the last 120 of 550 total units, then what was the time to build the first unit?
9. An automatic insertion machine costs $100,000. If the salvage value of the machine is $2,000 after 10 years and the interest rate is 15 percent, what is the equivalent annual cost of owning this machine?
10. If $30,000 is saved annually by a piece of automatic test equipment, what is the total lump savings at the end of 5 years with 10 percent interest?

BIBLIOGRAPHY

1. Bussey, Lynne E., and Ted G. Eschenbach, *The Economic Analysis of Industrial Projects* (2nd ed). Englewood Cliffs, N.J.: Prentice Hall, Inc., 1992.
2. Canada, John R., and William G. Sullivan, *Economic and Multiattribute Evaluation of Advanced Manufacturing Systems.* Englewood Cliffs, N.J.: Prentice Hall, Inc. 1989.
3. Cochran, E. G., *Planning Production Costs: Using the Improvement Curve.* San Francisco, Calif.: Chandler Publishing Company, 1968.
4. Grant, Eugene L., W. Grant Ireson, and Richard S. Leavenworth, *Principles of Engineering Economy* (8th ed.) New York: John Wiley & Sons, 1990.
5. Malstrom, Eric M., *What Every Engineer Should Know About Manufacturing Cost Estimating.* New York: Marcel Dekker, Inc., 1981.
6. Malstrom, Eric M., ed. *Manufacturing Cost Engineering Handbook.* New York: Marcel Dekker, Inc., 1984.
7. Malstrom, Eric M. and Richard L. Shell, "A Review of Product Improvement Curves," *Manufacturing Engineering,* 82, no. 5, 1979.
8. Niebel, Benjamin W., and Alan B. Draper, *Product Design and Process Engineering.* New York: McGraw-Hill Book Company, 1974.
9. Park, Chan S., and Gunter P. Sharp-Bette, *Advanced Engineering Economics.* New York: John Wiley & Sons, 1990.
10. Shupe, Dean S., *What Every Engineer Should Know About Economic Decision Analysis.* New York: Marcel Dekker, Inc., 1980.
11. Tarquin, Anthony J., and Leland T. Blank, *Engineering Economy.* New York: McGraw-Hill Book Company, 1976.

CREDITS

Figs. 3-1, 3-2, 3-3, 3-4, 3-5, 3-6, 3-7, 3-8, 3-9, 3-10, 3-12, 3-14, and 3-15, on pages 39–55; and Figs. 4-40, p. 126; 4-44, p. 129; 4-49, p. 133; 4-53, p. 137, and 4-56, p. 139; and Tables 3-1, p. 41; 3-2, p. 42; 3-3, p. 47; and 4-1, p. 135, reprinted from Thomas H. Jones, *Electronic Components Handbook*, © 1978. Reprinted by permission of Prentice-Hall, Inc., Englewood Cliffs, N.J. Fig. 3-11, p. 51, tubular ceramic capacitors courtesy of Tusonix, Inc. Fig. 3-13, p. 53, courtesy of Cornell-Dubilier Electric Co. Figs. 3-37, 3-38, pp. 73–74, reprinted, with permission, from *Microelectronics* by Clayton L. Hallmark. Copyright 1976 by TAB Books, a Division of McGraw-Hill Inc., Blue Ridge Summit, PA 17294-0850. (1-800-233-1128). Fig. 4-28, p. 116, Copyright of Motorola, Inc. Used by permission. Fig. 4-30, p. 120, © 1990 IEEE. Figs. 4-48, p. 132; 4-54, p. 138; 4-57, p. 139; 4-58, p. 140; 4-59, p. 141; and 4-60, p. 142 courtesy of ITT Canon. Figs. 4-61 and 4-62, pp. 142–43, courtesy of Aurora Optics, Inc. Blue Bell, PA, USA. Fig. 5-3, p. 154, courtesy of RTW-PCB Tooling Group, Inc., Rogers, Ark. Fig. 10-2, p. 290; Table 10-1, p. 291, reprinted with permission from *Industry Week*, 1986. Copyright Penton Publishing, Inc., Cleveland, Ohio. Figs. 10-3, p. 292; 10-5, p. 294; 10-7, p. 295; and 10-8, p. 296, reprinted from Geoffrey Boothroyd, Carrado Poli, and Laurence E. Murch, *Automatic Assembly*, Marcel Dekker, Inc., N.Y., 1982, by courtesy of Marcel Dekker, Inc. Figs. 10-9, p. 298; and 10-10, p. 299, excerpted from *Assembly Engineering*, January 1987. By permission of the Publisher © 1987. Hitchcock Publishing Co. All rights reserved. Fig. 10-14, p. 305, reprinted from John W. Priest, *Engineering Design for Producibility and Reliability*, Marcel Dekker, Inc., N.Y. 1988, by courtesy of Marcel Dekker, Inc. Fig. 11-1, p. 318, from R. M. Fortuna, "Beyond Quality: Taking SPC Upstream," *Quality Progress*, 21, no. 6 (1988). © 1988 American Society for Quality Control. Reprinted by permission. Figs. 11-12, 11-13, p. 335; and 11-15, p. 336, from G. E. P. Box and S. Bisgaard, "The Scientific Context of Quality Improvement," *Quality Progress*, 20, no. 6 (1987). © 1987 American Society for Quality Control. Reprinted by permission. Figs. 11-19, p. 345; 11-20, p. 347; Tables 11-1, p. 343, 11-2, p. 344; and 11-3, p. 347, reprinted from *AT&T Statistical Quality Control Handbook*. Reproduced with the permission of AT&T © 1956. All rights reserved. Fig. 11-21, p. 350; 11-22, p. 353; and 11-23, p. 354; and adapted tables 11-4, p. 351; and 11-5, p. 352, from M. S. Phadke, "Design Optimization Case Studies," *AT&T Technical Journal*, vol. 65, March/April 1986, pp. 51–67, copyright © 1986 AT&T. All rights reserved. Reprinted by permission. Fig. 11-27, p. 363, reprinted from Harold Ascher and Harry Feingold, *Repairable Systems Reliability: Modelling, Inference, Misconceptions, Causes*, Marcel Dekker, Inc., N.Y., 1984, by courtesy of Marcel Dekker, Inc. Fig. 12-13, p. 392, © 1987 IEEE. Fig. 14-6, p. 427, reprinted by permission of the publisher, Chapman & Hall. Fig. 14-10, p. 435, reprinted by permission of Stanley-Vidmar Automated Systems. Figs. 14-35–14-42, pp. 471–76, reprinted by permission of Texas Instruments. Figs. 15-5, p. 489; 16-3–16-6, pp. 524–28; and portions of chapter 16 reprinted from Eric M. Malstrom, *What Every Engineer Should Know About Manufacturing Cost Estimating*, by courtesy of Marcel Dekker, Inc. Fig. 16-1, p. 523, reprinted by permission of McGraw-Hill.

Index